Practical Lubrication for Industrial Facilities

Fourth Edition

RIVER PUBLISHERS SERIES IN ENERGY ENGINEERING AND SYSTEMS

Series Editors

BOBBY RAUF
Semtrain, LLC,
USA

The "River Publishers Series in Energy Engineering and Systems" is a series of comprehensive academic and professional books focussing on the theory and applications behind various energy-related technologies and control systems. The series features handbooks for related technology, as well as manuals on the fundamentals and theoretical aspects of energy engineering projects.

The main aim of this series is to be a reference for academics, researchers, managers, engineers, and other professionals in related matters with energy engineering and control systems.

Topics covered in the series include, but are not limited to:

- Energy engineering;
- Machinery;
- Testing;
- HVAC;
- Electricity and electronics;
- Water systems;
- Industrial control systems;
- Web based systems;
- Automation;
- Lighting controls.

For a list of other books in this series, visit www.riverpublishers.com

Practical Lubrication for Industrial Facilities

Fourth Edition

Kenneth E. Bannister

River Publishers

Routledge
Taylor & Francis Group
NEW YORK AND LONDON

Published 2024 by River Publishers
River Publishers
Alsbjergvej 10, 9260 Gistrup, Denmark
www.riverpublishers.com

Distributed exclusively by Routledge
605 Third Avenue, New York, NY 10017, USA
4 Park Square, Milton Park, Abingdon, Oxon OX14 4RN

Practical Lubrication for Industrial Facilities -- Fourth Edition / Kenneth E. Bannister.

Routledge is an imprint of the Taylor & Francis Group, an informa business

ISBN 978-87-7022-745-2 (hardback)
ISBN 978-87-7004-069-3 (paperback)
ISBN 978-10-0381-209-8 (online)
ISBN 978-1-032-63236-0 (ebook master)

Dedication

I dedicate this book to the memory of my Father, Douglas G.E. Bannister under whom I apprenticed from six years old to my teens, both willingly and unwillingly! Although his education was cut short by WWII, he was the most practical engineer I have ever met. He taught me much more than how to weld, form and fabricate metal; to take apart, repair, and rebuild engines with no leftover mystery parts. He also taught me the value of taking a disciplined and common-sense approach to problem solving, the value of practicality, and the value of recognizing and building on excellence in people and design. I truly miss him.

Contents

Preface **xxvii**

Acknowledgements **xxix**

List of Figures **xxxi**

List of Tables **xlv**

Section 1 Lubrication Fundamentals **1**

1 Friction and Wear **3**

1.1 Introduction 3
1.2 Friction 3
1.2.1 Causes of solid friction 6
1.2.2 Effect of friction 7
1.2.3 Lubrication 8
1.3 Machine Surface Break-in 11
1.4 Bearing Metals 12
1.5 Wear 13
1.5.1 Wear versus friction 13
1.6 Common Wear Mechanisms 13
1.6.1 Abrasive wear 13
1.6.2 Adhesive wear 14
1.6.3 Fatigue wear 15
1.6.4 Corrosive wear 15

2 Functions of a Lubricant **19**

2.1 Lubricants 19
2.2 Base Oil Types 22
2.2.1 Animal/Vegetable base oil 23
2.2.2 Petroleum base oil 23
2.2.3 Synthetic base oil 24

2.3 Base Oil Properties 25
2.3.1 Viscosity 25
2.3.2 Multi-grade viscosity oils 26
2.3.3 Viscosity index 28
2.3.4 Specific gravity 28
2.3.5 Flash point 28
2.3.6 Pour point 28
2.4 How Lubricants Fail in Service 29
2.4.1 Heat related lubricant failure 29
2.4.2 Oxidation failure. 30
2.4.3 Thermal failure 31
2.4.4 Practical prevention of lubricant temperature related failures 31
2.4.5 Additive depletion failure 32
2.4.6 Decomposition. 32
2.4.7 Separation 33
2.4.8 Absorption 33

3 Lubrication Regimes **35**
3.1 Lubricant Film Regimes 35
3.1.1 Regime 1—Hydrodynamic lubrication—(HDL) . . . 36
3.1.2 Regime 2—Hydrostatic Lubrication (HSL) 41
3.2 Fluid Friction 41
3.2.1 Laminar Flow 41
3.2.2 Shear stress 42
3.2.3 Effect of speed and bearing area 43
3.2.4 Regime 3—Partial film lubrication 43
3.3 Bearing Efficiency 45
3.3.1 Overall bearing friction 45
3.3.2 Coefficient of friction 45
3.3.3 ZN/P Curve 45
3.3.4 Effect of load on fluid friction 47
3.3.5 Efficiency factors 47
3.3.6 Temperature-viscosity relationships 47
3.3.7 Additives for heavier loads 50
3.3.8 Oiliness/lubricity agents 51
3.3.9 Anti-wear (AW) agents 51
3.3.10 Extreme-pressure (EP) agents 51
3.3.11 Multiple boundary lubrication. 52
3.3.12 Incidental effects of boundary lubricants 52

3.3.13 Stick-slip lubrication 52
3.3.14 Regime 4—elastohydrodynamic lubrication (EHDL) 53
3.3.15 Regime 5—zero film lubrication (ZFL) 55

4 Lubricant Selection **57**
4.1 Oil Versus Grease 57
4.1.1 Advantages and disadvantages of oil 58
4.1.2 Advantages and disadvantages of grease 58
4.2 Selecting a Suitable Lubricant 58
4.2.1 Machine decision factors 59
4.2.2 Environmental /working decision factors 59

Section 2 Lubricants **63**

5 Lubricant Categories **65**
5.1 Gaseous Lubricants 65
5.2 Liquid Lubricants 66
5.2.1 Animal/vegetable fatty oils 67
5.2.2 Mineral oils 67
5.2.3 Synthetic oils 69
5.3 Cohesive Lubricants 70
5.4 Solid Lubricants 70
5.5 Categorizing and Grouping Base Oils 71
5.5.1 Refining base stocks 72

6 Lubricant Properties **75**
6.1 Lubricating Oils 75
6.2 Lubricating Oil Tribotechnical Data 78
6.3 Additional Lubricant Additives 79
6.3.1 The additive role and function 80
6.3.2 The additive package 81

7 Lubricant Property Testing **85**
7.1 Air Entrainment 87
7.1.1 DIN 51 381 TUV impinger test 87
7.1.2 Significance of results 87
7.2 Aniline Point 88
7.2.1 ASTM D 611 and ASTM D 1012 88
7.2.2 Significance of results 88

7.3 Auto-ignition Temperature 90
7.3.1 ASTM D 2155. 90
7.3.2 Significance of results 90
7.4 Biodegradation and Ecotoxicity 91
7.4.1 Significance of results 93
7.5 Cloud Point 94
7.5.1 ASTM D 2500. 94
7.5.2 Significance of Results 94
7.6 Color Scale Comparison. 95
7.6.1 Color and color tests. 95
7.7 Composition Analysis of Petroleum Hydrocarbons 95
7.7.1 Types of analysis 96
7.7.2 General methods and instrumentation. 96
7.8 Consistency of Grease (Penetration) 98
7.9 Copper Strip Corrosion 98
7.9.1 ASTM D 130 98
7.9.2 Significance of results 99
7.10 Demulsibility 100
7.10.1 ASTM D 1401 and ASTM D 2711 100
7.10.2 Significance of results 101
7.11 Density. 102
7.12 Dielectric Strength. 102
7.12.1 ASTM D 877 and D 1816. 102
7.12.2 Significance of results 103
7.13 Dilution of Crank Case Oils. 104
7.13.1 ASTM D 322 104
7.13.2 Significance of results 105
7.14 Distillation. 106
7.14.1 Significance of Results 106
7.15 Dropping Point of Grease 108
7.15.1 ASTM D 566 and ASTM D 2265 108
7.15.2 Significance of results 108
7.16 Ecotoxicity. 108
7.17 Flash and Fire Points—Open Cup. 109
7.17.1 ASTM D 92 109
7.17.2 Significance of results 109
7.18 Flash Point-closed Cup 111
7.18.1 ASTM D 56 and D93 111
7.18.2 Significance of results 111
7.19 Foaming Characteristics of Lubricating Oils 112
7.19.1 ASTM D 892 112

7.19.2 Significance of results 112
7.20 Four-ball Wear Test—ASTM D 2266 113
7.21 Four—Ball EP Test—ASTM D 2596 113
7.22 Gravity 113
7.22.1 ASTM D 287 113
7.22.2 Significance of results 114
7.23 Grease Consistency 114
7.23.1 ASTM D 217 and D 1403 114
7.23.2 Significance of results 115
7.24 Interfacial Tension 116
7.24.1 ASTM D 971 116
7.24.2 Significance of results 118
7.25 Load-carrying Ability 119
7.26 Neutralization Number 121
7.26.1 ASTM D 664 and D 974 121
7.26.2 Acidity and alkalinity 122
7.26.3 Titration 122
7.26.4 pH 122
7.26.5 Potentiometric method 122
7.26.6 End points 123
7.26.7 Colorimetric method. 123
7.26.8 Reporting the results. 123
7.26.9 Significance of results 124
7.27 Octane Number 124
7.27.1 ASTM D 2699 and D 2700 124
7.27.2 Octane number in the laboratory 125
7.27.3 Road octane number. 125
7.27.4 Aviation gasoline knock rating 126
7.27.5 Significance of results 126
7.28 Oil Content of Petroleum Wax 127
7.28.1 ASTM D 721 127
7.28.2 Significance of oil content. 127
7.29 Oil Separation in Grease Storage 127
7.29.1 ASTM D 1742. 127
7.29.2 Significance of results 127
7.30 Oxidation Stability—Oils 128
7.30.1 ASTM D 943 128
7.30.2 Significance of results 129
7.31 Oxidation Stability—Greases 130
7.31.1 ASTM D 942 • 1P142, D 1402, and D 1261 130
7.31.2 Method of evaluation 130

7.31.3 ASTM D 942 • 1P142 130
7.31.4 Significance of results 130
7.31.5 ASTM D 1402 132
7.31.6 Significance of results 132
7.31.7 ASTM D 1261 132
7.31.8 Significance of results 132
7.32 Penetration 133
7.33 Pentane and Toluene Insolubles 133
7.34 Pour Point and Cloud Point 134
7.34.1 ASTM D 97 134
7.34.2 Significance of results 135
7.35 Power Factor 136
7.35.1 ASTM D 924 136
7.35.2 Significance of test results 136
7.36 Refractive Index 137
7.36.1 ASIM D 1218 137
7.36.2 Significance of results 137
7.37. Rotary Bomb Oxidation Test (RBOT) 138
7.37.1 ASTM D 2272 138
7.37.2 Significance of results 139
7.38 Rust-preventive Characteristics 139
7.38.1 ASTM D 665 139
7.38.2 Degrees of rusting 139
7.38.3 Reporting of results 140
7.38.4 Significance of results 140
7.39 Saponification Number 140
7.40 Timken Extreme Pressure Tests 141
7.40.1 ASTM D 2509—Lubricating Greases 141
7.40.2 ASTM D 2782—Lubricating Fluids 141
7.41 USP/NF Tests for White Mineral Oils 141
7.41.1 Significance of results 142
7.42 UV Absorbance 142
7.42.1 FDA method 142
7.42.2 Significance of results 143
7.43 Vapor Pressure 143
7.43.1 ASTM D 323 143
7.43.2 Significance of test results 144
7.44 Viscosity 145
7.44.1ASTM D 88, D 445, Redwood, and Engler 145
7.44.2 Significance of Results 145
7.45 Viscosity Classifications Comparison 147

7.46 Viscosity Index 147
7.46.1 ASTM D 567 and D 2270 147
7.46.2 The concept of viscosity index 148
7.46.3 The ASTM standards 150
7.46.4 Calculating viscosity index 150
7.46.5 Significance of viscosity index 150
7.47 Water Washout 153
7.47.1 ASTM D 1264 153
7.47.2 Significance of results 153
7.48 Water and Sediment 154
7.48.1 ASTM D 96, D 95, and D 473 154
7.48.2 Significance of results 154
7.49 Wax Melting Point 155
7.49.1 Melting point (Plateau) of petroleum wax (ASTM D 87) 155
7.49.2 Drop melting point of petroleum wax (ASTM D 127) 155
7.49.3 Congealing point of petroleum wax (ASTM D 938) 155
7.49.4 Significance of results 155
7.50 Wheel Bearing Grease Leakage 156
7.50.1 ASTM D 1263 156
7.50.2 Apparatus 156
7.50.3 Significance of results 156

8 General Purpose R&O Oils **159**
8.1 Are All "R&O" Oils the Same? 159
8.2 Additive Formulation 160
8.3 Designing R&O Lubricants 160
8.4 Extreme Pressure (EP) R&O Lubricants 163
8.4.1 Dependable turbine lubrication 163
8.4.2 Cleanliness levels 163
8.5 Superior R&O Oils Cover a Wide Range of Pumps 164
8.6 Hydraulic Applications for R&O Oils 165
8.7 Universal Application of R&O Oils 167
8.8 A Heat Transfer Fluid That Keeps its Cool 168
8.9 R&O Oil Use in Self-lubricating Bearings 169

9 Hydraulic Fluids **173**
9.1 Hydraulic Oils 174
9.2 Hydraulic Pumps 177

9.3 Maintaining Your Hydraulic Oil in Service 178
9.3.1 Exorcising the "Big Three" 179
9.3.2 Prevention control strategy 181
9.4 Environment-friendly Hydraulic Fluids 182

10 Fire Resistant Fluids **183**

11 Food Grade and Enviro-friendly/Biodegradable Oils **187**
11.1 Food Grade Lubricants 187
11.2 What Performance Features are Needed? 190
11.2.1 Anti-wear 190
11.2.2 Oxidation stability 190
11.2.3 Extreme-pressure protection 191
11.2.4 Rust protection 191
11.3 Environment Friendly Lubricants and Biodegradable Lubricants 192
11.3.1 Ambiguity of environmental claims for lubricants . 194
11.3.2 Natural base oils 195
11.3.3 Comparing natural efl hydraulic fluids to modern synthetic based fluids 195
11.3.4 Choosing the right EFL lubricant product to fit your needs 197
11.3.5 Practical tips for ensuring your EFL product stays that way in service 198

12 Automotive Lubricants **199**
12.1 Engine Lubricating Oils 199
12.1.1 Automotive oil viscosity grades 200
12.1.2 Automotive gasoline and diesel engine lubricant standards 200

13 Industrial Gear Lubricants **207**
13.1 Lubricant Selection for Closed Gears 207
13.2 Lubricant Types. 209
13.2.1 Mineral oils 209
13.2.2 Extreme pressure additives 210
13.2.3 Synthetic lubricants. 210
13.2.4 Compounded oils 211
13.2.5 Viscosity improvers. 211
13.2.6 Lubrication of high-speed units 211

13.3 Worm Gear Lubricants 212
13.4 Small Gear Lubricants 213
13.5 Testing the Performance of Gear Oils 214
13.5.1 Performance under pressure 214
13.6 Gear Coupling Lubrication 214
13.7 High Speed Coupling Grease 216
13.7.1 Grease filled gearboxes 217
References 218

14 Metal Working Fluids **219**

15 Synthetic Lubricants **221**
15.1 Synthetic Base Oils 222
15.2 Synthetic Formulations 223
15.3 Origin of Synthetic Lubes 224
15.4 Examining Synthetic Lubes 226
15.4.1 Synthetic hydrocarbon fluids 226
15.4.2 Organic esters 227
15.4.3 Polyglycols 227
15.4.4 Silicones 230
15.4.5 Synthetic lubricant blends 230
15.5 Properties and Advantages 232
15.5.1 Contaminant dispersion 233
15.5.2 Protecting the metal surface from rust and corrosion 233
15.6 Case Histories 234
15.6.1 Circulating oil system for furnace air preheaters 234
15.6.2 Right-angle gear drives for fin fan coolers 235
15.6.3 Plant-wide oil mist systems 235
15.6.4 Pulverizing mills in coal-fired generating plant . . 236
15.7 Synthetic Lubricants for Use in Extreme Pressure and Temperatures 241
15.7.1 Polyalphaolefins (PAOs) make the difference . . . 242
15.8 Case Histories Involving PAO-based Synthetic EP Oils . . . 244
15.9 Diesters: Another Synthetics Option 246
15.9.1 High film strength for better wear protection 252
15.9.2 Long-term oxidation resistance 253
15.9.3 Negligible carbon deposits 253
15.9.4 Low pour point advantage 254

15.9.5 Easy cold startup, low friction and energy savings 254
15.9.6 Reduced maintenance, fuel savings 255
15.10 Application Summary for Diester-base Synthetic Lubricants 257
15.10.1 Oil change procedures 258
15.11 Semi-synthetic Fluids. 260
Bibliography 260

16 Grease Lubrication **261**
16.1 Considering Grease as a Lubricant of Choice 262
16.1.1 Grease characteristics. 262
16.1.2 NLGI consistency rating 262
16.1.3 Appearance 264
16.1.4 Color 264
16.1.5 Pumpability 265
16.1.6 Slumpability 265
16.1.7 Dropping point 265
16.1.8 Operating temperature 266
16.1.9 Water resistance or washout 266
16.1.10 Shear stability. 266
16.1.11 Grease thickeners 266

17 Pastes, Waxes, and Tribo-systems **267**
17.1 Lubricating Pastes 267
17.2 Lubricating Waxes 268
17.2.1 Wax application. 269
17.2.2 Lubricating release agents 269
17.3 Tribo-system Materials 271
17.3.1 Tribo-system coatings 273
17.3.2 Dry lubricants for tribo-systems 274
17.3.3 Application 275

Section 3 Lubricant Delivery **277**

18 How Much and How Often? Calculating Bearing Requirements **279**
18.1 The Need for Precise Lubricant Delivery 280
18.2 How Much and How Often? 282
18.2.1 Method one 283

18.2.2 Method Two 284
References 286

19 Manual Lubrication Delivery Systems for Oil & Grease 287
19.1 Manual Oiling 287
19.2 Manual Greasing 287
19.2.1 Choosing the right grease fitting for your needs . . 290
19.2.2 Material choice 291
19.2.3 Thread choice. 291
19.2.4 Style choice. 292
19.2.5 Specialty fittings 292
19.3 Anatomy of a Grease Gun 294
19.3.1 Output and delivery matters 295
19.4 How to Load a Grease Gun. 296
19.4.1 Grease cartridge loading 297
19.4.2 Bulk grease loading. 299
19.4.3 Cleaning & storage practices 300
19.5 How to Grease a Bearing - In Seven Steps 301
19.5.1 Perform simple act of greasing in a consistent and controlled manner 302
19.6 Greasing Distribution Devices/Systems 306
19.7 Ultrasonic Greasing 308
19.8 Implementing a Manual Grease Program 312
Bibliiography 314

20 Automated Lubrication Delivery Systems for Oil & Grease 317
20.1 Automated Single Point Lubrication (SPL) Devices and Systems 317
20.1.1 Constant level oilers (Bottle oilers). 318
20.2 How a Modern SPL Functions 321
20.2.1 Chemical activated SPLs 322
20.2.2 Electro-chemical activated SPLs 323
20.2.3 Electro-mechanical activated SPLs. 325
20.2.4 SPL Pros and cons 325
20.2.5 Tips for set up and use of SPLs. 326
20.3 Automated Centralized Lubrication Systems 327
20.3.1 Cost studies prove favorable economics of automated lubrication systems 330
20.3.2 Elements of a quality dual-header lubrication system. 332

20.3.3 Key features of single-header lubrication systems . 332
20.3.4 Comparing manual and automatic grease lubrication provisions. 332
20.4 Circulating Lubrication Systems 333
20.5 Open, Centralized Oil Lubrication Systems 335
20.6 Air-oil Lubrication Systems 336
20.6.1 Pros and cons 339
20.7 Single Line Resistance (SLR) Systems. 341
20.7.1 How the system works 342
20.7.2 Pros and cons 344
20.8 Single Supply Line Injector System (Positive Displacement Injector) 344
20.8.1 How the system works 345
20.8.2 Pros and cons 346
20.9 Dual Line Parallel System 347
20.9.1 How the system works 347
20.9.2 Pros and cons 349
20.10 Series Progressive Divider Systems 350
20.10.1 How the system works 350
20.10.2 Monitoring the System 356
20.10.3 Pros and cons 358
20.11 Pump-to-point Lubrication System. 358
20.11.1 Pros and cons 359
20.12 Injector Pump Systems 360
20.13 Oil Mist Lubrication Technology and Applications 361
20.13.1 Benefits and description of an oil mist lubrication system. 363
20.13.2 How the system works 363
20.13.3 Conventional oil mist system. 366
20.13.4 New central mist generator design 366
20.13.5 Internal reservoir design 368
20.13.6 Distribution header system design 368
20.13.7 Mist manifold. 370
20.14 Lubrication Delivery System Pumps—Mechanical and Pneumatic Activated Types 371
20.14.1 Mechanical powered pump units 371
20.14.2 Pneumatic powered pumps 373
20.15 Maintaining Your Centralized Lubrication Delivery System 373
20.15.1 Cleanliness and contamination control 373

20.15.2 New installations 374
20.15.3 Existing Installation 374
20.15.4 The power of adjustability 375
20.15.5 Regular PM/Operator maintenance. 375
References 376
Bibliography 377

21 Lubrication Delivery System Design Components 379
21.1 Reservoirs. 379
21.1.1 Lubricant storage reservoirs 379
21.1.2 Lubricant working reservoirs. 380
21.2 Lubrication Lines and Hoses 384
21.2.1 Line size and material 384
21.2.2 Main causes of line problems. 386
21.3 Lubrication Seals 388
21.3.1 Labyrinth seals 389
21.3.2 Radial lip seal. 390
21.3.3 Bearing isolators 392
21.4 Lubrication System Controllers and Signal Devices 392
21.4.1 Controllers 393
21.4.2 Signal devices. 394
21.4.3 Broken line, or pressure loss protection 395
21.4.4 Reservoir fluid level gauge devices. 396
21.4.5 Planar sight glass 397
21.4.6 Columnar sight level gauge. 398
21.4.7 3-Dimensional sight glass 400
21.5 Reservoir Filters and Breathers Filters 402
21.5.1 Surface fluid filter—oil 402
21.5.2 Depth fluid filter—oil. 406
21.5.3 Grease filtration. 407
21.5.4 Hydraulic system bypass filter 408
21.5.5 Measuring filter efficiency and beta ratio 409
21.5.6 Portable filter carts 411
21.5.7 Breathers 416
21.5.8 The anatomy of a breather 417
21.5.9 Filter/Breather combination unit 417
21.5.10 Standard breather unit 418
21.5.11 Desiccant breather unit 418
21.5.12 IIOT sensing 418
Bibliography 422

Section 4 Applied Lubrication 423

22 Bearing Lubrication 425
22.1 Grease Lubrication 426
22.1.1 Lubricating greases 427
22.1.2 Relubrication 431
22.2 Oil Lubrication 435
22.2.1 Methods of oil lubrication 435
22.3 Tilting Pad Thrust Bearings 445
22.3.1 Flooded lubrication vs Directed lubrication 445
22.3.2 Bearing selection 450
22.4 Tilting Pad Radial Bearings 452
22.4.1 Instrumentation 454
22.5 Combination Thrust and Radial Bearings 455
22.6 Plain Bearings 455
22.6.1 Lubrication of hydrodynamic sliding bearings . . . 456
22.6.2 Sliding bearings in the mixed friction regime . . . 457
22.6.3 Lubrication of sintered metal sliding bearings . . . 460

23 Machine Element(s) Lubrication 461
23.1 Lubrication of Fastener Screws 461
23.2 Lubrication of Valves and Fittings 462
23.3 Lubrication of Electrical Switches and Contacts 466
23.4 Lubrication of Industrial Springs 468
23.5 Lubrication of Pneumatic Components 471
23.6 Lubrication of Shaft-hub Connections 472

24 Industrial Gear Lubrication 475
24.1 Lubricant Selection for Closed Gears 475
24.1.1 Film Thickness 479
24.1.2 Lubricant types 480
24.1.3 Lubrication of high-speed units 482
23.1.4 Types of lubrication regimes found in gear teeth . . 483
23.1.5 Methods of supplying lubricant 484
24.2 Lubrication of Large Open Gears 486
24.3 Lubrication of Worm Gears 487
24.4 Lubrication of Small Gears 488
24.5 Testing the Performance of Gear Oils 490
24.5.1 Performance under pressure 490
24.5.2 Superior oxidation stability 491
24.5.3 Corrosion resistance 491

24.6 Gear Coupling Lubrication 493
24.7 Lubrication of Small Geared Blowers 494
References 496

25 Electric Motor Lubrication **497**
25.1 Implementing a Motor Lubrication Strategy 499
25.1.1 Right lubricant 499
25.1.2 Right amount 500
25.1.3 Right place 501
25.1.4 Procedures for re-greasing electric motor bearings . . 505
25.1.5 Right time. 509
25.1.6 Right person 510
25.2 Oil Mist for Electric Motors 512
Bibliography 517

26 Pump Lubrication **521**
26.1 Type A - Lubrication Pumps 521
26.1.1 Centralized oil delivery 521
26.1.2 Hydraulic oil delivery 521
26.1.3 Centralized grease delivery 522
26.2 Type B – Process Fluid Pumps 522
26.2.1 Static oil bath bearing lubrication 522
26.2.2 Oil ring lubrication 524
26.2.3 Flinger disc lubrication 525
26.2.4 Oil spray/mist lubrication 526
26.2.5 Constant level lubricators. 527
26.2.6 Grease lubrication 528
26.2.7 Preferred pump lubrication method 528
References 529

27 Lubrication of Wire Ropes **531**
27.1 Wire Rope Failure 532
27.2 Wire Rope Lubrication 534
27.3 Wire Rope Lubricant Application 535
Bibliography 536

28 Lubrication of Chains **537**
28.1 Chain Failure 538
28.2 Chain Lubricants 539
28.3 Chain Lubricant Application 541
Bibliography 542

Section 5 Managing Lubricants 543

29 Lubricant Purchasing 545
29.1 Implementing a Cradle-to-Cradle Lubricant Management Program . . . 545
29.1.1 Introducing a lubricant purchase policy/ program . . . 546
29.1.2 Automotive assembly plant purchasing case study . . . 547
29.1.3 Introduce a service level agreement (SLA) . . . 550

30 Lubricant Consolidation 551
30.1 Why Consolidate? . . . 551
30.2 Choosing a Consolidation Partner . . . 553
30.3 Preparing for the Consolidation Process . . . 555
30.4 Implementing Change . . . 555
30.5 Program Monitoring . . . 557

31 Designing and Preparing a Lubricant Storage Area/Facility 559
31.1 Attributes of a Lubricant Storage Facility . . . 560
31.1.1 Location/size . . . 560
31.1.2 Ventilation . . . 561
31.1.3 Fixtures . . . 562
31.1.4 Transfer/Filtration equipment . . . 563
31.1.5 Spill control . . . 564
31.1.6 Safety . . . 564
31.1.7 Stock control . . . 564
31.1.8 Identification . . . 564
31.1.9 Processes and procedures . . . 564
31.2 Outdoor Storage . . . 564
31.3 Indoor Storage . . . 566
31.3.1 Lubricant storage policy . . . 568
31.3.2 Practical dispensing equipment . . . 569
31.3.3 Lubricant transfer policy . . . 571
31.3.4 Lubricant ID control systems . . . 573
31.3.5 Setting up and maintaining a lubricant ID control system . . . 576
31.3.6 Receiving lubricants . . . 578
31.3.7 Fluid cleanliness delivery/Acceptance procedure for all tote managed lubricants . . . 581

31.3.8 Stock rotation procedure 582
31.3.9 Why lubricants degrade 583
31.4 Guidelines for Designing a Lube Storage Facility 586
31.4.1 Design process 589
31.5 Used and Waste Oil Management 590
31.5.1 Identifying used oil 590
31.5.2 Identifying waste oil 591
31.5.3 Used or waste oil tanks must be clearly labeled and accessible 592
31.5.4 The cost of doing business 594
Bibliography 594

32 Contamination Control **597**
32.1 The Contamination Effect 597
32.1.1 Three types of lube oil contamination identified . . 598
32.1.2 Contamination sources 599
32.1.3 Built-in 599
32.1.4 Ingested 599
32.1.5 Generated 599
32.2 Solids Contamination 600
32.3 Water Contamination 603
32.3.1 Water ingression 606
32.3.2 Ideal water levels difficult to quantify 606
32.3.3 Methods employed to remove water 607
Bibliography 612

33 Lubricant Condition Testing – Oil Analysis **615**
33.1 Detecting Machine Faults and Abnormal Wear Conditions 615
33.2 Performing Condition-based Oil Changes 616
33.3 Monitoring and Proactively Responding to Oil Contamination 617
33.4 Oil Sampling Methods Examined 619
33.4.1 Maximize data density through sampling point selection 619
33.4.2 Typical procedure for extracting an oil sample. . . 621
33.4.3 Minimize sample data disturbance— don't contaminate the contaminant! 623
33.4.4 Maximize sample data consistency— oil sample frequency 626

33.5 Oil Sampling Frequency 627
33.5.1 Penalty of failure 627
33.5.2 Fluid environment severity 627
33.5.3 Machine age 628
33.5.4 Oil age 628
33.6 Selection of Oil Analysis Test 629
33.6.1 Fluid properties analysis 629
33.6.2 Contamination analysis. 629
33.6.3 Fluid wear debris analysis 631
33.7 Monitoring Changing Oil Properties 631
33.7.1 Viscosity stability. 632
33.7.2 Oxidation stability 634
33.7.3 Thermal stability and varnish tendency 636
33.7.4 Additive stability 637
33.8 Monitoring Oil Contamination 638
33.8.1 Particle contamination 639
33.8.2 Moisture contamination 643
33.9 Wear Particle Detection and Analysis 646
33.9.1 Elemental spectroscopy. 646
33.9.2 Ferrous density analysis 647
33.9.3 Analytical ferrography 648
33.10 Interpreting Test Results 650
33.11 Simple Steps to Implementing an Industrial
Oil Analysis Program. 651
33.12 Importance of Training 657
Bibliography 659

34 Safety **661**
34.1 Implementing a Lubrication Safety Program. 661
34.1.1 Purchasing 661
34.1.2 Lubricant use 662
34.2 Lube-application Safety Specifics 664
34.2.1 Pressure injury incidents 664
34.2.2 Breathing incidents 667
34.2.3 Eye and skin incidents 667
34.3 Safety Data Sheets (SDS) 670
34.3.1 Understanding SDS elements. 671
34.4 Disposal of Lubricants 674
Bibliography 674

Section 6 Implementing World Class Lubrication **677**

35 Implementing World-Class Lubrication **679**
35.1 Working to a Lubrication Standard 679
35.1.1 Interpreting the standard 681
35.1.2 ICML 55 Standard Series 682
35.2 Developing the Lubrication Program 682
35.2.1 Benefits of lubrication mapping 687
35.2.2 Implementing a lubrication mapping initiative . . . 688
35.2.3 Combating bearing failure 690
35.2.4 Passive monitoring control 694
35.2.5 Active monitoring control 695
35.2.6 Regular PM/Operator maintenance 697
Bibliography 698

36 Lubrication Work Management **699**
36.1 The Work Request 699
36.1.1 Stage One – The maintenance work request (MWR) 699
36.1.2 Stage Two – The maintenance work order (MWO) – Work 704
36.1.3 Breaking down the work order – Job plan section 704
36.1.4 Work order completion 712
Bibliography 715

37 Key Performance Indicators **717**
37.1 Key Performance Indicators (KPI) for Lubrication 718
Bibliography 720

Section 7 Lubrication Certification **723**

38 Lubrication Certification **725**
38.1 Lubrication Technician Certification 725
38.1.1 Why certify? 725
38.2 Certification Choices 726
38.3 Certification Body of Knowledge 732

Appendix **733**

Glossary of Terms 741

Index 761

About the Author 789

Preface

Lubrication truly is the lifeblood of all things mechanical, yet it continues to mystify and be taken for granted when it need not be the case. The year 2023, witnessed the centennial anniversary of the world's first centralized lubrication system, celebrating Joseph Bijur's introduction of the single line resistance (SLR) system first used to lubricate the chassis of many famous luxury cars of that era. The same system, virtually unchanged, is still the most copied and sold system for the lubrication of small machinery in the world today, and the company Joseph Bijur founded all those years ago still operates today as Bijur-Delimon International.

The question I have been continually asked over my career has always been "why is lubrication so often taken for granted, if it can cause over 70% of all bearing failures?" My take on this is simple; the advent of the zerk grease nipple and Gulberg's grease gun allowed engineered lubrication systems to be replaced by a simple "do it yourself" lubrication approach (think of the automobile) that cost virtually nothing for machine manufacturers to install. In abdicating responsibility to the end user, the manufacturer placed the burden of reliability and bearing life squarely in the hands of untrained individuals, who have continued to prematurely and unwittingly "kill" bearings for almost a hundred years.

The fault here is not the invention, because with quality education/training a grease gun is a very effective device; the fault lies in the education system for engineers and trades alike who receive virtually no, if any, formal education/training in the science of lubricants, their application methods, and their effects. In many plants I visit there is still an agrarian approach to lubrication that dictates "any oil or grease will do the job" where the lubrication specialist position is given to an apprentice or a soon to retire individual with no training. Sound familiar?

Fortunately, industrial lubrication IS changing! A century after Bijur's invention, 2023 also witnessed the unveiling of the ICML 55® Worldwide standard for industrial lubrication. Thanks to the tireless efforts of the International Council for Machinery Lubrication, a non-profit global organization dedicated to education, training and certification of individuals and corporate organizations, effective lubrication practices are now being recognized and advocated as an integral foundation to machine availability and reliability.

So congratulations and thank you! The fact that you have opened the book, and are reading this page shows that you care and are interested in the life-cycle preservation of bearings and machinery through effective lubrication practices.

This is the fourth edition of *Practical Lubrication for Industrial Facilities*, and is vastly different from previous editions. For one, you will notice the book illustrations and photos are in full color or colorized. The text has undergone a major rewrite making it more relevant for training purposes, easier to read, understand, and use. The book references most of the body of knowledge areas required for individual certification and corporate certification through ICML, STLE and ISO certification bodies.

It is my hope that you find the book appealing and interesting. I can promise that you that with this knowledge, you will be able to make a difference to the machinery you will design, build, and service, delivering world class availability, maintainability and reliability. In addition, application of this knowledge is a very effective and inexpensive way of cost reclamation / reduction that can have a true and measurable impact on the environment and your carbon footprint.

Effective lubrication practice truly does open the door to a new world, so I invite you to turn the page and make a difference to your world…

Kenneth E. Bannister, 2023

Acknowledgements

This book was completely rewritten with the help of individuals and companies whose assistance and cooperation are gratefully acknowledged.

I would especially like to mention Heinz Bloch who passed away in 2022. A true giant in the field of applied lubrication who first penned this book in its first and second edition. Heinz had agreed to write a foreword for the book, unfortunately this was not to be. He was however, very excited to hear of all the major changes proposed for the book and the introduction of color. Thank you, Heinz, it was a privilege to know you and to work with you.

I would like to thank my wife Jo for her encouragement and allowing me the space to write.

I would like to thank all the staff at Rivers Publishers for their patience, understanding, and professionalism in getting this book out there. A special thanks to Rajeev Prasad, Phillipa Jefferies, and Junko Nakajima, all of whom were instrumental in this effort.

The book would not have been complete without the input from numerous individuals and companies in the form of photos, illustrations, tables, and narrative. These include:

- ASTM
- Bijur-Delimon International
- Des-Case Corporation
- Efficient Plant magazine
- ENGTECH Industries
- Exxon-Mobil
- Farvel Lubrication Systems
- Fluid Defense Systems
- Glacier Metal Company
- Howard Marten Company

- ICML – International Council for Machinery Lubrication
- Kingsbury Inc.
- Noria Corporation
- Oil-Rite Corporation
- SKF
- STLE – Society for Tribologists and Lubrication Engineers
- The RAM Review E-zine

And to all other contributors in the book not mentioned here, I thank you!

List of Figures

Figure 1.1	How bearings and Machines lose their usefulness. . .	4
Figure 1.2	Forces on bodies at rest..	5
Figure 1.3	Solid friction—static versus kinetic.	6
Figure 1.4	Burnt out engine white metal bearing due to frictional heat caused by lubricant starvation in the bearing area.	8
Figure 1.5	Effects of fluid versus solid friction.	9
Figure 1.6	Three friction states.	10
Figure 1.7	Typical x2000 magnified ground steel surface.. . . .	11
Figure 1.8	Ground metal surface profilogram.	12
Figure 1.9	2 Body abrasive wear..	14
Figure 1.10	3 Body abrasive wear..	14
Figure 1.11	Adhesive wear.	15
Figure 1.12	Metal tear due to adhesion.	15
Figure 1.13	Surface cracking and pitting from metal fatigue.. . .	16
Figure 1.14	Acid etched corrosion on a rack gear..	16
Figure 2.1	Petroleum oils make excellent lubricating films. . . .	24
Figure 2.2	High viscosity fluids like molasses flow slower then low viscosity fluids like water.	26
Figure 2.3	Oil is thicker at lower temperatures, thinner at higher temperatures.	27
Figure 2.4	Relationship between oil operating temperature and the oil life..	30
Figure 3.1	Measuring surface roughness.	36
Figure 3.2	Oil film lambda oil thickness ratio curve..	36
Figure 3.3	Sliding load supported by a wedge-shaped lubricating film.	37
Figure 3.4	Shoe type thrust bearings..	39
Figure 3.5	Oil pressure distribution diagrams.	40
Figure 3.6	Fluid-bearing friction is the drag imposed by one layer of lubricant sliding upon another..	42

Figure 3.7 Factors that effect bearing friction under full-fluid-film lubrication. 43
Figure 3.8 Mixed film lubrication. 44
Figure 3.9 Boundary lubrication.. 44
Figure 3.10 Stribeck bearing performance curve. 46
Figure 3.11 Loose-fitting bearings require high-viscosity oils. . . 48
Figure 3.12 Relationships between oil viscosity, load, speed, and temperature.. 48
Figure 3.13 For the same temperature change, the viscosity of oil "B" changes much less than that of oil "A". 49
Figure 3.14 Sliding surfaces separated by a boundary lubricant of the polar type.. 50
Figure 3.15 EHD lubrication in a rolling-contact bearing. 54
Figure 5.1 Cross section of a gas lubricated air bearing. 66
Figure 5.2 Typical applications and industry sectors for lubricants.. 67
Figure 7.1 Grease classification and testing. 117
Figure 7.2 Schematic view of the four-ball wear test assembly in the tribometer.. 120
Figure 7.3 Bomb Oxidation Test—When pressure drop is plotted against time, the resulting curve will indicate a period of comparatively slow oxidation followed by a pronounced rise. The relatively flat portion at the beginning represents what is known as the "induction period," a phase during which oxidation is not ordinarily of serious magnitude. For practical reasons, it is not customary to continue the test beyond the induction period, its end being indicated by a sudden rise. Should the test be carried further, however, this rise would eventually taper off again as oxidation becomes complete. In some cases, test results have been expressed in terms of the duration of the induction period. 131
Figure 7.4 Fischer Style D445 Viscometer.. 146
Figure 7-5 Viscosity classification equivalents. 149
Figure 7.6 The concept of viscosity index.. 151
Figure 7.7 Chart for calculating V.I.'s above 100 from kinematic viscosity, based on ASTM D 2270. Dotted lines illustrate its use.. 152

Figure 8.1	Elementary circulating oil system.	161
Figure 8.2	Reciprocating compressor applications.	165
Figure 8.3	Rotary positive displacement applications.	166
Figure 8.4	Dynamic compressor applications.	166
Figure 8.5	The extraordinary loads placed on the self-lubricating bearings of high-speed power tools can lead to premature tool failures without the protection of quality oils.	171
Figure 9.1	Typical hydraulic circuit with vane pump.	174
Figure 9.2	Main types of pumps found in hydraulic systems. . .	178
Figure 11.1	Typical bottle filling line.	188
Figure 11.2	Sealed Flange Bearing (over) lubricated with H1 food grade grease..	189
Figure 11.3	Typical commercial bakery bread cooling conveyor system.	191
Figure 12.1	Typical monograde oil.	201
Figure 12.2	Typical multigrade oil.	201
Figure 12.3	API designation donut icon for gasoline engine vehicle SN designation..	202
Figure 13.1	Single-helical, high-speed gear unit being assembled..	208
Figure 13.2	Manual drill, double-stage gear and a Double stage spur gear.	213
Figure 13.3	Typical couplings requiring grease lubrication.. . . .	215
Figure 13.4	Section through a tooth-type coupling.	216
Figure 15.1	Synthetic oil versus mineral oil molecular structure. .	225
Figure 15.2	Bowl mill wear, 1,000 operating hours..	237
Figure 15.3	Viscosity stability of synthetic lubricant in bowl mill gear unit.	238
Figure 15.4	The arctic demands free-flowing synthetic lubricants.	242
Figure 15.5	Steel mill temperature environment calls for synthetic lubricants.	243
Figure 15.6	Paper machine rolls in severe duty service..	246
Figure 15.7	Overhead crane gear box exposed to severe ambient environment.	247
Figure 15.8	Viscosity increase due to oxidation, mineral oil vs. synthetic EP product.	248
Figure 15.9	A superior synthetic EP oil will give superior volatility control to provide long-term lubrication	

effectiveness. Plus, the low volatility and high flash point compared to conventional gear oils give an added margin of safety at high operating temperatures. 248
Figure 15.10 PAO-based synthetics have superior cold-weather performance. 249
Figure 15.11 High-quality polyalphaolefin base stock helps keep viscosity stable over a wide range of temperatures. The viscosity index for Spartan Synthetic EP gear oils ranges from 150 to 167, compared to 90 to 100 for a conventional petroleum-based oil. 250
Figure 15.12 Photo on left - No. 1 discharge valve after 4 months' service with ISO 150 mineral oil. Photo on right - No. 1 discharge valve after 6 months' service with Exxon SynESSTIC 100 diester synthetic oil.. 252
Figure 15.13 Laboratory studies with an instrumented bearing test rig demonstrate that a low viscosity SynESSTIC can provide the same protection as a higher viscosity mineral oil. In the test illustrated in this graph, the additive packages of all three lubricants were identical—only the base changed. Because a lower viscosity grade can achieve the same degree of protection, you may be able to save energy and reduce costs.. 253
Figure 15.14 Oxidation tests show diester-base synthetics excel over mineral oils. 254
Figure 15.15 At comparative test temperatures, the mineral oils solidify while the Diester 32 and 100 grades remain free flowing. 255
Figure 15.16 At –18°C, Exxon's Synesstic ISO 32 diester (top photo) has no wax crystals, while a comparable mineral oil (bottom photo) shows significant crystallization. The dark spots in the diester photo are air bubbles in the sample.. 256
Figure 15.17 Diester film strength offers improved bearing protection for all types of compressors.. 258
Figure 16.1 ASTM D217 Cone penetration test stand.. 265
Figure 17.1 Baking pan mold release lubrication rail 272
Figure 17.2 Tribo-system materials.. 273

Figure 17.3 Factors influencing service life of tribo-system coatings.. 274
Figure 17.4 Nuts coated with Klübertop. 276
Figure 18.1 Incorrectly set up pitman arm style auto lubricator creating a very messy and detrimental over-lubrication condition. 281
Figure 18.2 Manually over-lubricated pillow block bearing.. . . . 282
Figure 18.3 Lubricant delivery pattern. 283
Figure 18.4 Typical bearing company relubrication chart. 285
Figure 19.1 Plunger style oil can. 288
Figure 19.2 Lube poster showing oil cup. 288
Figure 19.3 Grease cup cutaway. 289
Figure 19.4 Arthur Gulborg's grease gun lubrication system invention. 289
Figure 19.5 Joseph Bystricky grease gun nozzle patent. 291
Figure 19.6 Assorted grease fitting styles.. 292
Figure 19.7 Button Head (DIN) fittings alongside regular zerk grease fitting and adaptor.. 293
Figure 19.8 Pressure relief and Hydraulic shut-off fittings. 293
Figure 19.9 The gold standard professional lever action grease gun design. 294
Figure 19.10 A lever action grease gun with see through barrel. . . 295
Figure 19.11 Marked syringe used to determine grease gun displacement. 297
Figure 19.12 End of grease barrel showing Tee handle and lock slot.. 298
Figure 19.13 Horizontal grease gun caddy. 301
Figure 19.14 Typical grease gun storage method particularly prone to dirt and lubricant cross contamination 301
Figure 19.13 Grease-point colored collar ID and dust protector tag. 303
Figure 19.14 Connecting a manual grease gun using a grease coupler and lever action grease gun (note gun lever in closed position.). 304
Figure 19.15 Grease injector needle. 307
Figure 19.16 Narrow needle nose dispenser. 307
Figure 19.17 Two styles of 90° - grease fitting adaptors. 308
Figure 19.18 Seal-off grease adaptor. 309
Figure 19.19 Recessed fitting grease extension adaptor 309

Figure 19.20 Grease gun adaptor kit. 310
Figure 19.21 Grease fitting multi-tool. 310
Figure 19.22 Operator performing manual greasing using a UE systems Ultraprobe 201 grease gun system. . . . 311
Figure 19.22 Non-engineered triple gang grease block. 313
Figure 19.23 Lubriquip/Trabon manual progressive divider valve block with cycle pin and over pressure indicators. . . 314
Figure 19.24 Simple manual centralized grease lubrication system. 315
Figure 20.1 Single-point spring loaded grease unit. 318
Figure 20.2 Filling a late 1800s original Single Point Lubricator with oil—still in use in 2016. 319
Figure 20.3 Piping "Air Vent" back into the equipment bearing housing produces a reliable pressure—balanced lubricator 320
Figure 20.4 Pressure-balanced lubricator with low level safety (warning or shut-down) switch.. 321
Figure 20.5 Pressure-balanced constant level oiler with integral sight glass. 322
Figure 20.6 Cutaway of Perma classic chemical activated SPL. 323
Figure 20.7 Programmable electro-chemical SPL. 324
Figure 20.8 Electro mechanical SPL. 325
Figure 20.9 Electro-chemical lubricator with date in service clearly indicated by the installer 327
Figure 20.10 Different versions of automated centralized grease/oil systems. 328
Figure 20.11 Safematic SG1-grease lubrication system 329
Figure 20.12 Graco/Trabon 250-point Centralized grease system using a drum pump with control station and progressive divider distribution installed on a large walking beam furnace. 331
Figure 20.13 Modern steel mills use automatic grease lubrication systems. 333
Figure 20.14 Circulating lubrication system comprised of filters/cooler/pumps and reservoir (1), oil flow metering modules (2), pressure piping (3), and return piping (4). 334
Figure 20.15 Large recirculating oil—Circoil™ pumping unit used of large paper mill applications. 334

Figure 20.16 Open, centralized oil lubrication system. 335
Figure 20.17 Forest product processing machinery using open, centralized lubrication. 336
Figure 20.18 Air Oil lubrication schematic. 337
Figure 20.19 Typical air-oil lubrication system. 338
Figure 20.20 Small air-oil lubrication system. 339
Figure 20.21 Four-section air-oil valve assembly. 340
Figure 20.22 Air-oil lubrication systems with coaxial delivery tubing.. 340
Figure 20.23 Typical application layout for air-oil lubrication. 341
Figure 20.24 Single line resistance (SLR) metering valve system lubrication cycle. 342
Figure 20.25 SLR metering units.. 343
Figure 20.26 Single line resistance system.. 344
Figure 20.27 Single line positive displacement injector lubrication cycle. 345
Figure 20.28 Single line parallel (PDI system).. 346
Figure 20.29 Lincoln style positive displacement injectors. 347
Figure 20.30 Dual line parallel injector system lubrication cycle. . 348
Figure 20.31 Dual supply line parallel system. 349
Figure 20.32 Dual line farval pump system in use at a small hydro generation facility 350
Figure 20.33 Series progressive system. 351
Figure 20.34 Series progressive divider system lubrication cycle 352
Figure 20.35 Piston A – Outlet one discharge (lubriquip slide card 412-C).. 353
Figure 20.36 Piston B – Outlet two discharge (Lubriquip slide card 412-C).. 354
Figure 20.37 Piston C – Outlet three discharge (Lubriquip slide card 412-C).. 355
Figure 20.38 Piston A – Outlet four discharge (Lubriquip slide card 412-C).. 356
Figure 20.39 Trabon progressive divider block with indicator pins.. 357
Figure 20.40 Victorian steam engine box cam pump to point lubricator (still in use today). 359
Figure 20.41 Modern multipoint pump-to-point lubricator. 360
Figure 20.42 Air-operated injector pump system. 361

Figure 20.43 Air-powered “purgex” grease injector pump.. 362
Figure 20.44 Repair cost comparison for 2 identical petrochemical facilities, with vs. without oil mist lubrication.. . . . 365
Figure 20.45 Typical mist system design.. 366
Figure 20.46 Central oil mist generator and supply tank.. 367
Figure 20.47 Oil mist distribution piping.. 369
Figure 20.48 Closed-loop oil mist system schematic. 370
Figure 20.49 Mechanically actuated grease pump. 372
Figure 20.50 Gearbox reservoir Bulls Eye sight gauge. 376
Figure 21.1 Typical on-site interim lubricant storage reservoirs. 380
Figure 21.2 Closed loop oil system with multiple pumps connected to single reservoir. 381
Figure 21.3 Total loss reservoir over pump style lubrication system. 383
Figure 21.4 Compare steel line routing versus nylon line routing. 386
Figure 21.5 How a compression nut system is put together showing the difference between the metal sleeve and nylon tubing sleeve.. 388
Figure 21.6 Steam turbine labyrinth seal. 390
Figure 21.7 Typical radial lip seal.. 391
Figure 21.8 Modern bearing isolator. 392
Figure 21.9 Progressive divider block signal indicators. 395
Figure 21.10 Gearbox level gauge system with Hi-Lo markers. . . 399
Figure 21.11 Modern 3-dimensional level sight gauge.. 401
Figure 21.12 Sight level indicator comparison chart 402
Figure 21.13 RAG sight level indication system on a total loss grease reservoir.. 403
Figure 21.14 Surface filter media.. 404
Figure 21.14 Different filter types found in a typical oil circulating lubrication system. 405
Figure 21.15 Typical depth filter media. 407
Figure 21.16 Wedge wire grease contaminant trap.. 408
Figure 21.17 Hydraulic filter bypass circuit. 409
Figure 21.18 Filter efficiency measured as a Beta Ratio. 410
Figure 21.19 Three different but similar filter cart designs 412

Figure 21.20 Filter cart with RAG indicators showing status of Filter cleanliness. 413
Figure 21.21 Desiccant style silica gel breather. 419
Figure 21.22 Hydraulic gearbox with damaged breather (Upper LH corner circled in red) c/w a planar style level window with upper (max) and lower level (min) markings . . 420
Figure 21.23 Breathing action of an adsorbent media style breather 421
Figure 21.24 Sensor activated IIOT breather. 421
Figure 22.1 Solidified grease thickener that has destroyed the bearing. 430
Figure 22.2 Grease relubrication intervals as a function of bearing type, size, and speed. 432
Figure 22.3 Advantageous, simple grease escape "valve." (Arrangements found in ASEA Electric Motors.) . . 434
Figure 22.4 Oil bath/oil spray on double row spherical roller bearings. In the oil bath configuration (left), the oil level reaches the center of the rollers at the bottom. In the spray configuration (right), the oil is conveyed to the tapered flinger by an oil ring which dips into the oil bath. An important feature of this application is that the air pressure on both sides of the bearing and enclosures is equalized by connecting ducts. This prevents leakage of the lubricant when the housings are located in an air stream.. 436
Figure 22.5 Circulating oil lubrication. 437
Figure 22.6 Oil-jet lubrication allows bearings to operate at lower temperature or at higher speeds than any other method of lubrication.. 438
Figure 22.7 Air-oil lubrication schematic.. 438
Figure 22.8 Kinematic viscosity requirement as a function of bearing mean diameter and speed. 439
Figure 22.9 The required ISO-grade of a lubricant can be obtained from this graph. 441
Figure 22.10 Oil mist lubrication applied to a set of split inner ring bearings 442
Figure 22.11 Simplified oil viscosity selection chart.. 444

Figure 22.12 20MW naval gearbox fitted with Glacier Series II flooded lubrication standard bearings for medium speed duties.. 446
Figure 22.13 Flooded lubrication: typical double thrust arrangement. 447
Figure 22.14 Glacier's standard 7 series bearings for both flooded and directed lubrication. 447
Figure 22.15 Glacier double thrust bearing size 10293 with directed lubrication installed in an ABB gas turbine. 448
Figure 22.16 Directed lubrication: typical double thrust arrangements designed to prevent bulk oil from contacting the collar. 449
Figure 22.17 Offset pivots: effect on thrust pad temperature.. . . . 450
Figure 22.18 Hydraulic thrust metering: arrangement diagram. . . 451
Figure 22.19 Glacier bearing featuring high pressure jacking oil for start-up and run down. Jacking oil annulus can be seen on the surface of each thrust pad. 451
Figure 22.20 Tilting-pad journal bearing, converging geometry. . . 452
Figure 22.21 Geometry and pressure.. 453
Figure 22.22 Glacier combination thrust/radial tilt pad bearing. . . 455
Figure 22.23 Lowering a large Glacier medium wall bearing into a marine pro pulsion gearbox. 457
Figure 22.24 Hydrodynamic bearing clearances: recommended minima against speed by shaft diameter. 459
Figure 22.25 Dry-running bearings: sheet bearing (left), having lubricating holes filled (right) with "Klüberdur" tribo-system material.. 459
Figure 23.1 Screw connections that are too tightly preloaded tend to fail. 462
Figure 23.2 Screw anti-seize lubricant shown in tube form.. . . . 463
Figure 23.2 Valve components may require lubrication. 465
Figure 23.3 Some electrical switches and contactors must be lubricated.. 467
Figure 23.4 Plug contacts. 468
Figure 23.5 Advantages of a lubricated, gold-plated plug contact as a function of the number of connections. 469
Figure 23.6 The performance of plate and annular springs can be improved with appropriate lubrication.. 470
Figure 23.7 Pneumatic cylinder.. 471

Figure 23.8 Pneumatic valves.. 472
Figure 23.9 Lubrication of a multiple spline shaft and formation of fretting corrosion. 473
Figure 23.10 Loose fits require different lubricants than interference fits.. 474
Figure 24.1 Single-helical, high-speed gear unit being assembled.. 476
Figure 24.2 Double-helical, low-speed gear unit being checked at Lufkin Gear Company, Lufkin, Texas. 477
Figure 24.3 Basic gear designs. 478
Figure 24.4 Open Girth gear drive of a drying cylinder.. 486
Figure 24.5 Efficiency of mineral hydrocarbon oil compared to synthetic oils. 487
Figure 24.6 Wear curve of mineral hydrocarbon oil compared to synthetic oils. 488
Figure 24.7 Electric hand drill, double-stage gear. 489
Figure 24.8 Gear motor, double-stage spur gear. 489
Figure 24.9 High viscosity index is mandatory for demanding marine applications.. 492
Figure 24.10 Typical couplings requiring grease lubrication.. . . . 494
Figure 24.11 Section through a tooth-type coupling. 495
Figure 24.12 Roots type positive displacement blower at a petrochemical plant.. 495
Figure 25.1 Temperature effect on fill level of grease packed bearings.. 498
Figure 25.2 Grease in motor winding. 498
Figure 25.3 Relief style grease fitting.. 499
Figure 25.4 Shielded, grease-lubricated bearing (no drain).. . . . 501
Figure 25.5 Single-shield motor bearing, with shield facing the grease cavity. 502
Figure 25.6 Double-shielded bearing with grease metering plate facing grease reservoir. 503
Figure 25.7 High load and/or high-speed bearings are often supplied without shield, as shown. 504
Figure 25.8 Open bearing with cross-flow grease lubrication. . . 504
Figure 25.9 EPRI NP-7502 motor greasing guide.. 510
Figure 25.10 Oil mist routed through electric motor bearings. . . . 513
Figure 25.11 Oil mist applied to the same side of a bearing is not providing optimal lubrication; much of the mist is simply flowing from entry to drain.. 514

Figure 26.1 Centralized oil lubrication pump.. 522
Figure 26.2 Large hydraulic oil pumping system. 523
Figure 26.3 Typical grease lubrication pumping unit. 524
Figure 26.4 Typical sump (oil bath) housing with 6-o'clock oil level. 524
Figure 26.5 An unrestrained oil ring can touch portions of the inside of the bearing housings and suffer abrasive damage. 525
Figure 26.6 Oil rings shown in correct position.. 526
Figure 26.7 Flinger discs avoid issues with oil rings; they can be accommodated in bearing housings fitted with cartridges designed to allow access and insertion. . . 526
Figure 26.8 Typical oil spray/oil mist) directed into the bearing cage at two entry points provides an optimum thickness oil film for lubrication and heat removal at any bearing orientation. Note the face type bearing seals used to prevent oil loss to atmosphere through the bearings.. 527
Figure 26.9 Constant level lubricator with pressure balance between bearing housing and lubricator body. 528
Figure 26.10 Typical direct motor driven centrifugal pump body. . 528
Figure 27.1 Typical open and closed style wire ropes configurations. 532
Figure 27.2 Wire rope in poor condition. 533
Figure 27.3 Typical wire rope clean/lubricate pressure boot system. 535
Figure 28.1 Duplex roller chain and sprocket drive system.. . . . 538
Figure 28.2 Cross section of a typical roller chain pin assembly. . 539
Figure 28.3 Opco OP-4A Automatic conveyor chain pin oiler. . . 542
Figure 29.1 Lubricant C2C cycle. 546
Figure 30.1 Wartime 3-R lubrication guide.. 552
Figure 31.1 Shipping container lube storage and dispensing unit. 562
Figure 31.2 Outdoor Storage exposes lubricants to the elements. 563
Figure 31.3 Twelve drum pallet rack showing bungs at 6 and 12 o'clock. The preferred bung location is 3 and 9 o'clock.. 566
Figure 31.4 Orderly lubricant storage with adequate spill containment. 567

Figure 31.5 Fully labelled and color coded dispensing system complete with a locable fireproof cabinet and fireproof garbage pail shown. 568

Figure 31.6 A variety of open style transfer containers-clearly contaminated. 570

Figure 31.7 State-of-the-art utility oil transfer container complete with color-coded top and ID label. 571

Figure 31.8 Transferring lubricant with the "new style" oil transfer container. 572

Figure 31.9 Typical lubricant transfer/filter cart. 573

Figure 31.10 Lube Label utilizes a yellow color, a shield shape and a description of the lubricant 578

Figure 31.11 Lube Labels hang off a dispensing tank utilizing a blue circle and red hourglass with their written descriptions to identify two different lubricants metered side by side. 579

Figure 31.12 400-gallon (1600 Ltr) refillable tote sitting on a dedicated transfer pump station. 580

Figure 31.13 Dirty Tote that should not be accepted for filling before it is cleaned 587

Figure 31.14 Outdoor used and waste oil storage facility example.. 592

Figure 32.1 Contamination ingression causes.. 600

Figure 32.2 The effect of solids contamination on MTBF in hours.. 601

Figure 32.3 Macpherson contamination curve. 601

Figure 32.4 Particulate count sample. 603

Figure 32.5 Fluids cleanliness under a microscope. 603

Figure 32.5 Oil reservoir sight gauge showing all three states of water presence. 604

Figure 32.6 Water contamination effect on bearing life 605

Figure 32.7 Portable oil centrifuge. 609

Figure 32.8 Schematic view of vacuum dehydrator.. 611

Figure 32.9 Portable vacuum distillation unit.. 612

Figure 33.1 Some maintenance strategies are more costly than others.. 618

Figure 33.2 Circulating system recommended oil sampling points.. 621

Figure 33.3 Low pressure sampling.. 622

Figure 33.4 High pressure sampling options. 622
Figure 33.5 Drain sampling with tap, can be connected to a vacuum sampler. 623
Figure 33.6 Drop tube static sampling arrangement using a vacuum sampler. 624
Figure 33.7 Zip-Lock® bags prevent contamination of samples. . 625
Figure 33.8 Absolute viscometer for plant use. 634
Figure 33.9 ISO contaminant code (ISO 4406). 640
Figure 33.10 Light scattering particle counter. 642
Figure 33.11 2 Pore blockage-type particle counter. 642
Figure 33.12 Particle count trend graphs. 643
Figure 33.13 Crackle test for water contamination.. 645
Figure 33.14 Three common categories of wear particle detection. 646
Figure 33.15 Combined detection and analysis process. 649
Figure 33.16 Examples of particles identified by analytical ferrography. 650
Figure 34.1 Purpose-built manual push lube truck. 663
Figure 34.2 LOTO station board. 664
Figure 34.3 Breathing hazard warning. 668
Figure 34.4 Breathing respirator. 668
Figure 34.5 Squeeze bottle eye-wash. 669
Figure 34.6 Eye station with shower. 669
Figure 34.7 Typical SDS manual containing current SDS sheets for ALL lubricants and Chemicals used in the plant. 670
Figure 35.1 Twelve ICML 55 Auditable areas. 683
Figure 35.2 A lubrication management system (LMS). 684
Figure 35.3 7-Step lubrication program journey. 685
Figure 35.4 Poorly set up auto lube system. 690
Figure 35.5 Lubrication system calibration savings example.. . . 691
Figure 35.6 Typical P-F curve.. 694
Figure 35.7 Manual marked gauge with needle showing outside safe area. 695
Figure 35.8 Commercial marked gauge with needle in safe area. 696
Figure 36.1 Typical air compressor PM instruction set. 710
Figure 36.2 Revised air compressor PM instruction set in a checklist format. 711

List of Tables

Table 1.1	Typical coefficient of friction values.	7
Table 2.1	8-Functions of a lubricant.	23
Table 2.2	Viscosity index rating chart.	28
Table 4.1	Advantages versus disadvantages of oil lubrication. .	58
Table 4.2	Advantages versus disadvantages of grease lubrication.	59
Table 4.3	Guidelines for choosing a suitable lubricant.	61
Table 5.1	Types of lubricant relationships.	68
Table 5.2	Paraffinic versus Naphthenic base oils.	69
Table 5.3	Properties of typical base oils for industrial lubricants.	71
Table 5.4	Typical base oil stock properties.	73
Table 6.1	Lubricating oil additives and their purpose.	76
Table 6.2	Typical lubricant additive formulations by lubricant type.	77
Table 6.3	Tribo-technical data pertaining to lubricating oils. . .	78
Table 6.4	ISO viscosity grades of fluid industrial lubricants. . .	79
Table 6.5	Oil additives by function.	81
Table 6.6	Oil type additive package reference table.	83
Table 7.1	Lubricant test chart.	86
Table 7.2	ASTM copper strip classification.	100
Table 7.3	NLGI grease grading system.	115
Table 7.4	Numerical relationships among viscosity classification systems..	148
Table 8.1	Typical viscosity grade range for R&O lubricants. . .	160
Table 8.2	Typical inspections for Exxon TERESSTIC® R&O lubricants.	162
Table 8.3	Temperature guidelines for premium grade R&O oils in heat transfer applications.	170
Table 9.1	Typical inspections for quality anti-wear hydraulic oils (Exxon NUTO®H).	176

Table 10.1	ISO 12922 test methods used to assess fire resistance.	184
Table 10.2	Comparison of fire resistant and mineral oil properties.	185
Table 11.1	Typical characteristics of a food-grade can seamer lubricant.	192
Table 11.2	Typical Characteristics of a food-grade grease formulated with aluminum-complex thickener.	193
Table 11.3	"Environment Friendly" hydraulic fluid comparison table.	196
Table 12.1	API gasoline engine oil service categories.	202
Table 12.2	API diesel engine oil service categories.	203
Table 12.3	ILSAC engine oil standard.	205
Table 13.1	Typical factors that affect lubricant selection.	209
Table 13.2	Equivalent viscosities of other systems.	210
Table 13.3	Shell 4 ball test—one minute wear load performance of four coupling greases.	217
Table 13.4	Oil separation (%) observed on four coupling greases.	217
Table 15.1	Base Oil Groups.	222
Table 15.1	Generalized properties of synthetic hydrocarbon lubes.	228
Table 15.2	Physical properties of ISO VG 320 gear oil.	239
Table 15.3	Cost benefit analysis table.	240
Table 15.4	Typical inspections for a synthetic EP oil.	245
Table 15.5	Typical inspections for a diester synthetic lubricant.	251
Table 15.6	Diester-base synthetic lubricants extend oil drain intervals.	257
Table 16.1	Grease versus oil comparison.	263
Table 16.2	NLGI grease classification # system.	264
Table 16.3	Grease thickener types.	266
Table 17.1	Typical machine elements and components lubricated by wax and wax emulsions.	269
Table 17.2	Selection criteria for lubricating pastes.	270
Table 17.3	Typical inspections for a balanced line of highly refined waxes (Exxon's PARVAN wax products).	271
Table 17.4	Application methods for dry lubricants.	276
Table 18.1	Lubricant replenishment rate (R) for normal operating conditions.	284
Table 19.1	Grease Gun comparison chart.	296

Table 21.1 ISO 4406:1999 solid contamination code suitability matrix.. 414
Table 22.1 Operating temperature ranges for greases used in rolling element bearings. 429
Table 22.2 Properties of bearing alloys. 454
Table 22.3 Bearing bore options. 458
Table 22.4 Porosity of sintered metal sliding bearings.. 460
Table 23.1 Typical Anti-seize thread lubricant compounds. 464
Table 24.1 Lubricant selection factors. 479
Table 24.2 Viscosity ranges for AGMA lubricants.. 480
Table 24.3 Equivalent Viscosities of other systems. 481
Table 24.4 Viscosity, SSU @ 100°F. 482
Table 24.5 Typical design attributes of a gearbox sump lubrication system. 485
Table 25.1 Baldor Electric Company guidelines for motor relubrication. 511
Table 26.1 Preferred lubrication method ranking table. 529
Table 28.1 Drive system comparison table.. 540
Table 30.1 Current lubricant listing chart. 555
Table 30.2 Consolidation data gathering form example. 556
Table 31.1 Design attributes of a lubricant storage and handling facility – aka "Lube Room". 561
Table 31.2 Lubricant ID control system color chart. 575
Table 31.3 Color-coded lubricant listing (partial). 576
Table 32.1 ISO particle concentration range table.. 602
Table 32.2 Water ingression effects on base oils.. 606
Table 32.3 Water ingression effects on lubricant additives. 607
Table 32.4 How guidelines on allowable water contamination vary. The table applies to turbomachinery lubricants. 608
Table 33.1 Application of lube oil analysis. 616
Table 33.2 Parameter change and its effect on remaining oil life.. 617
Table 33.3 7-Best practice principles of oil sampling. 620
Table 33.4 Conservatively recommended oil sampling intervals for different equipment categories. 628
Table 33.5 Selecting oil analysis tests by application. 630
Table 33.6 The origin of viscosity changes. 633
Table 33.7 Additive depletion tests. 636
Table 33.8 Contaminant severity factor (CSF).. 641

Table 33.9 Make-up of water concentration in oil at different temperatures. 644
Table 33.10 Oil analysis data interpretation/problem identification. 654
Table 38.1 Lubrication certification comparison guide. 727

SECTION 1

Lubrication Fundamentals

1

Friction and Wear

1.1 Introduction

It is often stated that in order to lubricate correctly, one must follow the "four" rights of lubrication, often referred to as "the law of the Four R's." The law simply states that "Good Lubrication Practice (GLP) depends on delivering the Right lubricant, in the Right place, in the Right amount, at the Right time." In doing so, friction and wear can be minimized.

In the 1970's, MIT Mechanical Engineering emeritus professor Ernest Rabininowicz published what has now become known as the "Robinowicz Law" that simply states "six percent of the US Gross Domestic Product (GDP) is lost through mechanical wear."

Rabinowicz performed ground breaking work during his term at MIT into how machines and their bearing surfaces lose their usefulness. Figure 1.1 shows clearly that 70% of [bearing] surface degradation is due to mechanical wear and corrosion—all combatable through effective lubrication practice, and the application of the four R's.

Consider that in 2021 the US GDP increased to $22.9 Trillion; the amount of loss due to friction and wear according to Rabinowicz's law translates to a whopping $1.38 Trillion!

1.2 Friction

In the mid-1960s, the world took note of a remarkable ground breaking study performed for the British government by Mr. H. Peter Jost who for the first time quantified the effects of ineffective lubrication practices on Britain's GNP (Gross National Product). In his initial report, now referred to as the *Jost Report*, Jost coined the word Tribology to describe the science of lubrication, friction and wear. For the first time, lubrication was internationally recognized as a bone-fide science and practice in the corporate world of asset reliability, and for its significant positive fiscal impact on industry, and the country's GNP when practiced effectively.

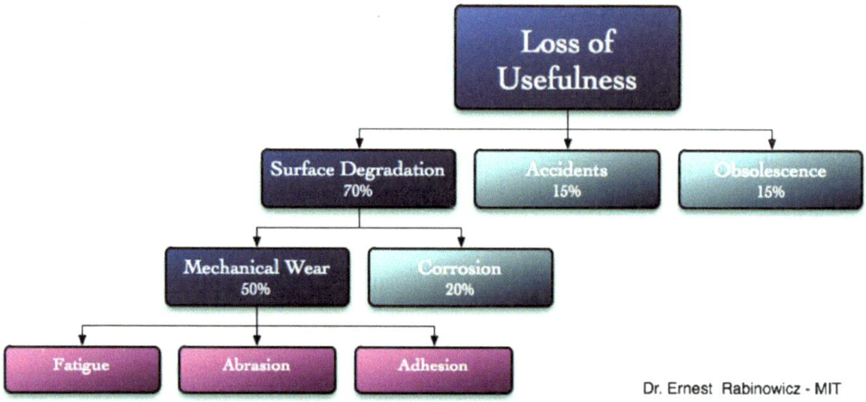

Figure 1.1 How bearings and Machines lose their usefulness.

Tribology can be broken down into the following four elements:

1. The design and manufacture of surface materials where two or more surfaces interact,
2. The correct combination of materials where surfaces interact,
3. The interactions of wear surfaces with lubricants,
4. The reduction of friction and wear.

The word Tribology is derived from the Greek word *tribos* meaning, "to rub". It is used to describe what happens when two mating surfaces move across one another. Sir Isaac Newton first recognized the resulting resistive force causing this "rubbing" action in his laws of motion, as *"an external force to motion that must be overcome for motion to occur"*. The force described became known as friction.

Webster's dictionary describes friction as *"the force, which opposes the movement of one surface sliding or rolling over another with which it is in contact."* Simply put, friction is the resistive force that retards motion.

Friction isn't necessarily a bad force. We employ frictional forces when we want to intentionally retard a body in motion. E.g. when slowing down a rotating machine, or slowing an automobile by applying a rough and soft consumable braking material with a high coefficient of friction against a smooth hard (less-consumable) surface. Friction becomes an undesirable force when it robs energy from an applied force used to intentionally move an object; estimates blame frictional forces for consuming over one third of the world's energy! When ignored in such cases, friction causes heat, wear, and sometimes, catastrophic failure of the bearing surfaces.

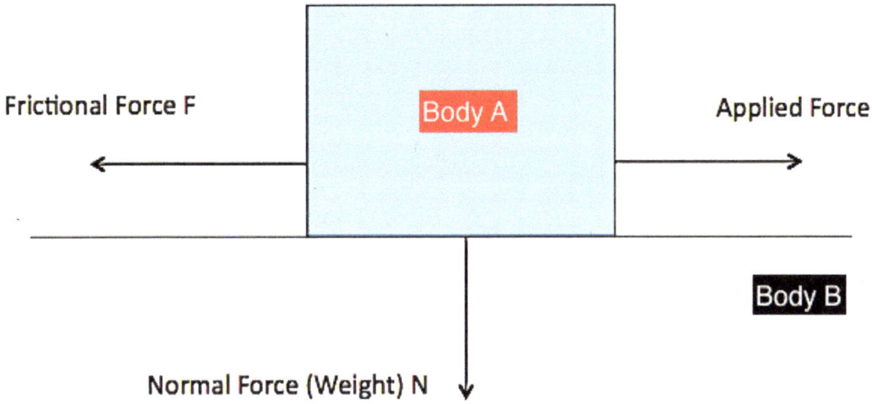

Figure 1.2 Forces on bodies at rest.

To better understand friction, we must recognize that there are two unchanging fundamental laws that govern friction:

1. Friction varies directly with load, and
2. Friction is independent of surface area

Figure 1.2 depicts the forces at play on two bodies at rest; to begin to move body A across body B we must first overcome its resistive frictional force in which the resistive force is a result of load N (the weight of body A) multiplied by the coefficient of friction. When bodies are rigid as shown, the frictional force is known as *solid* friction. Solid friction may be *static* or *kinetic*—the former encountered when initiating movement of body A from a rest position, the latter being a reduced force when body A is in motion. (Distinct from solid friction is *fluid friction*, a normally less resistive force that occurs between the molecules of gas or liquid in motion). As will be seen in later discussions, lubrication generally involves the substitution of high solid-to-solid friction for the much lower fluid friction.

Example: if body A depicted in Figure 1.2. represents a wooden steamer trunk full of books resting on a concrete floor (body B), using the formula $F = \mu N$, we can calculate the initial (static) resistive force (F) we need to overcome to start the trunk moving across the floor. Therefore, if we assume the normal force (N) or weight of the loaded trunk is 100lbs, and the coefficient of friction (μ) of wood on dry concrete is 0.62, the applied force required to start the trunk moving would be $0.62 \times 100 = 62$lbs.

As depicted in Fig 1.3, once the trunk has begun to move the static friction barrier has been broken and the kinetic force required to keep the trunk moving reduces somewhat as long as the body remains moving.

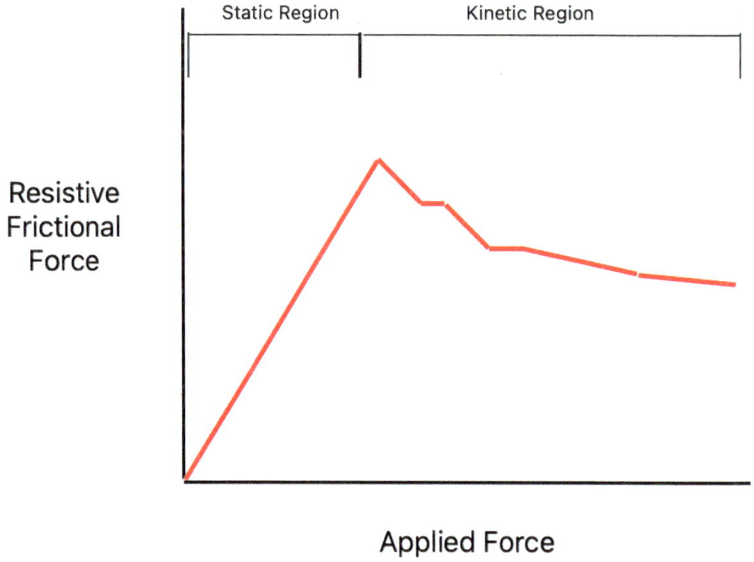

Figure 1.3 Solid friction—static versus kinetic.

The Coefficient of Friction (COF), denoted by the Greek letter μ, (mu), is different for every material and fluid with values ranging from almost 0 to well above 1. The lower the COF value, the lower the resistance and motion retardation effect—see Table 1.1 for typical COF values. Whenever we want to produce work from moving parts, lower COF values are preferred as they require less energy expenditure to achieve movement, or work. i.e., the motor requires less amperage draw, or the engine requires less fuel to achieve the desired work performance.

If industry were only to allow surface-to-surface contact on all moving machinery parts it would expend enormous amounts of energy! To offset/reduce the combined static and kinetic frictional forces present between moving surfaces, a protective fluid film must be introduced to separate the surfaces and lower friction.

1.2.1 Causes of solid friction

Solid friction, or *sliding friction*, as it is sometimes known, originates from two widely differing sources. The more obvious source is surface roughness. No machined surface, however polished, is ideally smooth. Although modern machinery is capable of producing finishes that approach perfection,

Table 1.1 Typical coefficient of friction values.

Materials		Dry μ	Lubricated μ
Brass	Steel	0.35	0.19
Sintered Bronze	Steel	0.16	Self lubricated
Cast Iron	Oak	0.49	0.075
Graphite	Steel	0.1	Self lubricated
Iron	Iron	1.0	0.2
Teflon	Steel	0.05-0.2	No lube required
Teflon	Teflon	0.04	No lube required
Wood	Concrete	0.62	
Copper Lead	Steel	0.22	
Tire	Dry Road	1.0	0.6 (wet)
Tire	Snow	0.3	

microscopic irregularities inevitably exist. Minute projections on a surface are called *asperities,* when two solids rub together, interference between opposing asperities will account for a considerable portion of the friction, especially if the surfaces are rough.

The second cause of sliding friction is the tendency for the flatest areas of the opposing surfaces to weld together, caused by the high temperatures produced by friction under heavy loads. Rupture of the tiny bonds created in this manner is responsible for the majority of the friction that occurs between machined parts. On finely ground surfaces, these minute welds constitute a major source of potential frictional resistance.

1.2.2 Effect of friction

Whenever friction is overcome, dislocation of the surface particles can generate heat. Excessive temperatures developed in this way can be destructive. The same frictional heat that ignites a match is similar to that which "burns out" the bearings of an engine, Figure 1.4.

Additionally, where there is solid friction (sometimes referred to as "dry" friction), there is wear, described as a loss of material due to the cutting action of opposing asperities, and to the shearing of infinitesimal welded surfaces. In extreme cases, welding may actually cause seizure of the moving

Figure 1.4 Burnt out engine white metal bearing due to frictional heat caused by lubricant starvation in the bearing area. Source: ENGTECH Industries Inc.

parts. Whether a piston ring, gear tooth, or journal is involved, the harmful effects of friction are difficult to overemphasize.

One of the tasks of the engineer is to control friction—to increase friction where friction is needed (brake linings) and to reduce it where it is objectionable (bearing surfaces). It has long been recognized that if a pair of sliding bodies are separated by a fluid or fluid-like film, the friction between them (sometimes referred to as "wet" friction) is greatly diminished. A barge can be towed through a canal (wet friction) much more easily than it can be dragged over, say, a sandy beach (dry friction). Figure 1.5 illustrates this fact.

1.2.3 Lubrication

The principle of supporting a sliding load on a friction-reducing film is known as lubrication. The substance of which the film is composed is a lubricant, and to apply it is to lubricate. When mechanical moving parts are present, the amount of lubrication required is dependent on the state of friction, which can manifest itself in three distinctive ways identified as:

Figure 1.5 Effects of fluid versus solid friction. Source: Exxon Corporation, USA, Practical Lubrication for Industrial Facilities – 2nd Edition (PLIF 2)

1. Sliding Friction
2. Rolling Friction
3. Combination Friction—Sliding and Rolling

Sliding friction is common where any plain surface moves over one another evident in plain bearings where a journal moves within a sleeve. Sliding friction arrangements require the most lubricant, as the friction is evident over a larger surface contact area.

Rolling friction is found in all rolling element bearings that at one time were described as "friction-less" bearings. The 'line" bearing contact surface is considerably smaller than in sliding friction bearings, thereby requiring much smaller amounts of lubrication to provide a protective full fluid film.

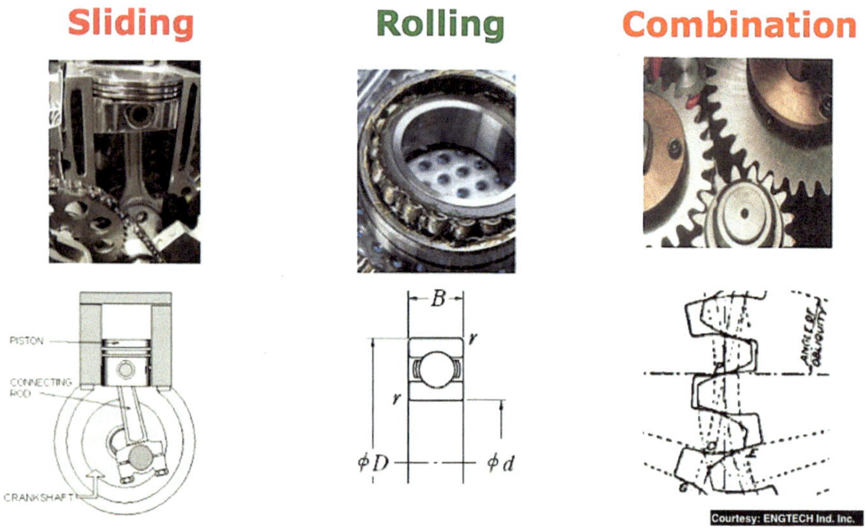

Figure 1.6 Three friction states. (Courtesy ENGTECH Industries Inc.)

Combination friction is unique to meshing gears due to the changing gear tooth profile when teeth mesh together and begin to slide on one another until the opposing pitch surfaces meet and rolling friction takes over as they disengage. Certain types of gears such as hypoid gears and worm gears produce much higher degrees of sliding friction. Figure 1.6 demonstrates the three friction states.

Whenever moving parts are employed, friction is ALWAYS present. Understanding friction helps us develop effective lubrication practices that in turn allow us to tame the harmful effects of friction and increase the component life cycle.

The principle of supporting a sliding load on a friction-reducing film is known as lubrication. The substance of which the film is composed is a lubricant, and to apply it is to lubricate. These are not new concepts, nor, in their essence, particularly involved ones. The ancient Egyptian warriors lubricated the axles of their war chariots with animal fat millenniums ago.

Since those days, modern machinery has become many times more sophisticated and complicated, and the demands placed upon lubricants have become proportionally more exacting. Though the basic principle still prevails—the prevention of metal-to-metal contact by means of an intervening layer of fluid or fluid-like material—modern lubrication has become a complex study.

In conclusion, understanding tribology, friction and wear is critical to:

Figure 1.7 Typical x2000 magnified ground steel surface.

- The reliability of mechanical machinery and electrical switching,
- The availability of mechanical machinery and electrical switching,
- The maintainability of mechanical machinery and electrical switching,
- Component life cycle,
- Corporate finances

1.3 Machine Surface Break-in

Though modern tools are capable of producing parts with close tolerances and highly polished surfaces, many machine elements are too rough, when new, to sustain the loads and speeds that they will ultimately carry. Frictional heat resulting from the initial roughness of mating parts may be sufficient to damage these parts even to the point of failure. This is why a new machine, or a machine with new parts, is sometimes operated below its rated capacity until the opposing asperities have been gradually worn to the required smoothness. Figure 1.7 shows a typical x2000 magnified ground metal surface with a

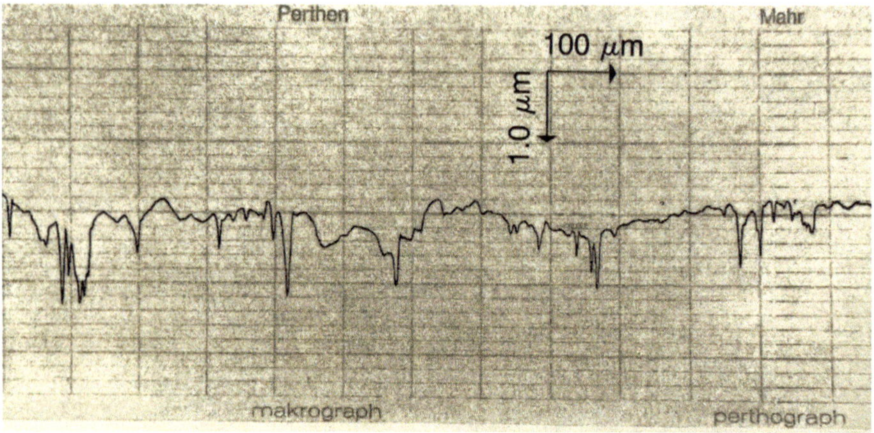

Figure 1.8 Ground metal surface profilogram.

surface roughness of 10 m. Figure 1.8 shows the same surface when traced with a laser beam to produce a cross sectional profilogram.

Under break-in conditions, it is sometimes necessary or advantageous to use a lubricant fortified with Extreme Pressure (EP) additives. The chemical interaction of these agents with the metal tends to remove asperities and leave a smoother, more polished surface. As the surface finish improves during initial run-in, the need for an EP lubricant may be reduced or eliminated, and it may then be appropriate to substitute straight mineral oil or EP oil with less chemical activity.

1.4 Bearing Metals

The break-in and operating characteristics of a journal bearing depend to a large extent upon the composition of the opposing surfaces. In the region of partial lubrication, friction is much less if the journal and bearing are of different metals. It is customary to mount a hard steel journal in a bearing lined with a softer material, such as bronze, silver, or babbitt.

There are several advantages in a combination of this sort. The softer metal, being more plastic, conforms readily to any irregularities of the journal surface, so that break-in is quicker and more nearly perfect. Because of the consequent closeness of fit, soft bearing metals have excellent wear properties. Moreover, in the event of lubrication failure, there is less danger of destructive temperatures. Friction is lower than it would be if steel, for example, bore directly against steel.

If temperature should rise excessively in spite of this protective feature, the bearing metal, with its lower melting point, would be the first to give way.

Yielding of the bearing metal often prevents damage to the journal, and replacement of the bearing lining is a relatively simple matter. However, composition of bearing metals has no effect upon performance under full-film lubrication.

1.5 Wear

Even with the most perfectly lubricated parts, some physical wear is to be expected. Sometimes wear is so slight as to be negligible, as in the case of many steam turbine bearings. Turbines used to generate power operate under relatively constant loads, speeds, and temperatures, a situation that leads to the most effective sort of lubrication.

Many other machines, however, operate under less ideal conditions. If they stop and start frequently, there will be interruptions of the lubricating film. Also, in any lubricating process, there is always the possibility of abrasive wear due to such contaminants as dirt and metallic wear particles. Wear is further promoted by overloading, idling of internal combustion engines, and other departures from optimal operating conditions.

1.5.1 Wear versus friction

Wear and friction generally go hand-in-hand, there are extreme situations in which this is not so. Some slow-speed bearings are so heavily loaded, that an oil of the highest viscosity is required for complete lubrication. Because of the greater fluid friction, this lubricant imposes more bearing friction than a lighter lubricant would.

On the other hand, the lighter lubricant, since it would provide only partial lubrication, could not be considered suitable from the standpoint of protection of the metal surface. Some frictional advantage must be sacrificed in favor of an improvement in wear characteristics. Contrary to popular thought, therefore, less wear actually means more friction under extreme conditions such as this.

1.6 Common Wear Mechanisms

There are four common wear mechanisms that cause surface degradation and eventual loss of usefulness in bearing surfaces.

1.6.1 Abrasive wear

Abrasive wear occurs when bearing surfaces run in a Mixed Film (MF) or a Boundary Layer (BL) Lubricant regime. Abrasive wear can occur as a result of

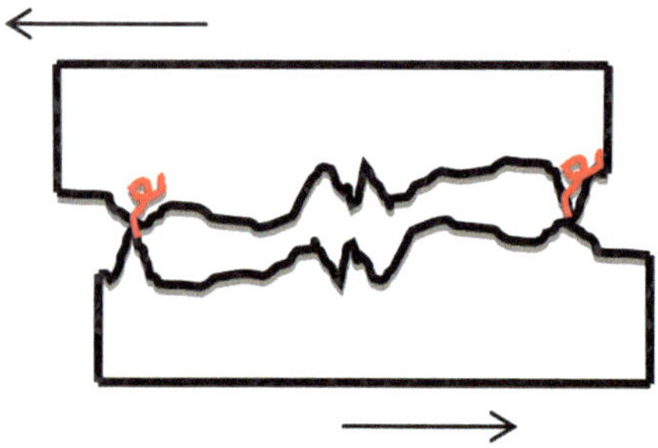

Figure 1.9 2 Body abrasive wear.

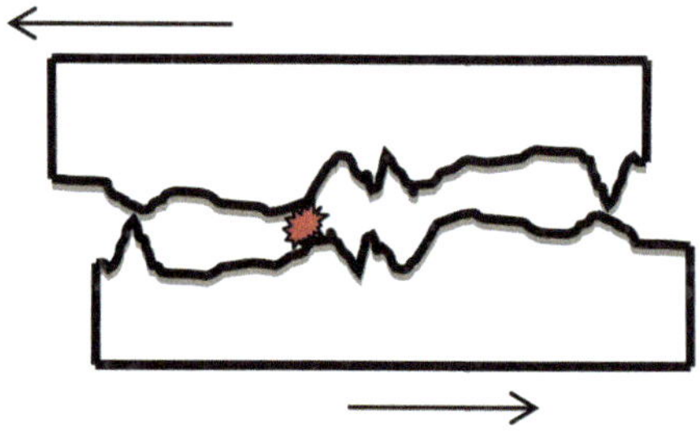

Figure 1.10 3 Body abrasive wear.

a 2-body surface interaction. As seen in Figure 1.9, the 2-body abrasion example depicts two surface points touching and cutting into the opposing sliding surface resulting in a scratched, grooved, or furrowed surface. In Figure 1.10 a third body metal cutting is shown being released into the lubricant. This third body is now free to also get caught between two surface points and add to the surface degradation. 3-body abrasion can also occur due to large particle (dirt) found or introduced into the lubricant from the lubrication process or machine operation. See Figure 1.4 for a physical example of abrasive wear. Erosive wear is a form of abrasive wear caused by particles impacting a surface.

Abrasive wear can be defined by the following:

- Most common and most preventable wear mechanism
- Also referred to as plowing, cutting and gouging wear,
- Identified by surface scratching, furrows and grooves,
- Primarily, caused by solid contamination (dirt), surface hardness level, and lubrication film thickness (regime),

1.6.2 Adhesive wear

Adhesive wear depicted in Figures 1.11, and 1.12, usually occurs under highly loaded sliding friction when an incorrect viscosity lubricant is used, when EP and AW additives have depleted, or under heavy shock loading. As the surfaces come together, they weld under the heat and load pressure, metal is transferred and torn apart under movement leaving a discreet, often jagged or smeared surfaces. Adhesive wear is also described as scuffing, shearing, or galling wear.

Adhesive wear can be defined by the following:

- Welding of surfaces, creating metal transfer and tearing,

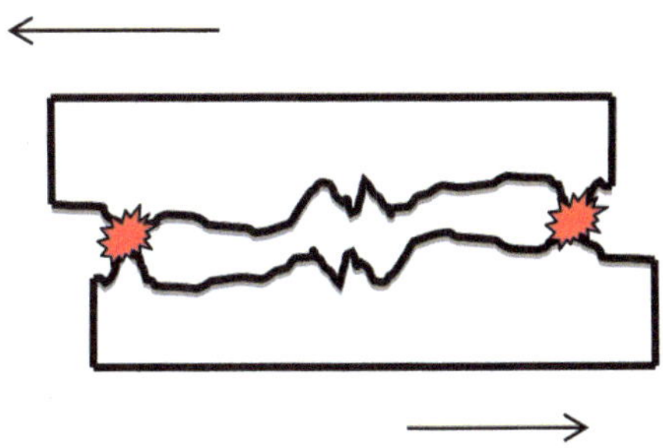

Figure 1.11 Adhesive wear.

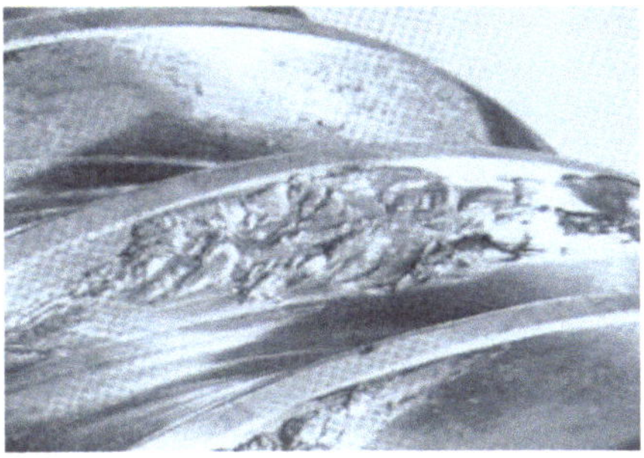

Figure 1.12 Metal tear due to adhesion.

- Also referred to as scuffing, shearing, galling and smearing,
- Primarily, occurs during highly loaded sliding friction,
- Leaves discreet, rough, jagged, and smeared surfaces,
- Caused by shock loads, incorrect lubricant use, lack of EW and/or Anti-Wear (AW) additives

1.6.3 Fatigue wear

Fatigue wear usually occurs in rolling friction surfaces that have experienced repeated long-term load cycles and stress that causes elastic deformation to the surfaces as in Elasto-hydrodynamic lubrication (EHDL) lubrication. This long-term action eventually results in small surface and sub-surface cracking – see figure 1.13, which eventually travel up to and across the bearing surface resulting in surface delamination and surface pitting.

Fatigue wear can be defined by the following:

- Surface and sub-surface cracking caused by repeated long term load cycles and stress due to elastic deformation at the surface.
- Surface delamination and pitting,
- Usually occurs in rolling friction surfaces that work in elasto-hydrodynamic fluid regimes,

1.6.4 Corrosive wear

Corrosive wear, depicted in figure 1.14 leaves an acid etched surface in the bearing contact surface area when surfaces corrode due to an oxidative chemical reaction that is accelerated in the presence of moisture contamination

Figure 1.13 Surface cracking and pitting from metal fatigue.

Figure 1.14 Acid etched corrosion on a rack gear.

and heat. Corrosive wear is also known as acidic pitting and is caused by a lubricant that is moisture contaminated, one that is additive depleted, or use of a lubricant with no corrosion inhibitor additive.

Corrosive wear can be defined by the following:

- Etched and pitted surfaces in the bearing contact surface area
- Surface damage caused by oxidative chemical reaction accelerated by moisture contamination,
- Also referred to as acidic pitting,
- Caused by lubricant that is corrosion additive depleted, or use of incorrect lubricant with no corrosion inhibitors

2

Functions of a Lubricant

2.1 Lubricants

The old adage "oil is oil, so any old oil will do!" may have had merit in a past agrarian society, but in today's world of sophisticated machinery and demand for asset reliability, choosing the correct lubricant is a highly important and informed decision.

Whether in liquid, solid or gas form, modern lubricants are pure liquid engineering. Through the blending of additives to a variety of different base stocks they can be designed to perform up to eight functions simultaneously whilst operating in a host of differing environments.

Webster's dictionary describes a lubricant as *"a substance (e.g. oil, grease, or soap) that when introduced between solid surfaces that move over one another, reduces resistance to movement, heat production, and wear (i.e. friction and its effects) by forming a fluid film between [mating] surfaces."*

Essentially, a lubricant's function is to control and minimize the sacrificial harmful effects of moving surfaces passing over one another under load, and at speed. It does this in the following eight definitive ways:

Function 1—Control and minimize friction:

The primary function of every lubricant is to control and minimize the effects of friction.

When two solid surfaces passing over one another are allowed to come into contact under load, they "rub" together and produce dry friction that requires considerable energy to keep the surfaces moving. With no lubricant present to separate the moving surfaces from one another, surfaces quickly degrade and can weld or lock together resulting in a "seize." The indiscriminate sacrifices of wear surfaces produces accelerated wear, temperature spiking, decreased asset performance, reduced reliability, and increased energy use.

The introduction of a lubricating film between the two wear surfaces serves to create a fluid barrier that prevents surface contact. Although a small amount of fluid friction is still present within the lubricant film, the energy required to move the surfaces over one another is but a small fraction of that required to overcome "surface to surface" *dry* friction.

Function 2—Control and minimize wear:

Knowing that a full lubricant film may not always be possible and that some metal-to-metal contact may occur under slow moving, heavy load, lubricant loss conditions, additives can be added to the lubricant that act as chemical "softening" agents on the metal surfaces.

The lubricant coats the surface with soft layers of metallic salts (sulfides and phosphate additives). As the two surfaces slide over one another alternating load cycles can cause the softened high points (asperities) on each surface to collide with one another when film thickness is reduced. When the unit loading exceeds the sulfur phosphide film, a rupture can occur creating a localized metal-to-metal contact area. Concentrated heat builds up causing the two surfaces to "weld and break" that results in small metal particulate, or asperity release into the lubricant film.

Many lubricants are designed to control wear by promoting minute surface degradation to allow asperity "tips" to be easily sacrificed and flushed away without "tearing" parent metal, thereby minimizing surface wear under varying lubricant film conditions.

Function 3—Control and minimize heat:

Whenever friction and wear levels are controlled and minimized, the amount of heat is also reduced. Excessive heat can "cook" the lubricant and cause it to oxidize rendering it less effective. To combat this, an anti-oxidant additive is added to the lubricant base stock.

Circulating oil system and air/oil system designs take advantage of a lubricant's ability to transfer localized heat build up at a bearing load point and prevent any thermal runaway at the bearing surfaces. To facilitate the heat transfer/cooling process the oil can be pumped through a heat exchange unit (oil cooler), and/or reservoir baffle system. In energy conservation environments, the heat exchanger can be employed to redistribute heat into the plant during the winter months.

Function 4—Control and minimize contamination:

As described above, a lubricant can become contaminated when wear asperities are introduced into the lubricant. Other forms of contamination, such as silica (dirt) can be introduced from the reservoir filling process when proper storage, transfer and cleanliness practices are not observed, or can enter the system when seals become compromised, or when breathers or fill caps are not firmly in place.

To remove contaminant solids, detergent and dispersant additives work together with lubricant flow to clean surfaces, neutralize acids and colloidal suspend particulate to ensure all contaminates are delivered to and extracted under pressure by the in-line oil filter.

Failure to refresh the oil filter on a regular basis can cause the contaminated lubricant to act as a "lapping" paste and accelerate the wear process in the bearing areas.

In the case of water or glycol contamination demulsifier additives are added to facilitate capture and release of moisture in the sump or filter. These additives types are more prevalent in automotive oils.

Grease lubricants can also act to seal out contamination ingress around shafts, this is especially so in the case of a labyrinth seal that depends on grease to fill up a series of annular grooves cut into the non-moving shaft housing, designed to act as a live shaft seal.

Function 5—Control and minimize corrosion:

Oxygen may be a basic human life force, but can be fatal to a lubricant. When present it acts as catalyst to combine metals and organics that can catalyze and generate corrosive acids harmful to the bearing surfaces. For ferrite (iron based) wear surfaces, acid(s) can attack the metal and form rust on the bearing surface.

A lubricant is designed to "cling" to the metal surfaces so as keep out moisture and oxygen from reacting with the surface. As not all lubricants are designed equal, if the bearing surfaces are iron based, a lubricant with anti-corrosive additives must be employed.

Function 6—Control and minimize shock:

Most will be familiar to the quieting effect of adding lubricant to a gear train or bearing area. This is due to the lubricant acting as a hydraulic shock absorber between mating gears and components as they mesh. When poorly lubricated, meshing gears set up shock waves as they start to mesh resulting in a "chattering" sound that can lead to physical fracture of the gear teeth.

The very word "shock absorber" is synonymous with automobile suspension systems that employ hydraulic oil to dampen and absorb the effects of road shock on the vehicle.

Function 7—Control and transmit power:

In a typical hydraulic system, hydraulic oil is used to transmit force and motion from a single source (usually a pump) into multiple sources, pistons, accumulators, etc. this occurs in real time due to the oil's incompressibility.

Hydraulic oil is also used to transmit power in soft start devices such as fluid couplings, automatic transmissions and torque converters.

Function 8—Control and minimize energy consumption: Effective lubrication practice dictates use of the Right lubricant, in the Right place, at the Right time, in the Right amount, using the Right method. Doing so, will ensure the asset is using the least amount of energy where moving parts are involved.

In studies[1] conducted on behalf of various electric power companies, effective use of lubricants, delivery systems and lubrication methods resulted in an energy reduction of 7.5% when a synthetic lubricant was used to replace a standard compressor oil, and a 17.92% energy reduction was achieved on a 1000-ton straight side stamping press when the automated "pump-to-point" oil delivery system was set up correctly and a more suitable oil viscosity was chosen!

All liquids will provide lubrication of a sort, but some do it a great deal better than others. The difference between one lubricating material and another is often the difference between successful operation of a machine and failure. Mercury, for example, lacks the adhesive and metal-wetting properties that are desirable to keep a lubricant in intimate contact with the metal surface that it must protect. Alcohol, on the other hand (Figure 1.6), wets the metal surface readily, but is too thin to maintain a lubricating film of adequate thickness in conventional applications. Gas, a fluid-like medium, offers lubricating possibilities—in fact, compressed air is used as a lubricant for very special purposes. But none of these fluids could be considered practical lubricants for the multitude of requirements ordinarily encountered.

2.2 Base Oil Types

All finished "ready to use" lubricants are manufactured to a proprietary formula, each designed to combat friction according to the differing design, function, and operating conditions under which bearing surfaces must be kept separated.

Lubricating oil, whether in its liquid form or thickened form known as grease, is a blended product consisting of base oil stock and additive package. The base oil percentage of the finished oil can range from 75% up to 99% depending on how much additive package is required to enhance, suppress,

Table 2.1 8-Functions of a lubricant.

CONTROL & REDUCE	• Friction • Wear • Shock • Temperature • Corrosion • Contamination • Power transmission • Energy

or contribute new properties to the oil. Modern Base oil stock is derived from three primary sources that classify the oil as an animal/vegetable, petroleum (mineral), or synthetic lubricant product. For more

2.2.1 Animal/Vegetable base oil

The industrial revolution ran on olive oil and rendered animal fat for a long time until the animal/vegetable oil's inability to arrest rapid acid formation under ever-increased speeds and loads was solved with the discovery of crude petroleum (mineral) oil. This oil classification is now generally reserved for cooking purposes only.

2.2.2 Petroleum base oil

The basic petroleum lubricant is lubricating oil, often simply referred to as "oil." Often defined as mineral based oil, petroleum base oil stock is overwhelmingly the most popular base oil stock in use today. For almost every situation, petroleum products have excelled as lubricants. Petroleum lubricants are high in metal-wetting capability and possess the body, or viscosity characteristic that a substantial oil film requires.

Petroleum oil must be refined to remove impurities that include aromatic hydrocarbons, sulfur compounds, acids and wax, so as to improve the base oil's desirable properties.

These properties include viscosity index, pour point, and stability. Refining crude oil also separates the crude oil molecules by size and weight to produce a variety of petroleum based products of which lubricating base

Figure 2.1 Petroleum oils make excellent lubricating films. Source: Exxon Corporation, USA, (PLIF 2)

stock oil constitutes between 1-2% of a barrel of crude oil's yield (gasoline is the highest at approximately 25% yield).

Where in the world a crude oil originates will establish its base oil properties and eventually what kind of service application the finished lubricating oil will be used in, determined by the levels of Paraffin and Naphthene present in the crude stock.

All petroleum-based oil is a complex mixture of hydrocarbon molecules representing one of the most important classifications of products derived from the refining of crude petroleum oils, and is readily available in a great variety of types and grades reviewed more closely in later chapters.

2.2.3 Synthetic base oil

Synthetic base oil stocks are man-made and designed to have an improved and more uniform molecular structure, allowing them to display more predictable fluid properties with the ability to work better under severe conditions less suited to petroleum oil stocks. Synthetic lubricants are reviewed more closely in Section 2 – Lubricants.

2.3 Base Oil Properties

The quality of any base stock oil is measured by its resulting properties that define how well the oil will perform in service, and what additives will be required to enhance its performance. There are five major properties identified in a base oil specification.

2.3.1 Viscosity

To understand how oil enters a bearing, picks up and carries the bearing load requires an understanding of viscosity. With lubricating oils, viscosity is arguably its most important property, and much of the story of lubrication is built around it.

Defined by the oil's molecule size, viscosity is recognized as the oil's "measure of resistance to flow". With larger molecule sizes, the oil's resistance to flow will increase while slowing down the flow rate. Higher viscosity, or "thicker" fluids like molasses will have a slow flow rate, whereas lower viscosity, or "thinner" fluids like water will flow faster and more easily as shown in Figure 2.2.

The viscosity of a particular fluid is not constant and can change based on the ambient temperature (as depicted in Figure 2.3), and load. When the temperature increases the lubricant becomes thinner and the viscosity decreases. Inversely, as the temperature decreases, the lubricant thickens and the viscosity increases, as illustrated in Figure 2.3, making it more difficult to flow. Therefore, a numerical number for viscosity is meaning-less unless accompanied by the temperature to which it applies.

When a lubricant is subjected to extreme load, its viscosity will increase. This is a phenomenon experienced in the elastohydrodynamic lubrication (EHDL) film state found in rolling element bearings when the ball or roller moves into the direct loading contact area known as the Hertzian contact area, causing it to elastically deform, trap and pressurize the lubricant momentarily raising its viscosity and causing it to change state from a fluid to a solid, and back again, as the roller or ball moves through the direct load area. For further details, reference the Lubrication Film Regime section later in this chapter.

Base oils are rated with a viscosity number according to a recognized viscosity index numbering system. The most common used imperial system for industrial lubricating oils is the SUS (Saybolt Universal Seconds) rating index, which charts the viscosity rating at two different temperatures of 100F and 210F. Its equivalent metric rating is the ISO VG rating index that follows

Figure 2.2 High viscosity fluids like molasses flow slower then low viscosity fluids like water. Source: Engtech Industries Inc.

the Kinematic viscosity rating measured in centistokes @ 40C and 100C. For example, ISO VG 220 gear oil is the equivalent viscosity to an SUS 1000 @100F gear. Different oil types have their own rating systems as shown in the Equivalence of Viscosity Grading System chart found in the Appendix section

2.3.2 Multi-grade viscosity oils

When it comes to automotive engine crankcase oils, choosing the right oil viscosity grade will depend on where you live and operate your vehicle. Northern climate use of vehicles in winter will subject oil to very cold temperatures on start up resulting in poor oil flow and lack of full film lubrication until the oil reaches operating temperature. The outcome is poor starting due to the oil thickness and drag on the engine, and excessive engine valve train wear during the start up/warm up stage due to momentary lack of lubricant being pumped into the valve train.

Figure 2.3 Oil is thicker at lower temperatures, thinner at higher temperatures. Source: Exxon Corporation, USA, (PLIF 2).

Up until 1952, all vehicle manufacturers recommended mono-grade (single viscosity grade) automotive crankcase lubricants for their vehicles and recommended the oil viscosity grade be changed to a lower viscosity number for winter driving. The viscosity grade choice would depend on the winter temperature the car would operate in. These viscosity grades were set by the SAE (Society of Automotive Engineers) grade viscosity system that designated oil viscosity with a number followed by the letter W to designate winter use oil, eg.10W, 15W, 20W weight. Or, just a number on its own to designate summer grade oils, eg. 20, 30, 40, or 50 weight oil. All that changed in 1952 when Esso (now Exxon Mobil) introduced its Esso "Uniflo" product, the world's first multi-grade oil.

Multi-grades offered motorists the advantage of single weight oil for all year round driving that delivered improved low temperature start-ups and high temperature performance.

The multi-grade designates its working range by identifying the base oil viscosity first that is designed for cold weather performance, followed by the viscosity the oil will perform as (or emulate) once it reaches operating temperature. For example, a 20W50 has relatively thin SAE20W winter base oil that thickens as it heats up so as to act similar to thicker, more viscous SAE50 weight summer oils at operating temperature.

This multi-state change provides full film protection over a wider temperature range and is achieved by blending polymeric viscosity improver additives to lower viscosity base oils. These additives are long string polymers that "curl up" like a ball at low temperatures, allowing them to move freely amongst the oil molecules. Once the oil heats up, the polymer strings then unfurl and expand to restrict the oil's flow and raise its apparent viscosity making the oil thicker.

Table 2.2 Viscosity index rating chart.

VI Rating	VI Group
<35	Low - **LVI**
35-80	Medium - **MVI**
80-110	High - **HVI**
>110	Very High - **VHVI**

2.3.3 Viscosity index

Viscosity Index, or VI, is a measure of oil's viscosity change due to temperature. Oils with higher VI rating are more desirable as they are more stable under changing temperature conditions, reflecting a narrower change in viscosity over a standard temperature range.

As previously noted, paraffinic oils have much higher VI ratings making them more stable and desirable where a wide operating temperatures range is experienced. Oils can be grouped and classified by their VI property as depicted in Table 2.2.

2.3.4 Specific gravity

Specific gravity rates the oil's density relative to water.

2.3.5 Flash point

Flash point is used to determine a lubricant's volatility, measured by the lowest temperature a lubricant can be heated to before its vapor, when mixed with air, will "flash" ignite but not sustain combustion. Paraffinic oils have higher flash points than naphthenic oils.

2.3.6 Pour point

Pour point defines the lowest temperature at which oil will still pour, or flow. Because of their level of wax content, paraffinic oils are known to have a wax pour point, which is not as low as a naphthenic base-oils that are described as having a viscosity pour point. Lower viscosity oils will have lower pour points.

An oil base stock is a canvas for the additive package that makes up the final lubricant blend designed for an intended application purpose.

2.4 How Lubricants Fail in Service

In the lubrication world, temperature presents an unusual paradox. Lubricants require heat to help them flow efficiently over and around their bearing surfaces, yet if the temperature gets too hot, the lubricant is likely to undergo a chemical change that can drastically reduce its life expectancy. Inversely, if the temperature gets too cold, the lubricant will thicken and loose its ability to flow and efficiently lubricate its bearing surfaces. In addition to this paradox, irony is also at work at work, as lubricating oil is not only designed to separate and lubricate a bearing surface, it is also designed to absorb and carry away frictional heat from the bearing surface.

The old adage "oil is oil and grease is grease" may have been true in our long ago agrarian society, but in today's high speed complex industrial environments lubricants must be tailored and managed to their machine host's specific needs and operating environment if they are to guarantee asset availability and reliability.

When a lubrication film is insufficient to separate moving surfaces the surfaces will collide and transfer energy, resulting in a rapid heat build-up and metal expansion further retarding motion until both surfaces eventually weld to one another and seize. To avoid this worst-case scenario the ideal intent is to ensure the correct lubricant is introduced into the bearing surface area in sufficient quantity to provide full separation of moving surfaces in a consistent manner. Lubricants also ensure the bearing temperatures stay relatively cool. Figure 2.4 illustrates the relationship between oil operating temperature and the oil life cycle expectancy. Note: at 100°C (210°F) oil has a life expectancy of 3 months, whereas operating at a temperature of 70°C (160°F) life expectancy now becomes 2 years.

In cold temperature climates, hydrocarbon-based oil can thicken to the point in which it will no longer pour, largely due to the wax content in the oil. More expensive hydro-treated and synthetic based oils will largely resolve this problem, or the user can choose to heat the reservoir containing the oil to a temperature that will allow it to flow again.

2.4.1 Heat related lubricant failure

In the late 19th century, the Swedish Nobel laureate Svante Arrhenius, discovered a direct relationship between temperature change and chemical reaction

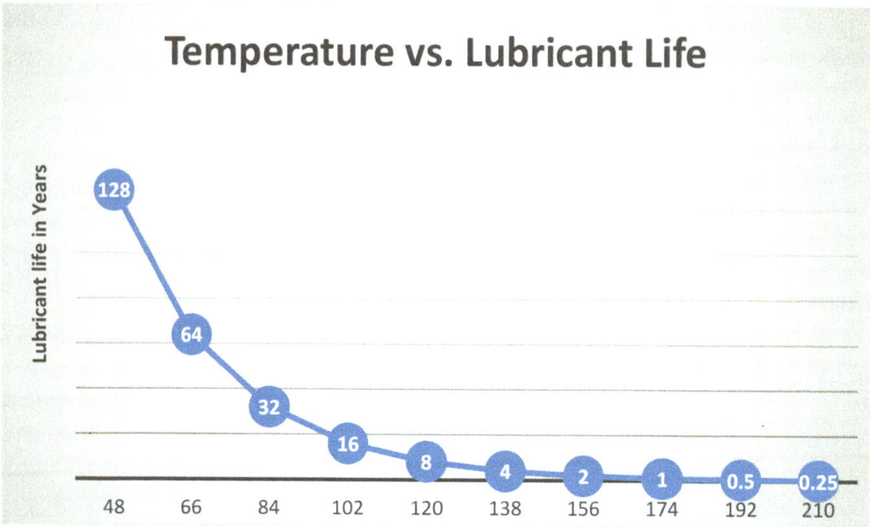

Figure 2.4 Relationship between oil operating temperature and the oil life. Source: Engtech Industries Inc.

rate in fluids, expressing this relationship in his Arrhenius equation. In the field of lubrication, we know this equation as the "Arrenhius rule", used to express the temperature change dependent failure rate of lubricating oils in the following statement, "for every 18°F (10°C) increase in oil temperature, a lubricant's lifecycle is reduced by half." Inversely, lowering oil temperature by the same rate will double its life cycle.

Two predominant failure mechanisms occur when oil is heated up, the most common categorized as oxidation failure and the lesser categorized as thermal failure.

2.4.2 Oxidation failure

Oxidation failure occurs when oxygen reacts with the lubricant base oil. Anti-foaming and anti-oxidant additive agents, if present in the oil, are designed to slow down the process but once depleted, the rate of oxidation will accelerate, especially in the presence of water and reactive bearing materials such as copper and iron, causing the oil to become oxidized. In an oxidized state, the hydrocarbon molecules in the oil will transform into a greasy sludge containing harmful corrosive acids. These will cause the oil to degrade and lose its lubricating properties and manifest themselves through an increase in lubricant viscosity, specific gravity, acidity (TAN), rapid additive depletion, a darkening of the oil, sour "rotten egg" odor and a varnishing of the bearing surfaces.

2.4.3 Thermal failure

Thermal failure can occur due to transference of a localized or external heat source to the lubricant, or through the adiabatic compression (the thermodynamic process that occurs when entrained air compresses and heats up according to Boyle's law) of entrained air bubbles in pumps, bearings, and even in pressurized hydraulic and lubricating systems.

The resulting heat causes the lubricant to decompose, and its corresponding hydrogen loss to create carbon rich particles in the form of sludge and carbon deposits. These effects manifest themselves as a *decrease* in lubricant viscosity, now a darkened fluid with greasy suspensions that smell of burned food, with evidence of coking and varnishing on the bearing surfaces.

Once a lubricant has failed, the molecular change state is usually irreversible requiring at the very least a lubricant change. If a lubricant failure is possible, this can be monitored through an oil analysis program. A lubricant's propensity to fail can be checked by subjecting a sample to a RPVOT (Rotary Pressure Vessel Oxidation Test) that simulates failure through a speeded-up oxidation process to give the maintainer an indication of a lubricant's suitability. This test can take a virgin oil sample or existing oil sample and predict its remaining life.

2.4.4 Practical prevention of lubricant temperature related failures

Whether your lubricant choice is grease or oil, once the correct one is chosen and employed it will require assistance from the maintainer to ensure the lubricant has a fighting chance to perform and deliver a reasonable life expectancy. This can be achieved by implementing some of the following best lubrication practices:

- Ensure the lubricant transfer delivery systems are immaculately clean so as not to allow the ingress of solid contamination that can create sludge, raise the lubricant viscosity and accelerate the oxidation process,
- Ensure the oil/grease delivery method/system is tuned to deliver the correct amount of lubrication in a timely manner. Over-lubrication will create fluid friction heat compounded by the bearing ball/rollers

working overtime to mechanically push through the excess lubricant, both causing the lubricant to heat up rapidly. Under-lubrication can allow the bearing to go into boundary lubrication creating surface interaction frictional heat that can "cook" the lubricant,

- Ensure the oil reservoir level is between the Min and Max fluid level so as not to create cavitation in the oil pump or oil churn in the reservoir that can result in air bubbles accelerating the oxidation process,
- Ensure oil reservoirs are clean and free of debris. Dirt, dust and debris can create the effect of a thermal blanket and raise the temperature of the oil inside the reservoir,
- Ensure the integrity of shaft seals. Poor shaft seals lead to excessive lubricant leakage that can quickly lead to an under-lubrication state,
- Implement a lubricant test and control process to ensure incompatible lubricants are not mixed together in the same bearing space that can lead to a variety of detrimental conclusions, including overheating,
- Where hydrocarbon base lubricants are employed in cold weather climates, use timed block heaters or blanket wrap element heaters for reservoir and drum/pail heating. If a lubricant is to protect a bearing surface it must readily flow across the bearing.

2.4.5 Additive depletion failure

Lubricants are an integral part of any mechanism or machine and are an engineered choice. Their life cycle is dependent on machine/mechanism load, speed, ambient condition factors (temperature, humidity, water, dirt) and the level of maintenance and operator interaction.

Every finished lubricant product is a combination of base oil plus additive package, making every lubricant product not only unique but also consumable due to the additive package design. By nature, additives are designed to deplete as the oil is employed in service and their rate of depletion is wholly dependent on the additive type and working conditions. Additives can deplete, or fail, in three different ways:

2.4.6 Decomposition

Although it is called decomposition, additives actually experience numerous metamorphic changes caused by a variety of causes listed below:

- **Hydrolysis**— Zinc dialkyldithiophosphate (ZDDP) additive + water and heat change to sulfuric acid and hydrogen sulfide
- **Neutralization**—Acid and Base (detergent additives) neutralize and change into salt and water
- **Oxidation**—ZDDP and hindered Phenol additives change into Polymers and precipitants
- **Thermal Degradation**—ZDDP and EP additives + heat change into Phosphates, Phosphides, Sulfates, Sulfides

2.4.7 Separation

Sometimes referred to as mass transfer, additives are lost, or transferred, due to the system design in the following manner:

- **Aggregate Adsorption**—Adsorptive filter media removes polar additives
- **Centrifugation**—heavy organic and metals separate under centrifugal forces
- **Condensation Settling**—additive becomes insoluble and settles
- **Evaporation**—vacuum dehydration
- **Filtration**—larger particles are captured in the oil filter. Better quality lubricants use sub-micron particle additives that more easily pass through most inline filters

2.4.8 Absorption

Through action within the tribo-system, or through contamination (dirt and water) ingression, additives can be attracted to system elements and be removed from the lubricant in the following manner:

- **Particle Scrubbing**—dirt particle attracts additive and traps additive in filter or drops out of suspension
- **Rubbing Contact**—Polarized EP and AW additives form soap like boundary films around themselves
- **Surface Adsorption**—Polarized additives attach to machine surfaces
- **Water Washing**—Water attracts the additive and settles on the reservoir bottom

As shown in the three depletion methods, additives deplete by performing their function as they were designed to do and due to the inefficiency of the system design and operating condition. Once the additives are depleted the oil is thought as consumed and must be changed or have its additives replenished. Failure to do so will cause a loss of protection to the machine surfaces and they will degrade rapidly.

3

Lubrication Regimes

3.1 Lubricant Film Regimes

To successfully combat friction and wear, a lubricant film must be present at all times between the mating bearing surfaces. The degree of protection, and subsequent bearing surface life, is directly relevant to the lubricant's working film thickness, load, speed, and lubricant viscosity The minimum working film thickness required to achieve full surface separation is also known as the Lambda—λ thickness ratio.

Because the degree of surface separation is dependent on the surface "roughness" (Ra), it must be determined by measuring the profile (peaks and valleys of the surface) of both mating surfaces and by defining a centerline through them so that the areas above and below the centerline are equal as illustrated in Figure 3.1. The lambda ratio is then defined as the ratio of lubricating film thickness to surface roughness, which is a lubricant film thicker than the combined height of both surface asperities enough to completely separate both surfaces.

Lubrication effectiveness is predominantly governed by two principles: *hydrodynamic lubrication* and *boundary lubrication.* Figure 3.2 shows the lambda λ ratio thickness curve that depicts the relationship between the working lubricant film thickness (between the two mating surfaces) and the resulting life expectancy of the lubricated component. Note that once the lambda ratio is greater than four, life expectancy remains constant. Figure 3.10 also makes reference to different film stages, or regimes, known as Hydrodynamic (Thick Film), Mixed Film and Boundary Layer (Thin Film).

Technically, there are five distinct lubricant film states or regimes, each one describing a different relationship between two interacting surfaces as they slide over one another.

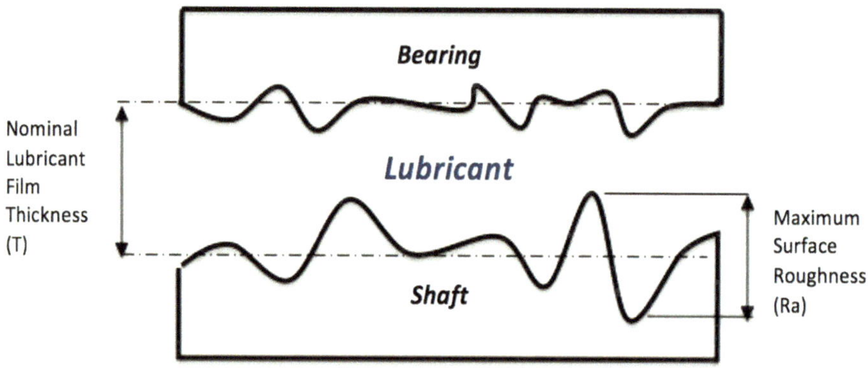

Figure 3.1 Measuring surface roughness.

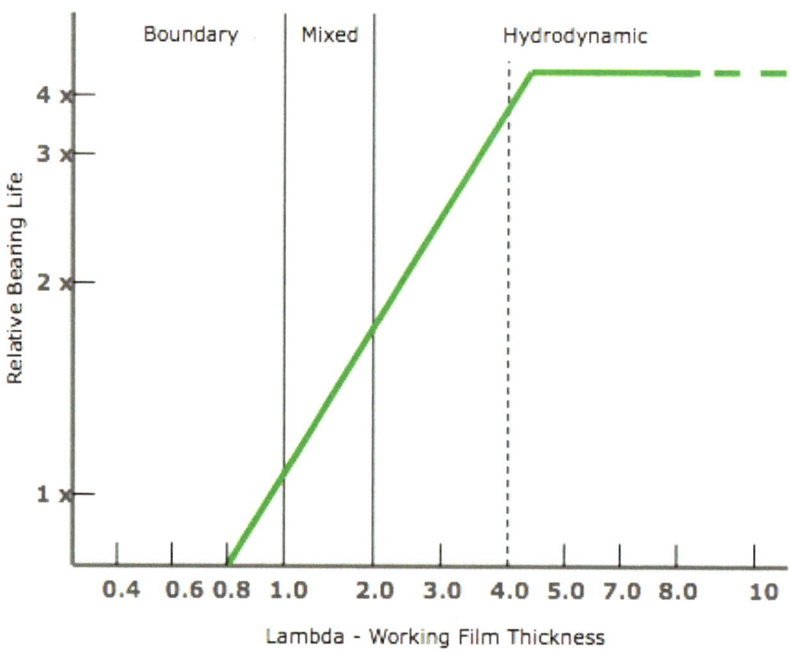

Figure 3.2 Oil film lambda oil thickness ratio curve.

3.1.1 Regime 1—Hydrodynamic lubrication—(HDL)

A hydrodynamic state exists when a continuous full-fluid film completely separates the sliding surfaces allowing them float over one another in the least hindered way. This is sometimes referred to as "Thick film" lubrication, and is applicable to nearly all types of continuous sliding action where extreme pressures are not involved.

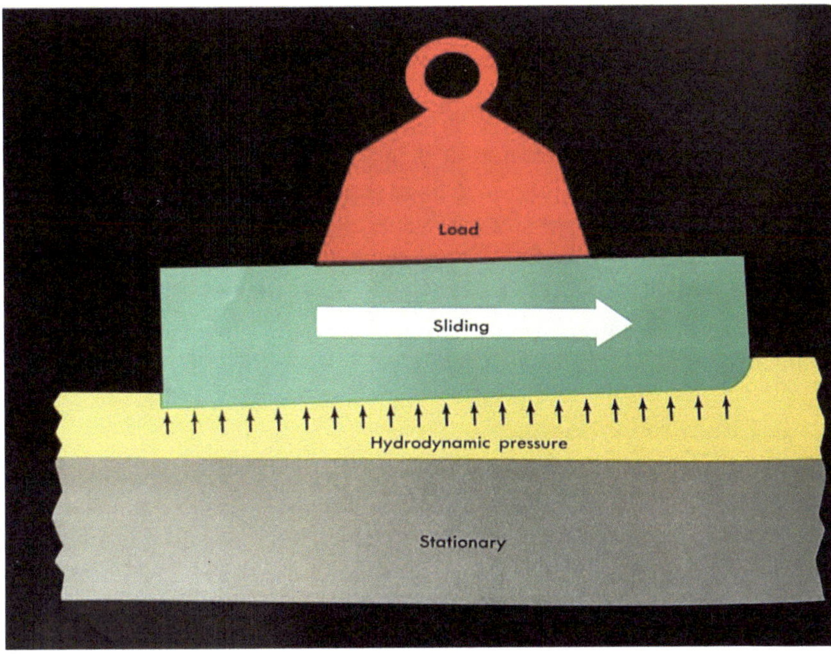

Figure 3.3 Sliding load supported by a wedge-shaped lubricating film. Source: Exxon Corporation, USA, (PLIF 2).

Whether the sliding occurs on flat surfaces, as it does in most thrust bearings, or whether the surfaces are cylindrical, as in the case of journal (plain or sleeve) bearings, the principle is essentially the same

3.1.1.1 Hydrodynamic lubrication of sliding surfaces

It would be reasonable to suppose that, when one moving part slides on/over another, the protective oil film between them would be scraped away. Except under some conditions of reciprocating motion, this is often not the case. With proper design, this very sliding motion constitutes the means of creating and maintaining that film.

Consider, for example, the case of a block that slides continuously on a flat surface as in Figure 3.3. If hydrodynamic lubrication is to be effected, an oil of the correct viscosity must be applied at the leading edge of the block, and three design factors must be incorporated into the block:

1. The leading edge must not be sharp, but must be beveled or rounded to prevent scraping of the oil from the fixed surface.
2. The block must have a small degree of free motion to allow it to tilt and to lift slightly from the supporting surface.

3. The bottom of the block must have sufficient area and width to “float” on the oil.

3.1.1.2 Full-fluid Film

Before the block is put in motion, it is in direct contact with the supporting surface. Initial friction is high, since there is no fluid film between the moving parts. As the block starts to slide, however, the leading edge encounters the supply of oil, and it is at this point that the significance of viscosity becomes apparent. Because the oil offers resistance to flow, the block does not wholly displace it. Instead, a thin layer of oil remains on the surface under the block, and the block, because of its rounded edge, rides up over it.

3.1.1.3 Effect of viscosity

As the sliding block rises from the surface, more oil accumulates under it, until the oil film reaches equilibrium thickness. At this point, the oil is squeezed out from under the block as fast as it enters. Again, it is the viscosity of the oil that prevents excessive loss due to the squeezing action of the block’s weight.

With the two surfaces completely separated, a full-fluid lubricating film has been established, and friction has dropped to a low value. Under these conditions, the block assumes of its own accord an inclined position, with the leading edge slightly higher than the trailing edge creating a wedge-shaped lubricating film.

3.1.1.4 Shoe-type thrust bearings

As illustrated in Figure 3.4, many heavily loaded thrust bearings are designed in accordance with the principle illustrated by the sliding block. A disk, or thrust collar, rotates on a series of stationary blocks, or shoes, arranged in a circle beneath it. Each shoe is mounted on a pivot, rocker, or springs, so that it is free to tilt and to assume an angle favorable to efficient operation. The leading edge of each contact surface is slightly rounded, and oil is supplied to it from a reservoir.

Bearings of the type described serve to carry the tremendous axial loads imposed by vertically mounted hydroelectric generators. Rotation of the thrust collar produces a flow of oil between it and the shoes so that the entire weight of the turbine and generator rotors and shaft is borne by the oil film. So closely does this design agree with theory, that it is said that the babbitt facing of the shoes may be crushed before the oil film fails.

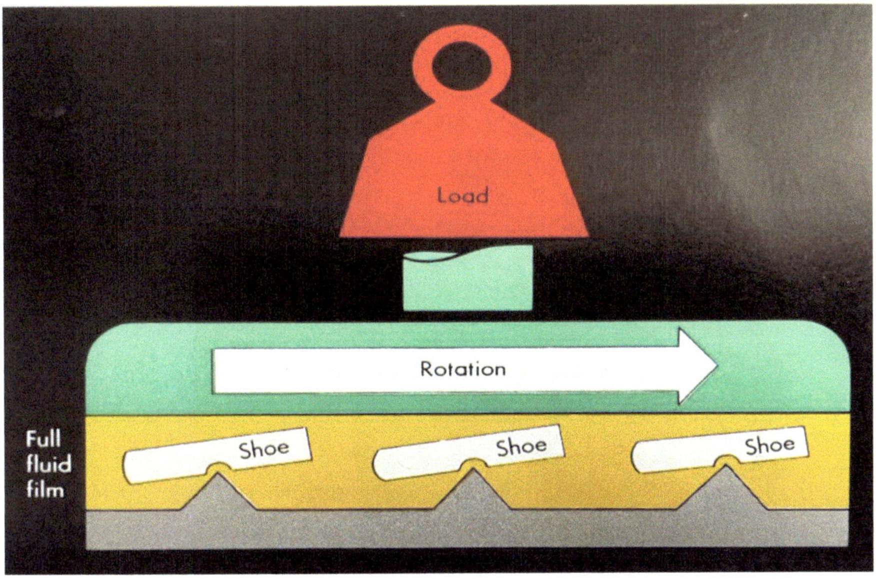

Figure 3.4 Shoe type thrust bearings. Source: Exxon Corporation, USA, (PLIF 2).

3.1.1.5 Journal bearings

The hydrodynamic principle is equally applicable to the lubrication of plain or journal type bearings. In this case, the load is radial, and a slight clearance must be provided between the journal and its bearing to permit the formation of the wedge-shaped film.

Let it be assumed, for example, that a journal supports its bearing, as it does in the case of a plain-bearing railroad truck. The journal is an extension of the axle and, by means of the bearing, it carries its share of the load represented by the car.

All of the force exerted by the bearing against the journal is applied at the top of the journal—none against the bottom. When the car is at rest, the oil film between the bearing and the top of the journal has been squeezed out, leaving a thin residual coating that is probably not sufficient to prevent some metal-to-metal contact.

As in the case of the sliding block, lack of an adequate lubricating film gives rise to a high initial friction. As the journal begins to rotate, however, oil seeps into the bearing at the bottom, where the absence of load provides the greatest clearance. Some of the oil clings to the journal and is carried around to the upper side, dragging additional oil around with it.

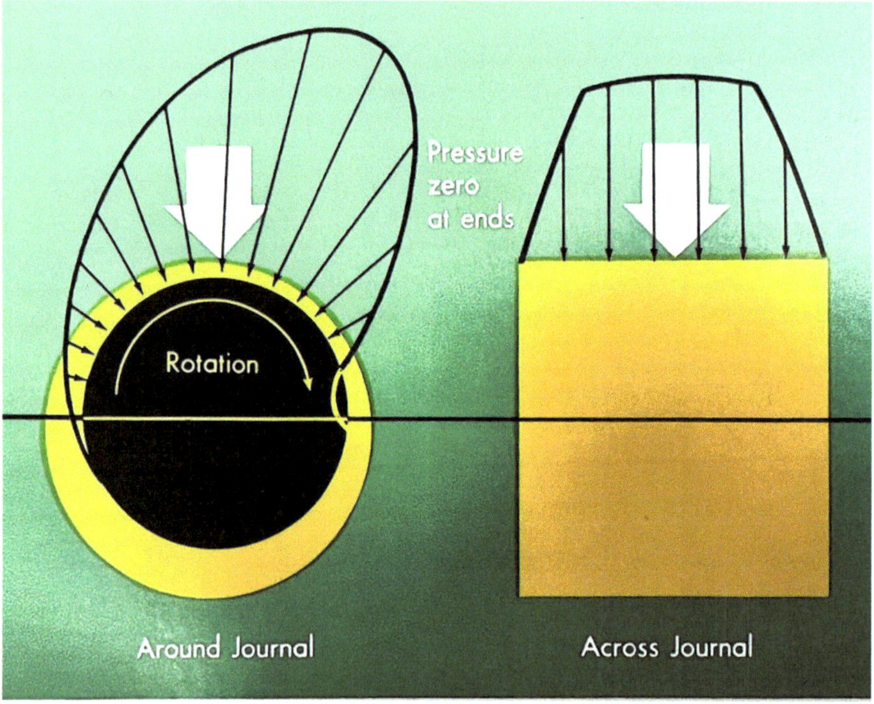

Figure 3.5 Oil pressure distribution diagrams. Source: Exxon Corporation, USA, (PLIF 2).

In this manner, oil is "pumped" into the narrowing clearance at the top of the journal, where there is greatest need. The consequent flow of oil from an area of low pressure through a converging channel to an area of high pressure produces a fluid wedge that lifts the bearing from the top of the journal, eliminating metal-to-metal contact.

When a state of equilibrium is reached, the magnitude of the entering flow displaces the bearing to one side, while the load on the bearing reduces the thickness of the film at the top. The situation is analogous to that of the inclined thrust-bearing shoe; in either case, the tapered channel essential to hydrodynamic lubrication is achieved automatically. The resulting distribution of hydrodynamic pressure is shown in Figure 3.5.

If the load were reversed, that is, if the bearing supported the journal, as is more generally the case, the relative position of the journal would be inverted. The low-pressure region would be at the top of the journal, and the protective film would be at the bottom.

3.1.1.6 Journal bearing design requirements

The performance of a journal bearing is improved by certain elements of design. In addition to the allowance of sufficient clearance for a convergent flow of oil, the edges of the bearing face should be rounded some-what, as shown in Figure 3.6, to prevent scraping of the oil from the journal. Like the leading edge of the thrust-bearing shoe, this edge should not be sharp.

Oil can enter the clearance space only from the low-pressure side of the bearing. Whatever the lubrication system, it must supply oil at this point. If the bearing is grooved to facilitate the distribution of oil across the face, the grooves must be cut in the low-pressure side. Grooves in the high-pressure side promote the discharge of oil from the critical area. They also reduce the effective bearing area, which increases the unit bearing load. No groove should extend clear to the end of the face.

3.1.2 Regime 2—Hydrostatic Lubrication (HSL)

Hydrostatic lubrication is a full film lubrication state set up when a lubricant is used to hydraulically separate a loaded surface and "float" one surface over another similar to that shown in Figure 3.3. This film regime is typically set up on precision machines and machine tools such as a plunge grinder in which the grindstone carriage is "floated" into the work piece on a controlled full film of oil to perform precision grinding on gears, etc.

3.2 Fluid Friction

It has been pointed out that viscosity, a property possessed in a greater or lesser degree by all fluids, plays an essential role in hydrodynamic lubrication. The blessing is a mixed one, however, since viscosity is itself a source of friction, commonly referred to as fluid friction.

Fluid friction is ordinarily but a minute percentage of the solid friction encountered in the absence of lubrication, and it does not cause wear. Nevertheless, fluid friction generates a certain amount of heat and drag that does require additional energy to overcome. Correct choice of lubricant viscosity is essential to minimize these affects and provide a strong enough lubricant film.

3.2.1 Laminar Flow

When two sliding surfaces are separated by a lubricating film of oil, the oil flows. Conditions are nearly always such that the flow is said to be *laminar,*

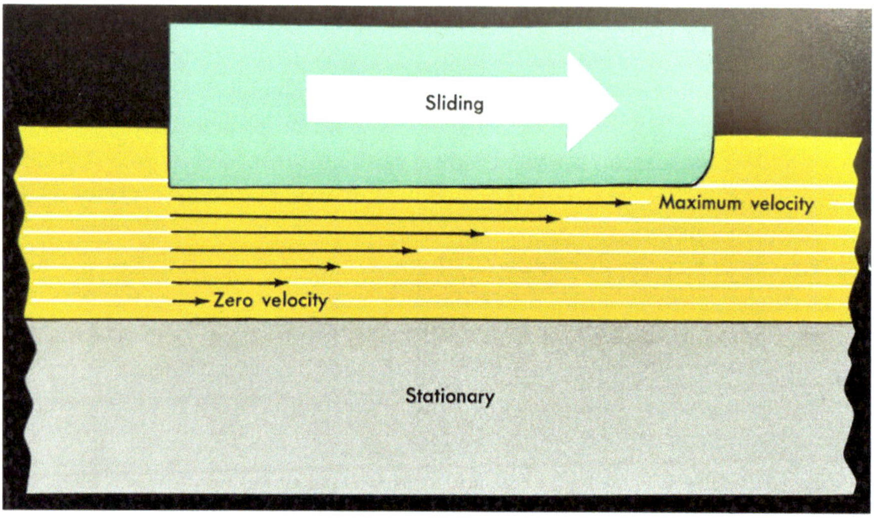

Figure 3.6 Fluid-bearing friction is the drag imposed by one layer of lubricant sliding upon another. Source: Exxon Corporation, USA, (PLIF 2).

that is, there is no turbulence. The film may be assumed to be composed of extremely thin layers, or laminae, with each moving in the same direction but at a different velocity, as shown in Figure 3.6.

Note: A practical example of this can be experienced by placing a new deck of playing cards between both hands, and by simulating movement by moving the hands back and forth whilst watching the cards move in relation to one another in a laminar flow manner.

Under these conditions, the lamina in contact with the fixed body is likewise motionless. Similarly, the lamina adjacent to the moving body travels at the speed of the moving body. Intermediate laminae move at speeds proportional to the distance from the fixed body, the lamina in the middle of the film moving at half the speed of the body in motion. This is roughly the average speed of the film.

3.2.2 Shear stress

Since the laminae travel at different speeds, each lamina must slide upon another, and a certain force is required to make it do so. Specific resistance to this force is known as *shear stress,* and the cumulative effect of shear stress is fluid friction. Viscosity is a function of shear stress, i.e., viscosity equals shear stress divided by shear rate. Therefore, fluid friction is directly related to viscosity.

Figure 3.7 Factors that effect bearing friction under full-fluid-film lubrication. Source: Exxon Corporation, USA, (PLIF 2)

3.2.3 Effect of speed and bearing area

In a bearing, however, there are two additional factors that affect fluid friction, both elements of machine design. One is the relative *velocity* of the sliding surfaces, the other, their effective *area.* Unlike solid friction, which is independent of these factors, fluid friction is increased by greater speeds or areas of potential contact.

Again, unlike solid friction, fluid friction is not affected by load, Figure 1.15. Other considerations being the same, a heavier load, though it may reduce film thickness, has no effect on fluid friction.

3.2.4 Regime 3—Partial film lubrication

Mixed Film (MF) and Boundary Lubrication (BL)

This discussion of friction has so far been limited to full-fluid film lubrication. However, formation of a full-fluid film may be precluded by a number of factors, such as insufficient viscosity, a journal speed too slow to provided the necessary hydrodynamic pressure, a bearing area too restricted to support the load, or insufficient lubricant supply.

Under these conditions, only partial lubrication in the form of mixed film lubrication or Boundary lubrication may be possible in which the resulting high bearing friction is a combination of fluid and solid friction, proportional to the severity of the operating condition.

Mixed film lubrication – Figure 3.8, is often referred to as partial lubrication and is classified as an Intermediate lubrication regime when

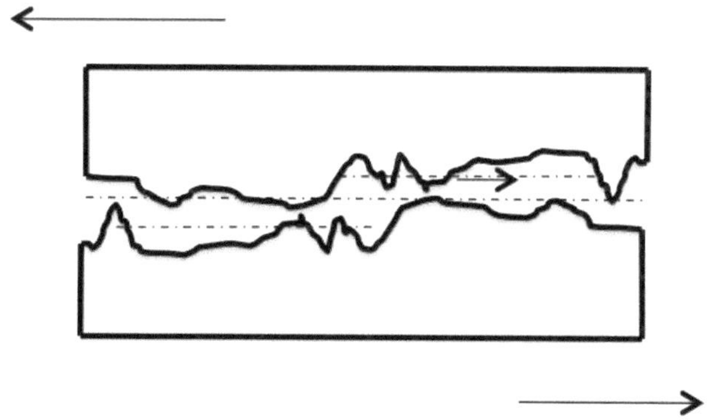

Figure 3.8 Mixed film lubrication.

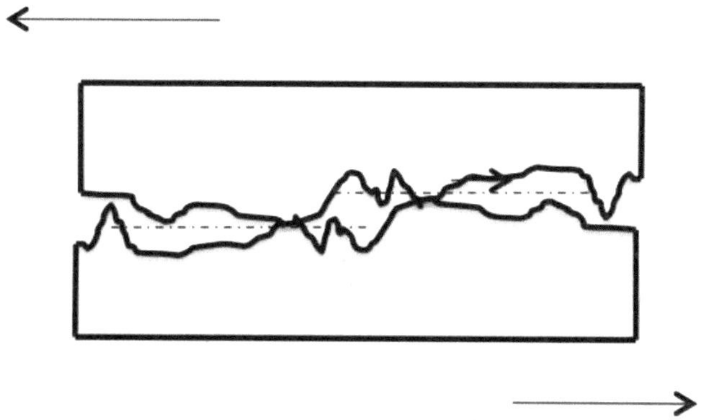

Figure 3.9 Boundary lubrication.

there is some lubricant present between two sliding surfaces, but not enough to fully separate the surface allowing intermittent contact between the highest bearing surface points. This is also thought of as an "unstable" regime.

Boundary Lubrication – Figure 3.9 is the least desirable lubrication regime with the highest coefficient of friction. Although a minimal amount of lubricant is present, the sliding surfaces are in full contact with one another at rest. With heavy load, slow moving machinery, boundary layer to mixed film regime may be the best condition achievable requiring a lubricant with EP (Extreme Pressure) and AW (Anti Wear) additives to offset the extreme bearing surface working condition. If insufficient lubricant is present, or an incorrect lubricant viscosity is used, a normally loaded bearing can stay in a boundary layer state when in full motion in which the surfaces will interfere with one another and cause rapid wear.

3.3 Bearing Efficiency

3.3.1 Overall bearing friction

It is thus possible to relate all bearing friction, regardless of lubricating conditions, to oil viscosity, speed, and bearing area. Engineers express the situation mathematically with the formula:

$$\mathbf{F = (f)\ ZNA} \tag{3.1}$$

where:

F is the frictional drag imposed by the bearing;
Z is oil viscosity;
N is journal speed;
A is the load-carrying area of the bearing.
(f) is a symbol indicating that an unspecified mathematical relationship exists between the two sides of the equation.

3.3.2 Coefficient of friction

It is customary to express frictional characteristics in terms of *coefficient of friction,* rather than friction itself. Coefficient of friction is more broadly applicable. It is *unit* friction, the actual friction divided by the force (or load) that presses the two sliding surfaces together. Accordingly, if both sides of equation (3.1) are divided by the load L:

$$\frac{F}{L} = (f)ZN\frac{A}{L} \tag{3.2}$$

Here, F/L is coefficient of friction and is represented by the symbol μ. Also, A/L is the reciprocal of pressure; or A/L 1/P, where P is pressure, the force per unit area that the bearing exerts upon the oil. By substitution, Equation (3.2) can therefore be written:

$$\mu = (f)\ \frac{ZN}{P} \tag{3.3}$$

This is the form that engineers customarily apply to bearing friction, the term ZN/P being known as a *parameter*—two or more variables combined in a single term.

3.3.3 ZN/P Curve

Equation (3.3) indicates only that a relationship exists; it does not define the relationship. In 1902, a Professor Richard Stribeck was the first to

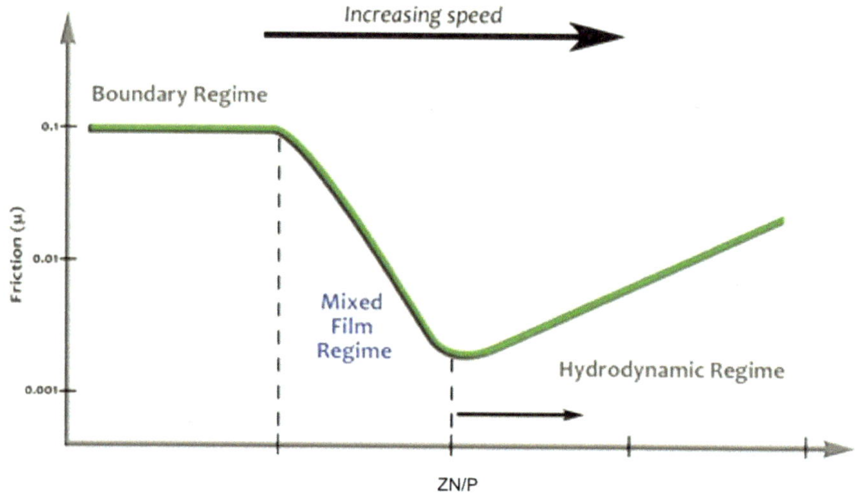

Figure 3.10 Stribeck bearing performance curve.

graphically define describe how the coefficient of friction changes for bearings experiencing different lubrication regimes as shown in Figure 3.10. The ZN/P curve shown depicts a typical example found in normally loaded sliding friction bearings such as those found in a shaft and sleeve bearing set up. At rest, the bearing surfaces are in a boundary or mixed film state prior to start up or shut down in which solid friction combines with fluid friction to yield generally high frictional values. As the shaft speed increases it will develop a fluid film reducing friction as it centrifugally centers in the bearing. Correspondingly, greater speed increases the value of the parameter ZN/P, driving operating conditions to a point on the curve farther to the right. A similar result could be achieved by the use of heavier oil, or by reducing pressure. Pressure could be reduced by lightening the load or by increasing the area of the bearing.

If these factors are further modified to increase the value of the parameter, the point of operation continues to the right, reaching the zone of perfect lubrication. This is an area in which a fluid film is fully established, and metal-to-metal contact is completely eliminated.

Beyond this region, additional increases in viscosity, speed, or bearing area reverse the previous trend. The greater fluid friction that they impose drives the operating position again to a region of high unit friction—now on the right portion of the curve.

3.3.4 Effect of load on fluid friction

Within the range of full-fluid-film lubrication, it would appear, from Figure 1.18, that increasing the bearing load or pressure could reduce bearing friction. Actually, as pointed out earlier, fluid friction is independent of pressure. Instead, the property illustrated by this curve is *coefficient* of friction—not friction itself.

Since the coefficient of friction μ equals F/L, then $F = \mu L$, and any reduction of μ due to greater bearing load under fluid-film conditions is compensated by a corresponding increase in the load L. The value of the actual bearing friction F remains unchanged.

In the region of partial lubrication, however, an increase in pressure obviously brings about an *increase* in μ. Since both μ and L are greater, the bearing friction F is *markedly* higher.

3.3.5 Efficiency factors

From this analysis, it is evident that proper bearing size is essential to good lubrication. For a given load and speed, the bearing should be large enough to permit the development of a full-fluid film, but not so large as to create excessive fluid friction. Clearance should be sufficient to prevent binding, but not so great as to allow excessive loss of oil from the area of high pressure. The relative position of the ZN/P curve for a loose-fitting bearing would be high and to the right, as shown in Figure 3.11, indicating the need for high-viscosity oil, with correspondingly high fluid friction.

Efficient operation also demands selection of an oil of the correct viscosity; oil just heavy enough to provide bearing operation in the low-friction area of fluid-film lubrication.

If speed is increased, heavier oil is generally necessary. For a given application, moreover, lighter oil would be indicated for lower ambient temperatures, while heavier oil is more appropriate for high ambient temperatures. These relationships are indicated in Figure 3.12.

3.3.6 Temperature-viscosity relationships

To a certain extent, lubricating oil has the ability to accommodate itself to variations in operating conditions. If speed is increased, the greater frictional heat reduces the operating viscosity of the oil, making it better suited to the new conditions.

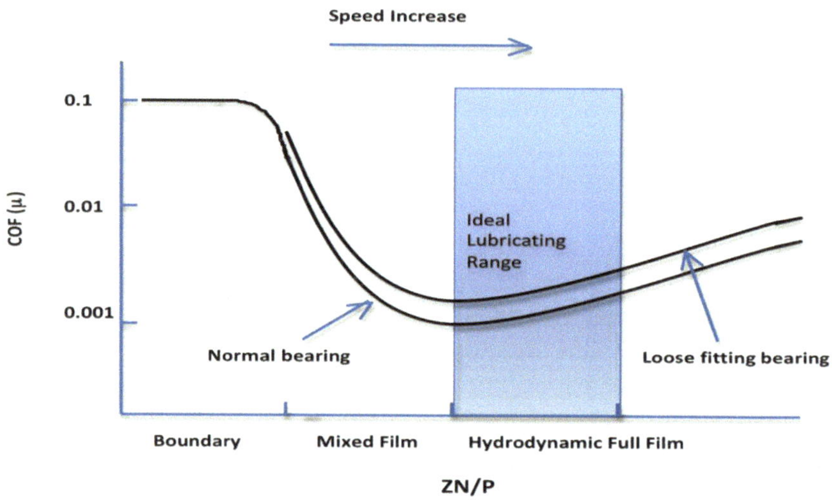

Figure 3.11 Loose-fitting bearings require high-viscosity oils.

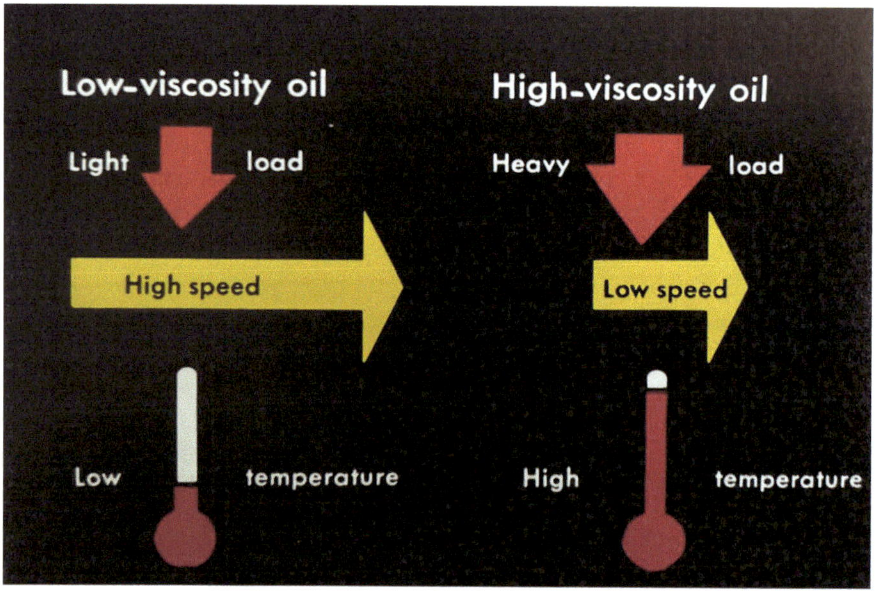

Figure 3.12 Relationships between oil viscosity, load, speed, and temperature. Source: Exxon Corporation, USA, (PLIF 2).

Similarly, an oil of excessive inherent viscosity induces higher operating temperatures and corresponding drops in operating viscosity. The equilibrium temperatures and viscosities reached in this way are higher, however, than if an oil of optimum viscosity had been applied, so the need for proper viscosity selection is by no means eliminated.

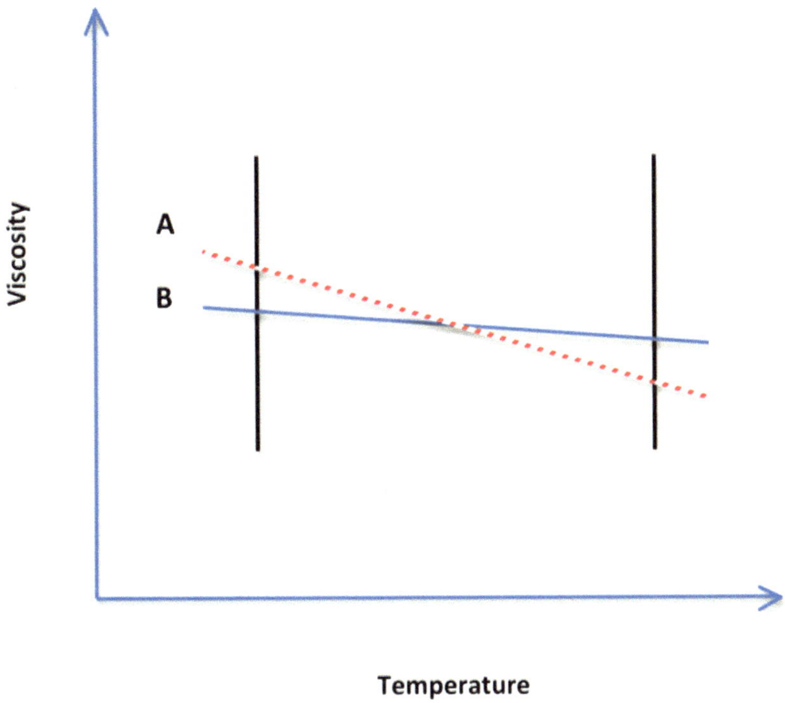

Figure 3.13 For the same temperature change, the viscosity of oil "B" changes much less than that of oil "A".

Oils vary, however, in the extent to which their viscosities change with temperature. Oil that thins out less at higher temperatures and that thickens less at lower temperatures is said to have a higher V.I. (viscosity index). For applications subject to wide variations in ambient temperature, a high-V.I. is more suited as depicted in Figure 3.13.

This is true, for example, of motor oils, which may operate across a 100°F temperature range. With an automobile engine, there is an obvious advantage in oil that does not become sluggishly thick at low starting temperatures or dangerously thin at high operating temperatures. So good lubrication practices include consideration of the V.I. of the oil as well as its inherent viscosity.

As previously stated, all factors that make hydrodynamic lubrication possible are not always present. Sometimes journal speeds are so slow or pressures so great that even heavy oil will not prevent metal-to-metal contact, as in for example, an oscillating linkage or heavily loaded, low rpm journal. Oil heavy enough to resist certain shock loads might be unnecessarily heavy for normal loads. In other cases, stop-and-start operation or reversals

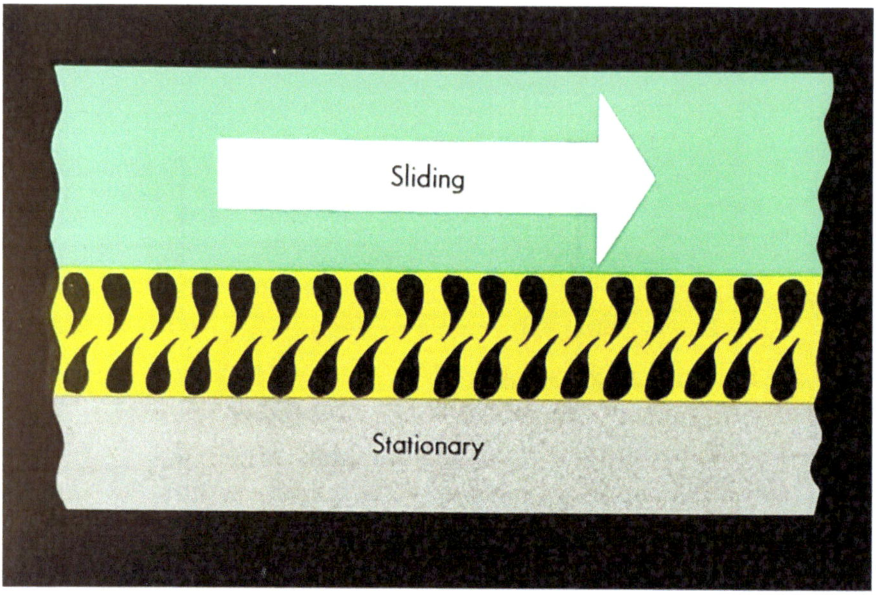

Figure 3.14 Sliding surfaces separated by a boundary lubricant of the polar type. Source: Exxon Corporation, USA, (PLIF 2)

of direction cause the collapse of any fluid film that may have been established. Also, the lubrication of certain heavily loaded gears—because of the small areas of tooth contact and the combined sliding and rolling action of the teeth—cannot be satisfied by ordinary viscosity provisions.

Since the various conditions described here are not conducive to hydrodynamic lubrication, they must be met with *boundary lubrication,* a method that is effective in the absence of a full-fluid film, Figure 3.14.

3.3.7 Additives for heavier loads

There are different degrees of severity under which boundary lubrication conditions prevail. Some are only moderate, others extreme. Boundary conditions are met by a variety of special lubricants with properties corresponding to the severity of the particular application. These properties are derived from various additives contained in the oil, some singularly, some in combination with other additives. Their effect is to increase the load-carrying ability of the oil.

Where loads are only mildly severe, an additive of the class known as *oiliness agents* or *film-strength additives* is applicable. Worm-gear and pneumatic-tool lubricants are often fortified with these types of agents. Where loads are moderately severe, *anti-wear agents* or *mild EP additives* are used. These additives are particularly desirable in hydraulic oils and engine oils. For more heavily loaded parts, a more potent class of additives is required; these are called *extreme pressure* (EP) *agents*.

3.3.8 Oiliness/lubricity agents

The reason for referring to oiliness agents as film-strength additives is that they increase the oil film's resistance to rupture. These additives are usually oils of animal or vegetable origin that have certain polar characteristics. A polar molecule of the oiliness type has a strong affinity both for the petroleum oil and for the metal surface with which it comes in contact. Such a molecule is not easily dislodged, even by heavy loads.

In action, these molecules appear to attach them-selves securely, by their ends, to the sliding surfaces. Here they stand in erect alignment, like the nap of a rug, linking a minute layer of oil to the metal. Such an array serves as a buffer between the moving parts so that the surfaces, though close, do not actually touch one another. For mild boundary conditions, damage of the sliding parts can be effectively avoided in this way.

Lubricity is another term for oiliness, and both apply to a property of oil that is wholly apart from viscosity. Oiliness and lubricity manifest themselves only under conditions of boundary lubrication, when they reduce friction by preventing breakdown of the film.

3.3.9 Anti-wear (AW) agents

Anti-wear agents, also known as *mild EP additives,* protect against friction and wear under moderate boundary conditions. These additives typically are organic phosphate materials such as zinc dialkyldithiophosphate (ZDDP) and tricresyl phosphate. Unlike oiliness additives, which physically plate out on metal surfaces, anti-wear agents react chemically with the metal to form a protective coating that allows the moving parts to slide across each other with low friction and minimum loss of metal. These agents sometimes are called "anti-scuff" additives.

3.3.10 Extreme-pressure (EP) agents

Under the extreme-pressure conditions created by very high loads, scoring and pitting of metal surfaces is a greater problem than frictional power losses, and seizure is the primary concern. These conditions require extreme-pressure (EP) agents, which are usually composed of active chemicals, such as derivatives of sulfur, phosphorus, or chlorine.

The function of the EP agent is to prevent the welding of mating surfaces that occurs at the exceedingly high local temperatures developed when opposing bodies are rubbed together under sufficient load. In EP lubrication, excessive temperatures initiate, on a minute scale, a chemical reaction between the additive and the metal surface. The new metallic compound is resistant to welding, allowing the metal peaks to easily detach without tearing the parent metal. This has the effect of "smoothing" out the surface, thereby minimizing friction that results from repeated formation and rupturing of tiny metallic bonds between the surfaces.

This form of protection is effective only under conditions of high local temperature. So an extreme-pressure agent is essentially an extreme-temperature additive.

3.3.11 Multiple boundary lubrication

Some operations cover not one but a range of boundary conditions. Of these conditions, the most severe may require oil with a chemically active agent that is not operative in the milder boundary service. Local temperatures, though high, may not always be sufficient for chemical reaction. To cover certain multiple lubrication requirements, therefore, it is sometimes necessary to include more than a single additive: one for the more severe, another for the less severe service.

3.3.12 Incidental effects of boundary lubricants

The question logically presents itself as to why all lubricating oils are not from the onset formulated with boundary-type additives? The answer is that this formulation is usually unnecessary; there is no justification for the additional expense of blending.

Additionally, the polar characteristics of oiliness agents may increase the emulsibility of the oil, making it undesirable for applications requiring rapid oil-water separation. Some of the more potent EP additives have a tendency to react with certain structural metals, a feature that might limit their applicability.

3.3.13 Stick-slip lubrication

A special case of boundary lubrication occurs in connection with stick-slip motion. Remember, a slow or reciprocating action, such as that of a machine way, can be destructive to a full-fluid film. Unless corrective measures are taken, the result is metal-to-metal contact, and the friction is solid, rather than fluid. Also remember that solid *static* friction is greater than solid *kinetic* friction, i.e., frictional drag drops after the part has been put in motion.

Machine carriages sometimes travel at very slow speeds. When the motive force is applied, the static friction must first be overcome, whereupon the carriage, encountering the lower kinetic friction, may jump ahead. Because of the slight resilience inherent in a machine, the carriage may then come to a stop, remaining at rest until the driving mechanism again brings sufficient force to bear. Continuation of this interrupted progress is known as *stick-slip* motion, and accurate machining may be difficult or impossible under these circumstances.

To prevent this chattering action, the characteristics of the lubricant must be such that kinetic friction is greater than static friction. This is the reverse of the situation ordinarily associated with solid friction. With a way lubricant compounded with special oiliness agents, the drag is *greater* when the part is in motion. The carriage is thus prevented from jumping ahead to relieve its driving force, and it proceeds smoothly throughout its stroke.

3.3.14 Regime 4—elastohydrodynamic lubrication (EHDL)

The foregoing discussion has covered what may be termed the classical cases of hydrodynamic and boundary lubrication. The former is characterized by very low friction and wear and dependence primarily on viscosity; the latter is characterized by contact of surface asperities, significantly greater friction and wear, and dependence on additives in the lubricant to supplement viscosity.

In addition to these two basic types of lubrication, there is an intermediate lubrication mode that is considered to be an extension of the classical hydrodynamic process. It is called *elasto-hydrodynamic* (EHD) *lubrication,* also known as EHL. It occurs primarily in rolling-contact bearings and in gears where non-conforming surfaces are subjected to very high loads that must be borne by small areas. An example of non-conforming surfaces is a ball within the relatively much larger race of a bearing (see Figure 3.15).

EHD lubrication is characterized by two phenomena:

1. The surfaces of the materials in contact momentarily deform elastically under pressure, thereby spreading the load over a greater area.
2. The viscosity of the lubricant momentarily increases dramatically at high pressure essentially transforming a fluid into a solid, thereby increasing load-carrying ability in the contact zone.

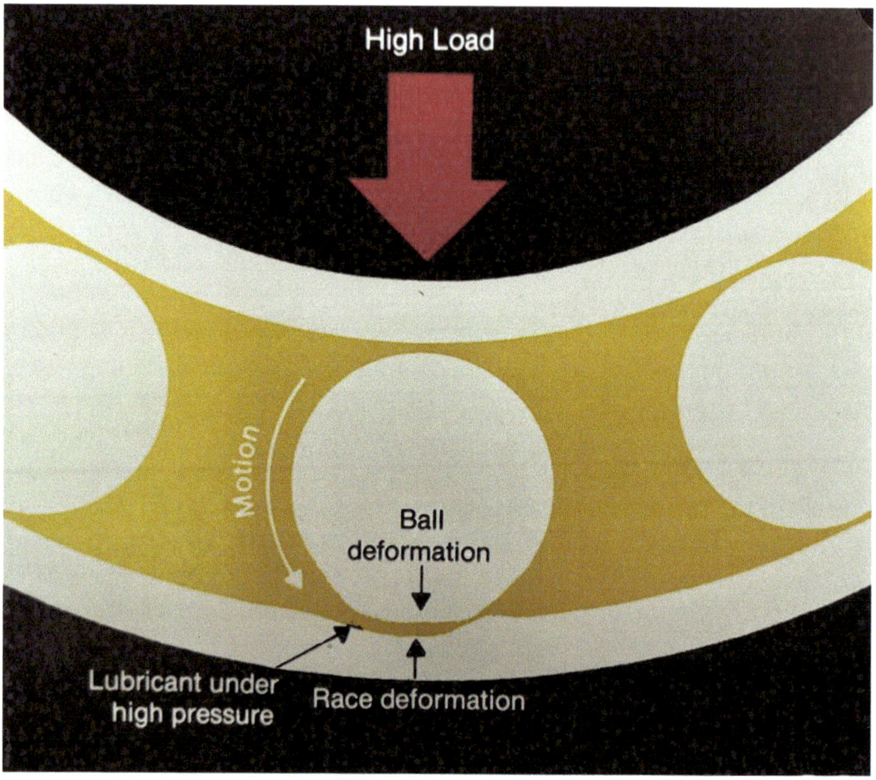

Figure 3.15 EHD lubrication in a rolling-contact bearing. Source: Exxon Corporation, USA, (PLIF 2).

The combined effect of greatly increased viscosity and the expanded load-carrying area is to trap a thin but very dense film of oil between the surfaces. As the viscosity increases under high pressures, sufficient hydrodynamic force is generated to form a full-fluid film and separate the surfaces.

The ball/race deformation zone shown in figure 3.15 is also known as the "Hertzian" contact area, or zone. This deformation characteristic of EHDL is analogous to the ever-changing traction zone of a modern-day automotive tire as it rotates, deforms where the tire contacts the ground, and restores to its original state as it rolls through the friction zone.

The repeated elastic deformation of bearing materials that occurs during EHD lubrication results in a far greater incidence of metal fatigue and eventual bearing failure than that witnessed in sliding, or plain, bearing operation. Even the best lubricant cannot prevent this type of failure.

3.3.15 Regime 5—zero film lubrication (ZFL)

The final regime is a lubricant-fluid-free type of lubrication known as zero film lubrication. This is typically employed in small sleeve (plain) type radial bearings manufactured from materials that have natural "self-lubricating" properties and therefore do not require fluid lubrication.

The bearing material is almost always captive in the bearing housing and manufactured from material softer (therefore sacrificial in nature) than the mating moving/rotating part. Typical bearing materials can include bronze alloys, nylon, PTFE, sintered or impregnated (graphite), which all offer a lower coefficient of friction without use of fluid lubricants.

4

Lubricant Selection

4.1 Oil Versus Grease

When specifying and designing new equipment, the engineering group must make a decision early on in the machine design and specification as to whether bearings are to be lubricated using oil, or grease. When making such a decision, a number of factors must be considered, that include:

- What are the operating conditions in which the equipment will be operated? For example, oil has a much higher operating temperature tolerance than that of grease.
- Will manufactured product contamination be an issue? Oil is more difficult to contain that grease where drippage may be considered detrimental to the manufactured product as in a food plant, or a paint line.
- Is the lubricant to be delivered to the bearing in an automated system or manually applied? Is the design conducive to a circulating oil system or total loss delivery system? Grease, because of its consistency can only be a total loss solution.
- Cost. Although oil is the preferred lubricant, it is often the more expensive lubricating delivery system to design and maintain.

When it comes to oil versus grease, oil is always the preferred lubricating medium. Purely due to the fact that it is always the oil that performs the job of lubrication. In grease, the oil is captured within a blend of soaps and fillers that allow it the grease to stay in place when applied. As the bearing heats up, the grease, acting like a sponge, allows the oil to wick out and lubricate the bearing. Upon cooling, the grease soap and fillers mop the oil back up. In essence the grease mix product acts as a live reservoir for the oil.

Table 4.1 Advantages versus disadvantages of oil lubrication.

Advantages	Disadvantages
Easy to apply	**Difficult to control**
Excellent cleaning and flushing properties	**Susceptible to leakage**
Can be recirculated	**Susceptible to temperature variations**
Centralized delivery systems are less expensive that grease	**Contamination when leakage occurs**
Can be used in highspeed, high temperature applications	
Can be filtered and cleaned in service	

4.1.1 Advantages and disadvantages of oil

Table 4.1 provides a comparison of Oil's advantages versus its disadvantages. Although oil is much easier to move around than grease, requiring lighter and often less expensive delivery system equipment, it is much more difficult to control.

In addition, use of oil as a lubricant is generally preferred over grease in closer tolerance bearings.

4.1.2 Advantages and disadvantages of grease

Table 4.2 compares the advantages and disadvantages of grease as a lubricating medium. Although grease is more difficult to move around, requiring higher pressure delivery pumps, it is relatively easy to control, generally requires less frequent applications, and is preferable in high load situations.

4.2 Selecting a Suitable Lubricant

The decision to use oil or grease is dependent on many factors that must be taken into account early on in the equipment design phase as it can be expensive to change at a later date.

Selection of a suitable lubricant is usually based on two main decision criteria, 1) machine condition factors, and 2) environmental decision factors.

Table 4.2 Advantages versus disadvantages of grease lubrication.

Advantages	Disadvantages
Easy to control, stays in place	**Excessive contamination will not settle out and could harm bearings**
Withstands heavy shock loads	**Greased bearings consume more energy**
Excellent seal forming ability	**Can cook in higher temperatures**
Generally, requires less frequent applications	**Depends on frictional heat to wick oil into bearing contact area**
Very inexpensive as a single point lube device	**Requires higher pressure delivery pumps**

4.2.1 Machine decision factors

Speed requires viscosity and application decisions. For example, a high-speed spindle machine can only be lubricated successfully by an air oil system using a very light viscosity oil.

Load, along with speed, requires additive (Anti Wear, Extreme Pressure additives) and viscosity decisions. Low speed, heavy loaded bearings will likely operate in a boundary regime requiring the shock load capability and film strength of a heavy grease.

Machine Age and wear may have bearing fits that will not accommodate oil unless the machine is overhauled and fitted out for oil lubrication.

Machine Design may already be in place requiring a review of the bearing types, bearing fits, bearing surface materials, seals, and bearing surface orientation to be taken into account.

Production Process use of raw material and/or finished product will dictate a preferred lubricant choice.

Warranty, if in effect, will require continued use of the existing lubrication delivery system and lubricants

4.2.2 Environmental /working decision factors

Environmental and working conditions relate to the in-situ conditions manifested by the production process and the geographical natural environment.

For example, machinery operating in a foundry, or a food processing plant environment would require modified approach to lubrication when compared to a "white-room" environment such as a pharmaceutical production line, or a precision bearing manufacturing line.

The foundry environment relies on silica(sand) as a casting material (or contaminant) with tremendous heat to complete the process (molten metal)—both enemies of lubrication. Food processing lines are generally hosed down on a daily basis, can easily bearing lubrication wash out unless preventive action is in place. On the other hand, a white room manufacturing suite environment is temperature and contamination controlled in a hepa-filtered environment ideal for effective lubrication.

Outside machinery such as facility roof fans, mining equipment, mobile equipment, wind turbines, and more are all at the mercy of the natural environment that will affect their lubrication approach. Coastal regions must deal with salt air and corrosion, northern regions must deal with both hot and freezing temperatures requiring changing viscosity requirements or use of lubricant heaters. Regions closer to the equator must deal with constant high temperatures. All of these climatic conditions must be factored into the lubrication program and system design and approach.

Sustainability programs and a corporation's carbon footprint are all affected by the efficiency of the lubrication approach. A tuned lubrication system using the correct viscosity lubricant(s) can produce tremendous energy savings and its associated carbon footprint reduction. The effect of lubrication on energy and the environment are discussed in numerous chapters later in the book.

Table 4.3 provides a guideline for choosing a suitable lubricant type in different situations. As lubricants are ever evolving, whenever making a final decision on a lubricant type it is always advisable to discuss with your lubrication partner(s). If a lubricant consolidation is in place, check first to see if your requirements can be met by the existing lubricants being used in the plant.

Table 4.3 Guidelines for choosing a suitable lubricant.

Condition	Oil	Grease	Solid
Clearances Designed for Oil	◆		◆
Clearances Designed for Grease		◆	◆
High Speed, Low Load	◆		
Low Speed, High Load		◆	
Low Speed Oscillating Load		◆	◆
High Temperature	◆		
Full Film Applications	◆	◆	
Mixed, Boundary Layer Applications		◆	◆
Contaminated Working Environment		◆	
Product NOT Leak Tolerant		◆	◆
Closed Gearbox	◆		
Isolated Bearings		◆	

SECTION 2

Lubricants

5

Lubricant Categories

By way of introduction, a brief overview of the principal lubricant categories is offered and illustrated in this chapter. Lubricants are divided into the following groups:

- Gaseous
- Liquid
- Cohesive
- Solid

Typical applications and industry sectors are given in Table 5.1.

5.1 Gaseous Lubricants

Gaseous lubricants are uncomplicated, low-viscosity lubricants that utilize industrial gases to form a gaseous barrier between the moving surfaces. Typical gases include air, nitrogen, oxygen, and helium.

Gas lubricated bearings are collectively referred to as "air-bearings". These bearings operate in a low-friction environment provided by a thin film of lubricated pressurized gas used to separate the components as they move over one another. Figure 5.1, demonstrates a typical cross-section of an air bearing in which the bearing "floats" on a cushion of lubricated gas.

Air bearings can be classified into two distinct groups, aerostatic, and aerodynamic. Aerostatic bearings require externally pressurized gas to be introduced into the bearing clearance (similar to the way a hovercraft's air skirt operates), whereas aerodynamic bearings use the relative motion of the internal moving parts to pressurize the self-contained gas.

Gaseous lubrication has certain advantages over fluid film lubrication, these include:

- Virtual frictionless contact between moving parts (note: a minute amount of airflow friction over the bearing surfaces may be present),

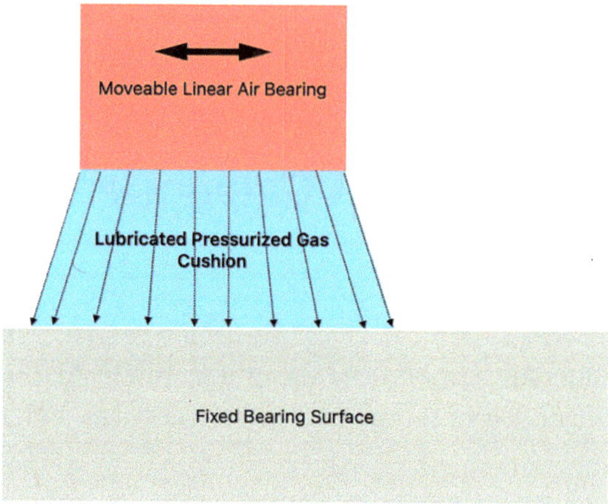

Figure 5.1 Cross section of a gas lubricated air bearing.

making them ideal for high-speed low load applications requiring precision measurement and processing. For example, 3-D printers, precision measuring equipment, gyroscopes, high speed spindles, dental drills, space craft simulators, air guided systems, and more

- Non-polluting and environmentally pure make them ideal for use in high-precision machinery required by pharmaceutical, medical, electronic and food industry applications.
- Gas lubricants operate well in extreme hot and cold temperatures.

Gas lubricants have limited appeal to mainstream manufacturing due to the cleanroom operating environment requirement, lack of ability to sustain high loads, and their need for very expensive close tolerance delivery components.

5.2 Liquid Lubricants

Liquid lubricants are by far the largest category of lubricants in general use, and also provide the lubrication component of all cohesive lubricants. This relationship is graphically depicted in Figure 5.2.

Liquid lubricants are generally catalogued into three main groups that include:

- Animal/Vegetable fatty oils,
- Mineral oils, and
- Synthetic oils.

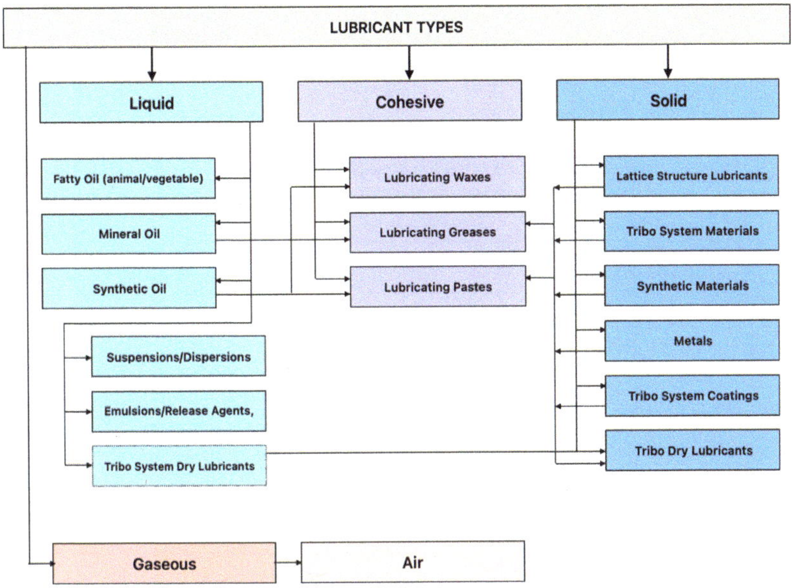

Figure 5.2 Typical applications and industry sectors for lubricants.

5.2.1 Animal/vegetable fatty oils

Although used sparingly in today's demanding industrial environment, animal/vegetable lubricants were the mainstay product well into the twentieth century. Olive oil and rendered animal fats lubricated both industrial revolutions with great success, but faltered due their lack of chemical inertness that caused oxidization and rapid acid formation under heat and load. As industry flourished and demanded more of its machinery, animal/vegetable lubricants were no longer viewed as a viable deterrent to prevent premature bearing failure, causing industry to move over to the newer, and more efficient, petroleum based mineral oils.

More recently, vegetable-based oil stocks have resurged in popularity, and are finding new life as base stocks for environmentally sensitive hydraulic lubricants (bio-lubes).

5.2.2 Mineral oils

Mineral oils are classified as petroleum products that are found deep under the earth's surface in geological fault formations. The product of trapped decayed vegetation metamorphosed by millions of years of heat and pressure

Table 5.1 Types of lubricant relationships.

● preferred application ○ possible application

Key words by lubricant type / Application	For machine elements and components ...														When exposed to				In the following industrial sectors ...																		
	Rolling bearings	Plain bearings	Chains	Seals	Ropes	Nuts and bolts	Electrical contacts	Springs	Valves/taps	Gears	Shaft/hub connections	Hydraulic systems	Pneumatic systems	Electrical switches	Fluids	Oxygen	Tribo-corrosion	Vacuum	Electrical engineering	Automotive	Precision mechanics	Transport	Base Materials	Rubber	Wood	Plastics	Food	Metal	Furniture	Mechanical engineering	Communications	Paper	Cableways	Textile	Machine Tools	Energy	Environment
Cohesive lubricants																																					
Sealing grease				○					●						●			●									○			○		●		○			
"Bio" lubricating grease	●	●						○			○									○		○	○					○		○			○			●	●
Damping grease	○									●				●					●	●	●								○	○	○						
Running-in grease		○	○	○						●	○						●			○			●			○				○					○	●	●
Running-in paste										●	○						●			○			●					○		○					○	●	●
Fluid grease		○								●												○					○								○		
Gas valve grease				○					●																			●								●	
Oil for plastics		●												●					●		○					●			○		○						
Long-term oil		○					○			●		○							○	●			○				○			○	○					○	○
Food grade oil		●	●		○			○		●		●	●														●										
Anticorrosion lacquer		○			○	●		○		○	○												○					●		●			○		○	○	○
Anticorrosion oil	●	○	○		○	○		○		○	●								○	○	○	○	○					●		●	○		○		●	●	○
Anticorrosion wax emulsion		●	○			○		○			○								○	●		○						●		○	○		○		●	●	○
Lubricating wax emulsion		●	○	●	○	○		○			○								○	○	○	○		●		○		○	●	●	○		○	○	○	○	○
Low-temperature oil		●	●		○		●			●									○	●	●	○								○	○		●				○
Impregnating fluid		●																	●	●	●									●	●						
Impregnating oil		●																●	●	●	●									●	●						
Sliding agent	○	○		●					●							●							●					●									
Tap grease		○							●						●												●	●								○	
High-speed grease	●																																	○	●		
High-temperature grease	●	●	●		○		●	○	○	○	○		○	●	○		○		○	●	○	●	○	○	●			○		○		○		○		○	○
High-temperature long-term grease	●						●												●	○		●	○		○		●	○		○		○		○		○	○
High-temperature paste		○		○		●				○	○						○			○			●					○		○		○		○		●	○
Grease for plastics		●		○				●	○	○		○	○	●					●	●	○					●			○		○						
Copper paste		○				●														○			●					○		○					○	●	○
Long-term grease	●				●		●		●	○	○		○	●			○		●	●	○	○	○					○	○	○	○	○	○	○	○	○	○
	●	●		●		○		○	●	○	○		○		●												●										
Smooth running grease	●												●							○	○					○			○		○		●		●		
Assembly grease		●		●		○		○			●	○	○		○		●			●			○			●	○	○	●	○			○		○	○	○
Assembly paste		●	●	●		●				○	●						●			●			●					●	○	●			○		●	●	○
Lubricating paste		●		○		●			○	○	●				●		●		○	●	○	○	○					○	○	○		○	○	○	○	○	○
Lubricating wax		●	●			○					○								●	○	○					○		○	●	●							
Heavy-duty grease	●																						●													●	●
Low-temperature grease	●	○		○	●	○	●	○	●	○	○	○	●	●					○	●	○	○								○	○		●				
Release agent																								●		●		●									
Solid lubricants																																					
Tribo-system materials																																					
Sheets		●		●					●					●	●		●	●	○	●	●	●	○		○		●		○	○	○			●	●	○	○
Granules		●	○			○			○	●		○	○	○	●		○		○	●	●	●					○		●	○	●		○	○			
Semi-finished parts		●		○					○						●		●	●	○	●	●	●	●				●		○	○	○		○	●		○	○
Bearings		●													●		●	●	●	●	●	●					●	○	○	●	●		○	○		○	○
Two-component systems		●													●		●			●								○		●					●		
Tribo-system coatings																																					
Heat-hardening	●	●	●			●		●	○	○			●		●		●		●	●	●	●	○		○	○	○	●	●	●	●	○		●		○	○
Air-drying	●	●	●	●	○	●	○	●					●	○			●	○	●	●	●	●	○		○	○	○	●	●	●	●	○		●	●	○	○
Dry lubricants for tribo-systems																																					
Heat-hardening		●	○		○	●		○	○	○	○	○	●		●		●	○	●	●	●	●						●		●							
Air-drying		●	○	●	○	●		○	○	○	●	○	●			●	●	○	●	●	●	●	○	○	○	○	○	●	○	●	○	○		●	●	○	○

into a viscous black liquid we call crude oil, from which, we refine amongst other petroleum products, lubrication grade mineral oil base stocks.

To make the different types of lubricating oils described in later chapters, many different additives are added to provide the lubricity, stability and protection required of a purpose built, modern fluid lubricating oil. Depending where in the world the crude oil is found, its refined oil characteristics will

Table 5.2 Paraffinic versus Naphthenic base oils.

Paraffinic	Naphthenic
Found in Mid USA, North Sea, Middle East	Found In Southern USA, South America
< 30% Naphthene	< 25% Paraffin
Viscosity Index 95-105	Viscosity Index 55-75
Wax-up to 10%	Wax - Trace
Automotive engine oils	Locomotive oils
Turbine Oils	Refrigerant Oils
Gear Oils	Compressor Oils
Bearing Oils / Hydraulic Fluids	Transformer oils

fall into two specific types based on whether the oil is paraffinic or naphthenic based.

As illustrated in Table 5.2, the majority of lubricating oils belong to the paraffinic family in which a higher Viscosity Index (VI) rating is more favorable. Their wax content levels do however, raise their cold temperature pour point, requiring additional processing at the refinery to extract them. Paraffinic stocks are refined using either a solvent extraction or hydroprocess refining method and are said to have a wax pour point.

Because naphthenic stocks have only a trace of wax and have good dielectric properties they make excellent candidates for low temperature applications and electrical switching equipment. Naphthenic stocks are refined using only a solvent extraction method and are said to have a viscosity pour point.

5.2.3 Synthetic oils

Synthetic oils are "man-made" synthesized hydro-carbon lubricants. As such, they can be custom designed with attributes that allow them to work more efficiently in conditions that regular mineral oils would find challenging.

Manufactured from chemical bases and compounds, synthetic lubricants contain no wax allowing them to operate in very low temperatures. Additionally, synthetics operate well above the temperature limits of mineral oils making them ideal for severe service.

Because they are more stable than mineral oils and have less internal fluid friction, synthetics not only deliver longer service life, they are generally more energy efficient.

The major drawback of synthetic lubricants is their comparative cost. To offset this, synthetic lubricants are now blended with mineral stocks and marketed as a less expensive "semi-synthetic" product. As the percentage blend is not always regulated, care must be taken when using these hybrid lubricants to ensure they are providing the right kind of protection.

The properties of all major synthetic lubricants can be compared to mineral oil in Table 2.3 below. For more detailed information on synthetic lubricants see Chapter 7.

5.3 Cohesive Lubricants

Cohesive lubricants are simply liquid lubricants blended with cohesive agents that allow them to be applied to a bearing surface and stay in place. This category of lubricants are identified as three main products:

- Lubricating greases,
- Lubricating pastes, and
- Lubricating waxes.

Lubricating pastes contain a high percentage of solid lubricants. They are used in the case of boundary and partial lubrication, especially for clearance, transition and press fits. Cohesive lubricants are used when the lubricant should not flow off, because there is no adequate sealing and/or when resistance against liquids is required. These lubricant types play an increasingly important role, since it is possible to achieve long-term or lifetime lubrication with minimum quantities. See Chapter 17 for a detailed review of Pastes, waxes and Tribo-systems.

5.4 Solid Lubricants

Solid lubricants include:

- Tribo-system materials,

Table 5.3 Properties of typical base oils for industrial lubricants.

Properties \ Oils	Mineral oils	Synth. hydro-carbon oils	Ester oile	Polyglycol oils	Polyphenyl ether oils	Silicone oils	Perfluoro alkylether
Density [g/m l] at 20°C	0.9	0.85	0.9	0.9 ... 1.1	1.2	0.9 ... 1.05	1.9
Viscosity Index (VI)	80 ... 100	130 ... 160	140 ... 175	150 ... 270	-20 ... -74	190 ... 500	50 ... 140
Pour point [°C]	-40 ... -10	-50 ... -30	-70 ... -37	-56 ... -23	-12 ... 21	-80 ... -30	-70 ... -30
Flash point [°C]	<250	<200	200... 230	150 ... 300	150 ... 340	150 ... 350	not flammable
Oxidation resistance	moderate	good	good	good	very good	very good	very good
Thermal stability	moderate	good	good	good	very good	very good	very good
Lubricity	good	good	good	very good	good to satisfactory	unsatistactory	good
Compatibility with elastomers, coatings, etc.	good	good	unsatis-factory	unsatis-factory to good	unsatis-factory	good	good

- Tribo-system coatings, and
- Dry lubricants for tribo-surfaces.

Solid lubricants include synthetic, metallic or mineral powders, such as PTFE, copper, graphite and Molybdenum Disulphide (MoS2). As powders are difficult to apply, they are mostly used as additives.

Solid lubricants are normally used as dry lubricants operating under boundary lubrication conditions. If liquid or cohesive lubricants are incorporated in the tribo-system materials there can even be partial lubrication. Solids are used when the application of liquid or cohesive lubricants is not ideal for functional reasons or risk of contamination and when, at the same time, the lubrication properties of solid lubricants are sufficient.

5.5 Categorizing and Grouping Base Oils

Base oil stocks are manufactured from crude oil extracted from many different oil fields throughout the world and refined using numerous methods based on the type of crude and the type of finished product required.

We can see from Table 5.3 that the majority of lubricating oils belong to the paraffinic family in which a higher Viscosity Index (VI) rating is more favorable. Their wax content levels do however; raise their cold temperature pour point, requiring additional processing at the refinery to extract them. Paraffinic stocks are refined using either a solvent extraction or hydro-processing refining method and are said to have a wax pour point.

Because naphthenic stocks have only a trace of wax and have good dielectric properties they make excellent candidates for low temperature applications and electrical switching equipment. Naphthenic stocks are refined using only a solvent extraction method and are said to have a viscosity pour point.

5.5.1 Refining base stocks

Solvent extraction process

Solvent extraction is performed using a multi-stage distillation/solvent extraction process. In stage one the crude is put through an atmospheric distillation process to remove the lighter and more volatile hydrocarbon fuel fractions that include benzene, gasoline, kerosene (jet fuel), and diesel. The stock is then charged into a vacuum distillation tower to take off (separate) the fluid by specific viscosity ranges. Next stage, a solvent is added to the stock to remove up to 85% of all unwanted large molecule aromatics, sulphur and nitrogen compounds. The primary solvent (usually N-Methyl pyrrolidone, or Furfural) is recovered and a second stage de-waxing solvent (usually Methyl-ethyl ketone or toluene) is added to facilitate the extraction of the wax crystals through cooling, solidifying, and filtering out unwanted wax. Once again, the solvent is recovered and the conventional base oil stock is now 70% to 85% pure and ready for additive formulation finishing, or further hydro-processing. Table 5.4 depicts the typical properties found in a base oil stock prior to blending its additive package.

These conventional base oil stocks are referred to as solvent refines, paraffinic **Group I base oil**.

Hydro-processing

Similar to solvent extraction, hydro processing employs a multi-stage process that begins with an atmospheric/vacuum distillation process. Then, depending on the finished product requirements, the base stock, now referred to as *feedstock*, can be put through a two-stage, severe hydrocracking process or a severe hydrocracking/hydro isomerization process.

Two-stage severe hydrocracking process

To remove unwanted polar compounds containing sulphur, nitrogen and oxygen, and convert aromatics hydrocarbons to more desirable saturated hydrocarbons, the distilled feedstock is charged into a hydrocracker. Here,

Table 5.4 Typical base oil stock properties.

Property	Explanation
Viscosity	Sets up base oil viscosity grade
Viscosity Index (VI)	Viscosity temperature relationship
Specific Gravity	Oil density relative to water
Flash Point	Volatility and flammability temperature
Pour Point	Lowest temperature oil stays fluid

the feedstock will react to a mixture of heat, pressure, and hydrogen gas in the presence of a catalyst.

The stock is once again distilled; the lube oil is chill de-waxed and then passed through a high-pressure hydro-treater to remove any last traces of aromatics and polar compounds.

This base stock is now better than 99% pure and ready for its additive package formulation and is referred to as a hydro process refined, paraffinic **Group II Base Oil**.

Severe hydrocracking/Hydro isomerization process

This process is identical to the severe hydrocracking process except that the chill de-waxing step is replaced with the more efficient hydro isomerization catalytic process. This efficiency yields very high VI base fluids up to 130 VI levels, with superior oxidation stability, excellent low temperature fluidity, High temperature stability, and low volatility/toxicity—similar characteristics to a Poly Alpha Olefin (PAO) synthetic lubricant!

These 99.9% pure base stocks are referred to as hydro-process refined, paraffinic **Group III Base Oil**. Due to their synthetic like characteristics, in 1999, the US National Advertising Division of the Council of Better Business Bureaus ruled that because of their molecular conversion, a Group III-Hydro Isomerized base oil can be referred to as a "synthetic" lubricant.

Group IV Base Oils are true synthetic PAO type base oils. (see Chapter 15 -Synthetic Lubricants.)

Group V Base Oils are all other oils not mentioned above. These would include solvent refined naphthenic and non-PAO synthetic lubricants (See Chapter 15- Synthetic Lubricants).

6

Lubricant Properties

6.1 Lubricating Oils

Lubricating oils consist of base oil and additives, which determine their performance characteristics. While the base oil is responsible for the typical properties of an oil as depicted in table 5.4, the additives determine its actual performance by influencing the base oil's:

- Oxidation stability,
- Anti-corrosion properties,
- Wear protection,
- Wetting behavior,
- Emulsibility,
- Stick-slip behavior,
- Viscosity-temperature behavior

Table 6.1 details the main lubricant additives and their purpose, while Table 6.2 shows additives common to the various lubricant types.

The advantages of lubricating oil, as compared to grease, are improved heat dissipation from the friction point, and its excellent penetrating and wetting properties. Its main disadvantage is that a complex design is required to keep the oil at the friction point and prevent the danger of leakage. Lubricating oils are used in a wide variety of elements and components, such as:

- Sliding bearings,
- Rolling bearings,
- Chains,
- Gears,

Table 6.1 Lubricating oil additives and their purpose.

Additive	Purpose
Anti-Oxidant	Prevents attack on base oil. Reduces formation of sludge, tar, varnish, and corrosive acid. These include:
Corrosion Inhibitor	Forms a protective shield against water on ferrous metals, tin, copper, and lead based bearing materials
Demulsifier	Prevents water from emulsifying with oil
Anti-Foam	Retards the formation of air bubbles in moving oil
Extreme Pressure (EP)	Chemical additives that soften the bearing surface to reduce adhesion when metal to metal contact is unavoidable
Anti-Wear	Solids additives to assist as additional sliding agents when metal to metal contact occurs under heavy loading
Detergent	Organic soaps that act as chemical cleaners to keep bearing surface deposit free
Dispersant	Work with detergents to suspend dirt particles in the oil to be easily extracted by the filter
Pour Point	Prevent wax crystals forming in paraffinic mineral oils at lower temperatures
Viscosity Improver	Long chain organic polymers that expand in heat to improve the oil's operating range and retard viscosity shearing

- Hydraulic systems,
- Pneumatic systems.

In addition to counteracting friction and wear, lubricating oils have other requirements to fulfill in various applications, e.g.,

- Corrosion protection,
- Neutrality to the applied materials,
- Meet food regulations,

Table 6.2 Typical lubricant additive formulations by lubricant type.

Additive	ATF	Bearing Oil	Compressor	Engine Oil	Gear Oil AGMA	Gear Oil SAE	Grease	Hydraulic	Turbine
Anti-Oxidant	✓	✓	✓	✓	✓	✓	✓	✓	✓
Corrosion Inhibitor	✓	✓	✓	✓	✓	✓	✓	✓	✓
Demulsifier	✓	✓	✓		✓	✓		✓	✓
Anti-Foam	✓	✓	✓	✓	✓	✓		✓	✓
Extreme Pressure EP	✓	Some			✓	✓	✓		
Anti-Wear	✓	Some	✓	✓	✓	✓	✓	✓	
Detergent	✓		✓	✓					
Dispersant	✓		✓	✓					
Pour Point	✓	✓	✓	✓	✓	✓		✓	✓
Viscosity Improver	✓			✓	Some	Some		Some	

- Resistance to temperatures,
- Biodegradability

Lubricating oil can also be applied in other primary or secondary applications that include:

- Running-in oils,
- Slideway oils,
- Hydraulic oils,
- Instrument oils,
- Compressor oils,
- Heat carrier oils

Their primary will always remain lubrication and protection against friction and wear, to protect surfaces, conduct electricity, and keep out foreign particles.

Lubricating greases are based on blended base oil and thickener imparting to them their cohesive structure. They can be used for elastohydrodynamic,

Table 6.3 Tribo-technical data pertaining to lubricating oils.

Parameters	Test	Notes
Ash Content	DIN 51 575 DIN EN 7	The residual oxide or sulphate ash remaining after the combustion of an organic compound; the sulphate ash content is determined only for metallic-organic additives or for used oil.
Color / Color Code	DIN 51 411 ISO 2049	The specific color of the lubricant determined using a colorimeter.
Demulsability	DIN 51 589	Oil's ability to separate from water.
Density	DIN 51 757	The amount of lubricant required for a certain friction point is generally indicated by volume. The volume in mm density is the weight in g of the required lubrication.
Flash Point	DIN ISO 2592	Lowest temperature at which an oil gives off vapors that will ignite when a small flame is passed over the surface of the oil.
ISO VG	DIN 51 519	ISO VG is the abbreviation of ISO Viscosity Grade, characterizing the classification of fluid lubricants in terms of viscosity.
Pour Point	DIN ISO 3016	Lowest point at which an oil will pour or flow under certain conditions.
Viscosity	DIN 51 561	Indicates the resistance of a fluid substance to flow Apparent Viscosity = Viscosity taking density into account 1 Pa s = 1 N s/m2 Kinematic Viscosity = Viscosity - Density ratio mm2/s
Viscosity - Temperature relation (VT)	DIN 51 563	Flow properties of a lubricant in relation to temperature
Viscosity Index (VI)	DIN ISO 2909	Non-dimensional figure characterizing the change of viscosity as a function of the temperature

boundary or partial lubrication. See Section 2, Chapter 16 for a detailed review of lubricating greases.

Lubricating waxes are based on hydrocarbons of high molecular weight and are preferably used for boundary or partial lubrication at low speeds.

6.2 Lubricating Oil Tribotechnical Data

Tribo-technical data relates to the characteristics of the lubricant; in this case, mineral oil. This data is shown in Table 6.3.

Within the framework of the intended application, they permit the selection of a lubricant suitable for the pertinent requirements (temperature,

Table 6.4 ISO viscosity grades of fluid industrial lubricants.

ISO viscosity grade	Mid-point viscosity at 40 °C, mm²/s (cSt)	Kinematic viscosity limits at 40 °C, mm²/s (cSt)	
		min.	max.
ISO VG 2	2.2	1.98	2.42
ISO VG 3	3.2	2.88	3.52
ISO VG 5	4.6	4.14	5.06
ISO VG 7	6.8	6.12	7.48
ISO VG 10	10	9.0	11.0
ISO VG 15	15	13.5	16.5
ISO VG 22	22	19.8	24.2
ISO VG 32	32	28.8	35.2
ISO VG 46	46	41.4	50.6
ISO VG 68	68	61.2	74.8
ISO VG 100	100	90.0	110
ISO VG 150	150	135	165
ISO VG 220	220	198	242
ISO VG 320	320	288	352
ISO VG 460	460	414	506
ISO VG 680	680	612	748
ISO VG 1000	1000	900	1100
ISO VG 1500	1500	1350	1650

load and/or speed). In this regard, the viscosity grade selection (Table 6.4) is of primary importance.

6.3 Additional Lubricant Additives

Earlier in this chapter we reviewed some of the different types and properties of base oils that constitute 75% to 99% of the finished oil that must then be blended with an additive package designed to meet the needs of the operating conditions within which the finished lubricant must perform. Designing a lubricant requires the tribology chemist to begin with a performance specification sheet outlining parameters and conditions the lubricant must meet and exceed in its finished commercial form.

Typical design parameters will include:

- **Lubricant application type:** gear oil, hydraulic oil, air tool oil, engine oil, etc.
- **Application requirements:** load, speed, bearing surface motion (sliding, rolling, combination), delivery method,
- **Lubricant quality:** viscosity index (VI), lubricant life expectancy, selling point,
- **Operating temperature range:** Viscosity,
- **Operating environment**: moisture, chemicals,
- **Biodegradability:** mineral or synthetic.

These operational design parameters make up what is technically known as the "***Tribological System***" in which the lubricant must perform. Lubricant type, application, quality, and operating temperature range are primarily used to determine the appropriate base stock, which is then supplemented with a variety of lubricant additives to strengthen or modify the lubricant's characteristics to meet the finished design specification. Interestingly, lesser quality base stocks can be significantly bolstered and modified to meet specifications with additive package.

6.3.1 The additive role and function

Additives play an important role in letting us know when the lubricant has lost it's efficacy, rendering it no longer useful in service. Additives are sacrificial by nature, and once depleted the oil must be replaced or replenished again with the necessary additive(s).

By monitoring and comparing the additive package "signature" of the virgin stock oil against a used oil sample, through the use of oil analysis, we can tell when the oil is degraded and ready to be changed/replenished.

Oil additives serve three functions: to enhance, to promote new properties, and to suppress undesirable base oil properties. Different lubricant types require different formulation packages. Table 6.5 - *Oil Additives by Function*, depicts which additive performs what function.

Additives can be organic or inorganic compounds and depending on their physical size, can dissolve in the oil (when sub-micron), or remain

Table 6.5 Oil additives by function.

Enhancers	New Properties	Suppressants
• Anti Foam	• Anti Wear	• Pour Point
• Anti Oxidant	• Detergent	• Viscosity Improver
• Demulsifier	• Dispersant	
• Film Strength	• Dye	
• Rust Inhibitor	• Extreme Pressure (EP)	
	• Friction Modifier	

as suspended solids. These solids are often visible to the naked eye when decanting new oils from one container to another.

6.3.2 The additive package

In the mixed film section of this chapter we reviewed in detail, Film strength (Oiliness), Anti-wear and Extreme pressure additives used to combat and alleviate boundary conditions. These additives, along with the ten additional additive types reviewed below make up the core additive package for most commercial lubricants.

Anti-Foam Agent—when a fluid is moved quickly through a pumping action, it can entrain small air bubbles in the lubricant. These air bubbles are detrimental to a lubricant as air contains oxygen that will attack the base oil (see Anti-Oxidant). Aerated fluids can also cause pump cavitation.

Also known as de-foamants, foam inhibitors, or anti-foam agents primarily work to increase the lubricant's surface tension so as to enlarge the foam bubble size, allowing them to collapse more easily.

Anti-Oxidant Agent—oxygen is base oil's primary enemy, especially at higher temperatures when in combination with contaminants such as debris and water. This can lead to sludge and viscosity thickening, tar, varnish, and eventually corrosive acid formation within the oil and on the bearing surfaces. Anti-oxidant agents, also known as oxidation inhibitors, can successfully improve oxidation stability by more than 10 times by deactivating catalytic metallic contaminants and by decomposing any formed reactive hydroperoxides within the oil. The most common anti-oxidant is Zinc dialkyldithiophophate, or ZDDP.

Demulsifier—also known as emulsion breakers, are used where water contamination is expected, and are designed to chemically prevent the formation of any water/oil emulsion by altering the surface tension of the oil allowing the water to separate easily and be drained off.

Detergents—used where combustion takes place as a chemical cleaner to keep combustion surfaces free from harmful deposits and to neutralize any combustion acids. Developed specifically for crankcase and compressor oils, detergent additives are made up from over base (alkaline) organic metallic soaps such as barium, calcium and magnesium.

Dispersants—again, used in crankcase and compressor oils, often in conjunction with detergents to chemically disperse and attach themselves to and suspend combustion and contaminant particles such as dirt, soot, glycol, depleted additives. The suspended contaminants can then be easily extracted by the oil filtration system.

Dye—used in transmission fluids and greases to better identify the product from other lubricants.

Friction Modifier—long chain polar additives that have an affinity for metal surfaces are added to crankcase and transmission oils to reduce the surface friction of lubricated parts in an effort to increase fuel economy

Pour Point Suppressant—used to prevent the formation of wax crystals in paraffinic mineral oils at low temperatures allowing the oil to pour at lower temperatures.

Rust Inhibitor—also known as a corrosion inhibitor, used to form a protective shield on the bearing surface against water and corrosive acids to stop the formation of corrosion and rust on ferrous, copper, tin, and lead based metals.

Viscosity Improver—these employ long string polymers that expand as the oil temperature increases and serve to "thicken" the oil, or increase its viscosity. Added to increase an oil's serviceability over a wider temperature range in multi-grade form and used to bolster lower quality base oils that have lower viscosity index (VI) ratings.

Table 6.6, Oil Type Additive Package Reference Table, details which additives are used in the different oil types. This is for guideline purposes only; always consult with your lubricant supplier to determine what additives are actually used in the lubricant you have chosen to use.

Table 6.6 Oil type additive package reference table.

Additive	Bearing Oil	Compressor Oil	Crankcase Oil	Gear Oil	Hydraulic Oil	Transmission Oil	Turbine Oil
Anti Foam	×	×	×	×	×	×	×
Anti Oxidant	×	×	×	×	×	×	×
Anti Wear	×	×	×	×	×	×	
Demulsifier	×	×		×	×		×
Detergent		×	×			×	
Dispersant		×	×			×	
Dye						×	
Extreme Pressure	×	×	×	×	×	×	
Friction Modifier			×			×	
Pour Point		×	×	×	×	×	
Rust Inhibitor	×	×	×	×	×	×	×
Viscosity Improver			×		×		

7

Lubricant Property Testing

Virtually every lubricant product is manufactured to specific performance standards and specifications. Each is designed to exhibit certain properties or characteristics described by the manufacturer in their sales literature, data sheets, or related documents.

Whenever industrial users are involved in the selection, or faced with the optimized application of their lubricant products, they will find themselves confronted by terms and descriptions that relate to these product properties and characteristics.

Sometimes termed "typical inspections," each lubricant property is an actual specification backed up by a physical test/standard, designed and overseen by an actual testing body.

Manufactured products must pass relevant test inspection requirements in order to claim their product meets or surpasses the specification. For example, a typical multi-purpose lubricating grease may advertise a grease specification that has passed such inspections items as *worked penetration, dropping point, viscosity, oil separation, wheel bearing leakage, Timken OK load, four-ball wear test, water washout,* and *corrosion prevention rating.* The question is, what do these terms mean, and how important are they?

Since the scope and intent of this book is aimed at conveying practical knowledge to the reader, it must enable the reader to make comparisons among products. With this in mind, this chapter is designed to describe the most important tests and their significance to the lubricant user.

Table 7.1, itemizes the major lubricant tests used to identify preferred properties typically found in industrial lubricants. All tests reviewed and described in this chapter are correctly referenced by their official test title (parameter), test number as catalogued by the referenced testing bodies. In this case, the European test body is represented by DIN – Deutsches Institut für Normung, (German National Institute for Standardization) as well as the North American body ASTM, or American Society for Testing and Materials, whose membership now represents 95% of all countries worldwide.

Table 7.1 Lubricant test chart.

Test Parameter	TEST	Notes
Air Entrainment	DIN 51 381	TUV Impinger test
Aniline Point	ASTM D 611/D 1012	
Ash Content	ASTM D 874	For sulphated residue
Auto Ignition Temperature	ASTM D 2155	
Cloud Point	ASTM D 2500	
Copper Strip Corrosion	ASTM D 130	
Demulsibility	ASTM D 1401/D 2711	
Grease Consistency	ASTM D 877/D 1816	
Dilution of Crank Case Oils	ASTM D 322	
Dropping Point of Grease	ASTM D 566/D 2265	
Flash / Fire Points - Open Cup	ASTM D 92	
Flash Point - Closed Cup	ASTM D 56 / D 93	
Foaming Characteristics - Oil	ASTM D 892	
Four Ball Wear Test	ASTM D 2266	
Four- Ball EP Test	ASTM D 2596	
Gravity	ASTM D 287	
Grease Consistency	ASTM D 217 / D 1403	
Interfacial Tension	ASTM D 971	
Neutralization Number	ASTM D 664 / D 974	
Octane Number	ASTM D 2699 / D 2700	
Oil Content of Petroleum Wax	ASTM D 721	
Oil Separation in Grease Storage	ASTM D 1742	
Oxidation Stability - Greases	ASTM D 942 / D 1402 / D 1261	
Pour Point and Cloud Point	ASTM D 97	
Power Factor	ASTM D 924	
Refractive Index	ASTM D 1218	
Rotary Bomb Oxidization Test	ASTM D 2272	RBOT
Rust Preventive Characteristics	ASTM D 665	
Timken Extreme Pressure Test	ASTM D 2509 / D 2782	
Vapor Pressure	ASTM D 323	
Viscosity	ASTM D 88 / D 445	Redwood, Engler
Viscosity Index	ASTM D 567 / D 2270	
Water Washout	ASTM D 1264	
Water and Sediment	ASTM D 95 / D 96 / D473	
Wax Melting Point	ASTM D 87 / D 127 / D 938	
Wheel Bearing Grease Leakage	ASTM D 1263	

7.1 Air Entrainment

7.1.1 DIN 51 381 TUV impinger test

Entrained air in lubricating oil can disrupt the lubricating film and cause excessive wear of the surfaces involved. In hydraulic systems, because entrained air is compressible, it can also cause erratic and inefficient system operation.

The term "air entrainment" refers to a dispersion of air bubbles in oil in which the bubbles are so small they tend to rise very slowly to the air-oil interface. The presence of the bubbles gives the oil a hazy or cloudy appearance.

Unfortunately, there is currently no standard method for testing the air entrainment characteristics of oil. The American Society for Testing Materials is still investigating various testing in. The German DIN 51 381 "TUV Impinger Test" is primarily used to test steam turbine oils and hydraulic fluids for their air release properties. It is the standard generally accepted in Europe and remains under consideration as the ASTM method.

7.1.2 Significance of results

Air entrainment consists of slow-rising bubbles dispersed throughout the oil and must be distinguished from foaming, which consists of bubbles that rise quickly to the surface of the oil. Both conditions are undesirable in lubricating systems. However, it is often difficult to distinguish between them because of high-flow rates and turbulence in the system.

Relatively small amounts of air are involved in air entrainment, with bubble diameters less than 0.04" dia. (1mm dia.), while larger amounts are involved with foaming. These conditions are considered separate phenomena and are measured in separate laboratory tests. Entrained air is not a normal condition; it is primarily caused by mechanical problems. Some of these are:

Insufficient Reservoir Fluid Level: Air can be drawn into the pump suction along with the oil.

Systems Leaks: Air can be introduced into the oil at any point in the system where the pressure is below atmospheric pressure.

Improper Oil Addition Methods: If make-up oil is added in a manner that causes splashing, it is possible for air to become entrained in the oil.

Faulty System Design: Design faults involve such things as placement of oil return so that the returning oil splashes into the reservoir, or placement of the pump next to the return opening.

The general trend in hydraulic oil systems, turbine oil systems, and industrial circulating oil systems of every kind is to decrease reservoir size and increase flow rates and system pressures. This trend increases the tendency for air entrainment, thereby making the air release property of oil more significant.

Some additives used to reduce foaming tend to increase the air entrainment tendency of oil. The choice of an anti-foam additive requires striking a balance between these two undesirable phenomena.

7.2 Aniline Point

7.2.1 ASTM D 611 and ASTM D 1012

The analine point is the measure of oil's aromaticy. Many petroleum products, particularly the lighter ones, are effective solvents for a variety of other substances. The degree of solvent power of the petroleum product varies with the types of hydrocarbons included in it. Frequently it is desirable to know what this solvent power is, either as a favorable characteristic in process applications where good solvency is important, or as an unfavorable characteristic when the product may contact materials susceptible to its solvent action.

The aniline point determination is a simple test, easily performed in readily available equipment. In effect, it measures the solvent power of the petroleum product for aniline, an aromatic substance. The solvent powers for many other materials are related to the solvent power for aniline.

Aniline is at least partially soluble in almost all hydrocarbons, and its degree of solubility in any particular hydrocarbon increases as the temperature of the mixture is increased. When the temperature of complete solubility is reached, the mixture is a clear solution; at lower temperatures, the mixture is turbid (cloudy). The test procedures make use of this characteristic by measuring the temperature at which the mixture clouds as it is cooled. The greater the solvent power of the hydrocarbon for aniline, the lower the temperature at which cloudiness first appears.

Usually, paraffinic hydrocarbons have the least solvency for aniline (and most other materials) and consequently have the highest aniline points. Aromatics have the greatest solvency and the lowest aniline points (usually well below room temperature), while naphthenic materials are intermediate between paraffins and aromatics.

7.2.2 Significance of results

Many internal combustion engine oils are formulated with detergent additives based on metallic derivatives such as barium and calcium. These additives help keep the engine clean. Being metallic, these materials appear in one chemical form or another in the ash.

For new oils of this type, sulfated ash may serve as a manufacturer 's check on proper formulation. Abnormal ash may indicate a change in additive content and, hence, a departure from an established formulation.

For new oils of unknown formulation, sulfated ash is sometimes accepted as a rough indication of detergent level. The principle is based on the dubious assumption that a higher percent of ash implies a stronger concentration of detergent and, hence, an oil of greater cleanliness properties. As a means of evaluating detergency, however, the test for ash is far less reliable than the usual engine and field tests, the primary advantage of ash content lying in the expediency with which it is determined.

There are several reasons why the relationship between sulfated residue and detergency may be extremely distorted:

- Detergency depends on the properties of the base oil as well as on the additives. Some combinations of base oil and additive are much more effective than others.
- Detergents vary considerably in their potency, and some leave more ash than others. Detergents have been developed, in fact, that leave no ash at all.
- Additives other than detergents may contribute some of the ash.
- There appears to be a limit to the effective concentration of detergent. Nothing is gained by exceeding this limit, and a superabundance of detergent may actually reduce cleanliness.

Sulfated ash has also been used to determine additive depletion of used diesel oils. The assumption has been that the difference between the ash of the used oil and that of the new oil is related to the amount of detergent consumed in service. Here again, results may be misleading. Consumption of the additive does not ordinarily mean that it has been disposed of, but that its effectiveness has been exhausted in the performance of its function. The metallic elements may still be present and may appear in the residue in the same concentration as in the new oil.

Sulfated ash of used diesel oils has significance only of a very general nature. If it runs higher than that of new oil, contamination with dirt or wear metals is suspected, and further analysis is required to identify the foreign material. If sulfated ash runs low, it may be attributed to faulty engine operation or a mechanical defect. With gasoline engines, a high sulfated ash may be caused by the presence of lead derived from the fuel.

7.3 Auto-ignition Temperature

7.3.1 ASTM D 2155

All petroleum products burn and, under certain conditions, their vapors will ignite with explosive force. For this to happen, however, the ratio of product vapor-to-air must be within certain limits.

When exposed to air, a certain amount of the liquid product evaporates, establishing a certain vapor-to-air ratio. As the temperature of the liquid increases, so does evaporation, and thus the vapor/air ratio. Eventually, a temperature is reached at which the vapor/air ratio will support combustion if an ignition source, such as a spark or flame, is present. This is the product's flash point.

If no ignition source is present, as the temperature increases above the product's flash point, a temperature is reached at which the product will spontaneously ignite, without any external ignition source. This temperature is the auto-ignition temperature of the fluid.

The auto-ignition temperature of a liquid petroleum product at atmospheric pressure is determined by the standard ASTM method D 2155 (which replaces the older ASTM D 286, discontinued in 1966).

7.3.2 Significance of results

The auto-ignition temperature of a petroleum product is primarily significant as an indication of potential fire and explosion hazards associated with the product's use.

The auto-ignition temperature may be used as a measure of the relative desirability of using one product over another in a high-temperature application. It is necessary to use a petroleum product with an auto-ignition temperature sufficiently above the temperature of the intended application to ensure that spontaneous ignition does not occur. Auto-ignition temperature thus places a—but by no means the only—limit on the performance of a product in a given application.

The auto-ignition temperature under a given set of conditions is the lowest temperature at which combustion of a petroleum product may occur spontaneously, without an external source of ignition. It is not to be confused with the flash point of a product, which is the lowest temperature at which a product will support momentary combustion, in the presence of an external ignition source.

The auto-ignition temperature of a product is a function of both the characteristics of the product and conditions of its environment. For example, the auto-ignition temperature of a substance is a function of such things as the pressure, fuel-to-air ratio, time allowed for the ignition to occur, and movement of the vapor-air mixture relative to the hot surface of the system container. Consequently, the auto-ignition temperature may vary considerably depending on the test conditions.

For a given product at atmospheric pressure, the auto-ignition temperature is always higher than the flash point. In fact, as a general rule for a family of similar compounds, the larger the component molecule, the higher the flash point, and the lower the auto-ignition temperature.

However, as the pressure of the system is increased, the auto-ignition temperature decreases, until a point is reached, usually at a pressure of several atmospheres, at which the auto-ignition temperature of a product under pressure may be less than the flash temperature of the product at atmospheric pressure. Thus, concern for the auto-ignition of a product increases as the pressure on the system increases.

As a general rule, the auto-ignition temperatures of many distillate products with similar boiling ranges can be related to hydrocarbon type. For example, aromatics usually have a higher auto-ignition temperature than do normal paraffins with similar boiling range. The auto-ignition temperatures of isoparaffins and naphthalenes normally fall somewhere in between those of the aromatics and normal paraffins.

However, care should be used in attempting to extend this guideline. For example, increasing the aromatics content of a lube or hydraulic oil tends to reduce the auto-ignition temperature of the oil. Conversely, increasing the aromatic nature of a solvent tends to increase the auto-ignition temperature of the solvent.

7.4 Biodegradation and Ecotoxicity

Biodegradation is the breakdown of a substance, e.g., hydraulic fluid, by living organisms into simpler substances, such as carbon dioxide (CO_2) and water. Most standard test methods for defining the degree of biodegradation of a substance use bacteria from a wastewater treatment system as the degrading organisms. This provides a relatively consistent source of bacteria, which is important, since the bacteria are the only variable in the test other than the test substance itself. A term that can be roughly defined as the opposite of biodegradability is *persistence*. A product is persistent if it does not degrade, or if it remains unchanged for long periods of time, i.e., years, decades.

There are many tests for biodegradation. Depending on the test design, it can measure *primary* biodegradability or *ultimate* biodegradability. *Primary biodegradability* is a measure of the loss of a product, but it does not measure the degree of degradation, i.e., partial or complete (to CO_2 and water), or characterize the by-products of degradation. It merely determines the percentage of the product that disappears over the term of the test or, conversely, determines the time required to reach a certain percentage of loss.

A popular primary biodegradation test in use today is the CECL-33-A-94, which measures disappearance of the test product and relates that to a biodegradation level. The assumption in this test is that all product that has disappeared is completely biodegraded. In actuality, this may not be the case, because the test does not measure *complete biodegradation,* but only the *loss* of the original product.

Ultimate biodegradability describes the percentage of the substance that undergoes complete degradation, i.e., degrades to CO_2 and water over the length of the test, or conversely, describes how long it takes to achieve a specified percentage of degradation.

Two tests that are designed to measure ultimate biodegradability are the Modified Sturm Test (OECD 301B) and the EPA Shake Flask Test, both of which quantify CO_2 generated over 28 days (a standard test duration).

Thus, the terms *primary* and *ultimate* describe the *extent of biodegradation*. The *rate of biodegradation* is defined by the term *ready biodegradation*. A product is considered to be *readily biodegradable* if shown to degrade 60-70%, depending on the test used.

Only a few tests measure ready biodegradability. The more commonly used include: the Modified Sturm Test (OECD 301 B); the Manometric Respirometry Test (OECD 301 F); and the Closed Bottle Test (OECD 301 D).

The final term to discuss here is *inherent biodegradability*. A product is considered *inherently biodegradable* if shown to degrade greater than 20%. However, unlike ready biodegradation tests, which run a specified 28 days, tests for inherent biodegradation have no defined test duration and are allowed to proceed as long as needed to achieve 20% degradation, or until it is clear that the product will never biodegrade to that extent. In the latter case, the product is then considered *persistent*.

Evaluating a substance's *environmental toxicity* (ecotoxicity) can involve examining its effect on growth, reproduction, behavior, or lethality in test organisms. In general, ecotoxicity is measured using aquatic organisms like fish, aquatic insects, and algae. The most common endpoint for expressing *aquatic toxicity* in the laboratory is the LC50, which is defined as the lethal concentration (LC) of a substance that produces death in 50% of the exposed organisms during a given period of time. Ecotoxicity data, properly developed, understood, and applied, is useful for evaluating the potential hazard of a material in the environment. Some of the most commonly used organisms for aquatic toxicity studies include rainbow trout, mysid shrimp, daphnids (water fleas), and green algae.

7.4.1 Significance of results

Both the test method and the intended use of the data must be considered when evaluating the biodegradability and environmental toxicity of a product. Data from different test methods are generally not comparable, and data developed on different products by different labs should be evaluated with strict caution by experts in the field.

Comparable biodegradation data should be developed using a consistent inoculum (bacteria) source, and in the same time frame, due to the variation of bacterial populations over time. As with most laboratory test procedures, results cannot be directly extrapolated to natural settings.

Similarly, for environmental toxicity tests, comparative data should be developed using the same test procedures and the same organisms. Exposures experienced in the laboratory will not be replicated in nature. The natural environment is a large dynamic ecostructure, while the laboratory environment is static and limited in size. Further, if a contaminant enters a natural aquatic system, the event will most likely be random in concentration and frequency, unlike the laboratory environment, which depends on constant, measured contamination.

Biodegradation and environmental toxicity data help us to begin to better understand how to protect the world around us. There are discrete, clearly defined methods for testing products for environmental toxicity, and there are many different test methods for evaluating a product's potential persistence in the environment. However, at this time, in the United States, there is no standard set of universally accepted test procedures defined by government or industry to measure the environmental performance of a product.

Standard biodegradability and environmental toxicity tests are very simplistic in their approach, and the usefulness of the data is generally limited in scope. The test systems typically used will never be able to consider the myriad variables that occur in the environment. In order to truly evaluate the "environmental friendliness" of a product, other investigative approaches, such as Life Cycle Assessment, in which manufacturing, delivery, useful life, and disposal undergo equal scrutiny, should be considered. This is particularly critical when comparing different classes of products, e.g., mineral oil-based versus vegetable oil-based hydraulic fluids.

Too much emphasis should not be placed on the quantitative results from these tests. Environmental studies cannot merely be represented by the simple numerical values that are often used to support claims regarding the "friendliness" of a product. Rather, they need to be understood in the context within which they were developed, i.e., how and why the tests were done. If not considered in that limited context, the information could improperly represent the "friendliness" or "unfriendliness" of a product.

In summary, biodegradation and environmental toxicity test results are not directly comparable in the same way as tests for physical characteristics, such as viscosity. Finally, although they are very important in the overall evaluation of a product, they represent only part of the data important to a product's complete evaluation.

7.5 Cloud Point

7.5.1 ASTM D 2500

If chilled to sufficiently low temperature, distillate fuels can lose their fluid characteristics. This can result in loss of fuel supply. The time when cold weather causes fuel stoppages is precisely the time when fuel is needed most in residential and commercial heating units.

Diesel powered equipment of all kinds is subject to failure due to poor low-temperature operability of the fuel. Consequently, it is generally necessary to know how cold a fuel can become before flow characteristics are adversely affected.

The most important indication of low-temperature flow characteristics of distillate fuels is *cloud point*. This is the temperature at which enough wax crystals are formed to give the fuel a cloudy appearance. (It should not be confused, however, with the turbid appearance that is sometimes caused by water dispersed in the fuel.)

Because of the effects of some additives on wax crystal formation, cloud point alone should not be used as an absolute minimum operating temperature. Some of these additives have been shown to lower the minimum operating temperature for specific fuels without affecting the base fuel's cloud point.

7.5.2 Significance of Results

The cloud point of a distillate fuel is related to the fuel's ability to flow properly in cold operations. Some additives may permit successful operation with fuels at temperatures below their cloud points; however, for distillate fuels without additives, clogging of filters and small lines may occur due to wax crystal formation at temperatures near the fuel's cloud point.

7.6 Color Scale Comparison

Several scales are used to measure color of petroleum products. Approximate conversion and comparison of the more common color scales can be accomplished through use of charts that are available from lubricant suppliers.

7.6.1 Color and color tests

Color is a term that is often misunderstood because it is a complex aggregate of human values and physical quantities. No two people have quite the same conception of color when it is allowed to assume its broader meanings. Hue, intensity, tone, purity, wavelength, opacity, and brightness are all directly or indirectly associated with color. It would be extremely difficult to depict mathematically all of these dimensions in a single index. Most attempts to define color do so in terms of only one or two factors, and any meaningful discussion of this index must be strictly confined to the dimensions it is able to represent.

Most of the color tests upon which these scales are based involve the same basic procedure. Light is transmitted simultaneously through standard colored glasses (or other standard reference material) and a given depth or thickness of the sample. The two light fields are compared visually and adjustments are made until a match is obtained. In some test, the volume of the sample is varied until the two fields match (as in the Saybolt test); in others, the light transmitted through a given depth or thickness of the sample is matched by using a series of glasses (as in the ASTM test). When the operator obtains a match, a color value is recorded. This color value corresponds to a point on the color scale associated with the particular color test.

These tests, by definition, involve only two qualities of the transmitted light—appearance, as compared with a standard, and intensity. These two dimensions are not sufficient to describe completely the color of the sample, and should be used only to indicate uniformity and freedom from contamination.

7.7 Composition Analysis of Petroleum Hydrocarbons

The analysis of a petroleum hydrocarbon involves the identification or characterization of various components of the substance. This can be accomplished through a variety of techniques. If the amount of information required is great, the analysis can be an extremely complex undertaking. For example, the API Project No. 6, an analysis of a single petroleum sample, continued for about 25 years.

The kind of analyses used in quality control and routine laboratory inspections of petroleum products are much faster, of course. These short-cut methods can be carried out in a variety of ways, using different test procedures and different types of instruments. The choice of method depends upon the nature of the substance to be analyzed, and upon the type of information required.

7.7.1 Types of analysis

The short-cut methods of analysis can generally be classified as either carbon-type or molecular-type. A carbon-type analysis is run when the distribution of the different sizes of molecules—as indicated by the number of carbon atoms in the nucleus—is required. For example, percentages of C_1, C_2, etc. molecules present in the substance can be determined by such an analysis. A molecular-type analysis is run when the object is to characterize the components according to the chemical *arrangement* of their molecules.

Since there are several different ways of classifying the chemical structure of hydrocarbons, several different approaches are possible in molecular-type analyses. For example, a molecular type analysis could be used to determine the relative percentages of the naphthenic, paraffinic, and aromatic components. Another analysis might simply determine the proportions of saturated and unsaturated compounds present.

7.7.2 General methods and instrumentation

The analysis of petroleum hydrocarbons is accomplished through use of a variety of instruments and techniques. The most common techniques go by such names as chromatography, mass spectrometry, ultraviolet and infrared absorption analysis, and precipitation analysis, according to the physical principle upon which each is based.

Chromatography is an analytical technique involving the flow of a gas or liquid, together with the material under analysis, over a special porous, insoluble, sorptive medium. As the flowing phase passes over the stationary phase, different hydrocarbon components are adsorbed preferentially by the medium. With some types of chromatography, these components are desorbed through a similar process, and they leave the chromatographic column in distinct individual patterns. These patterns can be detected and recorded, and with proper interpretation can provide an extremely accurate means of determining composition.

Chromatography is used in both carbon-type and molecular-type analyses. There are a number of chromatographic methods, each named according to the technique of analysis. *Gas chromatography* refers to the general method that uses a gas as the flowing, or mobile, phase; *gas liquid chromatography,* a more specific term, describes the techniques of using gas as the flowing phase and a liquid as the stationary phase; etc.

Silica gel analysis is a liquid chromatographic method that also involves physical separation of the components of a substance. The technique is based on the fact that polar compounds are adsorbed more strongly by silica gel than are non-polar saturated compounds. A sample of material under test is passed through a column packed with silica gel. Alcohol, which is more strongly adsorbed than any hydrocarbon, follows the sample through the column, forcing the hydrocarbons out—saturates first, unsaturated compounds next, then aromatic compounds. Small samples of the emerging material are taken periodically, and the refractive index of each sample is measured. From this information, relative percentages can be determined. (Clay/silica gel analysis, a method designed for rubber process oils, uses both activated clay and silica gel to determine the proportion of asphaltene, aromatic, saturated, and polar compounds present.)

Fluorescent indicator analysis (FIA) is a refinement of silica gel analysis in which a mixture of fluorescent dyes is placed in a small layer in the silica gel column. The dyes separate selectively with the aromatics, olefins, and saturates in the sample. Under ultraviolet light, boundaries between these different fractions in the column are visible; the amount of each hydrocarbon-type present can be determined from the length of each dyed fraction.

Mass spectrometry identifies the components of a substance by taking advantage of the difference in behavior exhibited by molecules of different mass when subjected to electrical and magnetic fields. A particle stream of the test material is first ionized, then directed in a curved path by a combination of the electrical and magnetic fields. The heavier ions, having greater inertia, tend toward the outside of the curve The stream of particles is therefore split up into a "mass spectrum"—they are distributed across the path according to their masses. This differentiated stream is played across a detecting slot on the "target," and a record of the analysis is thus made. (When the target is a photographic plate, the instrument is referred to as a Mass "Spectroscope").

As might be expected, the Mass Spectrometer is most useful, at least for hydrocarbon analysis, in the determination of carbon-number distributions.

However, because various types of material show distinct spectral patterns, the Mass Spectrometer is also used in molecular-type analysis.

Ultraviolet (UV) Absorption Analysis is a method in which the amount and pattern of ultraviolet light absorbed by the sample is taken as a "fingerprint" of the components. The analysis is carried out through use of a spectrophotometer, which measures the relative intensities of light in different parts of a spectrum. By comparing the UV-absorbance pattern of the test sample with patterns of known material, components of the sample may be characterized. Infrared (IR) Analysis is a similar method but utilizes a different radiation frequency range.

Precipitation Analysis is used primarily in the characterization of rubber process oils. Components are identified on the basis of their reaction with varying concentrations of sulfuric acid. Hydrocarbons are separated into asphaltenes, polar compounds, unsaturated compounds (which are further separated into two groups, First Acidiffins and Second Acidiffins), and saturated compounds.

7.8 Consistency of Grease (Penetration)

See "Grease Consistency."

7.9 Copper Strip Corrosion

7.9.1 ASTM D 130

Many types of industrial equipment use parts made of copper or copper alloys. It is essential that any oil in contact with these parts be non-corrosive to them.

Though modern technology has made great progress in eliminating harmful materials from petroleum oils, corrosion is still a possibility to be considered. Certain sulfur derivatives in the oil are a likely source. In the earlier days of the petroleum industry, the presence of active sulfur might have been attributable to inadequate refining. Today, however, practical methods have been developed to overcome this problem, and straight mineral oils of high quality are essentially free of corrosive materials.

On the other hand, certain oil additives, such as some of the emulsifying and extreme pressure (EP) agents, contain sulfur compounds. In the higher-quality oils, including those for moderate EP conditions, these compounds are of a type that is harmless to copper. For the more severe EP applications, however, chemically active additives are required for the prevention of scoring and seizure. Though oils containing these additives may not be desirable in the presence of copper or copper alloys, they are indispensable to many applications involving steel parts. Automotive hypoid rear axles are an example of this type of application.

To evaluate oil corrosiveness to copper, amd also to check for active sulfur-type EP additives, the copper strip corrosion test is the most widely accepted procedure. This test, described under the ASTM test method D 130 is applicable to the determination of copper-corrosive properties of certain fuels and solvents. Note: this is not to be confused, with tests for the rust-inhibiting properties of petroleum oils. The copper strip test evaluates the copper-corrosive tendencies of the oil itself, not the ability of the oil to prevent corrosion from some other source.

7.9.2 Significance of results

In the lubrication of bronze bushings, bearings that contain copper, and bronze wheels for worm-gear reduction units, corrosive oils must be carefully avoided. Because of the use of bronze retainers, manufacturers of anti-friction bearings insist on non-corrosive oils for their products. Hydraulic fluids, insulating oils, and aviation instrument oils must also be non-corrosive. In the machining of non-ferrous metals, moreover, cutting fluids must be of a non-corrosive type. The copper strip corrosion test helps to determine the suitability of these oils for the type of service they may encounter. In addition, it may help to identify oils of the active chemical type formulated for severe EP application.

This test may serve also in the refinery to check finished products for conformity with specifications. It may be applied, too, to solvents or fuels for assurance that these products will not attack cuprous metals with which they come in contact. In addition, there are certain special tests for corrosiveness, including the silver strip corrosion test of diesel lubricants. This test is applicable to crankcase oils for engines with silver bearing metals.

In conducting the copper strip corrosion test, there are 3 variables that may affect test results:

- Time of exposure of the copper to the sample,
- Temperature of the sample, and
- Interpretation of the appearance of the exposed sample.

It is reasonable to expect that these variables will be applied in such a way as to reflect the conditions to which the product is to be subjected.

Table 7.2 ASTM copper strip classification.

Classification	Designation		Description
1	Slight Tarnish	1a	Light orange, almost the same as a freshly polished strip
		1b	Dark orange
2	Moderate Tarnish	2a	Claret red
		2b	Lavender
		2c	Multi-colored with lavender blue and/or silver overlaid on claret red
		2d	Silvery
		2e	Brassy or gold
3	Dark Tarnish	3a	Magenta overcast on brassy strip
		3b	Multi-colored with red and green showing (peacock), but no gray
4	Corrosion	4a	Transparent black, dark gray or brown with peacock green barely showing
		4b	Graphite or lusterless black
		4c	Glassy or jet black

There is nothing to be gained, for example, by testing the oil at 212°F if test results at 122°F give better correlation with actual service conditions. If service conditions are more severe, however, test results at the higher temperature may give a more reliable indication of the oil's performance characteristics.

Similarly, selection of the critical ASTM classification must be based on experience gained in the type of service for which the product is formulated. A dark tarnish (Classification 3, Table 7.2) is wholly acceptable, for example, where it has been shown that this degree of copper discoloration is associated with safe performance of the tested product. The flexibility of the copper strip test makes it adaptable to a wide range of products and end uses.

7.10 Demulsibility

7.10.1 ASTM D 1401 and ASTM D 2711

In the petroleum industry, the term emulsion usually applies to an emulsion of oil and water. Though soluble to a degree, these substances can, under certain circumstances, be intimately dispersed in one another to form a homogeneous mixture. Such a mixture is an oil/water emulsion, usually milky or cloudy in appearance.

Commercial oils vary in emulsibility. A highly refined straight mineral oil will resist emulsification. Even after vigorous agitation with water, an oil of this type tends to separate rapidly from the water when the mixture is at rest. Emulsification can be promoted, however, by agitation and by the presence of certain contaminants or ingredients added to the oil. The more readily the emulsion can be formed and the greater its stability, the greater the emulsibility of the oil. Some products, such as soluble cutting fluids, require good emulsibility and are formulated with special emulsifying agents.

With many other products, such as turbine oils and crankcase oils, the opposite characteristic is desired. To facilitate the removal of entrained water, these products must resist emulsification. The more readily they break from an emulsion, the better their demulsibility.

Two tests for measuring demulsibility characteristics have been standardized by the ASTM. The older of the two is ASTM method D 1401, which was developed specifically for steam turbine oils having viscosities of 150-450 Saybolt seconds at 100°F. It can be used for oils of other viscosities if minor changes in the test procedure are made. This method is the one recommended for use with synthetic oils.

The second method, ASTM D 2711 is designed for use with R&O (rust and oxidation inhibited) oils. It can also be used for other types of oils, although minor modifications are required when testing EP (extreme pressure) oils.

7.10.2 Significance of results

In many applications, oil is exposed to contamination by water condensed from the atmosphere. With turbine oils, exposure is even more severe, since the oil tends to come in contact with condensed steam.

Water promotes the rusting of ferrous parts and accelerates oxidation of the oil. For effective removal of the water, the oil must have good demulsibility characteristics.

Steam cylinder oils that serve in closed systems require good demulsibility for the opposite reason: to facilitate removal of oil from the condensate, so that oil is kept out of the boiler. Hydraulic fluids, motor oils, gear oils, engine oils insulating oils, and many similar petroleum products must resist emulsification. Oil and water must separate rapidly and thoroughly.

For each of the 4 ASTM designations, there are two or more standards considered to cover the same degree of tarnish or corrosion. Differences in the type of chemical reaction produce differences in discoloration. Even a completely non-corrosive oil will alter the appearance of the freshly polished strip.

Either of the ASTM methods is suitable for evaluating the demulsification properties both of inhibited and uninhibited oils. However, correlation with field performance is difficult. There are many cases where the circulating oil is operating satisfactorily in the field, but fails the demulsibility tests in the laboratory. Hence, it must be recognized that these laboratory test results should be used in conjunction with other facts in evaluating oil's suitability for continued service.

7.11 Density

Density is a numerical expression of the mass-to-volume relationship of a substance.

Density is important in volume-to-mass and mass-to-volume calculations, necessary in figuring freight rates, fuel loads, etc. Although it is not directly a criterion of quality, it is sometimes useful as an indicator of general hydrocarbon type in lubricants and fuels. For a given volatility, for example, aromatic hydrocarbons have a greater density than paraffins, naphthenic hydrocarbons usually being intermediate. Density data is also used to monitor uniformity of composition.

Density may be determined by ASTM method D 1298, using a hydrometer graduated in units of density. Tables are available for conversion of observed density to that at 15°C.

See also Gravity.

7.12 Dielectric Strength

7.12.1 ASTM D 877 and D 1816

Dielectric is an electric insulating material, one that opposes a flow of current through it. There are two properties that contribute to this characteristic. One is resistivity, the specific resistance that a dielectric offers under moderate conditions of voltage. The other is dielectric strength, the ability to prevent arcing between two electrodes at high electric potentials. Though the two properties are not directly related, commercial insulating materials of high dielectric strength also possess adequate resistivity. In the insulation of high-voltage electrical conductors, it is dielectric strength that is of the greater concern.

Petroleum oil is an excellent dielectric and is used extensively in electrical equipment insulated with a liquid. Among the advantages that oil offers over solid insulation are the abilities to cool by circulation and to prevent corona. Corona is the result of ionization of air in the tiny voids that exist between a conductor and a solid insulating wrapper. Corona is destructive to certain types of solid insulation. By filling all of the space around a conductor, insulating oil eliminates the source of corona. Oil also has the high dielectric strength that good insulation requires.

At normal voltage gradients, conduction of electric current through a dielectric is negligible. The dielectric lacks the free charged particles that a conductor must have. If the voltage impressed on the dielectric is increased, however, the material becomes more highly ionized. Ions thus produced are free charged particles.

If a high enough voltage is applied, ions are produced in sufficient concentration to allow a discharge of current through the dielectric, and there is an arc. The minimum voltage required for arcing is the breakdown voltage of the dielectric incurred under the circumstances involved. When the dielectric breaks down, it undergoes a change in composition that temporarily permits it to conduct electricity.

The magnitude of the breakdown voltage depends on numerous factors, such as the shape and thickness of the electrodes and dielectric strength of the insulation between them. In accordance with the ASTM method D 877 or D 1816, the dielectric strength of insulating oil is evaluated in terms of its breakdown voltage under a standard set of conditions.

Because of the marked effect of contamination on test results, special care must be exercised in obtaining and handling the sample. The sample container and test cup must be absolutely clean and dry, and no foreign matter must come in contact with the oil.

In either case, the voltage noted at the specified end point is the breakdown voltage of the respective sample.

7.12.2 Significance of results

Insulating oils find wide application in transformers, cables, terminal bushings, circuit breakers, and similar electrical equipment. Depending upon the installation the purpose of these oils may be to prevent electrical leakage and discharge, to cool, to eliminate corona effects, or to provide any combination of these functions. High dielectric strength is obviously an important insulating-oil property.

When new, carefully refined petroleum oil can be expected to exhibit a high natural dielectric strength suitable for any of the conventional insulating purposes. Other properties of the oil, such as oxidation resistance, are now of greater significance.

In service, however, the oil eventually becomes contaminated with oxidation products, carbon particles, dirt, and water condensed from atmospheric moisture. Water is the the principal offender. Though small quantities of water dissolved in the oil appear to have little influence on dielectric strength, *free* water has a pronounced effect. The dispersion of free water throughout the oil is promoted, moreover, by the presence of solid particles. These particles act as nuclei about which water droplets form. Dielectric strength is impaired also by dirt and oxidation sludge that may accumulate in the oil.

A relationship exists, therefore, between a drop in dielectric strength and the deterioration of oil in service. Dielectric strength thus suggests itself as a method of evaluating the condition of used insulating oil. In this application, a significant drop in dielectric strength may indicate serious water contamination, oxidation, or both.

If water is the only major contaminant, the oil can generally be reclaimed by drying. But, if the drop in dielectric strength is attributable to oxidation, the oil may already have deteriorated beyond a safe limit. By itself, therefore, dielectric strength is not ordinarily considered a sufficiently sensitive criterion of the suitability of a batch of oil for continued service. Power factor, neutralization number, and interfacial tension are test values that have found greater acceptance for this purpose.

7.13 Dilution of Crank Case Oils

7.13.1 ASTM D 322

Excessive crankcase dilution is associated with faulty operation of an internal combustion engine. It is caused by the seepage of raw and partially burned fuel from the combustion chamber past the piston into the crankcase, where it thins the crankcase oil. It is often desirable to know the extent to which used oil has been diluted in this way. For motor oils from gasoline engines, dilution may be evaluated by the ASTM method D 322.

The procedure is to measure the percentage of fuel in the sample by removing the fuel from the oil. Since the fuel is considerably more volatile than the oil, the two can be separated by distillation.

To lower the distillation temperature and to make the test easier to run, a relatively large amount of water is added to the sample. Since the water and the sample are immiscible (incapable of mixing), the boiling point of the mixture is, at any instant, appreciably lower than that of the sample alone.

Because of its substantially higher volatility, the fuel will evaporate before the oil. A mixture of fuel vapor and water vapor passes into a condenser and is converted back to liquid. The fuel, which is lighter, floats on top of the water in a graduated trap. Here, the volume of condensed fuel can be observed before any significant distillation of the oil begins.

7.13.2 Significance of results

This test for crankcase dilution is applicable to used motor oils from gasoline engines. Excessive dilution, as determined by test, is harmful in is own right, as well as being indicative of faulty engine performance.

Dilution is an obvious source of fuel waste. Another effect is to reduce the viscosity of the oil, which may seriously impair its lubricating value. Diluted oil may lack the body required to prevent wear, and it may not make a proper seal at the piston rings. Pistons and cylinders are especially vulnerable, since the oil on their wearing surfaces is subject to the direct washing action of the raw fuel.

Fuel may also reduce the oil's oxidation stability and raise the oil level in the crankcase. Abnormally high levels cause an increase in oil consumption and give false readings as to the amount of lubricant present. Failure of motor oil to lubricate may be directly attributable to dilution.

As an indication of faulty performance, excessive crankcase dilution may be the symptom of unsuitable fuel. If the fuel's volatility characteristics are too low, the fuel does not vaporize properly, and combustion is incomplete. The unburned portion of the fuel finds is way into the crankcase.

A similar effect may be produced, however, by incorrect operation or poor mechanical condition of the engine:

Too rich a fuel mixture—maladjustment of the carburetor or excessive choking may admit more fuel to the combustion chamber than can be burned with the amount of air present.

Too low an engine temperature—defective temperature control or short operating periods may keep the engine too cold for proper vaporization.

Inadequate breathing facilities—insufficient venting of the crankcase vapors may interfere with normal evaporation of the fuel from the crankcase. With older cars, the trouble may be caused by stoppage of the crankcase breather. On cars built after 1963, the positive crankcase ventilation (PVC) system may be at fault.

Worn pistons, rings, or cylinders—excessive clearance between the pistons or rings and the cylinder walls facilitates the seepage of fuel into the crankcase.

Any of the deficiencies indicated by excessive crankcase dilution can be expected to jeopardize satisfactory engine performance.

With diesel engines, there is not the spread in volatility characteristics between fuel and lubricating oil that there is with gasoline engines. For this reason, there is no simple test for the crankcase dilution of diesel engine. The closest approximation is made by noting the reduced viscosity of the used oil as compared with that of the new oil and estimating what percentage of fuel dilution would cause such a viscosity reduction.

7.14 Distillation

A chemically pure hydrocarbon, like any other pure liquid compound, boils at a certain temperature when atmospheric pressure is constant. However, almost all commercial fuels and solvents contain many different individual hydrocarbons, each of which boils at a different temperature. If the petroleum product is gradually heated, greater proportions of the lower-boiling constituents are in the first vapor formed, and the successively higher-boiling constituents are vaporized as the temperature is raised.

Thus, for any ordinary petroleum product, boiling takes place over a range of temperature rather than at a single temperature. This range is of great importance in fuel and solvent applications, and is the property measured in distillation tests.

Several ASTM tests are used for measuring the distillation range of petroleum products. These tests are basically similar, but differ in details of procedure. The following tests are widely used:

- ASTM D 86-67: Distillation of Petroleum Products
- ASTM D 216-54: Distillation of Natural Gasoline ASTM D 850-70: Distillation of Industrial Aromatic Hydrocarbons
- ASTM D 1078-70: Distillation Range of Lacquer Solvents and Dilutants

7.14.1 Significance of Results

For both fuels and solvents, distillation characteristics are important.

Automotive Gasoline: The entire distillation range is important in automotive fuels. The distillation characteristics of the "front end" (the most volatile portion, up to perhaps 30% evaporated), together with the vapor pressure of the gasoline (see discussion on Vapor Pressure), control its ability to give good cold-starting performance. However, these same characteristics also control its vapor-locking tendency. An improvement in cold-starting can entail a decrease in vapor lock protection.

The temperatures at which 50% and 90% of the fuel are evaporated are indications of warm-up characteristics. The lower these points, the better the warm-up. A low 50% point is also an indication of good acceleration. A low 90% point is desirable for completeness of combustion, uniformity of fuel distribution to the cylinders, and less formation of combustion chamber deposits.

Usually, the volatility of a commercial gasoline is adjusted seasonally, and also in accordance with the climate in the region into which it is being shipped. In cold weather, a more volatile product is desired to provide better starting and warm-up. In warm weather, less volatility provides greater freedom from vapor lock.

Aviation Gasoline: In general, aviation gasoline has lower 90% points and final boiling points than automotive gasoline, but the significance of the various points on the distillation curve remain the same. A minimum limit on the sum of the 10% and 50% points is normally specified to control carburetor icing characteristics.

Diesel Fuel: Although diesel fuels have much lower volatility than gasoline, the effect of the various distillation points are similar. For example, the lower the initial boiling point for a given cetane number, the better its starting ability, but more chance of vapor lock or idling difficulties. Also, the higher the end point or final boiling point, the more chance there is of excessive smoking and deposits. The mid-boiling point (50% point) is related to fuel economy because, other things being equal, the higher the 50% point, the more Btu content and the better cetane number a diesel fuel possesses.

Burner Fuel: For burner fuels, ease of lighting depends on front end volatility. Smoking depends upon the final boiling point, with excessive smoke occurring if the final boiling point is too high.

Solvents: Many performance characteristics of solvents are related to distillation range. The initial boiling point is an indirect measure of flash point and, therefore, of safety and fire hazard. The spread between IBP and the 50% point is an index of "initial set" when used in a rubber or paint solvent.

The 50% point shows a rough correlation with evaporation rate; the lower the 50% point for certain classes of hydrocarbons, the faster the evaporation. If the dry point and the 95% point are close, there is little or no "tail" or slow-drying fractions. Also, useful indications of good fractionization of a solvent are the narrowness of the distillation range and the spread between the initial boiling point and 5% point and between 95% point and the dry point. Therefore, the smaller the spread, the better.

7.15 Dropping Point of Grease

7.15.1 ASTM D 566 and ASTM D 2265

It is often desirable to know the temperature at which a particular lubricating grease becomes so hot as to lose its plastic consistency. Being a mixture of lubricating oil and thickener, grease has no distinct melting point in the way that homogeneous crystalline substances do. At some elevated temperature the ordinary grease becomes sufficiently fluid to drip. This temperature is referred to as the dropping point, and is determined by the ASTM Method D 566-"Dropping Point of Lubricating Grease" or the ASTM Method D2265—"Dropping Point of Lubricating Grease of Wide Temperature Range."

7.15.2 Significance of results

Since both tests are held under static conditions, the results have only limited significance with respect to service performance. Many other factors such as time exposed to high temperatures, changes from high-to-low temperatures, evaporation resistance and oxidation stability of the grease, frequency of re-lubrication, and the design of the lubricated mechanism, are all influences that affect the maximum usable temperature for the grease.

Though both dropping point and consistency are related to temperature, the relationships follow no consistent pattern. The fact that grease does not liquefy at a particular temperature gives no assurance that its consistency will be suitable at that temperature. However, the dropping point is useful in identifying the grease by type and for establishing and maintaining benchmarks for quality control.

One of the weaknesses of either procedure is that a drop of oil may separate and fall from the grease cup at a temperature below that at which the grease fluidizes. This would then give an erroneous indication of the actual temperature at which the grease becomes soft enough to flow from the cup.

7.16 Ecotoxicity

(See "Biodegradation")

7.17 Flash and Fire Points—Open Cup

7.17.1 ASTM D 92

The flash point and the fire point of a petroleum liquid are basically measurements of flammability. The flash point is the minimum temperature at which sufficient liquid is vaporized to create a mixture of fuel and air that will burn if ignited. As the name of the test implies, combustion at this temperature is only of an instant's duration. The fire point, however, runs somewhat higher. It is the minimum temperature at which vapor is generated at a rate sufficient to *sustain* combustion. In either case, combustion is possible only when the ratio of fuel vapor to air lies between certain limits. A mixture that is too lean or too rich will not burn.

The practice of testing for flash and fire points was originally applied to kerosene (aircraft fuel) to indicate its potentiality as a fire hazard. Since then, the scope has been broadened to include lubricating oils and other petroleum products. Though it has become customary to report flash point (and sometimes fire point) in lubricating oil data, these properties are not as pertinent as they might appear. Only in special instances does lubricating oil present any serious fire hazard. Being closely related to the vaporization characteristics of a petroleum product, however, flash and fire points give a rough indication of volatility and certain other properties.

The fire point of a conventional lubricating oil is so closely associated with its flash point, that it is generally omitted from inspection data. For the ordinary commercial product, the fire point runs about 50°F above the flash point. Fire and flash points are not to be confused, however, with auto-ignition temperature, which is an entirely different matter.

Auto-ignition deals not so much with volatility, as with the temperature necessary to precipitate a combustive chemical reaction without an external source of ignition. Though a more volatile petroleum product may be expected to have lower flash and fire points than one that is less volatile, its ASTM auto-ignition temperature is generally higher.

7.17.2 Significance of results

To appreciate the significance of flash point and fire point test results, one must realize what the tests measure. It is necessary to understand how a combustible air-fuel mixture is created.

For all practical purposes, a petroleum liquid does not burn as such, but must first be vaporized. The vapor mixes with the oxygen in the air, and, when sufficient concentration of the vapor is reached, the mixture may be ignited, as by a spark or open flame. The mixture can be ignited only if the concentration of fuel vapor in the air is more than about 1% or less than about 6% by volume. A confined mixture containing more than 6% fuel vapor becomes a practical explosion hazard only if it is vented to admit a greater portion of air.

The significance of flash and fire-point values lies in the dissimilarity that exists in the volatility characteristics of different petroleum liquids. Even among lubricating oils of comparable viscosity, there are appreciable variations in volatility, and hence in flash and fire point. In general, however, the storage and operating temperatures of lubricating oils are low enough to preclude any possibility of fire. Among the exceptions to this situation are such products as quenching and tempering oils, which come in direct contact with high-temperature metals. Heat-transfer oils, used for heating or cooling, may also reach temperatures in the flash and fire-point ranges. Similarly, in the evaluation of roll oils, which are applied in steel mills to hot metal sheets from the annealing oven, fire hazard may likewise be a consideration. In many of these cases, however, auto-ignition temperature is of greater significance. At the auto-ignition temperature, as determined by test, fire is not merely a possibility—it actually occurs spontaneously, i.e., without ignition from any outside source.

Since flash and fire point are also related to volatility, however, they offer a rough indication of the tendency of lubricating oils to evaporate in service. It should be apparent that lower flash and fire points imply a greater opportunity for evaporation loss. The relationship between test results and volatility is by no means conclusive, however. The comparison is distorted by several additional factors, the most important of which is probably the manner in which the oil is produced.

The relationship between flash and fire point, on the one hand, and volatility, on the other, is further distorted by differences in oil type. For a given viscosity, paraffinic oil will exhibit higher flash and fire point than other types and may be recognized by these test results. Paraffinicity may also be indicated by a high viscosity index or by a high pour point.

Fire and flash points are perhaps of greater significance in the evaluation of used oils. If an oil undergoes a rise in flash or fire point in service, loss by evaporation is indicated. The more volatile components have been vaporized, leaving the less volatile ones behind; so an increase in viscosity is apparent. An excessive increase in viscosity may so alter lubricating properties that the oil is no longer suitable for its intended application.

If, on the other hand, the flash or fire points drop in service, contamination is to be suspected. This may happen to motor oils that become diluted with unburned fuel. Gasoline or heavier fuels in the crankcase reduce the viscosity of the oil, and bearings and other moving parts may be endangered by excessive thinning of the lubricant. These fuels, being more volatile than the oil, lower the flash and fire points of the mixture. So the flash or fire-point test on used oils constitutes a relatively simple method for indicating the presence of dilution.

7.18 Flash Point-closed Cup

7.18.1 ASTM D 56 and D93

All petroleum products will burn, and under certain conditions their vapors will ignite with explosive violence. However, in order for this to occur, the ratio of vapor to air must be within definite limits.

When liquid petroleum product is exposed to air, some of it evaporates, causing a certain vapor/air concentration. As the temperature of the liquid product is raised, more and more evaporates, and the vapor/air ratio increases. Eventually, a temperature is reached at which the vapor/air ratio is high enough to support momentary combustion, if a source of ignition is present. This temperature is the flash point of the product.

For fuels and solvents, the flash point is usually determined by a "closed cup" method, one in which the product is heated in a covered container. This most closely approximates the conditions under which the products are handled in actual service. Products with flash points below room temperature must, of course, be cooled before the testing begins.

Two closed cup methods of determining flash point are widely used. They differ primarily in details of the equipment and in the specific fields of application. However, the tests are basically similar, and may be grouped together for descriptive purposes. The two tests are:

- ASTM D 56 Flash Point by Means of the Tag Closed Tester
- ASTM D 93 Flash Point by Means of the Pensky-Martens Closed Tester

The D 56 test is used for most fuels and solvents, including lacquer solvents and dilutants with low flash points. Whereas the D 93 test is ordinarily used for fuel oils but can also be used for cutback asphalts and other viscous materials and suspensions of solids.

7.18.2 Significance of results

For a volatile petroleum solvent or fuel, flash point is primarily significant as an indication of the fire and explosion hazards associated with its use. If it is possible for any particular application to select a product whose flash point is above the highest expected ambient temperature, no special safety precautions are necessary. However, gasoline and some light solvents have flash points well below room temperature. When they are used, controlled ventilation and other measures are necessary to prevent the possibility of explosion or fire.

7.19 Foaming Characteristics of Lubricating Oils

7.19.1 ASTM D 892

Foaming in an industrial oil system is a serious service condition that may interfere with satisfactory system performance and even lead to mechanical damage.

While straight mineral oils are not particularly prone to foaming, the presence of additives and the effects of compounding can change the surface properties of oil and increase susceptibility to foaming when conditions are such as to mix air with the oils.

Special additives impart foam resistance to oils and enhance their ability to release trapped air quickly under conditions that would normally cause foaming.

The foaming characteristics of lubricating oils at specified temperatures are determined by the standard ASTM method D 892.

7.19.2 Significance of results

Foaming consists of large bubbles that rise quickly to the surface of the oil, and is to be distinguished from air entrainment, consisting of slow-rising bubbles dispersed throughout the oil. Both these conditions are undesirable, and are often difficult to distinguish due to high flow an equal volume of water at the same temperature.

In the petroleum industry, the API gravity scale is more widely used. This is an arbitrary scale, calibrated in degrees, related to specific gravity rates and turbulence in the system. These two phenomena are affected by different factors and are considered in separate laboratory tests. The primary causes of foaming are mechanical—essentially an operating condition that tends to produce turbulence in the oil in the presence of air. The current trend in hydraulic oil systems, turbine oil systems, and industrial oil systems of every kind is to decrease reservoir sizes and increase flow rates. This trend increases the tendency for foaming in the oils.

Contamination of the oil with surface-active materials, such as rust preventatives, detergents, etc., can also cause foaming.

Foaming in industrial oil is undesirable because the foam may overflow the reservoir and create a nuisance, and decrease lubrication efficiency of the oil, which may lead to mechanical damage.

Anti-oaming additives may be used in oils to decrease foaming tendencies of the oil. However, many such additives tend to increase the air entrainment characteristics of oil, and their use requires striking a balance between these two undesirable phenomena.

7.20 Four-ball Wear Test—ASTM D 2266

7.21 Four—Ball EP Test—ASTM D 2596

(See "Load Carrying Ability")

7.22 Gravity

7.22.1 ASTM D 287

Practically all, liquid petroleum products are handled and sold on a volume basis, e.g. by gallon, barrel, tank car, etc. Yet, in many cases, it is desirable to know the weight of the product. Gravity is an expression of the weight-to-volume relationship of a product.

Any petroleum product will expand when heated, and its weight per unit volume will therefore decrease. Because of this, gravity is usually reported at a standard temperature, although another temperature may actually have been used in the test. Tables are available for converting gravity figures from one temperature basis to another.

Gravity can be expressed on either of two scales. The "specific gravity" is defined as the ratio of the weight of a given volume of the product at 60°F to the weight of an equal volume of water at the same temperature.

In the petroleum industry, however, the API gravity scale is more widely used. This is an arbitrary scale, calibrated in degrees, and related to specific gravity by the formula:

$$\text{API Gravity (degrees)} = \frac{141.5}{\text{specific gravity } 60/60\,°\text{F}} \quad 131.5$$

As a result of this relationship, the higher the specific gravity of a product, the lower is its API gravity. It is noteworthy that water, with a specific gravity of 1.000, has an API gravity of 10.0°.

Gravity, specific or API, is determined by floating a hydrometer in the liquid, and noting the point at which the liquid level intersects the hydrometer scale. Corrections must then be made in accordance with the temperature of the sample at the time of test.

7.22.2 Significance of results

Gravity has little significance from a quality standpoint, although it is useful in the control of refinery operations. Its primary importance is in volume-to-weight and weight-to-volume calculations. These are necessary in figuring freight rates, aircraft and ship fuel loads, combustion efficiencies, etc.

To some extent, gravity serves in identifying the type of petroleum product. Paraffinic products have lower specific gravities (higher API gravities) than aromatic or naphthenic products of the same boiling range. Gravity data may also be used by manufacturers or by their customers to monitor successive batches of these products as a check on uniformity of product composition.

Gravity is important in process applications that depend on differences in gravity of the materials used. For example, petroleum products having higher specific gravities than 1.000 (that of water) are necessary in the field of wood preservation in order to permit separation of the materials involved. The specific gravity range of petroleum products is about 0.6 to 1.05.

Gravity is used in empirical estimates of thermal value, often in conjunction with the aniline point. With the exception of the above applications, gravity should not be used as an index of quality.

7.23 Grease Consistency

7.23.1 ASTM D 217 and D 1403

The consistency of lubricating grease is defined as its resistance to deformation under an applied force, i.e. its relative stiffness or hardness. The consistency of grease is often important in determining its suitability for a given application.

Grease consistency is given a quantitative basis through measurement with the ASTM Cone Penetrometer. The method consists of allowing a weighted metal cone to sink into the surface of the grease, and measuring the depth to which the point falls below the surface. This depth, in tenths of millimeters, is recorded as the penetration, or penetration number, of the grease. The higher its penetration, the softer the grease.

The ASTM D 217 method recognizes five different categories of penetration, depending on the condition of the grease when the measurement is

Table 7.3 NLGI grease grading system.

NLGI Grade	Worked Penetration Range mm/10
000	445-475
00	400-430
0	355-385
1	310-340
2	265-295
3	220-250
4	175-205
5	130-160
6	85-115

made. *Undisturbed* penetration is determined with the grease in its original container. *Unworked* penetration is the penetration of a sample which has received only minimum disturbance in being transferred from the sample can to the test cup. *Worked* penetration is the penetration of a grease sample that has been subjected to 60 double strokes in a standard grease worker (to be described). *Prolonged worked* penetration is measured on a sample that has been worked the specified number of strokes (more than 60), brought back to 77°F, then worked an additional 60 double strokes in the grease worker. *Block* penetration is the penetration of a block grease—a grease hard enough to hold its shape without a container.

All the above penetrations are determined on samples that have been brought to 77°F.

7.23.2 Significance of results

If grease is too soft, it may not stay in place, resulting in poor lubrication. If grease is too hard, it will not flow properly, and likely fail to provide proper lubrication or cause difficulties in dispensing equipment. These statements sum up the reasons for classifying greases by consistency. Penetration numbers are useful for classifying greases according to the consistencies required for various types of service, and in controlling the consistency of a given grade of grease from batch to batch.

The National Lubricating Grease Institute has classified greases according to their worked penetrations. These NLGI grades, shown in Table 7.3, are used for selection of greases in various applications.

In comparing greases, worked and prolonged worked penetrations are generally the most useful values. The change in penetration between the 60-stroke value and prolonged worked value is a measure of grease stability. Prolonged worked penetrations should report the amount of working (10,000 and 100,000 strokes are most common) in order to be useful. Unworked penetrations often appear in specifications and in grease product data, but are of limited practical value. No significance can be attached to the difference between unworked and worked penetration. Undisturbed penetration is useful mainly in quality control. Figure 7.1 further explores how greases are classified and tested.

7.24 Interfacial Tension

7.24.1 ASTM D 971

Molecules of a liquid have a certain attraction for one another. For some liquids, like mercury, this attraction is very great; for others, like alcohol, it is considerably less. Beneath the surface, the attractive forces are evenly distributed, since a molecule is drawn to the one above or below it as strongly as to the one at its side. But, at the surface, there are no similar molecules over it to attract the liquid upward; so the bonds between the molecules are concentrated in a lateral direction.

The strong mutual attraction between the surface molecules result in a phenomenon known as surface tension, and its effect is like that of a membrane stretched over the liquid face. Surface tension is an appreciable force, as anyone knows who has made the simple experiment of "floating" a needle on the top layer of still water.

Surface tension can be reduced by the introduction of materials that weaken the links between the original molecules. Differences in surface tension can be measured, and these measurements sometimes serve as a guide to the condition of used oil. Standard procedure for making these measurements is covered by the ASTM method D 971 for conducting the interfacial tension (IFT) test.

The IFT test measures the tension at the interface between two immiscible liquids: oil and distilled water. Ordinarily, oil and water do not mix, the oil floating on top of the water because it is less dense. At the interface, each liquid exhibits its own surface tension, the molecules of one having no great attraction for those of the other. To break through the interface, it is necessary to rupture the surface tensions both of the water and of the oil. However, if certain contaminants are added to the oil—such as soaps, dust particles, or

Grease Classification

1. Group I—General Purpose Greases
   Greases that are expected to give proper lubrication to bearings whose operating temperatures may vary from −40° C to 121° C (−40° F to 250° F).

2. Group II—High Temperature Greases
   Greases that are expected to give proper lubrication to bearings whose operating temperatures may vary from −18° C to 149° C (0° F to 300° F).

3. Group III—Medium Temperature Greases
   Greases that are expected to give proper lubrication to bearings whose operating temperatures may vary from 0° C to 93° C (32° F to 200° F).

4. Group IV—Low Temperature Greases
   Greases that are expected to give proper lubrication to a bearing whose operating temperature may go as low as −55° C (−67° F) or as high as 107° C (225° F).

5. Group V—Extreme High Temperature Greases
   Greases that are expected to give proper lubrication for comparatively short periods of time where bearing operating temperature may be as high as 232° C (450° F).

The requirements of the five groups of greases cover applications where the lubrication is not affected by extremely heavy loads, high speeds or excessive humidity. The tabulation of test requirements which follows is intended to establish for the five groups of greases those characteristics needed to meet the functional requirements of good rolling bearing greases.

Methods of Testing

Test Requirements	Method
ASTM Penetration Normal Worked	Method ASTM D-217
Oxidation	Method ASTM D-942
Low Temperature Torque	Method ASTM D-1478
Water Resistance	Method ASTM D-1264
Dropping Point	Method ASTM D-566
Evaporation	Method ASTM D-972
Dirt	Method Federal Test Method Standard No. 791-B Method 3005.3
Consistency, Stability	Method ASTM D-217
Abrasive Matter	Method ASTM D-1404
Free Acid and Free Alkali	Method ASTM D-128
Fillers	Method ASTM D-128
Moisture	Method ASTM D-128

Test Requirements

Tests	Group I	Group II	Group III	Group IV	Group V
ASTM Penetration Normal Worked[1]	250 to 350	200 to 300	220 to 300	260 to 320	250 to 310
Oxidation	10# max. drop in 500 hours	10# max. drop in 500 hours	10# max. drop in 500 hours	5# max. drop in 500 hours	5# max. drop in 100 hours at 121° C (250° F)
Low Temperature Torque	1 rev. in max. of 10 sec. at −40° C (−40° F)	—	—	1 rev. in max. of 5 sec. at −55° C (−67° F)	1 rev. in max. of 10 sec. at −40° C (−40° F)
Water Resistance	50% max. loss	—	50% max. loss	20% max. loss	50% max. loss
Dropping Point	149° C min. (300° F)	177° C min. (350° F)	149° C min. (300° F)	149° C min. (300° F)	232° C min. (450° F)
Evaporation	—	—	—	1.5% max. in 22 hours at 121° C (250° F)	4.0% max. in 22 hours at 204° C (400° F)
Dirt Max. Particles per cubic cm	5000/5-20 microns 2000/20-50 microns 50/50-75 microns None over 75 microns	Same as Group I	Same as Group I	Same as Group I	Same as Group I
Consistency Stability: max. ASTM Penetration after 100,000 strokes	Penetration shall not increase more than 100 points and in no case be more than 375	Penetration shall not increase more than 100 points and in no case be more than 350	Penetration shall not increase more than 100 points and in no case be more than 350	Penetration shall not increase more than 100 points and in no case be more than 375	Penetration shall not increase more than 100 points and in no case be more than 375

[1]National Lubricating Grease Institute Code — No. 0: 355-385; No. 1: 310-340; No. 2: 265-295; No. 3: 220-250; No. 4: 175-205

Figure 7.1 Grease classification and testing.

the products of oil oxidation—the situation is altered. These contaminants are said to be hydrophilic, i.e., they have an affinity for the water molecules—as well as for the oil molecules. At the interface, the hydrophilic materials extend bonds across to the water, so that any vertical linkage between the liquids is strengthened, and the lateral linkage is weakened. The interface is less distinct, and the tension at the interface is reduced. The greater the concentration of hydrophilic materials, the less the tension. Since oxidation products tend to be hydrophilic, IFT test results are related to the degree of the oil's oxidation.

7.24.2 Significance of results

For many purposes, large quantities of petroleum oil remain in service for very long periods. There is good reason, therefore, for checking the extent to which these oils have oxidized to determine their fitness for continued service. In some cases, neutralization number may provide a criterion by which used oils are evaluated. But, by the time the acid neutralization number has undergone a significant rise, oxidation has set in, and acids and sludges may be already formed.

It is felt by many people that the IFT test is more sensitive to incipient oxidation, that it anticipates oxidation before deterioration has reached serious proportions. Since the oxidation of an oil increases at an accelerated rate, an early warning of impending deterioration is advantageous.

The IFT test is frequently applied to electric transformer oils, where oxidation is especially harmful. Acids formed by oxidation may attack the insulation, and oxidation sludges interfere with circulation and the cooling of the windings. Because of the importance of good quality, transformer oils may be checked periodically for IFT to determine the advisability of replacement. The critical IFT value is based on experience with the particular oil in service, and testing conditions must be uniform in all respects. Tests for power factor and dielectric strength are also related to the condition of the oil.

For new oils, IFT values have little significance, though they may be used for control purposes in oil manufacture. Additives added to the oil to improve its performance may grossly distort IFT test results, so that they bear no apparent relation to the oil's quality. For this reason, special consideration must be given if attempts are made to evaluate the condition of an inhibited steam turbine oil by the IFT method.

The IFT test itself is an extremely delicate one, and consistent results are not easily obtained. The test should be conducted only by an experienced person, and the apparatus must be scrupulously clean. A minute quantity of foreign matter can cause a tremendous increase in the oil's hydrophilic properties. The sample must be carefully filtered to remove all solid materials, which reduce the IFT value appreciably. Even under the most meticulous conditions, however, good reproducibility is difficult to achieve.

It has been found, moreover, that test results are affected by conditions outside of the laboratory. Prolonged storage of the oil sample may cause a drop in its IFT value; so may exposure to sunlight. Similarly, agitation of the sample may increase its hydrophilic properties, and, if the sample is not tightly sealed against air, there may be an increase in the oxidation products. While indifferent handling practices would not be expected to affect a new oil, they may cause the deterioration of a used oil to appear greater than it actually is.

7.25 Load-carrying Ability

For machine parts that encounter high unit loadings, the lubricant must be capable of maintaining a film that prevents metal-to-metal contact under the extreme pressures involved. Otherwise, scoring of the surfaces and possible failure of the parts will result. Special extreme pressure (EP) lubricants are required for such applications.

Several test machines have been constructed and test procedures established in attempts to approximate closely the conditions a lubricant will meet in field applications. Four widely used tests are the Timken machine, the FZG test, the 4-Ball EP test, and the 4-Ball friction and wear test.

Timken Machine—In the Timken test (ASTM D 2509), a rotating member is brought to bear against a stationary member with lubrication provided by the lubricant under test. The lubricant is evaluated on the basis of its ability to prevent scoring of the metal surfaces. The maximum load that can be applied without scoring is reported as the Timken OK load. The minimum load required to cause scoring is reported as the score load.

In addition to the OK and score loads, the actual pressure at the point of contact is sometimes reported. The area of the wear scar is determined using a Brinnell microscope; the unit loading in megapascals (or in psi) on the area of contact can then be calculated using the OK load.

Only very general conclusions can be drawn from the Timken EP test. Results should be related to additional information about the lubricant, such as anti-wear properties, type of additive, and corrosion characteristics. Used in this way, Timken EP results can provide an experienced engineer with valuable information about the performance of one lubricant relative to another. In addition, the Timken EP test is often used in quality control of lubricants whose performance characteristics have already been established.

FZG Test—The FZG test is used in Europe to evaluate EP properties. Two sets of opposing gears are loaded in stages until failure of gear tooth surfaces

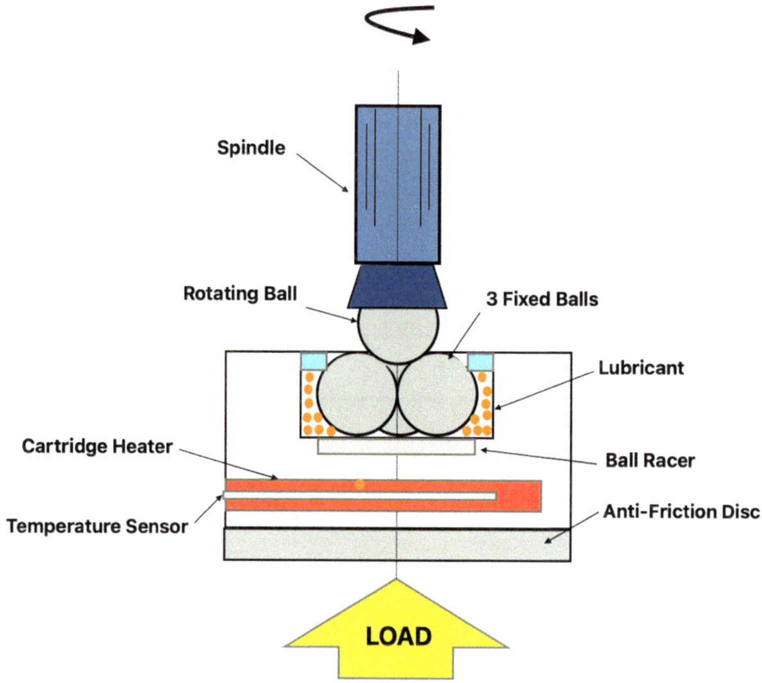

Figure 7.2 Schematic view of the four-ball wear test assembly in the tribometer.

occurs. Results are reported in terms of the number of stages passed. Two standard sets of temperature and gear speed are used and should be stated along with the number of stages passed.

Four-Ball Wear Test and Four-Ball EP Test—Each of the four-ball tests is designed to evaluate a different load-carrying characteristic of lubricating oil or grease. Both use similar equipment and mechanical principles. Four 1/2-inch steel balls are arranged with one ball atop the three others. The three lower balls are clamped together to form a cradle, upon which the fourth ball rotates on a vertical axis. See Figure 7.2.

The *four-ball wear test* (ASTM D 2266) is used to determine the relative wear-preventing properties of lubricants on sliding metal surfaces operating under boundary lubrication conditions. The test is carried out at specified speed, temperature, and load. At the end of a specified period the average diameter of the wear scars on the three lower balls is measured and reported.

Under standardized conditions, the four-ball wear test provides a means for comparing the relative anti-wear properties of lubricants. Results of two tests run under different conditions cannot be compared, so operating conditions should always be reported. No correlation has yet been established with field service, so individual results should not be used to predict field performance.

The *four-ball EP test* (ASTM D 2596) is designed to evaluate performance under much higher unit loads than applied in the wear test, hence the designation EP (extreme pressure). The EP Tester is of slightly different design and construction than the Wear Tester. One steel ball is rotated against the other three at constant speed, but temperature is not controlled. The loading is increased at specified intervals until the rotating ball seizes and welds to the other balls. At the end of each interval, the scar diameters are measured and recorded.

Two values from the EP test are generally reported: load wear index (formerly called Mean Hertz Load) and weld point. *Load Wear Index (LWI)* is a measure of the ability of a lubricant to prevent wear at applied loads. *Weld point* is the lowest applied load at which either the rotating ball seizes and then welds to the three stationary balls, or at which extreme scoring of the three stationary balls occurs. It indicates the point at which the extreme pressure limit of the lubricant is exceeded.

The four-ball EP test is used in lubricant quality control, and to differentiate between lubricants having low, medium, or high extreme pressure qualities. Results do not necessarily correlate with actual service and should not be used to predict field performance unless other lubricant properties are also taken into consideration.

For comparison of the capabilities of various lubricants, the results of both four-ball tests should be considered, particularly if additives or grease thickeners are unknown or widely dissimilar. Lubricants with good extreme pressure properties may not be equally effective in reducing wear rates at less severe loads, and conversely.

7.26 Neutralization Number

7.26.1 ASTM D 664 and D 974

Depending on its source, additive content, refining procedure, or deterioration in service, a petroleum oil may exhibit certain acid or alkaline (base) characteristics. Data on the nature and extent of these characteristics may be derived from the product's neutralization number, or "neut number," as it is commonly known. The two principal methods for evaluating neut number are ASTM D 664 and ASTM D 974. Although respective test results are similar, they are not identical, and any reporting of results should include the method by which they are obtained.

7.26.2 Acidity and alkalinity

Acidity and alkalinity are terms related to dissociation, a phenomenon of aqueous solutions. Dissociation is a form of ionization, the natural breaking up of some of the molecules into positive and negative ions. If the chemical composition of the aqueous solution is such that it yields more hydrogen ions (positive) than hydroxyl ions (negative), the solution is considered acid; an excess of hydroxyl ions on the other hand results in a solution that is considered to be basic or alkaline. The greater the excess, the more acid or alkaline the solution, as the case may be. If the hydrogen and hydroxyl ions are in equal concentration, the solution is by definition, neutral.

7.26.3 Titration

Since acidity and alkalinity are opposing characteristics, an acid solution can be neutralized (or even made alkaline) by the addition of a base. The converse is also true. In either case, neutralization can be accomplished by titration, the gradual addition of a reagent until a specified end point is reached. The amount of acid or base materials in a solution can thus be measured in terms of the quantity of added reagent. Being non-aqueous, however, petroleum oils cannot truly be said to be acid or alkaline. Nevertheless, they can be modified to exhibit these properties by addition of water—plus alcohol to extract oil-soluble acid or alkaline compounds from the sample, and to dissolve them in the water. This principle is utilized in the determination of neutralization number.

7.26.4 pH

Actual acidity or alkalinity, on the other hand, can be expressed in accordance with the pH scale, where zero represents maximum acidity, 14 maximum alkalinity, and 7 neutrality. The pH value of a solution can be determined electrolytically. When two electrodes of different materials are immersed in the solution, a small electric potential (voltage) is generated between them, and the magnitude and polarity of this potential can be related directly to pH value.

7.26.5 Potentiometric method

The potentiometric method for determining neut number (ASTM D 664) is based on the electrolytic principle, pH, as indicated potentiometrically, is recorded against added reagent. If the initial pH reading of the specially prepared sample lies between 4 and 11 (approximately), the sample may contain weak acids, weak bases, or an equilibrium combination of the two. It may be titrated to one end point with base to yield a total acid number, and then may be titrated to another end point with acid to yield a total base number.

If, on the other hand, the initial pH reading lies below 4 (approximately), the sample may be titrated with base up to this point to yield a strong acid number. It may also be titrated up to 11 (approximately) to yield a total acid number. Similarly, a sample whose initial pH reading lies above 11 (approximately) can be titrated with acid down to this value to yield a strong base number, and it can be titrated down to 4 (approximately) to yield a total base number.

7.26.6 End points

Titration end points are not at fixed pH readings but at inflections that occur in the curve: reagent versus pH. Whether or not an end point represents a strictly neutral condition is of little significance. With test procedure carefully standardized, the results obtained in reaching an end point can be compared on an equal basis with other results obtained in the same way. A result reported simply as "neut number," moreover, may be assumed to be a total acid number. Although it is not provided for by ASTM procedure, the initial pH reading may also be reported.

7.26.7 Colorimetric method

Under the colorimetric method for determining neut number (ASTM D 974), end point is identified by the change of a color indicator. This indicator exhibits one color above a specified pH value, another below. By this means, a total acid or strong base number can be determined with a *p*-naphtholbenzene indicator, while a strong acid number can be determined with a methyl orange indicator. Obviously this method is not suitable for the investigation of dark-colored liquids.

7.26.8 Reporting the results

Whatever the method, all acid numbers are expressed in milligrams of potassium hydroxide (KOH), a base, required to "neutralize" a gram of sample. For reasons of uniformity, base numbers, which are obtained by titrating with hydrochloric acid (HCl), are expressed in the same units, the HCl being converted to the number of KOH units that it would neutralize.

7.26.9 Significance of results

Because acidity is associated with corrosiveness, there has been a tendency to attribute undesirable properties to an oil that exhibits a high acid number or a low pH reading. This attitude is fostered by the fact that deterioration of an oil in service (oxidation) is ordinarily accompanied by an increase in acid test results. While this attitude is not in actual disagreement with fact, its oversimplification may be conducive to harmful misconceptions.

Firstly, petroleum oil is not an aqueous solution, and conventional interpretations of acidity and alkalinity do not apply. Secomdly, the test results, while involving certain acid or alkaline implications, do not distinguish between those that are undesirable and those that are not. The ASTM Standards themselves include the statement that the test "method is not intended to measure an absolute acidic or basic property that can be used to predict performance of an oil under service conditions. No general relationship between bearing corrosion and acid or base number is known."

This is not to say, however, that neut number or pH reading are without significance. They are applied widely and effectively to turbine oils, insulating oils, and many other oils in critical service. With new oils, neutralization test results provide a useful check on consistency of product quality. With used oils, they may serve as a guide to mechanical condition, change in operating conditions, or product deterioration. A rise in acid number and/or a drop in base number or pH reading are generally indicative of increasing oxidation. They may also be related to depletion of an additive, many of which are alkaline.

It is impossible, however, to generalize about the limits to which the neutralization values of an oil in service may safely be allowed to go. Each combination of oil, machine, and type of service follows a pattern of its own. Only through experience with a particular set of conditions can it be determined at what neutralization value an oil is no longer suitable for service.

7.27 Octane Number

7.27.1 ASTM D 2699 and D 2700

The octane number of a gasoline is a measure of its anti-knock quality; that is, its ability to burn without causing the audible "knock" or "ping" in spark-ignition engines. While octane number is a common term, it is also widely misunderstood, primarily because there are several different methods of measuring this property. Motor Octane Number, Research Octane Number, and Road Octane Number are the three basic procedures. Each assesses antiknock quality of a given fuel under a particular set of conditions.

7.27.2 Octane number in the laboratory

In principle, the octane number of a fuel is a numerical expression of its tendency to prevent engine knock relative to a standard fuel. In the laboratory, this quantity is determined through use of the ASTM engine, a special, single-cylinder engine whose operating characteristics can be varied.

The fuel to be rated is first burned in the engine, and the air-fuel mixture is adjusted to produce *maximum knock,* which is measured by a sensing device known as a *knock-meter.* Next, the compression ratio of the engine is varied until a knock intensity of 55 is recorded on the knock-meter. The knock intensity of the test fuel under these conditions is then compared (by referring to charts) to the knock intensities of various reference fuels. The reference fuels are normally blends of two hydrocarbons: iso-octane, which resists knocking, and normal heptane, which knocks severely. Iso-octane is arbitrarily assigned an octane number of 100, while heptane is rated as zero.

The percentage, by volume, of iso-octane in the blend that matches the characteristics of the test fuel is designated as the Octane Number of the fuel. For example, if a blend of 90% iso-octane and 10% heptane will match the knock intensities of the "unknown" fuel, under the same conditions, the fuel would be assigned an octane number of 90. For fuels having octane numbers above 100, the gasoline under test is compared with blends of iso-octane and tetraethyl lead, an effective anti-knock agent. Such blends can have octane numbers considerably above 100.

The general test procedure outlined above is the basis for two distinct laboratory methods of determining octane number, *Motor Octane Number* and *Research Octane Number.* Motor Octane Number, ASTM D 2700, is the name given to the octane rating as determined with the ASTM engine and a standard set of operating conditions that became widely known during the 1930's. Research Octane Number, ASTM D 2699, is a more recent method, and is determined under another set of conditions, the chief difference being the slower engine speed. The research method is therefore less severe than the motor method, and most gasolines have a higher octane number by the research method.

7.27.3 Road octane number

Laboratory octane ratings do not always provide an accurate prediction of how a fuel will behave in an automobile engine. A more reliable means of predicting antiknock quality is to test the gasolines in automobiles under varying condition of speed and load. There are several methods of determining this rating, which is known as *Road Octane Number;* each method compares the test fuel with various blends of iso-octane and heptane.

The Uniontown Procedure, one of the most common Road Octane methods, records the knock intensity at various speeds during acceleration. Knock ratings are recorded, either with instruments or by the human ear. The procedure is repeated, using various blends of iso-octane and heptane, until a reference fuel that produces the same knock characteristics is found. The test fuel is then assigned the same octane number as the reference blend.

The Modified Uniontown Procedure, another common method, depends on the human ear to establish where "trace knock" first occurs. A series of test runs, using various reference fuels of known octane numbers, is first made. For each blend, the spark advance setting that produces trace knock is determined, and the various settings are plotted into a curve. The Road Octane Number of the test fuel can then be assessed by referring to the curve to determine the octane number associated with the spark advance setting that produced trace knock with the test fuel.

7.27.4 Aviation gasoline knock rating

The anti-knock level of aviation gasoline is indicated by composite grade numbers, i.e., 80/87, 100/130, 115/145. In each case, the first number is the knock rating determined under conditions of lean air-fuel ratio by ASTM method D 614, while the second number is the rating under the supercharged-rich method, ASTM D 909. Values above 100 are expressed as Performance Numbers, which are related to the number of milliliters of tetraethyl lead in iso-octane.

7.27.5 Significance of results

Motor Octane Number is normally taken as an indication of a fuel's ability to prevent knocking at high engine speeds, while Research Octane Number measures low-speed knocking tendencies. It is the Road Octane Number, however, that an automobile engine will actually "see" in a given fuel with regard to knock characteristics.

The amount of technical literature devoted to octane numbers is immense, and many correlations exist among the three methods of determining octane numbers which, in the hands of experts, can be meaningful. For the motorist, however, the Road Octane Number of a gasoline offers the most practical prediction of whether the fuel is going to knock in his engine under the conditions to which the car is subjected.

7.28 Oil Content of Petroleum Wax

7.28.1 ASTM D 721

A major step in wax refining is the removal of oil; fully refined paraffin waxes usually contain less than 0.5% oil. Therefore, a measure of the oil content of a wax is also an indirect measure of the degree of refinement, and is a useful indicator of wax quality.

The ASTM D 721 test method is based on the low-temperature insolubility of wax in methyl ethyl ketone. A sample of the wax is dissolved in the solvent under heat, the solution is cooled to precipitate the wax and then filtered. The oil content of the filtrate is determined by evaporating the methyl ethyl ketone and weighing the residue. By definition, the oil content of a wax is that portion which is soluble in methyl ethyl ketone at –25°F.

7.28.2 Significance of oil content

The oil content of a petroleum wax is a criterion of purity and degree of refinement. Highly refined petroleum waxes have high purity and low oil content. This renders them suitable for many applications in the manufacture of drugs, other pharmaceuticals, and food packages. Crude scale waxes are not so highly refined and consequently contain more oil. Such waxes are suitable for applications where some odor or taste can be tolerated, and where higher oil content is permitted.

Oil content of microcrystalline wax can also be determined by ASTM D 721.

7.29 Oil Separation in Grease Storage

7.29.1 ASTM D 1742

ASTM Method D 1742, "Oil Separation From Lubricating Grease During Storage," provides an indicator of the tendency of greases to separate oil while in containers in storage. The separation of a few ounces of oil at the top of a 35-lb container of grease may represent less than 1% of the total oil in the grease and is not detrimental to the performance of the product. However, this may produce housekeeping problems as well as cause loss of the user's confidence.

7.29.2 Significance of results

ASTM states that the test correlates directly with the oil separation that occurs in 35-pound grease pails in storage. The test is also indicative of the separation that may occur in other sizes of containers. This method is not suitable for greases softer than NLGI No. 1 consistency and is not intended to predict the bleeding tendencies of grease under dynamic service conditions.

Due to improved grease technology the problem of grease separation in containers rarely occurs today. Therefore, the relevance of this test to the service performance of modern greases is questionable. The test is primarily of value as a means of assuring batch-to-batch uniformity.

7.30 Oxidation Stability—Oils

7.30.1 ASTM D 943

Oxidation is a form of deterioration to which all oils in service are exposed. It is a chemical reaction that occurs between portions of the oil and whatever oxygen may be present, usually the oxygen in the atmosphere. The oxidation of lubricating oils is accelerated by high temperatures, catalysts (such as copper), and the presence of water, acids, or solid contaminants. The rate of oxidation increases with time.

Oxidation tends to raise the viscosity of an oil. The products of oxidation are acid materials that lead to depositing of soft sludges or hard, varnish-like coatings. Paraffinic oils characteristically have greater oxidation resistance than naphthenic oils, although naphthenic oils are less likely to leave hard deposits. Whatever the net effect of oxidation, it is undesirable in any oil that lubricates on a long-term basis. Much has been done to improve oxidation resistance by the use of selected base stocks, special refining methods, and oxidation inhibitors. As might be expected, moreover, a great deal of study has been devoted to the means by which oxidation resistance of an oil may be evaluated.

A number of oxidation tests are in use. Some may be better related to a particular type of lubrication service than others. All are intended to simulate service conditions on an accelerated basis. At an elevated temperature, an oil sample is exposed to oxygen or air, and sometimes to water or catalysts, usually iron and/or copper. All speed up oxidation. Results are expressed in terms of the time required to produce a specified effect, the amount of sludge produced or oxygen consumed during a specified period.

One of the more common methods of examining steam turbine oils is the ASTM method D 943. This test is based on the time required for the development of a certain degree of oxidation under accelerated conditions; the greater the time, the higher the oil's rating. Here, oxidation is determined by an increase in the oil's acidity, a property measured by its acid neutralization number (See discussion on *"Neutralization Number."*)

7.30.2 Significance of results

Oxidation stability is an important factor in the prediction of an oil's performance. Without adequate oxidation stability, the service life of an oil may be extremely limited. Unless the oil is constantly replaced, there is a serious possibility of damage to lubricated parts. Acids formed by oxidation may be corrosive to metals with which the oil comes in contact. Sludges may become deposited on sliding surfaces, causing them to stick or wear; or they may plug oil screens or oil passages.

Oxidation stability is a prime requisite of oils serving in closed lubrication systems, where the oil is recirculated for extended periods. The higher the operating temperature, the greater the need for oxidation stability, especially if water, catalytic metals, or dirt are present. Resistance to oxidation is of special importance in a steam-turbine oil because of the serious consequences of turbine bearing failure. Gear oils, electric transformer oils, hydraulic fluids, heat-transfer oils, and many crankcase oils also require a high degree of oxidation stability.

Obviously, the ability to predict oxidation life by a test, and to do it with reasonable accuracy is highly desirable. There are certain factors, however, that make reliable test results difficult to obtain. In the first place, the tests themselves are very time-consuming; a method such as ASTM D 913 may require the better part of a year to complete. Prolonged though the test may be, moreover, its duration usually represents but a small fraction of the service life of the oil under investigation. A steam turbine oil, for example, may well last for a decade or more without serious deterioration. It is impossible to reproduce service conditions of this sort in the laboratory with a test even of several hundred hours' duration. In addition to the time factor, there are many other operational variables that cannot be duplicated under test conditions. Results can be distorted also by the presence of certain additives in the oil.

For these reasons, the correlation between oxidation test results and field experience leaves much to be desired. Test results are subject only to broad interpretations. It would be difficult to show, for example, that an oil with a 3000-hour ASTM test life gives better service than an oil with a 2500-hour test life. In evaluating the oxidation stability of an oil, primary consideration should be given to the record that it has established over the years in the type of service for which it is to be used.

7.31 Oxidation Stability—Greases

7.31.1 ASTM D 942 • 1P142, D 1402, and D 1261

The bomb oxidation test was developed in 1938 by the Norma-Hoffman Bearing Corporation. Its purpose was to evaluate the oxidation stability of a grease during the storage of machine parts to which it had been applied. It was not intended to predict either the stability of greases in service or their shelf life in commercial containers.

7.31.2 Method of evaluation

Oxidation is a form of chemical deterioration to which no petroleum product is immune. Petroleum products vary appreciably in their resistance to oxidation, a property that can be evaluated in many ways for many purposes. Evaluation is related to the quantity of oxygen that reacts with a grease sample during a specified period under standard conditions. The oxidation rate is plotted as pressure drop vs. time, Figure 7.3.

7.31.3 ASTM D 942 • 1P142

Procedure

Four grams of the grease to be tested are placed in each of five Pyrex dishes. These samples are then sealed and pressurized with oxygen (at 110 psi) in a heated bomb (210°F). The pressure is observed and recorded at stated intervals. The decrease in oxygen pressure determines the degree of oxidation after a given time period.

7.31.4 Significance of results

A relationship exists between the pressure lost during this time and the amount of oxygen that has entered into chemical reaction with the grease. However, the drop in pressure is the *net* change resulting from absorption of oxygen and the release of gaseous products by the grease. This is a basic weakness of the test, since a grease that is being oxidized, and at the same time releasing gaseous products, would appear to have greater oxidation resistance than is actually the case. This is a static test and is not intended to predict the stability of grease under dynamic conditions. Nor does it reflect oxidizing influence on bulk quantities in the original container. It more closely represents the conditions in a thin film of grease, as on pre-lubricated bearings or machine parts subjected to extended storage.

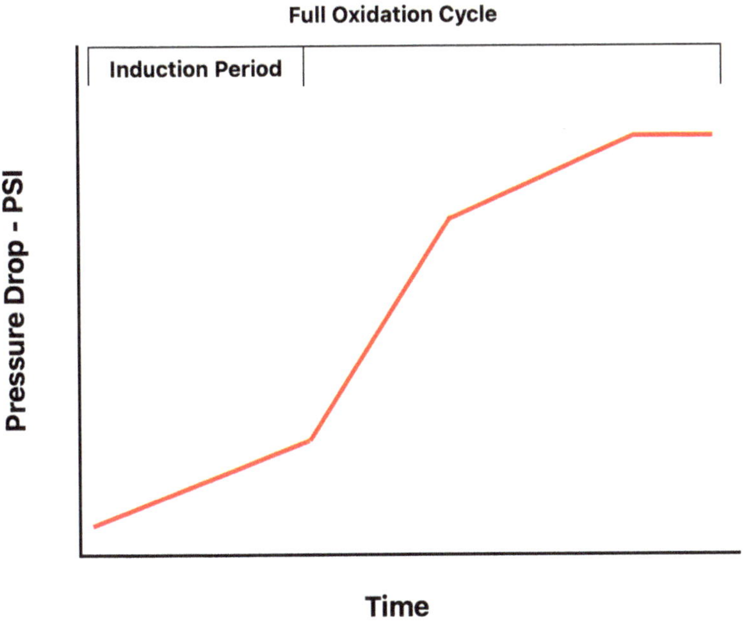

Figure 7.3 Bomb Oxidation Test—When pressure drop is plotted against time, the resulting curve will indicate a period of comparatively slow oxidation followed by a pronounced rise. The relatively flat portion at the beginning represents what is known as the "induction period," a phase during which oxidation is not ordinarily of serious magnitude. For practical reasons, it is not customary to continue the test beyond the induction period, its end being indicated by a sudden rise. Should the test be carried further, however, this rise would eventually taper off again as oxidation becomes complete. In some cases, test results have been expressed in terms of the duration of the induction period.

Certain machine parts are stored after an application of lubricating grease has been applied. This is particularly true of lubricated-for-life anti-friction bearings, which are greased by the manufacturer and then sealed. It is a common practice to make up these parts in advance and then stockpile them against future requirements.

There is hardly need to point out the damage that can be inflicted by a grease that deteriorates rapidly during this type of storage. The acidity associated with grease oxidation is corrosive to the highly sensitive bearing surfaces, and oxidation deposits may bind the bearing's action even before it has been put in operation. At best, a grease that has undergone significant deterioration in storage can hardly be in a condition to yield the long service life expected of it.

7.31.5 ASTM D 1402

Procedure

This test is run the same as ASTM D 942 is except that prepared copper strips are immersed on edge in each grease sample. Pressure readings are taken at 2-hour intervals over the duration of the test, until the pressure drops to 55 psi or for a specified time period if the pressure hasn't dropped to 55 psi during this time.

7.31.6 Significance of results

The same limitations exist with this test as with ASTM D 942 since the determination of oxygen absorption rate as an indication of oxidation reaction is affected by the release of gaseous products from the grease.

Results from this test indicate the catalytic effect of copper and its alloys) in accelerating the oxidation of greases under static conditions. The results are not applicable to greases under dynamic conditions or when stored in commercial containers.

7.31.7 ASTM D 1261

Procedure

Each of two Pyrex dishes is filled with 4 grams of grease and a prepared copper strip is partially immersed in each grease sample. The procedure from this point on is the same as in ASTM D 942. At the end the 24 hour test period, the copper strips are removed, washed, and examined for evidence of discoloration, etching, or corrosion.

7.31.8 Significance of results

The effect of grease on copper parts of bearing assemblies with which the grease comes in contact is determined from the results of this test. In addition, some indication of the storage stability of greases which are in contact with copper may be found by visual inspection of the grease at the end of the test. The results do not apply to greases in contact with copper under dynamic service conditions.

In spite of their limitations, these tests do have significant value. Many concerns find that the bomb test serves as an accurate check on uniformity of grease composition. Though test results may mean little by themselves, they are highly reproducible and highly repeatable. Results that are consistent from batch to batch give a good indication of product uniformity.

7.32 Penetration

(See "Grease Consistency.")

7.33 Pentane and Toluene Insolubles

When a used oil is diluted sufficiently with pentane, certain oxidation resins that it normally holds in solution are precipitated out. In addition, the dilution helps to settle out materials suspended in the oil. Among the latter are insoluble oxidation resins and extraneous matter such as dirt, soot, and wear metals. All of the contaminants that can be separated from the oil by precipitation and settling are referred to as *pentane insolubles.*

The pentane insolubles may then be treated with a toluene solution which dissolves the oxidation resins. The extraneous matter left behind is called the *toluene insolubles.* The difference between the pentane insolubles and the toluene insolubles represents the quantity of oxidation resins in the oil. This is termed the *insoluble resins,* meaning insoluble in pentane.

Toluene has replaced benzene as the aromatic solvent in ASTM D 893 because of concern about the potential toxicity of benzene. Insoluble sludges are generally similar with the two solvents.

With detergent engine oils, a pentane-coagulant solution is customarily used instead of pentane. This precipitates material held in suspension by the detergent-dispersant which would not otherwise separate out.

As with other tests, interpretation depends on the type of oil, the service to which it has been put, and the results of other tests on the oil. In general, however, low pentane insolubles indicate an oil in good condition and little is to be gained by continuing with other phases of the test.

High pentane insolubles, on the other hand, indicate oxidation or contamination. The point at which an oil change is called for depends on many factors which must be evaluated by experience.

A relatively high value for toluene insolubles indicates contamination from an outside source such as soot from partially burned fuel; atmospheric dirt, the result of inadequate air filtration; tiny particles of metal produced by extreme wear. Emission spectrometry is often used to reveal the makeup of metal contamination: excessive lead, copper, or silver implies bearing wear; aluminum, piston wear; silicone, atmospheric dirt.

High insoluble resins mean a highly oxidized oil, which may result from excessive engine temperatures, contamination, an unsuitable oil, or excessive crankcase dilution. Loose-fitting piston rings, faulty injection, or low-temperature operation may allow raw fuel to enter the crankcase, where its oxidation adds to the amount of insoluble resins.

7.34 Pour Point and Cloud Point

7.34.1 ASTM D 97

It is often necessary to know how cold a particular petroleum oil can become before it loses its fluid characteristics. This information may be of considerable importance, for wide variations exist in this respect between different oils, even between oils of comparable viscosity.

If a lubricating oil is chilled sufficiently, it eventually reaches a temperature at which it will no longer flow under the influence of gravity. This condition may be brought about either by the thickening of the oil that always accompanies a reduction in temperature, or by the crystallization of waxy materials that are contained in the oil and that restrain the flow of the fluid portions. For many applications, an oil that does not flow of its own accord at low temperatures will not provide satisfactory lubrication. The extent to which an oil can be safely chilled is indicated by its *pour point,* the lowest temperature at which the undisturbed oil can be poured from a container.

The behavior of an oil at low temperature depends upon the type of crude from which it is refined, the method of refining, and the presence of additives. Paraffinic base stocks contain waxy components that remain completely in solution at ordinary temperatures. When the temperature drops, however, these waxy components start to crystallize, and they become fully crystallized at a temperature slightly below the pour point. At this temperature, the undisturbed oil will not generally flow under the influence of gravity.

Crystallization of the waxy component does not mean that the oil is actually solidified; flow is prevented by the crystalline structure. If this structure is ruptured by agitation, the oil will proceed to flow, even though its temperature remains somewhat below the pour point.

An oil that is predominantly naphthenic, on the other hand, reacts in a somewhat different manner. In addition to having a comparatively low wax content, a naphthenic oil thickens more than a paraffinic oil of comparable viscosity does when it is cooled. For these reasons, its pour point may be determined by the actual congealing of the entire body of oil instead of by the formation of wax crystals. In such a case, agitation has little effect upon fluidity, unless it raises the temperature.

The pour point of a paraffinic oil may be lowered substantially by a refining process that removes the waxy component. For many lubricating oils, however, these components impart advantages in viscosity index and oxidation stability. Good performance generally establishes a limit beyond which the removal of these waxy component is inadvisable. It is possible, nevertheless, to lower the pour point of a paraffinic oil by the introduction of a pour depressant. Such an additive appears to stunt the growth of the individual crystals so that they offer less restriction to the fluid portions of the oil. It is hardly necessary to point out, however, that a pour depressant, as such, can have little, if any, effect upon a naphthenic oil.

Cloud point is the temperature, somewhat above the pour point, at which wax crystal formation gives the oil a cloudy appearance. Not all oils exhibit a cloud point and, although this property is related to pour point, it has little significance for lubricating oils. It is significant, however, for distillate fuels, and it is measured by ASTM D 2500.

7.34.2 Significance of results

The pour point of an oil is related to its ability to start lubricating when a cold machine is put in operation. Agitation by the pump will rupture any crystalline structure that may have formed, if the oil is not actually congealed, thereby restoring fluidity. But oil is usually supplied to the pump by gravity, and it can not be expected to reach the pump under these conditions, if the temperature is below the pour point. Passenger car engines and many machines that are stopped and started under low-temperature conditions require an oil that will flow readily when cold.

What is true of circulating lubrication systems, moreover, is equally true of gravity-feed oilers and hydraulic systems. A low pour point oil helps to provide full lubrication when the equipment is started and is easier to handle in cold weather. Low pour point is especially desirable in a transformer oil, which must circulate under all temperature conditions. The control of large aircraft is dependent upon hydraulic oils that must remain fluid after being exposed to extreme temperature drops. For these and similar applications, pour point is a very important consideration.

If the temperature of an oil does not drop below its pour point, the oil can be expected to flow without difficulty. It sometimes happens, however, that oil is stored for long periods at temperatures below the pour point. In some cases, the waxy crystalline structure that may be formed under these circumstances will not melt and redissolve when the temperature of the oil is raised back to the pour point. To pour the oil under these conditions, it is necessary to put the waxy crystals back in solution by heating the oil well above its pour point.

7.35 Power Factor

7.35.1 ASTM D 924

Petroleum oils serve extensively as dielectrics for electrical power transmission equipment. Their primary functions are the cooling of coils (by circulation) and the prevention of arcing between conductors of high potential difference. In serving these purposes, any dielectric tends to introduce a degree of *dielectric loss,* a form of leakage equivalent to a flow of current through the dielectric from one conductor (wire or cable) to another. It is a leakage peculiar to a-c circuits. Though the loss associated with insulating oils is ordinarily a minor consideration, it could, under unusual conditions, assume a significant magnitude. In such a case, it would not only reduce the efficiency of the unit, but could cause a harmful rise in the unit's temperature.

Dielectric loss depends, among other things, on the nature and magnitude of the insulation's *impedance*, its opposition to the flow of alternating current through it. This is the current that is related to dielectric loss, and it increases as the impedance decreases. Only a portion of this current, however, is directly involved: a component equivalent to an *active* current. In a given a-c circuit, the loss is directly proportional to this active current. The ratio of active to *total* alternating current may vary—theoretically—from one to zero. This ratio is known as the *power factor* of the dielectric, and it can be considered to be an inherent dielectric property. Because of its effect on dielectric loss, the power factor of the dielectric should be as low as possible.

Though the power factor of an insulating oil is defined by the same mathematical expression as that of an a-c circuit, the two concepts should not be confused The overall power factor of a power-circuit affect *line* losses, rather than local dielectric losses, and the reduction of line losses requires a *high* power factor for the circuit.

7.35.2 Significance of test results

In a-c transmission cables, conductors of opposite polarity may extend for long distances in close proximity to each other. There is abundant opportunity for dielectric loss associated with the insulating material between the conductors. The higher the power factor of the insulation, the greater this loss will be.

For other applications, as in the insulation of transformers, dielectric loss is not appreciable, and small differences in power factor have little significance. A high-quality oil that is free of contamination can be expected to exhibit the low power factor that good performance requires.

In the evaluation of a used insulating oil's condition, however, power factor may be more meaningful. Here, the principal criterion is freedom from water and oxidation products, water that promotes the tendency to arc and oxidation sludges that interfere with cooling. Oxidation of the oil and contamination with water, dirt, or carbonized particles cause the power factor to rise.

Many engineers consider power factor to be a highly sensitive index of the oil's deterioration. If sufficient data on the performance of a particular oil in a particular service is available, it is possible to relate in-creases in power factor to degradation of the oil. In this way, power factor tests on a used oil may be helpful in estimating its remaining service life.

Tests for power factor frequently serve a useful purpose in the refinery as a check on uniformity of product quality. Consistent test values are indicative of consistent performance characteristics.

7.36 Refractive Index

7.36.1 ASIM D 1218

Uniformity of composition of highly refined petroleum products is of importance, especially in process applications such as those involving solvents or rubber process oils. Refractive index is one test often used either alone or in combination with other physical tests as an indication of uniformity.

Refractive index is the ratio of the velocity of light of a specified wavelength in air to its velocity in a substance under examination. When light is passed through different petroleum liquids, for example, the velocity of light will be different in each liquid. Several sources of light of constant wavelength are available, but the yellow D line of sodium (5893 Å) is the one most commonly used in this test. Since the numerical value of the refractive index of a liquid varies with wavelength and temperature, it must be reported along with the wavelength and temperature at which the test was run.

This test is intended for transparent and light colored hydrocarbon liquids having refractive indices between 1.33 and 1.50. The method is capable of measuring refractive index with a reproducibility of ± 0.00006. It is generally not this accurate with liquids having ASTM colors darker than 4 (ASTM D 1500), or with liquids that are so volatile at the test temperature that a reading cannot be obtained before evaporation starts.

7.36.2 Significance of results

The refractive index is easily measured and possesses good repeatability and reproducibility. It is sensitive to composition. This makes it an excellent spot test for uniformity of composition of solvents, rubber process oils, and other petroleum products. A general rule for petroleum products of equivalent molecular weight is that paraffins have relatively low refractive indices (approximately 1.37), aromatics have relatively high indices (approximately 1.50), and naphthalenes have intermediate indices (approximately 1.44).

Refractive index may be used in combination with other simple tests to estimate the distribution of carbon atom types in a process oil. Empirical refractive index charts relating viscosity, specific gravity, and refractive index have been prepared, and they make it possible to estimate the percent naphthenic, aromatic, and paraffinic carbon atoms present. This is an inexpensive and quick set of tests to run, in contrast to the more time consuming clay/silica gel analysis, which is also used to determine hydrocarbon composition directly.

7.37. Rotary Bomb Oxidation Test (RBOT)

7.37.1 ASTM D 2272

Oxidation is a form of chemical deterioration to which petroleum products are subject. Even though oxidation takes place at moderate temperatures, the reaction accelerates significantly at temperatures above 200°F.

In addition to the effect of high temperatures, oxidation may also be speeded by catalysts (such as copper) and the presence of water, acids, or, solid contaminants. Moreover, the peroxides that are the initial products of oxidation are themselves oxidizing agents, so that oxidation is a chain-reaction—the further it progresses the more rapid it becomes.

Even though subject to oxidation, many oils (such as turbine oils) give years of service without need for replacement. Petroleum products can be formulated to meet service and storage life requirements by: (1) proper selection of crude oil type; (2) thorough refining, which removes the more-oxidation-susceptible materials; and (3) addition of oxidation inhibitors.

A number of oxidation tests are currently being used. Some may be better related to a particular type of service than others. All are intended to simulate service conditions on an accelerated basis. The most familiar method is the "Oxidation Characteristics of Inhibited Steam-Turbine Oils," ASTM D 943. The long time (over 1000 hours) required to run this test makes it impractical for plant control work. ASTM Method D 2272, "Continuity of Steam-Turbine Oil Oxidation Stability by Rotating Bomb," on the other hand, allows rapid evaluation of the resistance of lubricants to oxidation and sludge formation, using accelerated test conditions that involve high temperatures, high-pressure oxygen, and the presence of water and catalytically active metals.

The "rotary bomb oxidation test" (RBOT) does not replace ASTM D 943, but is intended primarily as an aid in quality control during the manufacture of long-life circulating oils.

7.37.2 Significance of results

The ASTM D 2272 procedure allows relative oxidation life of a turbine oil to be determined rapidly. Results are obtained by the RBOT test up to 1000 times faster than by the D 943 method. This speed makes the test practical for use as a product quality control measure, permitting decisions to be made within a matter of a few hours. The RBOT is also distinguished among oxidation tests by its good repeatability and reproducibility.

It should be remembered that the test is essentially a quality control device and no direct correlation has been established with other oxidation tests currently being used. For two oils of similar composition, both base stock and additive package, the RBOT test can be used as an indication of their relative oxidation stability.

7.38 Rust-preventive Characteristics

7.38.1 ASTM D 665

This test method was originally designed to indicate the ability of steam-turbine oils to prevent the rusting of ferrous parts, should water become mixed with the oil. While still used for this purpose, its application is now often extended to serve as an indication of rust preventive properties of other types of oils, particularly those used in circulating systems. It is a dynamic test, designed to simulate most of the conditions of actual operation.

In the method, a standard steel specimen is immersed in a mixture of the test oil and water under standard conditions and with constant stirring. At the end of a specified period, the steel specimen is examined for rust. Depending on the appearance of the specimen, the oil is rated as passing or failing.

7.38.2 Degrees of rusting

An indication of the degree of rusting occurring in this test is sometimes desired. For such cases, the following classification is recommended:

- **LIGHT RUSTING**—Rusting confined to not more than six spots, each of which is 1 mm or less in diameter.

- **MODERATE RUSTING**—Rusting in excess of the preceding, but confined to less than 5% of the surface of the specimen.
- **SEVERE RUSTING**—Rust covering more than 5 % of the surface of the specimen.

7.38.3 Reporting of results

Results obtained with a given oil are reported as "pass" or "fail." Since the test may be conducted with either distilled water or with synthetic sea water, and for varying periods of time, reports of results should always specify these conditions. For example: "Rust Test, ASTM D 665, Procedure B, 24 Hours—Pass."

7.38.4 Significance of results

When the lubricating oil of a turbine or other system is contaminated with water, rusting can result. Particles of rust in the oil can act as catalysts that tend to increase the rate of oil oxidation. Rust particles are abrasive, and cause wear and scoring of critical parts. In addition, rust particles can add to other contaminants in a circulating system, increasing the tendency toward the clogging of low-clearance members, such as servo valves, and increasing the probability of filter plugging.

In many cases, the rusting characteristics of the system in service are better than is indicated by testing a sample of the used oil, because the polar rust inhibitor "plates out" on the metal surfaces (which are therefore adequately protected). The sample of oil, being somewhat depleted of the inhibitor, will then allow greater rusting in the test than would occur in service.

The relative ability of an oil to prevent rusting can become a critical property in many applications. As noted, this test method was originally applied exclusively to steam turbine oils. However, the test is now frequently applied to other oils in different types of applications, whenever undesirable water contamination is a possibility.

7.39 Saponification Number

Many lubricating oils are "compounded" with fatty materials to increase their film strengths or water-displacing qualities. The degree of compounding is indicated by the saponification number of the oil, usually called its *sap number*. Sap number is commonly determined by ASTM D 94 or D 939, methods based on the fact that these fatty materials can be saponified—that is, converted to soap—by reaction with a base (alkali), usually at an elevated temperature.

A specified quantity of potassium hydroxide (KOH) is added to the prepared oil sample and the mixture heated to bring about reaction. The excess KOH is titrated to neutralization with hydrochloric acid, either colorimetrically (D 94) or potentiometrically (D 939). The sap number is reported as milligrams of KOH assimilated per gram of oil.

Other factors being the same, a higher degree of compounding will result in a higher sap number. For a given degree of compounding, however, some fatty materials show a higher sap number than others.

Even with a new oil, therefore, sap number cannot be translated directly into percentage of fatty materials unless their exact nature is known.

Considerable experience with a particular set of conditions and types of oil is needed to properly interpret the sap number of a used oil. While loss or decomposition of fatty materials is reflected in a drop of the sap number, oxidation of the mineral oil base may cause the sap number to rise. Test results may be further distorted by acid or metallic contaminants picked up in service.

It is advisable, therefore, to consider sap number in relation to neut number. Comparison of the two indicates what portion of the sap number is due to the presence of fatty materials and what portion to acids in the oil.

Sap number has little relevance for oils for internal combustion engines.

7.40 Timken Extreme Pressure Tests

7.40.1 ASTM D 2509—Lubricating Greases

7.40.2 ASTM D 2782—Lubricating Fluids

(See "Load Carrying Ability.")

7.41 USP/NF Tests for White Mineral Oils

The *US Pharmacopeia* and the *National Formulary*, publications by two independent associations of physicians and pharmacologists, contain specifications for white mineral oils. The US Pharmacopeia (USP) specification covers the more viscous "Mineral Oil," which is used primarily as a pharmaceutical aid or levigating agent. The National Formulary (NF) sets specifications for the less viscous "Light Mineral Oil," which is used as a vehicle in drug formulations. Both compendia have legal status, being recognized in federal statutes, especially the Federal Food, Drug and Cosmetic Act.

White mineral oils have certain physical properties that distinguish them from other petroleum products. Both the USP and NF describe them as colorless, transparent, oily liquids free or nearly free from fluorescence. When cold they are odorless and tasteless and develop only a slight petroleum odor when heated. They are insoluble in water and alcohol, soluble in volatile oils, and miscible with most fixed oils with the exception of castor oil.

7.41.1 Significance of results

These tests are designed to establish standards that assure that the oils involved are pure, chemically inert, and free from potentially carcinogenic materials. Oils that meet these standards find use, not only in pharmaceuticals and cosmetics, but also in chemical, plastics, and packaging applications where they are considered as direct or indirect food additives as defined by the Federal Food and Drug Administration (FDA)

7.42 UV Absorbance

7.42.1 FDA method

Petroleum product applications often extend into areas other than the obvious ones. One such area is the direct or indirect application of a petroleum product to food. A direct food additive is one that is incorporated, in small quantities, into or onto food meant for human or animal consumption. An example would be the use of white mineral oil to coat raw fruits to protect them or to coat animal feeds to reduce dustiness. An indirect additive is one that has only incidental contact with food, as through contact with a packaging material.

The use of petroleum products as food additives falls under the jurisdiction of the Food and Drug Administration. A major concern in the regulation of food additives of petroleum origin is the potential contamination of food by polynuclear aromatic hydrocarbons, some of which are considered to be carcinogenic. In an attempt to assure the absence of carcinogens, the FDA has sanctioned the use of ultraviolet (UV) absorbance as a test for monitoring the polynuclear aromatics content.

Ultraviolet absorbance is a measure of the relative amount of ultraviolet light absorbed by a substance. Types of compounds can be characterized by the wavelength range of UV light that they absorb. As the wavelength of ultraviolet light is varied, broad peaks of absorbance occur at the wavelengths that are characteristic of the compounds present. Most polynuclear aromatics have their principal absorbances at wavelengths between 280 and 400 millimicrons. Most carcinogens absorb UV light in this range, but not all materials with UV absorbance between 280 and 400 millimicrons are carcinogenic.

7.42.2 Significance of results

These results are compared with the corresponding UV absorbance limits set by the FDA for the specific regulation that applies to each UV. When UV absorbance of a petroleum product falls within these limits, the product is considered acceptable for the particular food application involved. The UV method described here represents the simplest case. The method becomes more complex as the aromatic concentration of the oil increases.

The UV test is not the only criterion the FDA has established for food additives of petroleum origin. There are often additional requirements for boiling range, color, odor, method of manufacture, US Pharmacopeia quality, and non-volatile residues.

7.43 Vapor Pressure

7.43.1 ASTM D 323

All liquids are disposed to vaporize, that is, to become gases. This tendency is a manifestation of the material's liquid vapor pressure, the pressure exerted by molecules at the liquid surface in their attempt to escape and to penetrate their environment. For a given liquid, this pressure is a function purely of temperature. The liquid vapor pressure of water at boiling temperature (212°F) is 147 psi, the pressure of the atmosphere. The more volatile the liquid, the higher the liquid vapor pressure at a specified temperature, and the faster the vaporization. In the same dry atmosphere and at the same liquid temperature, gasoline evaporates much more readily than heating oil.

For a given temperature, therefore, the vapor pressure of a liquid is a measure of is volatility. This applies only to vapor pressure exerted by a *liquid.* Pressures exerted by vapor disassociated from the liquid are functions of *volume,* as well as temperature, and they cover a wide range of values less directly related to volatility. As used in engineering circles, the term vapor pressure means *liquid* vapor pressure.

Unlike water, a petroleum product usually comprises many different fractions, each with a composition and a vapor pressure of its own. The vapor pressure of the product is therefore a composite value that reflects the combined effect of the individual vapor pressures of the different fractions in accordance with their mole ratios. It is thus possible for two wholly different products to exhibit the same vapor pressure at the same temperature—provided the cumulative pressures exerted by the fractions are also the same. A narrow-cut distillate, for example, may exhibit the same vapor pressure as that of a dumbbell blend, where the effect of the heavy fractions is counterbalanced by that of the lighter ones.

When a petroleum product evaporates, the tendency is for the more volatile fractions to be released first, leaving a material of lower vapor pressure and lower volatility behind. This accounts for the progressive rise in distillation-curve temperature, boiling point being related to volatility. Distillation, which is another measure of volatility, was described earlier.

Vapor pressure is commonly measured in accordance with the ASTM method D 323 (Reid vapor pressure), which evaluates the vapor pressures of gasoline and other volatile petroleum products at 100°F.

7.43.2 Significance of test results

Reid vapor pressure has a special significance for gasoline, which contains a portion of high-volatility fractions such as butane, pentane, etc. These fractions exert a major influence on vapor-pressure test results. A high vapor pressure is an indication of the presence of these high-volatility fractions, components required for satisfactory starting in cold weather. Without them, it would be difficult or impossible to vaporize gasoline in sufficient concentration to produce a combustible air-fuel mixture at low temperatures.

On the other hand, vapor pressure may be too high. An excess of high-volatility actions in hot weather can lead to vapor *lock,* preventing delivery of fuel to the carburetor. This is the result of the partial vacuum that exists at the suction end of the fuel pump and that, along with high temperatures, increases the tendency of the fuel to vaporize. If the fuel vapor pressure is too high, vapors formed in the suction line will interrupt the flow of liquid fuel to the pump, causing the engine to stall.

While Reid vapor pressure is the principal factor in determining both the vapor-lock and the cold-starting characteristics of a gasoline, they are not the only criteria. Distillation data, which defines the overall volatility of the fuel, must also be considered.

The higher the vapor pressures of automotive and aviation gasolines, solvents, and other volatile petroleum products, the greater the possibility of evaporation loss and the greater the fire hazard. Sealed containers for high-vapor-pressure products require stronger construction to withstand the high internal pressure. In the refinery, moreover, vapor pressure tests serve as a means of establishing and maintaining gasoline quality.

7.44 Viscosity

7.44.1 ASTM D 88, D 445, Redwood, and Engler

Viscosity is probably the most significant physical property of a petroleum lubricating oil. It is the measure of the oil's flow characteristics. The thicker the oil, the higher its viscosity, and the greater its resistance to flow. The mechanics of establishing a proper lubricating film depend largely upon viscosity.

To evaluate the viscosity of an oil numerically, any of several standard tests may be applied. Though these tests differ to a greater or lesser extent in detail, they are essentially the same in principle. They all measure the time required for a specified quantity of oil at a specified temperature to flow by gravity through an orifice or constriction of specified dimensions. The thicker the oil, the longer the time required for its passage.

Close control of oil temperature is important. The viscosity of any petroleum oil increases when the oil is cooled and diminishes when it is heated. For this same reason, the viscosity value of an oil must always be accompanied by the temperature at which the viscosity was determined. The viscosity value by itself is meaningless.

The two most common methods of testing the viscosity of a lubricating oil are the Saybolt and the kinematic. Of these, the Saybolt (ASTM D 88) is the method more frequently encountered in conjunction with lubricating oils. However, the kinematic method—STM D 445 is generally considered to be more precise, see Figure 7.4. There are also the Redwood and the Engler methods, which are widely used in Europe, but only to a limited extent in the United States. Each test method requires its own viscosimeter (or viscometer) apparatus.

7.44.2 Significance of Results

Viscosity is often the first consideration in the selection of a lubricating oil. For most effective lubrication, viscosity must conform to the speed, load, and temperature conditions of the bearing or other lubricated part. Higher speeds, lower pressures, or lower temperatures require an oil of a lower viscosity grade. An oil that is heavier than necessary introduces excessive fluid friction and creates unnecessary drag.

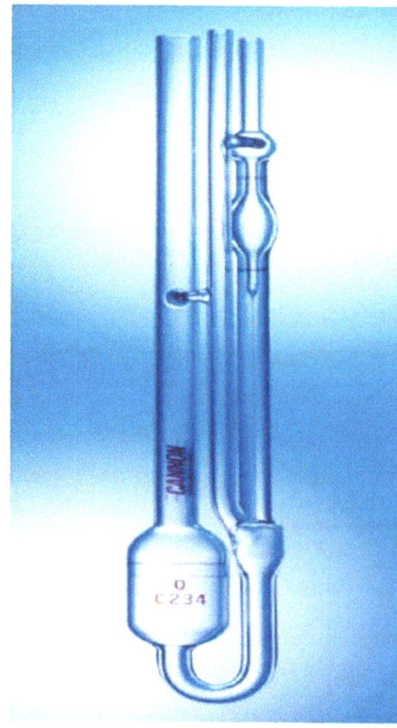

Figure 7.4 Fischer Style D445 Viscometer. Source: Fischer scientific catalog.

Lower speeds, higher pressures, or higher temperatures, on the other hand, require an oil of a higher viscosity grade.

An oil that is too light lacks the film strength necessary to carry the load and to give adequate protection to the wearing surfaces. For these reasons, viscosity tests play a major role in determining the lubricating properties of an oil.

In addition to the direct and more obvious conclusions to be drawn from the viscosity rating of an oil, however, certain information of an indirect sort is also available. Since, to begin with, the viscosity of the lube oil cut is determined by its distillation temperature, it is apparent that viscosity and volatility are related properties. In a general way, the lighter the oil, the greater its volatility—the more susceptible it is to evaporation. Under high-temperature operating conditions, therefore, the volatility of an oil, as indicated by its viscosity, should be taken into consideration.

Though the significance of viscosity test results has been considered from the standpoint of new oils, these tests also have a place in the evaluation of used oils. Oils drained from crankcases, circulating systems, or gear boxes are often analyzed to determine their fitness for further service or to diagnose defects in machine performance.

An increase in viscosity during service may often indicate oxidation of the oil. Oxidation of the oil molecule increases its size, thereby thickening the oil. When oxidation has progressed to the point of causing a material rise in viscosity, appreciable deterioration has taken place.

7.45 Viscosity Classifications Comparison

There are four common systems for classifying the viscosities of lubricating oils. It is frequently desirable to compare a grade in one system with a grade in another system, but this is often impossible because the standards in the different systems are not based on viscosities at the same temperature. The charts presented in Table 3.4 and Figure 3.4 are designed to overcome this problem by comparing the systems on the basis of viscosities at a single temperature—100°F, which is the base temperature for the ASTM viscosity grade system.

In order to convert all viscosities to 100°F, it is necessary to assume appropriate viscosity indices (VI's) for the oils involved. (Viscosity index of an oil is a measure of its resistance

These values are representative for the products involved in the respective classifications. Close comparison should not be attempted if the VI of the product differs appreciably from the values used.

Table 7.4 shows the numerical relationships; Figure 7.5 shows the graphical equivalents to change in viscosity as temperature changes.) The VI's assumed here are:

- 110 VI for crankcase oils (SAE)
- 90 VI for automotive gear oils (SAE)

7.46 Viscosity Index

7.46.1 ASTM D 567 and D 2270

Liquids have a tendency to thin out when heated and to thicken when cooled. However, this response of viscosity to temperature changes is more pronounced in some liquids than in others.

Often, as with petroleum liquids, changes in viscosity can have marked effects upon the performance of a product, or upon its suitability for certain applications. The property of resisting changes in viscosity due to changes in temperature can be expressed as the viscosity index (V.I.). The viscosity index is an empirical, unitless number. The higher the V.I. of an oil, the less its viscosity changes with changes in temperature.

Table 7.4 Numerical relationships among viscosity classification systems.

ASTM Industrial Oils	Viscosity, SSU at 100°F	SAE Crankcase Oils	Viscosity, SSU at 100°F (assumes 110 VI)	AGMA Industrial Gear Oils	Viscosity SSU at 100°F	SAE Gear Oils (a	Viscosity at 100°F Assumes 90 VI)
32	33-34						
36(A)	36-38						
40	40-43						
50(A)	46-50	5W	140 Max.			75	211 Max.
60	55-62						
75	72-83						
105	97-116						
150	136-165	10W	140-220				
215	193-235	20	168-366	1	193-234		
315	284-347	20W	202-500	2,2EP	284-347	80	211-575
465	417-510	30	366-560	3,3EP	417-510		
700	625-764	40	560-812	4,4EP	626-765		
1000	917-1121	50	812-1272	5,5EP	916-1122	90	768-1844
1500	1334-1631	60(B)	1272-1561	6,6EP	1335-1632		
2150	1918-2344	70(B)	1561-2085	7EP,7 Comp.	1919-2346		
						140	1844-4345
3150	2835-3465			8EP,8 Comp.	2837-3467		
4650	4169-5095			8A Comp.	4171-5098		
						250	4345 Min.
7000	6253-7643						

Notes: (A) Currently is not considered a standardized grade in the USA. (B) Not officially recognized by SAE.

*ASTM/BSI (American Society for Testing and Materials/British Standards Institution) "Viscosity Classification System for Industrial Fluid Lubricants."

SAE (Society of Automotive Engineers) "Crankcase Oil Viscosity Classification" and "Transmission and Axle Lubricant Classification."

AGMA (American Gear Manufacturers Association) "Specification— Lubrication of Industrial Enclosed Gear Drives."

7.46.2 The concept of viscosity index

One of the things that led to the development of a viscosity index was the early observation that, for oils of equal viscosities at a given temperature, a naphthenic oil thinned out more at higher temperatures than did a paraffinic

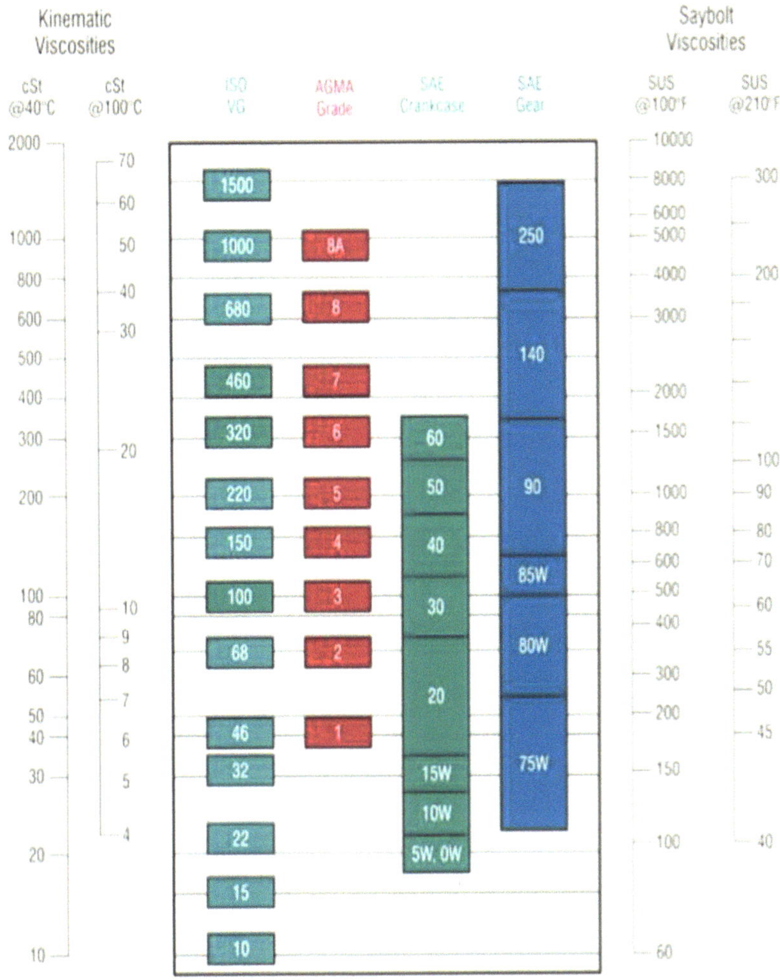

Figure 7-5 Viscosity classification equivalents. Source: Exxon Corporation, USA, (PLIF 2).

oil. However, there existed no single parameter that could express this type of response to temperature changes.

The viscosity index system that was developed to do this was based upon comparison of the viscosity characteristics of an oil with those of so-called "standard" oils. A naphthenic oil in a series of grades with different viscosities at a given temperature, and whose viscosities changed a great deal with temperature, was arbitrarily assigned a V.I. of zero. A paraffinic series, whose viscosities changed less with temperature than most of the oils that were then available, was assigned a V.I. of 100. With accurate viscosity data on these two series of oils, the V.I. of any oil could be expressed as a percentage factor relating the viscosities at 100°F of the test oil, the zero-V.I. oil, and the 100-V.I. oil, all of which had the same viscosity at 210°F. This is illustrated by Figure 7.6, and is the basis for the formula,

$$\text{V.I.} = \frac{L-U}{L-H} \times 100$$

where L is the viscosity at 100°F of the zero-V.I. oil, H is the viscosity at 100°F of the 100-V.I. oil, and U is the viscosity at 100°F of the unknown (test) oil.

7.46.3 The ASTM standards

This viscosity index system eventually became the ASTM standard D 567, which has been used for years in the petroleum industry.

ASTM D 567 is a satisfactory V.I. system for most petroleum products. However, for V.I.'s above about 125, mathematical inconsistencies arise which become more pronounced with higher V.I.'s. Because products with very high V.I.'s are becoming more common, a method (ASTM D 2270) that eliminates these inconsistencies has been developed.

7.46.4 Calculating viscosity index

The viscosity index of an oil can be calculated from tables or charts included in the ASTM methods. For V.I.'s below 100, ASTM D 2270 and ASTM D 567 are identical, and either method may be used. For V.I.'s above 100, ASTM D 2270 should be used. Since ASTM D 2270 is suitable for all V.I.'s, it is the method now preferred by the leading petroleum companies.

The V.I. of an oil may also be determined with reasonable accuracy by means of special nomographs or charts developed from ASTM tables. A chart for V.I.'s above 100, as determined by ASTM D 2270, is shown in Figure 7.7.

7.46.5 Significance of viscosity index

Lubricating oils are subjected to wide ranges of temperatures in service. At high temperatures, the viscosity of an oil may drop to a point where the lubricating film is broken, resulting in metal-to-metal contact and severe wear. At the other extreme, the oil may become too viscous for proper circulation, or may set up such high viscous forces that proper operation of machinery is

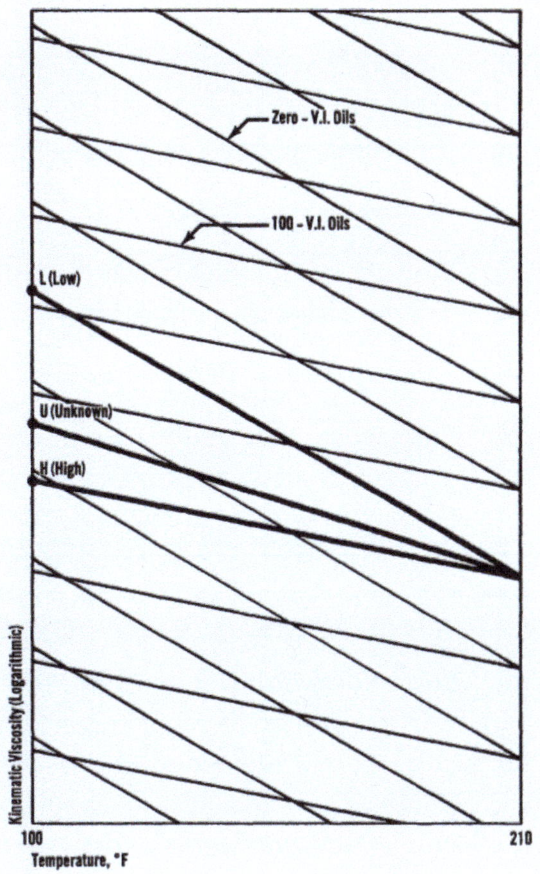

Figure 7.6 The concept of viscosity index. Source: Exxon Corporation, USA, (PLIF 2).

difficult. Consequently, many applications require an oil with a high-viscosity index.

In an automobile, for example, the crankcase oil must not be so thick at low starting temperatures as to impose excessive drag on the engine and to make cranking difficult. During the warm-up period, the oil must flow freely to provide full lubrication to all engine parts. After the oil has reached operating temperature, it must not thin out to the point where consumption is high or where the lubricating film can no longer carry its load.

Similarly, fluid in an aircraft hydraulic system may be exposed to temperatures of 100°F or more on the ground, and to temperatures well below zero at high altitudes. For proper operation under these varying conditions,

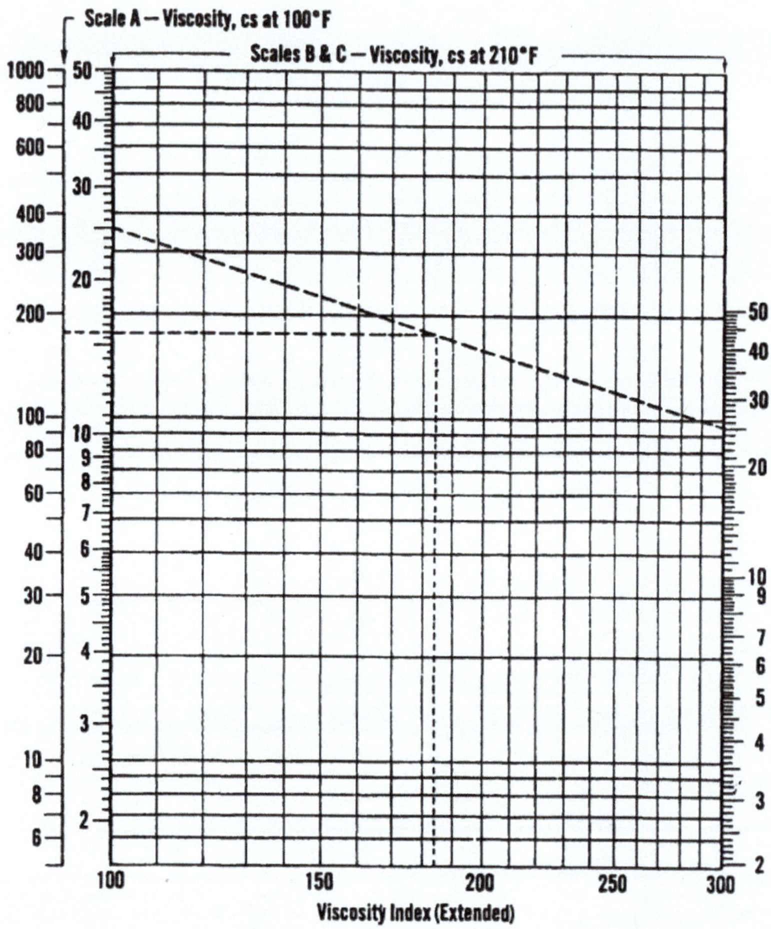

Figure 7.7 Chart for calculating V.I.'s above 100 from kinematic viscosity, based on ASTM D 2270. Dotted lines illustrate its use. Source: Exxon Corporation, USA, (PLIF 2).

the viscosity of the hydraulic fluid should remain relatively constant, which requires a high viscosity index.

As suggested by the relationship between naphthenic and paraffinic oils, the viscosity index of an oil can sometimes be taken as an indication of the type of base stock. A straight mineral oil with a high V.I., 80 or above, is probably paraffinic, while a V.I. below about 40 usually indicates a naphthenic base stock.

In general, however, this relationship between V.I. and type of base stock holds only for straight mineral oils. The refining techniques and the additives that are available today make it possible to produce naphthenic oils with many of the characteristics, including V.I., of paraffinic oils. V.I., then, should be considered an indication of hydrocarbon composition only in the light of additional information.

7.47 Water Washout

7.47.1 ASTM D 1264

Lubricating greases are often used in applications that involve operation under wet conditions where water may enter the mechanism and mix with the grease. Therefore, the ability of a grease to resist washout becomes an important property in the maintenance of a satisfactory lubricating film, and tests for evaluating the effect of water on grease properties are of considerable interest.

Greases can be resistant to water in several ways. Some greases completely reject the admixture of water or may retain it only as occluded droplets with little change in structure. Unless these greases are adequately inhibited against rusting they may be unsuitable for lubrication under wet conditions since the "free" water could contact the metal surface and cause rusting.

Yet other greases that absorb water may be satisfactory under wet conditions. These types of grease absorb relatively large amounts of water by forming emulsions of water in oil. This absorption has little effect on the grease structure and leaves no "free" water to wet and rust the metal. Therefore, the grease continues to supply the proper lubrication while also acting as a rust preventive.

Other water-absorbing greases form thin fluid emulsions so that the grease structure is destroyed. These are useless for operation under wet conditions, and can be considered to have poor water resistance.

There are many effects that water has on grease, and no single test can cover them all. Many of the tests are useful tools; however, the results are subject to the personal judgment of the test operator and much skill is needed to interpret their meaning. ASTM Method D 1264, "Water Washout Characteristics of Lubricating Greases," is one method of evaluating this property.

7.47.2 Significance of results

Test results are useful for predicting the probable behavior of a grease in a shielded (not positively sealed) bearing exposed to the washing action of water. They are a measure of the solubility of a grease in water and give limited information on the effect of water on the grease structure. They say nothing about the rust preventive properties of the grease.

The test is a laboratory procedure and should not be considered equivalent to a service evaluation. Results on greases tested by this method may differ from service results because of differences in housing or seal design. Therefore, a grease that proves unsatisfactory according to this test, may be satisfactory under service conditions if the housing or seal design is suitable.

7.48 Water and Sediment

7.48.1 ASTM D 96, D 95, and D 473

Whether a petroleum fuel is burned in a boiler or in an engine, foreign matter in the fuel is undesirable. Excessive quantities of such impurities as water or solid contamination may interrupt the operation of the unit, and may also damage it.

The two most common impurities found in fuel oils are water and sediment, and several test procedures are available for measuring their concentrations. Water and sediment may be determined together by a centrifuge procedure. Water alone may be determined more accurately in most cases by distillation, and sediment alone may be determined with good accuracy by solvent extraction or by hot filtration.

The tests referred to are the following:

- ASTM D 96 Water and Sediment in Crude Oils
- ASTM D 95 Water in Petroleum Products and Other Bituminous Materials
- ASTM D 473 Sediment in Crude Petroleum Fuel Oil by Extraction

7.48.2 Significance of results

Like many other tests, the determination of water and sediment gives results that must be interpreted in the light of a great deal of previous experience. It is obvious that large quantities of water and/or sediment can cause trouble in almost any application. However, different applications can tolerate different concentrations of impurities. In addition, the quantities of water and sediment determined by the various test procedures are not identical.

Therefore, for any particular application, it is necessary to determine the relation between the tolerance of that application and the results of one or more tests. When this has been done, these tests may be used as controls for that application.

It should be remembered that although a petroleum product may be clean when it leaves the refinery, it is possible for it to pick up contamination from the storage and handling equipment and practices, or as a result of condensation. Water and sediment are often picked up in tanks of ships and in other types of transportation or storage facilities.

7.49 Wax Melting Point

7.49.1 Melting point (Plateau) of petroleum wax (ASTM D 87)

7.49.2 Drop melting point of petroleum wax (ASTM D 127)

7.49.3 Congealing point of petroleum wax (ASTM D 938)

Each of the three test methods discussed here provide information about the transition between the solid and liquid states of petroleum waxes. The tests differ, however, by procedure and in the types of material to which they are applicable.

Both ASTM D 87 and ASTM D 127 are designed to determine the temperature at which most of the wax sample makes the transition between the liquid and the solid states. ASTM D 87 is applicable only to materials that show a "plateau" on their cooling curve. This plateau occurs when the temperature of a material passing into the solid phase remains constant for the time required to give up heat of fusion.

ASTM D 127 determines the temperature at which the material becomes sufficiently fluid to drip. The melting points of high-viscosity waxes that do not show a plateau can be determined by this method.

As determined by ASTM D 938, the congealing point is the temperature at which molten wax ceases to flow.

7.49.4 Significance of results

All three of the test methods are found in common specifications and buying guides among industries using large volumes of wax. The choice of a particular test method depends on the nature of the wax and the application.

Petroleum waxes are mixtures of hydrocarbon materials having different molecular weights. If these materials crystallize at about the same temperature, the cooling curve of the wax will show a plateau. ASTM D 87 is applicable to such waxes. Microcrystalline waxes, however, do not show a plateau in their cooling curve. The melting point of these waxes is usually reported by ASTM D 127. In general, ASTM D 127 is best suited for high-viscosity petroleum waxes.

The congealing point of a wax is usually slightly lower than either of its melting points. Congealing point (ASTM D 938) is often used when storage or application temperature is a critical factor, since it will give a more conservative estimate of the level at which temperatures should be maintained.

7.50 Wheel Bearing Grease Leakage

7.50.1 ASTM D 1263

Under actual service conditions automotive front wheel bearings frequently operate at high temperatures. This is caused by a combination of heavy loads, high speeds, and the heat generated by braking. Because of this, greases used to lubricate these wheel bearings must be resistant to softening and leaking from the bearing. ASTM Method D 1263, "Leakage Tendencies of Automotive Wheel Bearing Greases," is an evaluation of leakage tendencies under prescribed laboratory test conditions.

7.50.2 Apparatus

The test apparatus consists of a special front wheel hub and spindle assembly encased in a thermostatically controlled air bath. Grease that leaks from the bearings is collected in the hub cap and leakage collector. Means for measuring both ambient and spindle temperatures are provided.

7.50.3 Significance of results

This test is an accelerated leakage test and is mainly a measure of the ability of a grease to be retained in the bearings at the test temperature. However, experienced operators can observe other changes in grease condition such as softening or slumping, but these are subjective judgments and not readily expressed in quantitative terms. There is no load or vibration applied to the bearings such as exists in normal wheel bearing service, and the test temperature of 220°F may be considerably lower than encountered in modern vehicles equipped with disc brakes. These factors are recognized and other test devices and procedures are under study by ASTM. The test is of primary value as a screening procedure to be used in conjunction with other stability tests in the development and evaluation of new grease formulations. Because of its limited sensitivity and precision, it permits differentiation only among products of distinctly different leakage characteristics.

8

General Purpose R&O Oils

The very first non-aqueous lubes were base oils—plain, non-additive base oils. But when machinery was subjected to moisture, heat and oxygen, the oil oxidized. This introduction of moisture also led to rust, which began its corrosive creep. The result: breakdowns… blowouts… and, finally, replacement of expensive machinery.

However, the discovery of how to add certain ingredients to the base oil to help control rust and resist oxidation, lead to the broad, universal use of lubricants.

These rust and oxidation inhibited lubricants became universally known as R&O oils, and have now become part of the language of lubrication for industrial machinery. R&O oils have become the workhorse lubricants in thousands of applications.

Ongoing improvements to these products through the decades have delivered industry an outstanding record of dependable service, whether as a hydraulic fluid, gear oil, heat transfer fluid or self-lubricating bearing oil.

8.1 Are All "R&O" Oils the Same?

A common misconception is: "R&O oils are really all the same because they're mostly just oil." R&O products do contain "mostly oil," but the small concentration of carefully selected additives in proprietary base stocks that make up some superior lines of lubricants provide several key advantages:

- Base oils made from dedicated crude sources for most grades,
- Refining for optimized hydrocarbon composition,
- Advanced systems for additive treatment, based on knowledge and understanding of the fundamentals of additive and lubricant behavior, and
- Reliable quality control procedures to ensure a highly consistent, superior products

Table 8.1 Typical viscosity grade range for R&O lubricants.

Grade	32	32	46	68	77	100	150	220	320	460
ISO viscosity grade	32	32	46	68	–	100	150	220	320	460
Viscosity: cSt @ 40°C, ASTM D 445	32.3	30.0	44.2	65.1	78.7	105.0	157.5	214.0	306.0	432.0
cSt @ 100°C	5.3	5.2	6.5	8.4	9.5	12.1	15.7	18.7	23.8	30.3
SUS @ 100°F, ASTM D 2161	166	158	228	339	410	546	840	1123	1620	2284
SUS @ 210°F	44.2	43.8	48.1	58.5	66.5	80.0	93.0	115.0	149.0	

8.2 Additive Formulation

The combination of premium quality base oils plus advanced additive systems in the lubes provides qualities essential to premium R&O oils:

- Rust protection with film tenacity for persistent action
- Oxidation resistance for long service life of the oil without acid formation or sludge, even in the presence of catalytic metals
- Thermal stability to minimize deposit formation during prolonged exposure to high temperatures
- Demulsibility for rapid separation of water that becomes entrained in the lubrication system
- Foam and air entrainment control to ensure maximum lubrication efficiency
- High VI to avoid wide viscosity swings when variations in temperature occur
- Full range of viscosities to satisfy the wide range of machine condition-Table 8.1.
- Low pour points to ensure oil flow at startup

8.3 Designing R&O Lubricants

Designed to lubricate and protect industrial machinery in a wide variety of applications, premium R&O oils must be comprised of a broad range

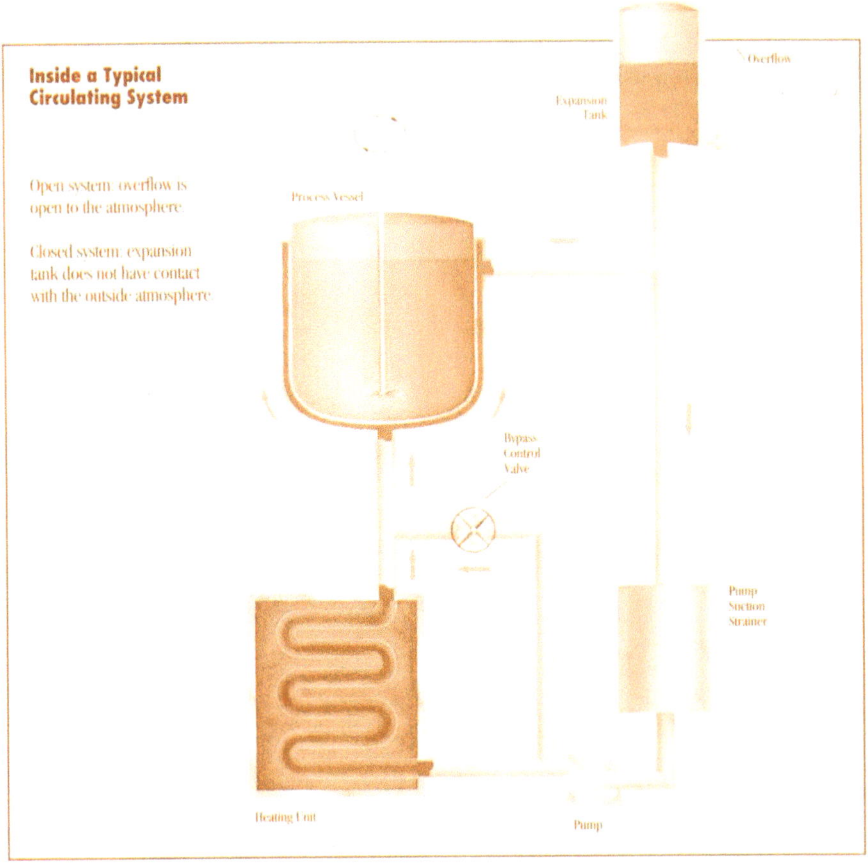

Figure 8.1 Elementary circulating oil system.

of viscosity grades to meet the varied conditions of use in a modern plant environment.

ISO grades 32 through 100 are used primarily in turbine and other circulating oil systems (Figure 8.1), hydraulic systems, compressors, pumps and general-purpose applications. ISO grades 150 through 460 are used primarily in light-duty geartrains and higher temperature applications. Viscosity data on these products are shown in Table 4.2.

Before the creation of the ISO viscosity grading system, it was customary to use the SAE grading system when selecting lubes for industrial applications. Refer to the appendix for a review of the approximate relationships between the grading systems.

Table 8.2 Typical inspections for Exxon TERESSTIC® R&O lubricants.

Specification	R&O Lubricant Grades				
Grade	32	33	46	68	77
ISO viscosity grade	32	32	46	68	—
AGMA lubricant grade	—	—	1	2	—
Viscosity:					
cSt @ 40°C, ASTM D 445	32.2	30.0	44.2	65.1	78.7
cSt @ 100°C	5.3	5.2	6.5	8.4	9.5
SUS@100°F, ASTM D 2161	166	158	228	339	410
SUS @ 210°F	44.2	43.8	48.1	54.4	58.5
Viscosity index	97	97	97	97	97
Gravity, specific @ 15.6°C(60°F)	0.877	0.873	0.877	0.879	0.880
°API		30.6	29.9	29.5	29.3
Flash point, ASTM D 92 COC, °C(°F)	206(403)	207(405)	212(414)	224(435)	227(441)
Pour point, ASTM D 97 °C(°F)	-37(-35)	-21(-6)	29.9	-21(-6)	-21(-6)
Rust test ASTM D 665 A/B					
Distilled water	pass	pass	pass	pass	pass
Synthetic sea water	pass	pass	pass	pass	pass
Copper strip corrosion, ASTM D 130 3hrs@ 100°C(212°F)	1	1	1	1	1
FZG Gear test, A/8.3/90, DIN 51354, Failure load stage (FLS)					
Foam, ASTM D 892	pass	pass	pass	pass	pass
Water Separability, ASTM D 1401	pass	pass	pass	pass	pass
Neutralization No.	0.06	0.06	0.06	0.06	0.06
Oxidation life, ASTM D 943, hrs	5000+	2500	3000	2500	2000
Four-ball wear test	—	—	—	—	—
scar diameter, 10 kg, 1800 rpm, 150°C(302°F), 30 min, mm					

8.4 Extreme Pressure (EP) R&O Lubricants

Certain geared turbines, steam and gas, are subject to shock loads and occasional overloading. This creates extreme pressure (EP) that can force the normal lubricating film out from between meshing gear teeth. The resulting grind of metal-to-metal contact can cause excessive wear.

Most R&O EP lubricants are formulated with non-zinc anti-wear additives to help reduce the possibility of metal-to-metal contact. In addition, quality EP lubricants will contain oxidation inhibitors to help assure long, dependable operating life, as well as the usual rust-inhibiting and anti-foam agents. These lubricants should exhibit no rusting in either distilled water or synthetic sea water when tested against the ASTM D 665 rust test procedure.

Quality EP oils will usually have extremely good demulsibility: any condensed moisture collecting in the lubricating system is readily shed by the oils. They should also have a high viscosity index (VI), which allows more uniform operation of the system throughout a wide range of ambient and operating temperatures.

8.4.1 Dependable turbine lubrication

A few products have achieved a long record of reliable lubricant performance in the lubrication of steam turbines and gas turbines. For many years, the power industry has recognized a quality R&O lubricants ability to provide:

- Long life with infrequent change out
- Prevention of acidity, sludge, deposits
- Excellent protection against rust and corrosion, even during shutdown
- Good demulsibility to shed water that enters the lubrication system
- Easy filterability without additive depletion
- Good foam control

8.4.2 Cleanliness levels

Compressor lubrication can be one of the most demanding jobs for a lube oil because all compressors generate heat in the compressed gas. This heat directly impacts lube oil life. The degree of impact depends upon the compressor type and the severity of operation. In some units, reciprocating or rotary type, the lubricant is *directly* exposed to the compressed gas.

This stress can cause rapid oxidative degradation and resultant formation of deposits and corrosive by-products leading to increased maintenance needs. But superior oils meet and beat the compressor lubrication challenge.

What makes a line of lubricants successful in compressor applications?

- Special resistance to oxidation under conditions of high temperature and intimate exposure to air
- Good demusibility during water condensation
- Long-lasting rust and corrosion inhibition
- Anti-foam properties

Proper lubricant selection is crucial to compressor life and service. The main types of compressors and typical R&O lubricant grades used at some of the most profitable refineries, utilities, and petrochemical plants are shown in Figures 8.2 through 8.4. Specific grade selections are discussed in the text segment dealing with compressor applications and depend both on manufacturer recommendations and on the expected operational severity.

8.5 Superior R&O Oils Cover a Wide Range of Pumps

Like compressors, industrial pumps come in many shapes and sizes and serve thousands of industries. The selection of the proper pump depends on: the nature of the liquid being pumped (its viscosity, lubricating value, density, volatility, corrosivity, toxicity, solids content), the pumping rate, desired pressure and the type of lubrication system to be used.

The centrifugal pump is the most widely used in the chemical and petroleum industries for transferring liquids of all types: raw materials, materials in process and finished products. Characterized by uniform (non-pulsating) flow and large capacity, the centrifugal pump is also used for water supply, boiler feed and condensate circulation and return.

Reciprocating and rotary pumps are particularly well adapted to low-capacity, high-pressure applications. They can deliver constant capacity against variable heads. Very close tolerances are required between internal rubbing surfaces in order to maintain volumetric efficiency, so the use of reciprocating or rotary pumps is generally restricted to liquids that have some lubricating qualities. Rotary and reciprocating pumps are used in fuel, lube circulating and hydraulic oil systems.

Quality R&O oil products are well-suited to many pump operations. They can serve as the external lubricant for the pump itself, drive motor

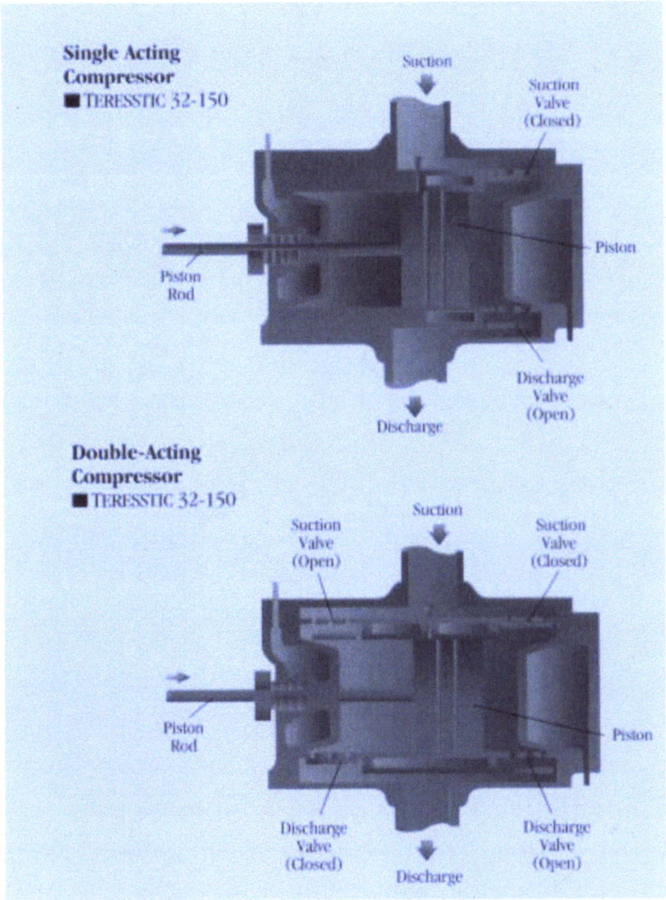

Figure 8.2 Reciprocating compressor applications.

bearings or other lubricated parts in the pump system. However, some R&O oils should not be used in portable water systems.

Recommendation of the right R&O lubricant product for a given pump can be determined by the pump manufacturer 's specification or by contacting the appropriate sales office.

8.6 Hydraulic Applications for R&O Oils

In hydraulic power applications, quality R&O oils, especially grades 32-150, are particularly well suited for general-purpose hydraulic systems that do not require anti-wear protection but do require a premium quality oil

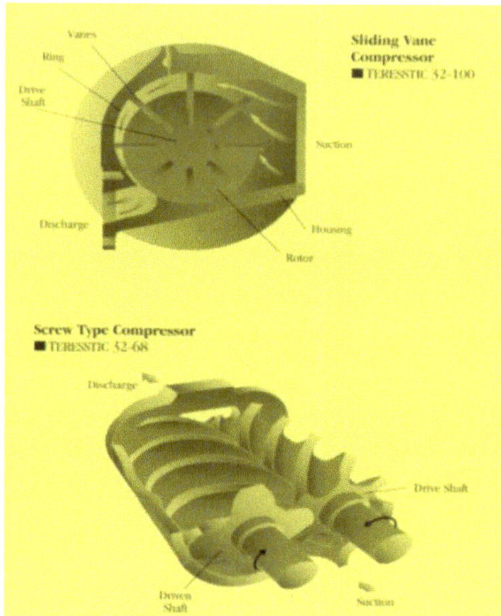

Figure 8.3 Rotary positive displacement applications.

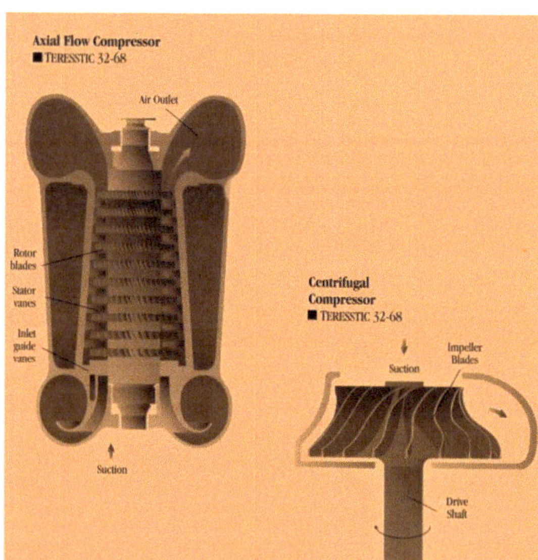

Figure 8.4 Dynamic compressor applications.

with long life, and for pump components requiring full hydro-dynamic film lubrication.

Selecting the proper viscosity grade is important for the most effective hydraulic system lubrication. Items to consider during selection: the expected ambient environment and the extremes of oil temperature expected in the system.

When choosing the correct lubricant, remember to choose one with a viscosity high enough to:

- Maintain sufficient hydrodynamic film thickness
- Prevent excessive wear of moving parts
- Avoid excessive pump slippage or case drain and loss of pressure system response

Also, choose an oil with a viscosity low enough to:

- Eliminate startup problems at low temperatures and avoid high wear during startup
- Reduce pump cavitation tendency
- Achieve good response in hydraulic system devices
- Ensure good defoaming and good demulsibility
- Save energy

8.7 Universal Application of R&O Oils

The almost universal applicability of superior R&O lubricants is best demonstrated by their use for gear lubrication, as heat transfer fluids, or in self-lubricated bearings.

For lubrication of enclosed gear drives in industrial equipment, the manufacturer's recommendation for the proper lubricant grade is usually indicated on the gear case as the American Gear Manufacturers Association (AGMA) number. As discussed later in our chapter dealing with gears, these AGMA specifications cover a wide range of viscosity and load-carrying grades, from light-duty to severe applications. The most widely used are the grades of circulating lubricants. They can easily be distributed to gears and bearings for lubrication and heat removal, and are readily filtered and cooled.

For light-duty applications where extreme-pressure properties are not required, AGMA recommends the use of R&O lubricants.

Typical gear types suitable for quality R&O oil use include: spur gears, helical gears, double helical (herringbone) gears, bevel gears and spiral bevel gears. With their long-lasting rust inhibitor system, excellent oxidation prevention, good demulsibility and foam control, this class of lubricants can provide excellent gear protection and R&O performance.

8.8 A Heat Transfer Fluid That Keeps its Cool

R&O oils are made from petroleum components that are vacuum-fractionated to selectively remove lower volatility elements. The components are then fully refined and additive-treated to provide outstanding oxidation resistance. These characteristics make most R&O oil products excellent heat transfer oils.

In many process applications that include industrial heat-treating, chemical manufacturing and food processing there are clear advantages to carrying heat by means of fluid transfer systems rather than using direct-fired heating systems. Using fluids for heat transfer allows closer control of the process temperatures, eliminates hot spots in vessel walls, permits heating several process vessels using one primary heat source, and provides better economy in the overall heating operation.

The working fluid used in the system must have a high degree of oxidation stability, especially in open circuit systems where the expansion tank is open through a breather tube and air can contact the hot oil. The fluids must not develop sludge or allow carbonaceous deposits on the primary heat exchanger walls, pipes or the heated reaction kettle. For safety reasons, the fluid should have volatility, i.e., low vapor pressure at the heat transfer temperatures. It should also have high flash and fire point properties.

Several grades, because of their excellent resistance to oxidation and thermal degradation, are ideally suited for use as high-quality heat transfer fluids. They are capable of providing long service in a heat transfer circulating system. The grades are listed in Table 8.3, which also shows general guidelines on:

- The maximum operating temperature—based upon safety considerations such as flash and fire points —for the fluid in the expansion tank, where some exposure to air may occur in open systems. (In less critical closed systems, considerably higher expansion tank temperatures may be used because no air contact occurs.)
- The maximum bulk oil temperature as it leaves the primary source heat exchange unit. (The actual film temperature or skin temperature next to the furnace tubes can be even higher, limited only by the ultimate thermal degradation property of the fluid.)

A brief tabulation of good practices in using heat transfer fluids might include a listing of general requirements, followed by a few memory joggers. Thus, in order to:

- Maximize the efficiency of the overall heat transfer operation
- Maximize the useful life of the fluid
- Minimize formation of deposits on walls and tubes

Tips for using R&O oils for Fluid Transfer:

- Select a fluid of lowest feasible viscosity and the highest coefficient of heat transfer
- Use fluids that are effectively inhibited with non-volatile additive systems
- Avoid catalytic metals (i.e., copper alloys) in the system
- Do not use clay filters (i.e., avoid additive removal)
- Minimize air contact, avoid air leaks at pumps, etc.
- Do not mix different types of heat transfer fluids
- Do not use heating oil or solvent for flushing; use a mineral oil such as coray or Faxam.
- Be watchful for oil leaks onto hot surfaces
- Use an oil with flash point at least 15°C (27°F) above expansion tank temperature
- Never operate either open or closed systems at temperatures near the auto-ignition point; keep the system below 360°C (680°F).

It is important to give close attention to the flash points and maximum temperatures, depicted in Table 8.3.

8.9 R&O Oil Use in Self-lubricating Bearings

Development of lubricated-for-life bearings has brought us electric shavers, kitchen appliances, power tools (Figure 8.5), automotive electric motors and a host of other conveniences we take for granted. The bearings, lightweight and complex in design, never have to be relubricated.

Self-lubricating bearings have evolved from the steadily emerging technology of powder metallurgy. Their success depends on having an

Table 8.3 Temperature guidelines for premium grade R&O oils in heat transfer applications.

Product	Flash Point	Maximum expansion tank temperature for open systems, COC,°C(°F)	Maximum bulk oil temperature for open or closed systems, °C(°F) °C(°F)
TERESSTIC 32	206(403)	185(365)	316(601)
TERESSTIC 46	212(414)	197(387)	316(601)
TERESSTIC 68	224(435)	209(408)	316(601)
TERESSTIC 77	227(441)	212(414)	316(601)
TERESSTIC 100	257(495)	242(468)	316(601)
TERESSTIC I50	266(511)	251(484)	316(601)

Courtesy: Exxon-Mobil Corporation

oil impregnated into the metal of the bearing that is capable of lifelong, trouble-free operation.

A self-lubricating bearing is typically made in four steps:

1. A selected metal or alloy is reduced to powder— usually by atomization of a molten stream—to a specified particle size and distribution.
2. The powder is compacted in a mold to the dimensions of the intended finished part.
3. The part is heated to sinter the powder, while leaving a controlled amount of internal space and capillary passageways.
4. The part is impregnated with a lube-for-life lubricant.

The oil absorbed into the pores of the sintered metal acts as a reservoir for lubricant when the part is in use. At startup a thin film on the surface provides initial lubrication. As the bearing warms up, the oil expands and is forced out of the pores into the space for journal/bearing lubrication. When rotation stops and the bearing and oil cool, the oil is drawn back into the pores of the bearing by capillary action.

Conventional powder metallurgy (P/M) bearings can absorb about 10-30% by volume of oil. The thin film of oil and the small reservoir within the pores do the whole job of lubrication. There is no circulating system or oil reservoir in the conventional sense.

There are significant advantages to this technology:

- The finished part is less dense than a conventionally machined part; weight savings are of great importance in many applicators.

Figure 8.5 The extraordinary loads placed on the self-lubricating bearings of high-speed power tools can lead to premature tool failures without the protection of quality oils.

- Controlling of powder metallurgy adds strength and durability to the part.
- The P/M system eliminates the need for conventional machining, saving time and allowing fabrication of more complex shapes.
- Self-lubricating bearings simplify design considerations in the unit of machinery.
- High-quality self-lubricating bearings eliminate the need for maintenance or repair service and reduce the cost of warranty claims.

P/M parts are particularly effective where relatively light loading is present: home appliances, automotive accessory equipment, power tools, business machines and the like. Selection of high-quality oil is extremely important:

- The oil must be fully refined to provide maximum stability without forming gums, varnish or sludge that would block the porous structure.

- Additives in the oil should have high permanence but should not interact with the sintered metals or otherwise cause corrosion or deposits.
- The viscosity grade must provide the proper hydrodynamic lubrication under the conditions of use both at startup and at maximum operating temperature.

For these types of self-lubricating bearing applications great success has been achieved using quality R&O ISO grades 68 and 77 lubricants.

9

Hydraulic Fluids

Hydraulic fluids are primarily petroleum-based fluids used as a power transfer medium in industrial hydraulic system and are generally lighter viscosity that most mineral oils, found offered in the 32 to 150 cSt @ 40 viscosity range. Although, technically, water can be uses as a hydraulic fluid, the majority of hydraulic fluid is formulated from Group 1 mineral base oil stocks. However, in a high temperature operating environment

In an ever, eco-conscious environment, eco fluids using canola vegetable base oils stocks are becoming more popular (see environment fluids chapter 11 for more information). Although water is technically a viable hydraulic fluid medium, this chapter will focus on mineral oil hydraulic fluids.

Hydraulic fluid has many critical functions. It must serve not only as a medium for energy transmission, it must also act as a lubricant, sealant, and heat transfer medium. The fluid must maximize power and efficiency by minimizing wear and tear on the equipment. Figure 9.1 shows a typical "closed loop" hydraulic system in which the lubricant is pumped from a reservoir tank via a suction screen under pressure to a pressure control valve using a vane style pump. Pressurized lubricant is then held in check by a pressure control valve until released into the circuit via a directional control valve to perform work (in this case a piston style actuator), or return to the control valve or reservoir via an in-line oil filter.

Working in such systems as previously described, a hydraulic oil can be called upon to fulfill many functions that include:

- facilitate seal mounting,
- improve the sealing effect,
- reduce adhesive and starting friction,
- reduce wear during operation,
- be neutral to NBR, EPDM, FMP and PU material,
- have excellent load-carrying capacity,

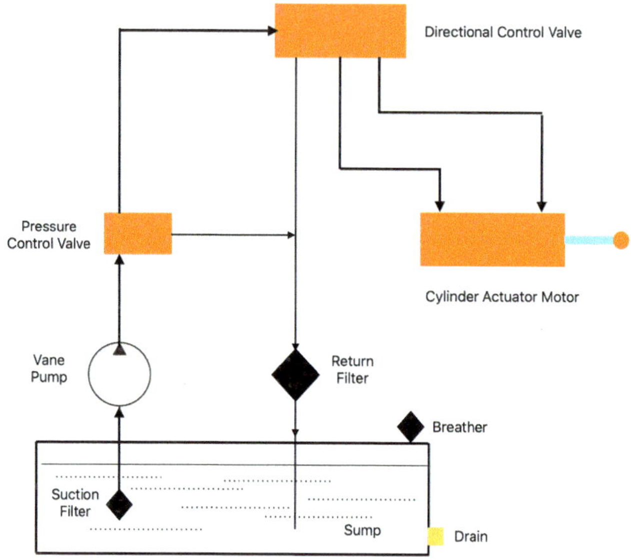

Figure 9.1 Typical hydraulic circuit with vane pump.

- show high affinity towards materials such as steel, plastics and elastomers
- protect components against rust and oxidization

As previously mentioned, hydraulic fluids not only act as the fluid-power medium, they lubricate systems components. Modern hydraulic pump units subjected to high system pressure and pump speeds can create conditions thin-film lubrication that cause unwanted mechanical wear unless the fluid contains protective additives.

Specific needs of hydraulic systems and their components, may differ. Some require fluid with greater oxidative or thermal stability; some need tougher anti-wear protection; some require extra lubricant stability in extreme-temperature environments; and some require the assurance of fire-resistant fluids.

Often, many manufacturers will add a color dye in their hydraulic fluids so they are more noticeable when troubleshooting for hydraulic system leaks.

9.1 Hydraulic Oils

Many lubricant manufacturers offer a family of mineral based hydraulic oils that are essentially the same product with minor variations suited to specific applications and working environments. For example, Exxon Corporation offers a premium quality line of hydraulic lubricants under the NUTO® trade name.

NUTO® H is the trademark for a line of premium-quality anti-wear hydraulic oils designed to meet the stringent requirements of most major OEM manufacturers and users of hydraulic equipment. The five grades listed meet the viscosity requirements of essentially all hydraulic systems. NUTO®H is very effective in reducing vane and gear pump wear in systems operating at high loads, speeds, and temperatures. Its specialized additive makeup also allows the use of NUTO® H in severe-service hydraulic systems employing axial and radial piston pumps. This lubricant's typical inspections listed in Table 9.1 are the marque of a premium quality product.

This family of oils are designed to provide outstanding rust protection, low deposit formation, good demulsibility, low air entrainment, oxidation resistance, low pour points, and good anti-foam properties, typical of all high-quality mineral hydraulic oils. These lubricants are also non-corrosive to metal alloys, except those containing silver, and are fully compatible with common seal materials. With a Viscosity Index in the high 90's, this lubricant specification is perfect for performing within extended temperature ranges.

NUTO®HP is a line of high-performance, ashless (metal free), mineral-oil-based anti-wear hydraulic oils formulated with additives that provide reduced environmental impact in the case of an accidental release into the environment. This formulation makes these oils suitable for applications in woodland, marine, construction, mining, pulp and paper, and farming, as well as general industrial hydraulic applications when environmental concerns exist.

NUTO®FG is a food-grade hydraulic oils designed for use in food processing and packaging operations. It incorporates the following unique combination of features.

- Compliance with FDA 21 CFR 178.3570, "Lubricants with Incidental Food Contact (see Chapter 6)
- USDA H-1 approved
- Outstanding anti-wear (AW) properties, for pump protection
- Excellent extreme-pressure (EP) properties, for bearing protection
- Superior oxidation stability, for long, trouble-free life
- Suitable for hydraulic systems up to pressures of 3000 psi

Table 9.1 Typical inspections for quality anti-wear hydraulic oils (Exxon NUTO®H).

Typical Inspections - Grade	32	46	68	100	150
Gravity, °API					
specific @ 15.6°C (60°F)	29.9	29.6	29.2	28.8	28.6
Density, kg/m³	0.877	0.878	0.881	0.883	0.884
lb/gal	7.303	7.311	7.336	7.352	7.360
Viscosity, cSt @ 40°C	32.2	46.0	66.1	99.0	154.1
cSt @ 100°C	5.3	6.5	8.5	11.1	14.9
SUS @ 100°F	166	230	343	516	810
SUS @ 210°F	44.2	48.1	54.8	64.3	79.3
Viscosity index	97	96	97	97	95
Pour point, °C/°F	-37/-35	-29/-20	-27/-17	-23/-9	-21/-6
Flash point, °C/°F	206/403	216/421	222/432	241/466	266/511
Color, ASTM D 1500	2.0	2.5	2.5	3.0	3.5
Neutralization number, ASTM D 974	0.40	0.40	0.40	0.40	0.40
Rust test, ASTM D 665					
A. Distilled water	———	———	no rust—	———	———
B. Synthetic sea water	———	———	no rust—	———	———
Hydrolytic stability, ASTM D 2619					
Cu mass loss, mg/cm²	0.12	0.12	0.12	0.12	0.12
Denison HF-0 pump test -	———	———	Approved	———	———
Cincinnati Milacron Spec. No.	P68	P70	P69	–	–
Vane pump test, ASTM D 2882,					
total ring and vane wear, mg	25.0	25.0	25.0	25.0	25.0
35VQ25 vane pump test	———	———	Pass—	———	———
Oxidation life					
Turbine Oil Stability Test,ASTM D 943, hrs	3000+	2500+	2500+	2500+	2500+
Rotary Bomb Oxidation Test,ASTM D 2772, min	350	350	350	350	350
FZG Spur Gear Test (FLS)	–	11	11	–	–

9.2 Hydraulic Pumps

Modern day hydraulic pump units are subject to high system pressures and pump speeds. This can create conditions of thin-film lubrication and cause eventual mechanical wear unless the fluid viscosity is sufficient, and the fluid contains special protective additives.

Three main types of pumps (Figure 9.2) are found in hydraulic systems: gear pumps, piston pumps, and vane pumps.

Vane pumps (as depicted in Figure 9.1 and 9.2) are most common and require the most anti-wear protection, due to the high contact pressures developed at the vane tips. Vanes are designed to be in continual contact with the inner surface of the cam ring. This allows the vane pump to deliver a constant flow rate, though surface interference does create a high wear rate operating environment requiring an AW additive in the oil. A pressure-compensated variable displacement pump can reach pressures up to 2900 psi before a pump pressure compensator cuts in to reduce internal pressure. Oil viscosity range in normal operating conditions can range between 14 and 160 centistokes (cSt).

The anti-wear properties of a hydraulic fluid are typically tested by operation in an actual vane pump under overload conditions. Results are measured in terms of hours to failure or as the amount of wear (weight loss of the vanes and ring) after a specified number of hours of operation. Experience has shown that a good anti-wear hydraulic fluid can reduce wear by 95% or more when compared to conventional R&O oils.

Gear pumps are positive displacement devices that work by drawing oil through the meshing action of the gear teeth at delivery pressures up to 7000psi (external gear pump). Gear drive pumps can be designed with external (shown in Figure 9.2) or internal gear (Gear within a gear) gear sets. Whereas external gear pumps viscosity range is limited somewhat to below 300cSt, internal gear pumps offer a much wider range is use that can exceed 2000cSt.

Gear and piston pumps don't usually require anti-wear oils. If uncertain, always consult with the pump manufacturer, or the OEM maintenance and operations manual for specific requirements.

Piston pumps are offered in both variable and fixed displacement designs and are often lauded for their versatility and relatively trouble-free operation. Operating at pressures greater that 6000psi they are the obvious choice for higher pressure systems. Piston pumps typically operate using a fluid viscosity between 15 to 160cSt.

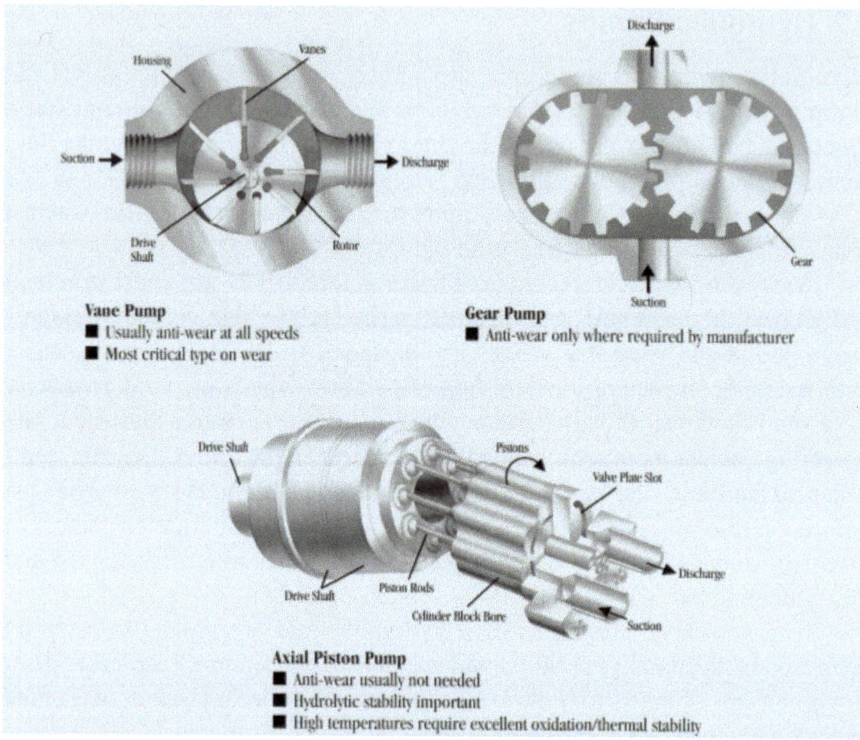

Figure 9.2 Main types of pumps found in hydraulic systems. Source: Exxon Company, U.S.A., Houston, Texas PLIF 2.

When looking to determine a pump's operating viscosity range always refer to the manufacturer's data.

9.3 Maintaining Your Hydraulic Oil in Service

Hydraulic systems rely on hydraulic fluids to transfer and amplify power, lubricate critical components, and dissipate frictional heat build-up in the system.

Despite their complexity, hydraulic systems are very forgiving—often to their detriment. They can continue to perform inefficiently for a long time before catastrophic failure occurs. Unfortunately, this forgiving nature can foster apathy toward failure prevention, efficiency, and fluid/component life cycle management. Hydraulic fluid is the most important part of any hydraulic system and when systems fail, the cause is most often related to fluid contamination. Fluid failure almost always results in critical equipment and manufacturing process failure.

To perform efficiently, hydraulic fluids must be clean and free of contamination as much as possible, making contamination prevention the primary focus of any hydraulic system preventive maintenance approach.

9.3.1 Exorcising the "Big Three"

Hydraulic fluid contamination presents in three forms: solid particulate, water, and air. All of these can seriously affect the hydraulic fluid and the equipment it serves. Maintaining hydraulic fluid in optimum condition requires measuring, controlling and preventing the introduction of contaminants. Hydraulic fluid is the most important component in any hydraulic system; when systems fail, the cause is most often related to fluid contamination. Understanding how contaminants are introduced into the system is important in developing effective preventive strategies.

Solid Particulate Contamination. Hydraulic system components are designed to operate with very close tolerances that can be as close as 1.5 microns. Solid particulate most often manifests itself as grit or dirt and if allowed into the system can be very damaging to bearing surfaces and hydraulic seals. The solid particles, which can be over 100 microns in size, will set up in a three-body abrasion state and easily score the mated machined surfaces, creating rapid bearing and component surface wear. This in turn leads to unwanted fluid bypass that reduces operating efficiency. Solids contamination can also cause valve stiction, fluid viscosity increase and unwanted fluid leakage through nicked and scored cylinder seals.

If your equipment is new or rebuilt, solids contamination in the form of leftover dirt or swarf from the manufacturing/rebuild process can be present in the hydraulic lines. Prior to initial start-up lines must be internally wad cleaned, existing oil flushed from the system. And new correct viscosity fluid added.

Many users are unaware that solids contamination levels can be excessive in new oil supplied from the manufacturer or introduced by the supplier if bulk transferred with dirty transfer hose and equipment. When receiving new oil—especially bulk oil—always perform an oil analysis test to detect for solids and water contamination. New oil can often be found at a 19/17/14 cleanliness level on the ISO solid contamination code, which is not clean enough for high-pressure hydraulic systems. These systems require fluids with a minimum 16/14/11 cleanliness level. See Chapter 32 Contamination control.

Other sources of solids contamination, all of which are preventable, can include:

- Improperly stored oil

- Dirty transfer equipment used to transfer the oil into the equipment reservoir
- Reuse of dirty "leaked" oil
- "Open to air" reservoir due to missing fill port or reservoir breathers
- Lack of filter maintenance causing the dirty oil to bypass into the system
- Poor housekeeping practices

Water Contamination. Water, a universal contaminant, will saturate a hydraulic fluid at a mere 300ppm or 0.04% concentration level. Water can be present in the oil in:

- a free state separated from the oil in an unstable form,
- an emulsified state in a stable form that appears cloudy
- in a saturated dissolved form that appears invisible

Water depletes vital oil additives or, even worse, can react with additives to create corrosive acids that attack system components. It can also reduce lubricant film strength and its ability to release air, which can increase wear, corrosion and cavitation.

Some hydraulic fluids such as brake fluids are designed to be hygroscopic and entrain moisture in the fluid until its saturation point. In high-heat applications, water can boil off and create huge inefficiency in the hydraulic power transfer motion.

Typical water contamination sources include:

- Incorrect outdoor lubricant storage that cause hot-cold cycle condensation
- Open-to-air reservoirs into which washdown and/or [process water can infiltrate

Water can be detected visually in both its free and emulsified state. To remove water contaminant use:

- A polymeric style filter media designed to absorb the water as it passes through the filter
- Using vacuum distillation to boil off the oil
- Dehumidification in the reservoir headspace.

Air Contamination. Air contamination presents in a number of forms, of which entrained air can be the most problematic. In this form, air bubbles

(< 1mm dia.) dispersed throughout the fluid, reduce fluid viscosity and, thus, film strength. This situation, in turn, can lead to premature component wear; reduction in the oil's bulk modulus, causing a lack of efficiency and control due to sponginess in the oil condition; an increased heat load, resulting in fluid deterioration; and system erosion, due to cavitation.

Air bubbles greater than 1mm create foam, which can quickly deplete any antifoam additive and cause fluid oxidization. Typical air contamination causes include:

- Over/under filled lubricant reservoir
- Clogged inlet/suction filters
- Clogged reservoir breather
- Restricted inlet lines
- Loosely clamped inlet lines
- Pump shaft seal failure

9.3.2 Prevention control strategy

Fortunately, all of the contamination problems discussed are easily preventable—in most cases at minimal cost. While the following advice is not all encompassing, it provides an excellent starting point for a successful hydraulic fluid PM strategy:

- Implement a fluid cleanliness standard
- Store all lubricants in a cool, dry place and practice FIFO (First in-First Out) stock rotation
- Practice good house-keeping, keeping reservoirs clean of all dirt and debri
- Cap all hoses and manifolds during handling and maintenance
- Wad-clean all lines prior to initial system start up
- Flush all lube systems and change oil prior to start up
- Use a dedicated filter cart for each hydraulic fluid type, with quick connect fittings to transfer /clean hydraulic fluid before it enters reservoir
- Install external sight gauges marked with a high and low fluid level mark in reservoirs to check for fluid levels and water presence

- Use polymeric oil filters and dessicant reservoir air breathers
- Specify rod wipers and replace all worn actuator seals

With a little care and common sense, your plant's hydraulic systems—and ythe vital hydraulic fluids that keep them running—will be efficient, reliable and long lived!

9.4 Environment-friendly Hydraulic Fluids

Hydraulic fluids, by tradition, are petroleum-based products. In recent years, industry has witnessed a move toward environment sustainability, especially where lubricants are concerned. Hydraulic systems are inherently susceptible to leakage due to their system design (high pressure versus seals) and require diligent maintenance to ensure they remain leak free. Unfortunately, hydraulic systems are often employed in environmentally sensitive situations, especially in mobile equipment and ship transportation. Government regulations have forced equipment and lubricant manufacturers to take a closer look at less toxic and more environmentally friendly hydraulic fluids. Refer to Chapter 11— Food Grade and Enviro-friendly/Bio-degradable oils Lubricants section.

10

Fire Resistant Fluids

Mineral-oil hydraulic leaks in high-temperature operations can be costly, and potentially disastrous. If hot fluid sprays or drips onto a hot surface, it can burst into flames and quickly propagate a fire. In these working conditions, the designer must investigate use of fire-resistant hydraulic lubricants.

Currently there is no legal definition for Fire-resistant hydraulic fluids. Selection is the responsibility of the user. As these fluids are primarily formulated according to ISO 12922 requirements their test methods depicted in Table10.1 can be used to determine a suitable application choice.

Fire resistant fluids are designed to resist combustion0, and are primarily formulated to meet the stringent safety requirements of the mining and steel industries These premium products help reduce fire hazards while providing excellent lubrication, foam resistance, and protection against rust and corrosion.

Fire resistant fluids are typically offered as Water-based fluids that do not have a flash point or combustion point, or Water-free fluids.

Fore resistant fluids are offered in four specific formulation categories that offer a range of fire-resistant capabilities and lubricant properties.

HF-A (Oil-in-Water Fluid): For low-pressure applications in which anti-wear properties are not critical and cost is a major concern. HF class lubricants have the same viscosity as water (between 1-5mm2/sec). Where viscosity is an issue, HFB or HFC lubricants must be considered.

HFAS fluids are formulated using various chemicals mixed in water. These formations can be found in use in foundries, mining equipment, and hydro-forming applications.

HFAE fluids are oil in water emulsions designed to provide very good fire resistance and emulsion stability. Typical applications include mine support and hydrostatic drive systems.

HF-B (Water-in-Oil Fluid): A pre-mixed invert (water-in-oil) emulsion, the HFB product provides superior anti-wear and anti-corrosion properties when compared to oil-in-water emulsions.

Table 10.1 ISO 12922 test methods used to assess fire resistance.

ISO Test	Testing for:	Test Purpose
ISO 15029-1	Spray flame persistence – Hollow cone nozzle method	Spray Ignition characteristics
ISO 15029-2	Spray flame persistence – Stabilized heat release method	Spray Ignition characteristics
ISO 14935	Average flame persistence	Wick flame persistence of fluids
ISO 20823	Ignition temperature and flame spread	Flammability on hot surfaces

HFB products can contain up to 60% mineral-based fluid that may discount them for use as a fire-resistant fluid in some applications. Always check the lubricant specification against the regulated requirement before use. Typical applications include hydrostatic control systems and drives

HFB products typically provide protection up to 650

HF-C (Water Glycol Fluid): This water/glycol fluid combines outstanding fire resistance with excellent performance at low temperatures and good resistance to corrosion.

HFC products contain approximately 35% water and are used in applications where water free hydraulic fluid is not permitted due to fire risk. Typical applications include injection molding, pressure die-casting, forming presses, foundries, coking plants, steel plants.

HFCE products are similar to HFC but contain less water, usually 20% in solution. These products can be used in higher temperature applications found in coal mines, and provide addition wear protection. These fluids are not standardized according to ISO 12922.

HFC products typically provide protection up to 600

HF-D (Synthetic Fluid): HF-D fluids are 100% synthetic fluids containing no petroleum-based products or water. The least hazardous of all fire-resistant fluids utilize a synthetic base oil that offers outstanding lubrication properties and pump protection.

HFDR products are phosphate-ester based and offer the best protection with temperatures up to 700. This fluid can, however, rapidly deteriorate most seal materials.

HFDU products are the most commonly used synthetic fluid that are manufactured predominantly from polyol-ester based fluids that offer protection up to 400

Table 10.2 Comparison of fire resistant and mineral oil properties. (Source: Exxon).

	NUTO H 68	HFA	HFB	HFC 46	HFD 68
Type of fluid	Premium mineral oil	Oil-in water emulsion	Water- in-oil emulsion	Water /glycol	Synthetic (polyester)
Appearance		Milky	Off-white	Clear red	Dark amber
Cost -mineral=1	1	0.2-0.4	1.5	3	5
Fire resistance	Poor	Excellent	Excellent	Excellent	Good
Pump wear, ASTM D 2882	25	Pump failure	164	<200	5.7
Wear protection	1	0.1	0.7	0.5	1.5
Pump guidelines	HP- All pump types	LP- Piston, plunger Flooded suction Max. inlet size	MP- Most pump types Flooded suction preferred	MP- Most pump types Flooded suction preferred	HP- All pump types
Operating Range40-85 °C (°F)	(103-183)	4-54 (40-130)	4-54 (40-130)	-18 to 60 (0-140)	-18 to 93 (0-200)
Freeze/thaw stability	Excellent	Poor	Moderate	Good	Excellent
Water content %	0	95	39	42	0
Water hardness limit	N/A	250 ppm $CaCCO_3$	N/A	N/A	N/A
Ease of recovery from water	Good	Good to Poor	Good	Poor	Good
Anti-Corrosion	Excellent	Fair	Good	Good	Excellent
ISO viscosity	68	N/A	N/A	46	68
Viscosity index	97	N/A	120-150	150	200
Specific gravity	0.88	1.005	0.928	1.09	0.90
Paint compatibility	Excellent	Very Good	Good	Moderate	Moderate
Seal compatibility	BN,N,V,T	NY,BN,V,T,P,N	BN,N,V,T,S	BN,N,V,T,S,B,E	NY,BN,N,V,T,S,P
Pour Point, °C /°F	-27 (-17)	0 (32)	-37 (-35)	-45 (-50)	-26 (-15)
Flash Point, °C/ °F	222 (430)	—	—	—	306 (585)
Auto-ignition °C/°F	372 (700)	—	410 (770)	440 (825)	483 (900)

SEALS: BN—acrylonitrile-butadiene copolymer; N—polychloroprene; V—fluoroelastomer; T—PTFE; NY—nylon; P—polyurethane, B—butyl E—ethylene propylene copolymer; S—silicones

Table 10.2 provides a fluid property data comparison of the different fire fluid offerings versus a straight mineral oil.

Note: occasionally, lubricants are marketed in different regions of the world with brand designations that differ from each other. If a lubricant is no longer available under its previous, widely used designation, the reader is encouraged to: (a) ask the supplier for the current name of a particular lubricant, and (b) compare the current "inspections"—the collective name for properties and ingredients--to those found widely accepted and described in this text. The next step would be to investigate the significance of these deviations by consulting Chapter 7, Lubricant Properties.

11

Food Grade and Enviro-friendly/ Biodegradable Oils

The food/beverage and pharmaceutical industries manufacture products that are not compatible with most lubricants. Product safety is key at all times, recognized in both GMP (Good Manufacturing Practices) and HACCP (Hazard Analysis and Critical Control Points) manufacturing control programs. These programs are designed to assure the consumer public and regulatory bodies such as the USDA (United States Department of Agriculture), and the FDA (Food and Drug Administration) that a quality preventive maintenance program focused on product contamination control is in place.

Both programs recognize a three-tier classification system for food grade lubricants. This tier system classifies lubricants based on how they are used in a food or pharma plant environment. All lubricants must comply with food safety regulations and be registered as a NSF (National Sanitation Foundation) lubricant. NSF International is a non-profit international third-party verification body founded in 1944 that specializes in providing lubricant certifications to protect food, water, consumer products and the environment.

11.1 Food Grade Lubricants

Food grade lubricants are designed for use where food is processed. These specialty lubricants are primarily classified according to how they are to be used in the plant environment. Understanding the differences between the three grade classifications H1, H2 and H3 allows the end user to determine the correct lubricant for use in the different plant areas.

H1 classified lubricants are the only true food grade lubricants designed for use when there is potential for incidental food contact. H1 lubricant are tasteless, odorless, physiologically inert and according to NSF international, are deemed suitable for “incidental, technically unavoidable contact with a food-products up to 10 parts per million (0.001%). These lubricants are used for lubrication of equipment or applications in which a lubricant *may have*

Figure 11.1 Typical bottle filling line.

incidental (direct)contact with edible or pharma products. These can include conveyors, transfer machines, cutting machines, can seamers, mixers, bottling machines, and more. H1 lubricants must also meet the FDA 21 CFR 3570 regulation that outlines allowed ingredients for use in H1 classified lubricants.

H2 classified lubricants are general lubricants that can be used where there is no possibility of contact with food. These lubricants are considered food grade but are not specific food safe products. NSF H2 certified lubricants are strictly screened for toxins and cannot be deemed safe for use if they contain known carcinogens, mutagens, teratogens, mineral acids, or heavy metals. Typical applications would include fork lifts, HVAC equipment, etc.

H3 classified lubricants are generally edible oils used as additives in the formulation of food-grade products, and can be used to clean and prevent rust

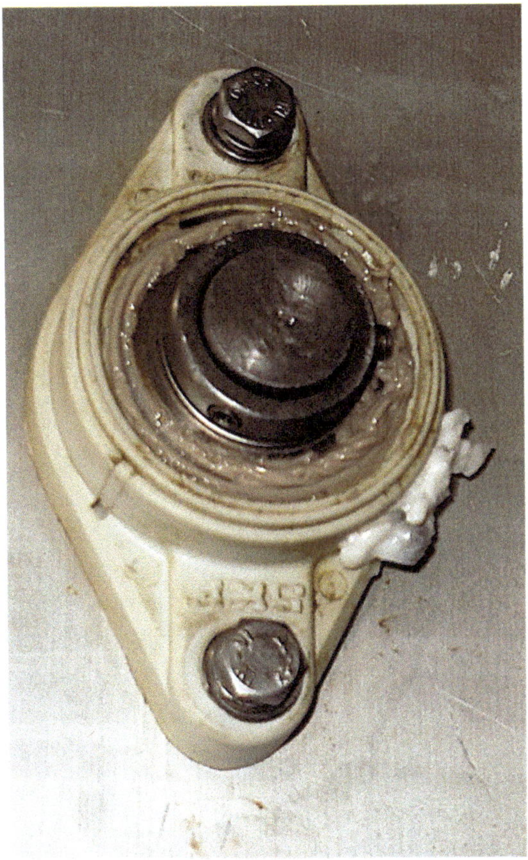

Figure 11.2 Sealed Flange Bearing (over) lubricated with H1 food grade grease (Courtesy ENGTECH Industries Inc.).

on hooks, trolleys, racks etc. Typical oils include all vegetable oils (corn, sunflower, soybean, cottonseed, etc.) that are biodegradable and meet the FDA 21 CFR 172.860 and 172.878 regulations.

Additional FDA title 21 classification regulations include:

- FDA 21 CFR 178 3620: use of white mineral oils as a component of non-food articles intended for use in contact with food
- FDA 21 CFR 178 182: substances generally recognized as safe

Figure 11.2 shows a bearing greased with a synthetic H1 lubricant. Note the typical neutral coloring and that the bearing is grossly over-lubricated by evidence of grease purged through the seal.

11.2 What Performance Features are Needed?

The challenge in formulating lubricants for the food processing industry is to meet the necessary FDA, USDA, and NSF food-grade requirements while also meeting the performance features needed to adequately protect food processing machinery. The required performance characteristics of a lubricant vary depending on the application, but key parameters often necessary for outstanding equipment protection are anti-wear, oxidation stability, extreme-pressure characteristics, and rust protection.

11.2.1 Anti-wear

Oils used in hydraulic systems are often subjected to high pressures and velocities. These forces can create conditions of thin-film lubrication and accelerated mechanical wear unless the fluid contains special protective additives. Each competent lubricant manufacturer has their own additives formulation. The anti-wear additives in these oils are selected to provide peak performance in the equipment they are designed to lubricate. In addition, all anti-wear additives selected for these products must meet the stringent requirements specified within the H-1 approval rating requirements.

11.2.2 Oxidation stability

Oxidation stability is a measure of an oil's ability to resist oxidation, i.e., chemical deterioration, in the presence of air, heat, and other influences. This is an important quality in a lubricant. Insoluble oil and sludge resulting from oil oxidation can interfere with the performance of moving parts. Varnish and sludge can plug lines, screens, and filters and prevent equipment from operating efficiently. In addition, removing these contaminants can be very expensive and time-consuming.

Oxidation accelerates with time and increasing temperature. The deterioration process begins slowly, but speeds up as the oil nears the end of its useful life. Equipment metallurgy can also affect oxidation. Catalytic metals, such as copper and iron, which are commonly used in food and beverage equipment, can also accelerate oxidation. The service life of an oil depends upon its ability to resist these influences.

The majority of H-1 rated lubricants have natural oxidation stability because they are formulated with extremely stable base-stocks. In addition, the oxidation stability of many of these oils and greases is further enhanced with carefully selected additives.

Figure 11.3 Typical commercial bakery bread cooling conveyor system (Courtesy ENGTECH Industries Inc.).

11.2.3 Extreme-pressure protection

Extreme-pressure (EP) protection is a measure of an oil's ability to protect metal surfaces under heavy loads when the oil film has been pushed away or squeezed out by the mechanical action of gears or bearings. EP additives react with the metal surface to prevent welding, scuffing, and abrasion. Such additives have to be carefully selected because they can act as pro-oxidants, thereby reducing the useful life of the oil.

11.2.4 Rust protection

It is often difficult to keep lubrication systems free of water, particularly in the food industry where many machines are constantly washed down to keep the surface free of dirt and contaminants. Even under the most favorable conditions, rust is a possibility… and a potential problem.

Rust can score mating surfaces, form scale in piping, plug passages and damage valves and bearings. Ram shafts are sometimes exposed directly to the elements, and any pitting of their highly polished surfaces is likely to rupture the packing around them.

Table 11.1 Typical characteristics of a food-grade can seamer lubricant (Source: Exxon Company).

Typical Inspections for an ISO 150 Can seamer lubricant	
Color, ASTM D 1500	0.5
Viscosity,	
cSt @ 40°C	146.8
cSt @ 100°C	14.3
Viscosity Index	97
Specific gravity @ 15.6°F (60°F)	0.879
Flash point, °C(°F)	200(392)
Rust test, ASTM D 665B	Pass

Table 11.1 identifies typical inspections that can be found in a H1 classification FG 150 cSt viscosity can seamer oil. A competent supplier formulates all of its USDA H1 food-grade lubricants with rust inhibitors to give extra protection against the destructive effects of water.

Table 11.2 identifies the typical inspections found in a FG1 premium industrial food-grade grease formulated with an aluminum-complex thickener and USP white oil base-stock. This FG 1 Exxon lubricating grease product provides excellent water resistance and outstanding pumpability. It is white in color, has a smooth-tacky appearance and contains an extreme-pressure additive for carrying heavy loads.

11.3 Environment Friendly Lubricants and Biodegradable Lubricants

In these days of heightened sustainability and eco awareness, the search for an "environmentally friendly lubricant", or EFL, has become a common quest. The search results however, can differ greatly depending on your understanding of the term and how the lubricant is to be used. There are many user interpretations of the term "environmentally friendly lubricant", the following five being the most common:

1. The lubricant is friendly or non-toxic to the natural environment when in use, and will quickly biodegrade without harm if spilled or disposed of (the most common interpretation used by oil companies)

Table 11.2 Typical Characteristics of a food-grade grease formulated with aluminum-complex thickener (Source: Exxon).

Typical Inspections for an NLGI 1 Food Grade Grease	
Penetration, ASTM D 217, worked 60X, mm/10	325
Thickener type	Aluminum-Complex
Color White	
Texture	Smooth, tacky
Dropping point, ASTM D 2265, °C(°F)	232(450)
Base oil viscosity	
SUS @ 100°F	800
SUS @ 210°F	75
cSt @ 40°C	60
cSt @ 100°C	14
Base oil viscosity index ASTM D 2270	90
Four-ball EP test, ASTM D 2596	
Weld point, kg	315
Load wear index, kg	39
Rust test, ASTM D 1743	Pass

2. When exhausted of its additive packages through extended use, the lubricant base stock is renewable and is therefore a sustainable resource based on reduced oil consumption
3. The lubricant is capable of extended drain intervals—long a claim of synthetic versus mineral based lubricants—and is therefore more friendly to the environment, again based on reduced oil consumption
4. Lubricants used to reduce energy and a user's carbon footprint. Due to their lubricity and performance under extreme operating condition many synthetic lubricants play an active role in efficiently reducing machine energy consumption whilst delivering a correlative CO^2 emission saving—excellent for the environment.
5. Specialty lubricants designed specifically to address issues related to specific "working" environments. For example, an automotive assembly paint line overhead conveyor chain lube designed with tactification

additives that ensure the lubricant that does not drip on to the newly painted automobile surface, or a food manufacturing / processing plant's requirement for specialty lubricants that will not cause harm to humans or significantly degrade its products if they come into contact with the product.

A strong environmental commitment is a basic obligation that any manufacturing operation has to its customers and the community. This interest is best served by expanded use of EFL lubricating products. Traditionally, when searching for an EFL, many will commence their search with lubricants used by the food and beverage or pharmaceutical industry in the belief that a food grade lubricant is an environmentally safe product. Beware; as this is not automatically the case!

Performance wise, EFL's formulated with vegetable oils can offer features such as high viscosity index and high flash points, and offer an elevated degree of friction and wear protection due to the natural lubricity of the vegetable-based oil. On the down side, these lubricants may not work well in adverse temperatures and may have a shorter shelf life than standard or synthetic lubricants.

True EFL's are designed to degrade quickly and naturally with non-toxic decomposed fractions and are generally based on renewable sources. If you are looking for a guaranteed non-toxic lubricant you are best to start your search with a USDA certified bio-based products. These lubricants are formulated with base products that originate from renewable biological sources such as vegetable oils or organoclay, and are readily biodegradable and free from heavy metals and other toxic ingredients.

11.3.1 Ambiguity of environmental claims for lubricants

The terms most often discussed with respect to environmentally friendly lubricants are "biodegradable" and "non-toxic." Both of these terms are ill-defined and severely situation dependent. There are many variables and little standardization in biodegradation testing. A given material may be found to be highly biodegradable under one set of test conditions and only moderately biodegradable under another. Thus, when a material is said to be "biodegradable," it is important to know the specific test circumstances. (For a detailed discussion, see the entry "Biodegradation and Ecotoxicity," Chapter 7.4)

The term "non-toxic" is similarly ambiguous. A material found to be non-toxic to one species may be toxic to another. Nevertheless, some lubricant manufacturers have claimed lubricants to be "non-toxic" on the basis of tests with only a single type of organism. The ambiguity of the term "non-toxic" is further exemplified by a past case in which a lubricant had been reformulated to eliminate an EPA-identified y component. However, the reformulated product, claimed to be less toxic, was found to contain an additive suspected of causing skin reactions in humans. This made the product OSHA hazardous.

Even assuming the validity of the environmental claims made for a product, there are important potential trade-offs that must realistically be considered. For example, the performance and useful life of a "green" lubricant may be significantly inferior to that of an alternative product, as discussed below.

11.3.2 Natural base oils

At first glance, the use of "natural" base oils, such as vegetable oils, as lubricants appears to be appropriate and prudent from an environmental perspective. Undoubtedly, the more rapid biodegradability of such oils versus petroleum oils is desirable in the event of accidental, routine, or excessive environmental exposure.

However, for most industrial lubricant applications, a number of additional factors must be considered. Foremost among these considerations is product performance. Vegetable oils have poor hydrolytic and oxidative stability; this may necessitate more frequent oil changes and result in significant disposal problems that may outweigh any environmental advantages. They also have relatively high pour points, which can impair low-temperature performance. Additives such as pour depressants and anti-oxidants may help compensate for these drawbacks, but they tend to reduce biodegradability and may increase the toxicity of the overall product to humans and the environment.

Additionally, vegetable oil-base lubricants are more susceptible to microbial action, which can both limit their storage life and rapidly degrade their performance in use.

As for the recyclability of "natural" oils, there is a practical problem here as well. Because these oils are not compatible with mineral oils, it may be difficult to find a recycler that will accept "natural" oils.

11.3.3 Comparing natural efl hydraulic fluids to modern synthetic based fluids

The original "environment friendly" hydraulic oil offerings were vegetable-based products that utilized base oils derived from such products as

Table 11.3 "Environment Friendly" hydraulic fluid comparison table.

Lubricant	Type	Base Oil	Advantages	Disadvantages
HETG Hydraulic Environmental Triglyceride	Vegetable	Soy-Bean, Canola, Sunflower	Excellent biodegradability Non toxic Excellent Lubricity High VI (>200) High Flash Point (>225°)	Thickens at high and low temperatures Rapid oxidization Hydrolyze in water Poor nitrile seal compatibility Miscible with mineral oil (lowers biodegradability)
HEES Hydraulic Environmental Ester Synthetic	Synthetic	Synthetic Esters	Excellent service life Excellent thermal stability Excellent oxidative stability Good low temp flow	Miscible with mineral oil (lowers biodegradability Hydrolyze in water Poor nitrile seal compatibility Expensive
HEPR Hydraulic Environmental Polyalphaolefin & Related	Synthetic	Polyalohaolefin (PAO)	Excellent service life Excellent oxidization stability Good corrosion protection Good lubricity Good VI Low pour point (-5 to - 40°F)	Only low PAOs are considered environment friendly Poor nitrile seal life Limited operating temperature range (-20 to 212°F Expensive
HEPG Hydraulic Environmental Poly Glycol	Synthetic	Polyalkylene Glycol (PAG)	Excellent fire resistance Good viscosity range available Can be used as an anhydrous (water-free) lubricant	Poor polyurethane seal compatibility Poor nitrile seal compatibility Limited operating temperature range (-5 to 180°F) Biodegradability dependent on propylene/ethylene ratio

soybean, canola and sunflower seeds classified as Hydraulic Environmental Triglyceride (HETG) fluids.

There are four generally accepted "environment friendly" hydraulic fluids available, one vegetable-based product and three synthetic based products. As with all synthetics each type offers different characteristics suitable for different operating environments giving the equipment designer and end user some flexibility in their lubricant of choice. Table 11.3 defines the advantages and disadvantages of the four fundamental types of available "environment friendly" lubricants.

As with all lubricants, EFL's require careful consideration and testing before placing them in service. Their seal and hose compatibility, and operating temperature will play are big part in their choice. As with all lubricants, when replacing an already existing lubricant the correct choice of flushing oil is crucial to operational success requiring consultation with the oil company prior to use.

Use of any EFL still requires regular system maintenance to ensure a leak or spill does not occur in the first instance, and if a major leak or spill does occur, most authorities still require the spill to be reported. The difference with using EFL's is how the spill or leak is managed.

11.3.4 Choosing the right EFL lubricant product to fit your needs

The development of environmentally friendly lubricants is an extremely worthwhile goal. However, consumers would be well-advised to ask the following questions before purchasing any product claimed to be "environmentally friendly" or more environmentally responsible:

- Are the environmental claims made for the product valid and well-documented?
- What performance debits or other trade-offs are associated with the product?
- Do these trade-offs outweigh the environmental advantages of using the product.

Making the correct lubricant choice requires the purchaser to provide the prospective lubricant vendor with some basic information on its intended use that should include:

- **WHAT** is the reason for your choice to move toward an environmentally friendly product? Are you looking to implement or comply with a legislative, corporate or department mandate/program?
- **WHERE** do you intend to use the product(s)? You will need to explain the type of equipment being lubricated and its manufactured product, e.g. is it a conveyor system for painted products; is it a bake oven drive system or is it a release agent for baking pans, etc.? At this time, you should also describe the working environment/conditions the lubricant must operate in, for example: extreme hot or cold (oven/freezer, indoor/outdoor), continual or occasional water presence (process water, wash down water, humidity), etc.;

  The lube supplier will also want to know if the intention is to apply the EFL product on a single machine, production line or plant wide and the current products you are intended to substitute
- **HOW** the lubricant is to be dispensed to the bearing surface or point will determine if it is to be oil or grease and if it can contaminate the product. You will need to answer if the lubricant will reside in a closed

reservoir? Will it be delivered through a total loss automated distribution system? Will it be misted? Will it be manually applied?

11.3.5 Practical tips for ensuring your EFL product stays that way in service

Purchasing the right environmentally friendly lubricant that meets your needs is only the first step of your journey. Your environmentally friendliness is only as good as your lubrication program. The following application tips will ensure your environmentally friendly products stay that way throughout their service life!

Tip #1: when changing from a less environmentally friendly product, work with your lube supplier to correctly purge all old (previous) product out of the lube system reservoir, lines, bearings, etc. This will usually require the use of an interim flushing or cleaning product

Tip #2: review your receiving practices to ensure your environmentally friendly lubricants are stored correctly and separate from other lubricants in your plant

Tip #3: use only new and dedicated transfer equipment for each new lubricant introduced into the plant to ensure cross contamination with other lubricants does not occur

Tip #4: Ensure all reservoirs and transfer equipment is product labeled with the correct lubricant type and viscosity to ensure the right lubricant is being used in the right place

Tip #5: DO NOT OVER LUBRICATE! Over lubricating not only overheats and reduces the life of both bearing and lubricant, it increases lubricant use and exposes the lubricant to the environment unnecessarily

Tip #6: Follow the manufacturer's recommendations for spillage and lubricant disposal.

Purchasing and using environmentally friendly lubricants is not only good for a healthy environment, is also provides a good reason to update and modernize your lubrication program and ensure machine health by placing the RIGHT lubricant, in the RIGHT place, at the RIGHT time, in the RIGHT amount.

12

Automotive Lubricants

When speaking with non-lubrication personnel about the subject of practical lubrication, the standard "go to" analogy explanation almost always involves a discussion in terms of automobiles and their use and need for lubricants. Most people have at some time in their life driven and maintained an automobile requiring a old fashioned "oil and lube job."

Worldwide, the transportation sector consumes more lubricants than any other. Although this book is about industrial lubrication, automotive lubricants are occasionally employed for both gasoline (petrol) and diesel fuel powered internal combustion engines in the operation of portable power tools, pumps, generators, and transport vehicles.

12.1 Engine Lubricating Oils

An automotive lubricating oil is similar to that of any industrial lubricant (refer to Chapter one and two), with the additional purpose of acting as a sealant to control and prevent the blow-by of combustion gases in the internal combustion chamber.

Because of this volatile, explosive operating environment, automotive lubricants are designed with additional detergent and dispersant additives to prevent destructive varnish buildup on metal surfaces. Detergents are utilized to capture and suspend moisture in the oil, while dispersants work alongside to capture and suspend varnish, soot and mechanical debris in the oil. These parasite materials are then captured in the lubrication system's oil filter. As additives deplete, the oil and filter will quickly saturate; if too dirty, the oil will thicken and transform into an acidic gel commonly referred to as "sludge" that will dramatically raise the oil's viscosity and internal friction. Because of this action, automotive oils are to be changed out on a much more regular basis.

12.1.1 Automotive oil viscosity grades

When choosing a viscosity grade to use in your automotive application, always chose based on the manufacturer's recommendation found in the operating manual. Because an engine is often used to power a portable tool or machine, the original equipment manufacturer (OEM) has no control over its use, yet must design it to provide useful life in all operating climates and condition.

If an engine oil is to properly protect all engine wear surfaces, at any operating temperature, the oil viscosity must be "thin" enough to be pump around an engine in sub-zero temperatures, yet "thick" enough to provide a protective lubricating film in the plus 200 operating temperature of an IC engine. Because of this changing temperature, mono-grade oils such as those used in early vehicle engine design, cannot provide the range of viscosity demanded by modern high-powered automotive engines that require the use of a multigrade viscosity oil to perform successfully in seasonal temperature conditions.

There still is a market for monograde engine oils with special anti-wear additives for older automobile, marine and aircraft piston engines, as shown in Figure 12.1.

Multi-grade viscosity oils like shown in Figure 12.2, are easily recognized by their double number designation separated by the letter W, e.g., SAE 0W20, SAE 10W30. This example typifies the automotive oil and packaging found in the late 1970-80's

The first number designation **10W** represents the oil thickness viscosity when the engine is cold. In the example shown in figure 12.2, the **30** is the viscosity value when the oil has reached engine operating temperature. Having a thinner oil in cold atmospheric temperatures allows the oil to be 1) pourable, and 2) thin enough to quickly move through the engine galleries and reach the valve train to provide vital start up protection.

With new vehicle engines operating with tighter engine tolerances and higher power output, it is essential to follow the manufacturer's recommendation regarding oil viscosity for warranty and engine performance. Often a synthetic oil with of lower viscosity such as 0W5 is required. In older cars, traditional 10W30 to 20W50 thicker oils are preferred as they compensate for engine wear and larger engine fit tolerances to create a better piston ring seal, help reduce oil blow-by, and reduce oil consumption.

12.1.2 Automotive gasoline and diesel engine lubricant standards

When dealing with automotive lubricants there are currently three governing bodies that set standards for automotive motor oils. These are known

Figure 12.1 Typical monograde oil.

Figure 12.2 Typical multigrade oil.

as the American Petroleum Institute (**API**), the International Lubricant Standardization Committee (**ILSAC**), and the Association des Constructuers Europeens d'Automobiles (**ACEA**).

The API categorization, arguably the most common standard used to categorize an automotive oil service category, classifies oils based on their additive formulation and ability to provide protection based on the latest gasoline and diesel engine power and emission requirements.

Automotive oils are formulated specifically for either spark ignition engine (**S** designation) used in gasoline fueled engines, or Compression ignition engines (**C** designation) found in diesel fueled engines. The API uses a donut icon (see Figure 12.3) that is prominently displayed on oil packaging to identify that the oil has been formulated to meet the performance requirements set by US and International vehicle and engine manufacturers and the lubricant industry. Table 12.1 identifies the current and past S designation categories for gasoline engine oil, while Table 12.2 identifies the current and past C designation requirements for diesel engine oils.

Figure 12.3 API designation donut icon for gasoline engine vehicle SN designation.

Table 12.1 API gasoline engine oil service categories (Information Source: API).

Category	Status	Service Description
SP	Current	Current Introduced in May 2020, designed to provide protection against low-speed pre-ignition (LSPI), timing chain wear protection, improved high temperature deposit protection for pistons and turbochargers, and more stringent sludge and varnish control. API SP with Resource Conserving matches ILSAC GF-6A by combining API SP performance with improved fuel economy, emission control system protection and protection of engines operating on ethanol- containing fuels up to E85.
SN	Current	Current For 2020 and older automotive engines.
SM	Current	Current For 2010 and older automotive engines
SL	Current	Current For 2004 and older automotive engines
SJ	Current	Current For 2001 and older automotive engines
SH	Obsolete	CAUTION: Not suitable for use in most gasoline-powered automotive engines built after 1996. May not provide adequate protection against build-up of engine sludge, oxidation, or wear
SG	Obsolete	CAUTION: Not suitable for use in most gasoline-powered automotive engines built after 1993. May not provide adequate protection against build-up of engine sludge, oxidation, or wear
SF	Obsolete	CAUTION: Not suitable for use in most gasoline-powered automotive engines built after 1988. May not provide adequate protection against build-up of engine sludge
SE	Obsolete	CAUTION: Not suitable for use in most gasoline-powered automotive engines built after 1979
SD	Obsolete	CAUTION: Not suitable for use in most gasoline-powered automotive engines built after 1971. Use in more modern engines may cause unsatisfactory performance or equipment harm
SC	Obsolete	CAUTION: Not suitable for use in most gasoline-powered automotive engines built after 1967. Use in more modern engines may cause unsatisfactory performance or equipment harm
SB	Obsolete	CAUTION: Not suitable for use in most gasoline-powered automotive engines built after 1951.
SA	Obsolete	CAUTION: Contains no additives. Not suitable for use in most gasoline-powered automotive engines built after 1930. Use in more modern engines may cause unsatisfactory performance or equipment harm

Table 12.2 API diesel engine oil service categories (Information Source: API).

Category	Status	Service Description
CK-4	Current	API Service Category CK-4 describes oils for use in high-speed four-stroke cycle diesel engines designed to meet 2017 model year on-highway and Tier 4 non-road exhaust emission standards as well as for previous model year diesel engines. These oils are formulated for use in all applications with diesel fuels ranging in sulfur content up to 500 ppm (0.05% by weight). However, the use of these oils with greater than 15 ppm (0.0015% by weight) sulfur fuel may impact exhaust aftertreatment system durability and/or oil drain interval. These oils are especially effective at sustaining emission control system durability where particulate filters and other advanced aftertreatment systems are used. API CK-4 oils are designed to provide enhanced protection against oil oxidation, viscosity loss due to shear, and oil aeration as well as protection against catalyst poisoning, particulate filter blocking, engine wear, piston deposits, degradation of low- and high-temperature properties, and soot-related viscosity increase. API CK-4 oils exceed the performance criteria of API CJ-4, CI-4 with CI-4 PLUS, CI-4, and CH-4 and can effectively lubricate engines calling for those API Service Categories. When using CK-4 oil with higher than 15 ppm sulfur fuel, consult the engine manufacturer for service interval recommendations..
CJ-4	Current	For high-speed four-stroke cycle diesel engines designed to meet 2010 model year on-highway and Tier 4 non-road exhaust emission standards as well as for previous model year diesel engines. These oils are formulated for use in all applications with diesel fuels ranging in sulfur content up to 500 ppm (0.05% by weight). However, the use of these oils with greater than 15 ppm (0.0015% by weight) sulfur fuel may impact exhaust aftertreatment system durability and/or drain interval. API CJ-4 oils exceed the performance criteria of API CI-4 with CI-4 PLUS, CI-4, CH-4, CG-4 and CF-4 and can effectively lubricate engines calling for those API Service Categories. When using CJ-4 oil with higher than 15 ppm sulfur fuel, consult the engine manufacturer for service interval.
CI-4	Current	Introduced in 2002. For high-speed, four-stroke engines designed to meet 2004 exhaust emission standards implemented in 2002. CI-4 oils are formulated to sustain engine durability where exhaust gas recirculation (EGR) is used and are intended for use with diesel fuels ranging in sulfur content up to 0.5% weight. Can be used in place of CD, CE, CF-4, CG-4, and CH-4 oils. Some CI-4 oils may also qualify for the CI-4 PLUS designation.
CH-4	Current	Introduced in 1998. For high-speed, four-stroke engines designed to meet 1998 exhaust emission standards. CH-4 oils are specifically compounded for use with diesel fuels ranging in sulfur content up to 0.5% weight. Can be used in place of CD, CE, CF-4, and CG-4 oils.
CG-4	Obsolete	**CAUTION**: Not suitable for use in most diesel-powered automotive engines built after 2009.
CF-4	Obsolete	**CAUTION**: Not suitable for use in most diesel-powered automotive engines built after 2009.
CF-2	Obsolete	**CAUTION**: Not suitable for use in most diesel-powered automotive engines built after 2009. Two-stroke cycle engines may have different lubrication requirements than four-stroke engines, so the manufacturer should be contacted for current lubrication recommendations.

Table 12.2 *(Continued)*

Category	Status	Service Description
CF	Obsolete	**CAUTION**: Not suitable for use in most diesel-powered automotive engines built after 2009. Later "C" category oils are usually suitable or preferred for diesel automotive engines for which "CF" oils were specified. Older equipment and/or two-stroke diesel engines, especially those calling for monograde products, may however require "CF" category oil.
CE	Obsolete	**CAUTION**: Not suitable for use in most diesel-powered automotive engines built after 1994.
CD II	Obsolete	**CAUTION**: Not suitable for use in most diesel-powered automotive engines built after 1994.
CD	Obsolete	**CAUTION**: Not suitable for use in most diesel-powered automotive engines built after 1994.
CC	Obsolete	**CAUTION**: Not suitable for use in most diesel-powered engines built after 1990.
CB	Obsolete	**CAUTION**: Not suitable for use in most diesel-powInformation Source ered engines built after 1961
CA	Obsolete	**CAUTION**: Not suitable for use in most diesel-powered engines built after 1959.

The ILSAC standard is a consortium standard developed by Chrysler, GM, Ford, and the Japan Automobile Manufacturers Association in 2010. The standard was developed specifically for high-performance automobile engines that operate at temperatures greater than 100°C and for vehicles running on E85 ethanol gasolines viscosity grades are limited to 0W-16, as shown in Table 12.3.

ACEA standards are predominantly used by European automobile manufacturers. These standards are divided into four classes: A represents gasoline engine oils, B represents Diesel engine oils, C is reserved for Catalyst compatible oils, and E representing Heavy duty diesel oils.

Unlike API, the ACEA category designations represent specific performance requirements for the oil(s) and later designations do not necessarily supersede earlier designations. Therefore, designations will usually have a revision date in their designation title, e.g., ACEA A3/B3-16. ACEA rated oils are designed to be stable, stay-in-grade oils intended for use at extended drain intervals in gasoline and diesel engines, and are specifically designed to be low-friction, low-viscosity oils with a high-temperature/high-shear (HTHS) rate viscosity for increased fuel mileage.

Table 12.3 ILSAC engine oil standard (Information Source – API).

Category	Status	Service Description
GF-6A	Current	Introduced in May 2020, designed to provide protection against low-speed pre-ignition (LSPI), timing chain wear protection, improved high temperature deposit protection for pistons and turbochargers, more stringent sludge and varnish control, improved fuel economy, enhanced emission control system protection and protection of engines operating on ethanol-containing fuels up to E85.
GF-6B	Current	Applies only to oils having an SAE viscosity grade of 0W-16. Introduced in May 2020, designed to provide protection against low-speed pre-ignition (LSPI), timing chain wear protection, high temperature deposit protection for pistons and turbochargers, stringent sludge and varnish control, improved fuel economy, emission control system protection and protection of engines operating on ethanol-containing fuels up to E85.
GF-5	Obsolete	Use GF-6A where GF-5 is recommended
GF-4	Obsolete	Use GF-5 where GF-4 is recommended
GF-3	Obsolete	Use GF-5 where GF-3 is recommended
GF-2	Obsolete	Use GF-5 where GF-2 is recommended.
GF-1	Obsolete	Use GF-5 where GF-1 is recommended

13

Industrial Gear Lubricants

Gears are one of man's oldest mechanical devices. In the public mind, gears are one of the most well recognized kinds of machinery. They create the impression of positive action, coordinated-interlocked-precise application of effort to secure a desired result. The primary early uses of gears were for navigation, timekeeping, grinding, etc. The automobile transmission is probably the most common use of gearing recognized by the everyday citizen.

13.1 Lubricant Selection for Closed Gears*

The factors affecting lubricant selection for the various arrangements are shown below and should be reviewed and evaluated to determine the required properties

Viscosity is probably the single most important factor in lubricant selection and relates to load, speed, and temperature.

Table 13.2 relates to the AGMA 250.04 standard and gives starting recommendations for gearbox oil viscosity ranges based on gear center distance. (Note: use recommendations with caution as they are only approximations.)

However, required viscosities can also be calculated from two empirical expressions:

1Vg = 420 (K/V)0.43

Where:

Vg = Viscosity, Centistokes

K = Operating '"K" Focor

$$k = \frac{W}{dF}\pi\left(\frac{m_G - 1}{m_G}\right)$$

* James R. Partridge, Lufkin Industries, Lufkin, Texas; also Exxon Company, USA, Houston, Texas.

Figure 13.1 Single-helical, high-speed gear unit being assembled. (Source: Maag Gear Company, Zurich, Switzerland).

W = Tangential Load
d = Pitch Diameter of Pinion
F = Effective Face Width
m_G = Ratio
V = Pitch Line Velocity, ft./min.

and 2:

$$SSU = \frac{V_g + \left(V_g^2 + 158.4\right)^{.5}}{0.44}.$$

This formula was first published by Shell Oil, and all values were converted to SSU.

It should be noted that by this formula the high-speed gears (above 5000 FPM) require a heavier oil than the 150 SSU @100°F usually used, but compromises are made for bearings and sometimes seals.

Normally, as a general rule for high-speed gearing, the minimum viscosity at supply temperature should be 100 SSU.

Table 13.1 Typical factors that affect lubricant selection.

Factor	Lubricant Property
Load	Viscosity P Additives
Speed	Viscosity EP Additives
Temperature (Operating and Ambient)	Viscosity Viscosity Index Fluidity Oxidation Stability
Contamination	Demulsibility Corrosion Protection Oxidation Stability
Life	Oxidation Stability Additive Depletion
Compatibility	Synthetic (Paint and Seals) EP Additives

13.2 Lubricant Types

13.2.1 Mineral oils

Although synthesized hydrocarbons (diesters and PAOs) are rapidly gaining more wide-spread industry acceptance, mineral oils are still the most commonly used type of gear lubricant. Containing rust and oxidation inhibitors, these oils are less expensive, readily available, and have very long life. When gear units operate at high enough speed or low enough load intensity, the mineral oil is probably the best selection.

Table 13.2 Equivalent viscosities of other systems (for reference only).

AGMA LUBRICANT NO.	EQUIVALENT ASTM-ASLE GRADE NO.	EQUIVALENT ISO GRADE NO.	METRIC EQUIVALENT VISCOSITY RANGES cSt AT 37.8°C (100°F)
1	S215	46	41.4 to 50.6
2, 2 EP	S315	68	61.2 to 74.8
3, 3 EP	S465	100	90 to 110
4, 4 EP	S700	150	135 to 156
5, 5 EP	S1000	220	198 to 242
6, 6 EP	S1500	320	288 to 352
7 comp., 7 EP	S2150	460	414 to 506
8 comp., 8 EP	S3150	680	612 to 748
8A comp.	S4650	1000	900 to 1100

13.2.2 Extreme pressure additives

Extreme pressure additives of the lead-napthenate or sulphur-phosphorus type are recommended for gear drives when a higher load capacity lubricant is required. As a general rule, this type of oil should be used in low speed, highly loaded drives, with a medium operating temperature.

It should be remembered that EP oils are more expensive and must be replaced more often than straight mineral oils. Some of these oils have a very short life above 160°F temperature.

A good gear EP oil would have a Timken OK load above 60 pounds and pass a minimum of 11 stages of the so-called FZG test.

Boron compounds, as EP additives, are more recent on the market and have the added ability to work with extremely high loaded gearing. The compounds tested show Timken OK loads greater than 100 pounds, and 14 stages of the FZG test. This additive is **nontoxic, highly stable, but sensitive to water.**

13.2.3 Synthetic lubricants

Not to be confused with the highly desirable synthesized hydrocarbons (typically diesters and PAOs), true "synthetic" lubricants (Type V) are not recommended for general gear applications due to their cost and availability. Synthetics are excellent when used in extreme applications of high, or low temperature and for added fire protection. However, the user must be careful when selecting these lubricants since some of them remove paint and attack rubber seals.

The more recent synthesized hydrocarbons (SHC) have many desirable features such as compatibility with mineral oils and excellent high and low temperature properties. They should be an excellent selection when EP lubricants are required along with high temperature operation. Refer to Chapter 15 for further details.

13.2.4 Compounded oils

Compounded oils are available with many different additives. The most commonly available is a molybdenum-disulfide compound that has been successfully used in some gear applications. It is difficult for a gear manufacturer to recommend these oils since some of these additives have a tendency to separate from the base stock.

13.2.5 Viscosity improvers

Viscosity improvers in gear-drives should be used with great care. These polymer additives do great textbook things for the viscosity index and extend the operating temperature range of an oil. What must be remembered is that polymers are non-Newtonian fluids, and the viscosity reduces with shearing. A gear drive is a very heavy shear application; and as a result, the viscosity reduces rapidly if too much polymer is used.

Lubricants in gear units have basically two functions: (1) to separate the tooth and bearing surfaces and (2) cooling. On low speed gear units, the primary function is lubrication; on high speed units, the primary function is cooling. This does not mean that both are not important but relates to the relative quantity of oil.

On low speed units, the amount of oil is determined by what is required to keep the surfaces wetted. On high speed units, quantity is generally determined by heat loss (or inefficiency). As a general rule, one GPM must be circulated for each 100 HP transmitted which results in a temperature rise of approximately 25°F. Higher HP units use a 40°F to 50°F temperature rise and require .5 to .6 GPH per 100 HP transmitted. This is based on a 98% efficiency.

13.2.6 Lubrication of high-speed units

The oil furnished to high-speed gears has a dual purpose: Lubrication of the teeth and bearings; and cooling. Usually, operating where only 10% to 30% of the oil is for lubrication, and 70% to 90% is for cooling.

A turbine type oil with rust and oxidation inhibitors is preferred. This oil must be kept clean (filtered to 40 microns maximum, or preferably 25 microns), cooled, and must have the correct viscosity. Synthetic oils should not be used without the manufacturer's approval.

For some reason, the high-speed gear makes all the compromises when oil viscosity for a combined lube oil system is determined. Usually, a viscosity preferred for compressor seals or bearings is selected and gear life is probably reduced. The bearings in a gear unit can use the lightest oils available, but gear teeth would like a much heavier oil to increase the film thickness between the teeth.

When selecting a high-speed gear unit, the possibility of using an AGMA No. 2 Oil (315 SSU @ 100°F) should be considered. In most cases, the sleeve bearings in the system can use this oil and, if not, a compromise 200 SSU at 100°F oil should then be considered.

When 150 SSU at 100°F oil is necessary, inlet temperatures should be limited to 110°F to 120°F to maintain an acceptable viscosity. Oil should be supplied in the temperature and pressure range specified by the manufacturer.

Up to a pitch line speed of approximately 15,000 feet per minute, the oil should be sprayed into the out mesh of the mating gears. This allows maximum cooling time for the gear blanks and applies the oil at the highest temperature area of the gears. In addition, a negative pressure is formed when the teeth come out of mesh pulling the oil into the tooth spaces.

Above approximately 15,000 feet per minute, 90% of the oil should be sprayed into the out-mesh and 10% into the in-mesh. This is a safety precaution to assure the amount of oil required for lubrication is available at the mesh.

Furthermore, in the speed ranges from 25,000 to 40,000 feet per minute, oil should be sprayed on the sides and gap area (on double helical) of the gears to minimize thermal distortion.

13.3 Worm Gear Lubricants

As compared to other gears (e.g., bevel gears), the efficiency of worm gears is relatively poor. However, they make high transmission ratios possible in a single step.

Suitable synthetic lubricants, can reduce the friction and power loss in worm gears by up to 30 %. The operational wear of the worm wheels, which usually consist of copper bronze, can be reduced substantially with synthetic lubricants containing suitable additives.

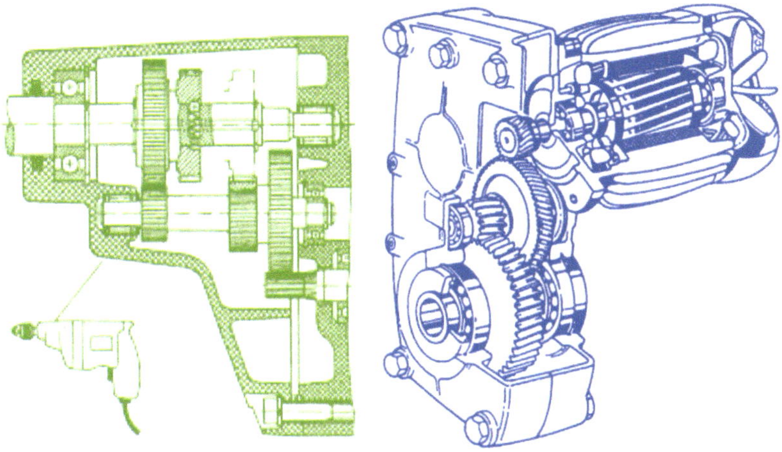

Figure 13.2 Manual drill, double-stage gear and a Double stage spur gear.

Properly formulated special synthetic gear lubricants will not only improve gear efficiency; their anti-wear additives can optimize the gear 's wear behavior.

The synthetic lubricants' excellent resistance to aging allows extended lubricant change intervals. These can be three to five times longer than intervals recommended with mineral hydrocarbon oils, which means lifetime lubrication in many cases.

13.4 Small Gear Lubricants

Small gears comprise spur, bevel and worm gears of an open, semi-closed or closed design (often not oil-tight). Small and miniature gears are used in adjusting and control drives in the automotive industry, office machines, household appliances and machines for do-it-yourselfers, (see Figure 13.2). Their main task is to transfer movements and power,

Due to the construction of these gears and the materials that are used—steel/steel, steel/bronze, steel/ plastic and plastic/plastic components—the lubricants have to meet various requirements, including

- lifetime lubrication
- noise damping
- low starting torque
- low and high temperature operation

- resistance to ambient media
- compatibility with the materials used

As small gears often are not oil-tight they are mainly lubricated with greases applied by dip-feed lubrication or one-time lubrication of the tooth flanks.

Dip-feed lubrication is preferred for gears in continuous operation or gears used for power transmission. One-time lubrication is suitable for gears used for the transmission of movements or gears only operating for short intervals or intermittently.

Lubricating greases of NLGI grade 000 to 0 are used for dip-feed lubrication, and grade 0-2 for life-time lubrication. To avoid the lubricant being thrown off the gear, pastier greases are preferred in case of increased peripheral speeds. In case of dip-feed lubrication at higher peripheral speeds, however, the grease should be softer to avoid channeling.

13.5 Testing the Performance of Gear Oils

In order to perform under pressure for extensive time periods there are three major characteristics a superior EP gear oil should exhibit, 1) extreme pressure capability, 2) cleanliness, and 3) demulsibility. A gear oil must also keep machine systems (and reservoirs) clean.

Gear oils undergo extensive testing for EP retention, oxidation, panel coking, copper strip corrosion and demulsibility. The tests measure EP capability, the degree of cleanliness and the demulsibility lubricants exhibit.

13.5.1 Performance under pressure

Superior EP lubricants are long lived. They hold up well under pressure and over time and as a result do not have to be replaced as often. This saves money and extends gear unit life.

The reserve EP capability test measures a lubricant's load-carrying ability over time. EP—extreme pressure—oils are specially formulated to lubricate under heavy-load conditions. The longer a lubricant can maintain its load-carrying ability, the less often it must be replaced...which once again, reduces operating costs.

13.6 Gear Coupling Lubrication

When users elect to use grid or gear couplings instead of non-lubricated coupling types, they must be aware of their vulnerability. The gear coupling,

Grid Type

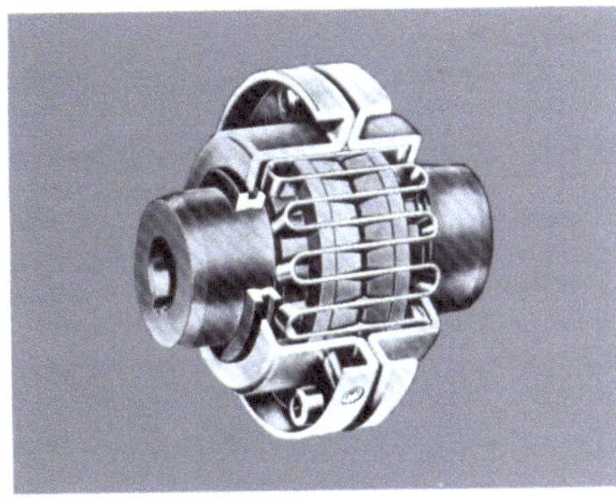

Gear Type

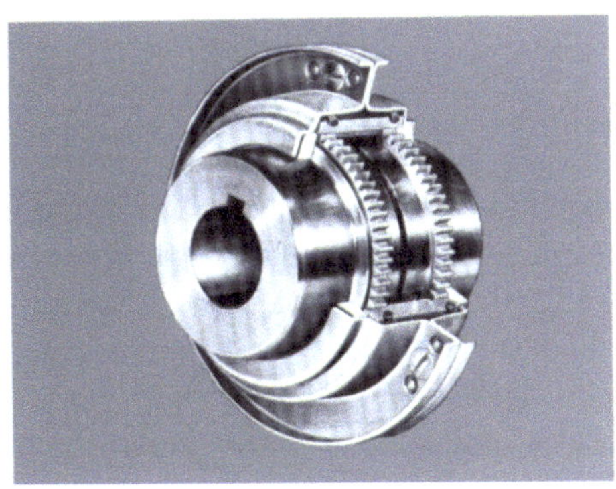

Figure 13.3 Typical couplings requiring grease lubrication..

Figures 12.16 and 12.17, is one of the most critical components in a turbomachine and requires special consideration in regards to lubrication.

There are two standard methods of gear coupling lubrication: batch and continuous flow. In the batch method the coupling is either filled with grease or oil; whereas the continuous flow type uses only oil, generally a light oil from the circulating oil system.

The grease-filled coupling requires special quality grease. The importance of selecting the best quality grease cannot be overemphasized. A good coupling grease must prevent wear of the mating teeth in a sliding load environment and resist separation at high speeds. It is not uncommon for centrifugal forces on the grease in the coupling to exceed 8,000Gs.

Testing of many greases in high-speed laboratory centrifuges has proved a decided difference exists between high quality grease, and inferior quality

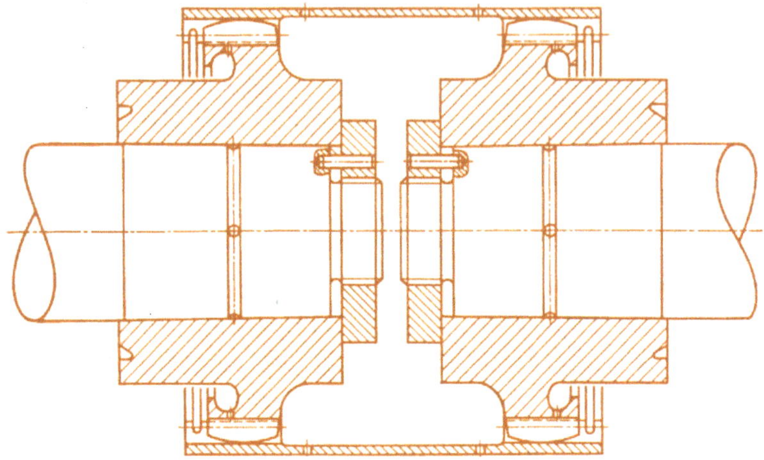

Figure 13.4 Section through a tooth-type coupling. (Source: ASEA Brown-Boveri, Baden, Switzerland)

grease used for coupling service. Testing has also shown separation of oil and soap to be a function of G level and time. In other words, oil separation can occur at a lower centrifugal force if given enough time. The characteristics of grease that allow it to resist separation are high viscosity oil, low soap content, with the soap thickener and base oil as near the same density as possible.

In the late 1970s, a number of greases were tested for separation characteristics in a Sharples high speed centrifuge and for wear resistance on a Shell 4 Ball Extreme Pressure Tester. (see figure 7.2). It was found that the high-quality grease B exceeded all other greases tested in separation characteristics. Zero separation was recorded at all speeds up to and including 60,000 Gs. The remaining three greases A, C and D were rated much lower in separation characteristics at all speeds tested.

Table 13.3 illustrates how these four greases performed on the Shell 4 Ball Extreme Pressure Tester in comparison with a typical Extreme Pressure gear oil. Based on these data, Greases A and B should provide excellent wear protection in severely loaded service. However, only Grease B passes the oil separation test (Table 13.4) and would qualify for long-term service at a modern facility.

13.7 High Speed Coupling Grease

utilizes a state-of-the-art calcium sulfonate thickener system. This thickener system has several unique performance advantages over conventional

Table 13.3 Shell 4 ball test—one minute wear load performance of four coupling greases.

Grease	Kilogram Load Passed
Grease "A"	80
Grease "B"	90
Grease "C "	50
Grease "D"	20
Typical E.P. 1 40 Oil	90

Table 13.4 Oil separation (%) observed on four coupling greases.

G Level	60,000	32,000	9,700
GREASE	2.5hr	6hr	6hr
GREASE "A"	53	35	
GREASE "B"	0	0	0
GREASE "C "	79	75	40
GREASE "D"	51	—	—

lithium-polymer thickener systems used in many competitive coupling greases. Some of the performance benefits of the calcium sulfonate thickener are:

- Excellent corrosion prevention, even in the presence of salt mist
- Inherent EP and anti-wear protection
- High dropping points
- Superior oxidation resistance
- Excellent shear stability

13.7.1 Grease filled gearboxes

Small, slow running /compact gearboxes, especially those with porous cast casings or tired /damaged seals can be filled and lubricated with grease.

A fluid or semi fluid, NLGI EP grade 000 to 0 grease is preferred in such instances. The grease has an advantage in that it is thixotropic (softens when worked and stiffens when at rest) and is far less likely to leak when compared to filling with oil. In such applications, the recommended oil viscosity should be between ISO VG 320 to 460

References

[1] Bloch, Heinz P.; "Improving Machinery Reliability," (1998), Third Edition, Gulf Publishing Company, Houston, TX, 77520 (ISBN 0-88415-661-3)

[2] Bloch, Heinz P.; "Consider Dual Magnetic Hermetic Sealing Devices for Equipment In Modern Refineries," (Pumps & Systems, September 2004)

[3] Bloch, Heinz P.; "Counting Interventions Instead of MTBF," (Hydrocarbon Processing, October 2007)

14

Metal Working Fluids

Metal working lubricants are often referred to as "Cutting Fluids" due to their use in the metal cutting industry. These lubricants see most use in metal removal machine tools; these typically include metal lathes, milling machines, broaches, shaping machines, surface grinders, and metal cut off saws.

The lubricant's primary purpose is to lubricate the metal surface, dissipate heat buildup produced by the metal-to-metal friction, and wash away metal cutting swarf from the work area. Metal working lubricants are for the most part applied using a continuous circulating delivery system.

The metal working fluid is sometimes mistakenly referred to as just "coolant." Depending on the metals being cut, metal working fluids can be straight undiluted mineral or synthetic oils, or a soluble emulsion of mineral/synthetic oil and water. These miscible fluids are the preferred lubricant/fluid for most applications.

In virtually all metal working scenarios the metal working fluid is both lubricant and coolant; in the case of precision grinding, a non-metal grindstone is used for metal removal. In this case, use of a mineral oil lubricant will likely result in the grindstone becoming clogged. As the work still needs to be cooled, a synthetic based water miscible coolant is applied to the work.

The oil type and additive package will vary greatly depending on the metal being processed.

15

Synthetic Lubricants

Most of today's industrial oils use either a mineral or synthetic base oil. These base oils are categorized into five groups according to their refining or manufacturing process. Groups I, II, and III (see table 15.1 - Base Oil Groups) represent conventional mineral based lubricants while groups IV and V are reserved for man made synthetic base oils.

- Group I base oils are known as "conventional" base oils made from solvent refined crude stock and have a Viscosity Index of between 80 to 120. Their sulfur content is above 0.03% and their saturated hydrocarbon levels are less than 90%.
- Group II base oils are refined using a hydro processing method known as "Hydrotreating" designed to add hydrogen to the base oil at temperatures above 600F. This is performed using a catalyst and by applying moderate pressure over 500psi to convert the base stock and reduce its sulfur content to less than 0.03% and increase its hydrocarbon saturation to levels of 90% and above.
- Group III base oils are known as "bright stock" and are primarily manufactured using a severe hydro processing conversion method known as "hydrocracking" that employs a catalyst at a temperature above 650F combined with pressure exceeding 1000psi to take out undesirable elements such as sulfur and nitrogen and replace them with hydrogen to attain a more stable base oil with a VI - viscosity index above 120 and a low pour point. In addition, remaining wax compounds are often removed to reduce the pour point even further. Due to this more complex refining process, Group III base oils perform in a similar manner to Group IV pure synthetic base oils and in most countries around the world, including North America, are allowed to be classified as a synthetic lubricant, even though it is hydrocarbon based.

Table 15.1 Base Oil Groups.

Group	VI	% Saturates	% Sulphur	Description
I	80-120	<90	>0.03	Solvent Processed
II	80-120	>90	<0.03	Hydro processed
III	>120	>90	<0.03	Severe Hydro processed
IV	>140		-	PAO – Polyalphaolefin - Synthetic
V	>150		-	All other base oils not in I-IV and all other Synthetics

- Group IV base oils are reserved for Polyalphaolephin – (PAO for short) synthetically manufactured base oils made up of very small, synthesized hydrocarbon molecules.
- Group V base oils represent all other synthetic base oil configurations

15.1 Synthetic Base Oils

Synthetic lubricants can trace their inception to the early jet engine, developed to cope with the extreme temperatures encountered when operating a jet aircraft. Using a polymerization process similar to that used in the plastic manufacturing industry, synthetic base oils are designed with specific and consistent molecular structures that result in highly stable base oils with a very high Viscosity Index (VI) rating

Synthetic lubricants offer many advantages over mineral based oils, the largest being their ability to operate reliably in both extremes of heat and cold at temperature ranges much wider than mineral oils. In addition to increased VI levels and improved thermal stability, synthetics also demonstrate improved oxidation stability (major reduction of sludge and acid buildup) and lower volatility resulting in extended lubricant life and reduced oil consumption.

The disadvantages of synthetics are primarily their cost, which, depending on the synthetic lubricant type can be as low as two times the cost of mineral base oil to exponentially more! Certain synthetics can also cause seal swelling, and many are incompatible with any other base oil type.

Judicious application of properly formulated synthetic lubricants can benefit a wide spectrum of process machinery. This informed usage will drive down overall maintenance and downtime expenditures and can markedly improve plant profitability.

However, although synthetic lubricants have gained considerable acceptance in many forward-looking process plants worldwide, there are still misconceptions that impede even wider acceptance many of these fluids so richly deserve. One of the most erroneous misunderstandings is that for a synthetic lubricant costing $60.00 per gallon to be justified, the drainage or replacement interval should be five times that of a mineral oil costing $12.00 per gallon. This reasoning does not take into account such savings as labor, energy, downtime avoidance, disposal of spent lubricants and equipment life extension.

15.2 Synthetic Formulations

The most knowledgeable formulators use a polyalphaolefin/diester blend. Additives are more readily soluble in diesters than in a straight PAO, making PAO/diester blends stable over a very wide temperature range. These superior synthetic base oils can then be blended with additives to obtain the highest level of performance required.

It should be emphasized that additives represent by far the most important ingredients of properly formulated, high performance synthetic lubricants. Often, additives used in synthetic oil formulations are the same conventional additives used to formulate mineral oils, which result in only marginal performance improvements. Truly significant performance improvements are obtained when superior synthetic base oil stocks are blended with superior synthetic additive technology.

The various proven PAO/diester blends contain synergistic additive systems identified with proprietary trade names, e.g. Synerlec. The synergism obtained in a competent additive blend combines all of the desirable performance properties, plus the ability to ionically bond to bearing metals to reduce the coefficient of friction, and greatly increase the oil film strength. The resulting tough, tenacious, slippery synthetic film results in longer lasting equipment that runs cooler, quieter, smoother and more efficiently. Synergistic additive systems, when in service will "micro-polish" bearing surfaces, thereby reducing bearing vibration, reducing friction, and minimizing energy consumption.

The most valuable synthetic lubricant types excel in high film strength and oxidation stability. However, while there are many high film strength oils on the market, these may not be appropriate for some process machine applications. High film strength oils based on extreme pressure (EP) technology intended for gear lubrication may typically incorporate additives such as sulfur, phosphorus and chlorine which are corrosive at high temperatures and/or in moist environments. Sensitive to this fact, a reputable lubricant manufacturer would not offer an EP industrial oil with corrosive additives as a bearing lubricant for pumps, air compressors, steam turbines, high speed gear reducers and similar machinery.

At least one U.S. manufacturer of synthetic lubricants can lay claim to having pioneered the development of non-corrosive high film strength industrial oils with outstanding water separation properties. Although such oils may not be critically important to the operating success of vast numbers of pumps, air compressors and turbines, which quite obviously have been running without high film strength oils for years, there still remain compelling reasons to look into the merits of superior lubricants. There is an exhaustive body of evaluated evidence that supports the fact that *properly formulated* synthetic lubricants based on diesters, PAO, or a combination of these base stocks will result in significantly reduced bearing and gear operating temperatures.

15.3 Origin of Synthetic Lubes

Synthetic-based fluids, used in the production of synthetic lubricants, are manufactured from specific chemical compounds that are usually petroleum derived. The base fluids are made by chemically combining (synthesizing) various low molecular weight compounds to obtain a product with the desired properties. Thus, unlike petroleum oils which are complex mixtures of naturally occurring hydrocarbons, synthetic base fluids are man-made and have a controlled molecular structure with predictable properties as depicted in figure 15.1.

Manufactured from chemically modified petroleum constituents, or from a number of chemical bases and compounds, there are five common types of Synthetic base oils:

1. **Poly-Alpha-Olefin (PAO)** – are often described as a man made mineral oil formulated through the synthesis of ethylene gas molecules into a polymerized uniform structure similar to pure paraffin. With a pour point down to – 90F, a VI above 140, good seal and mineral oil compatibility, Class IV PAO oils are widely used for automotive crankcase oil, industrial gear oils, compressor oils and turbine oils. Negative features include poor additive solubility and biodegradability. PAO's are fully oil soluble but not water soluble, are non-polar and recommended for rolling wear

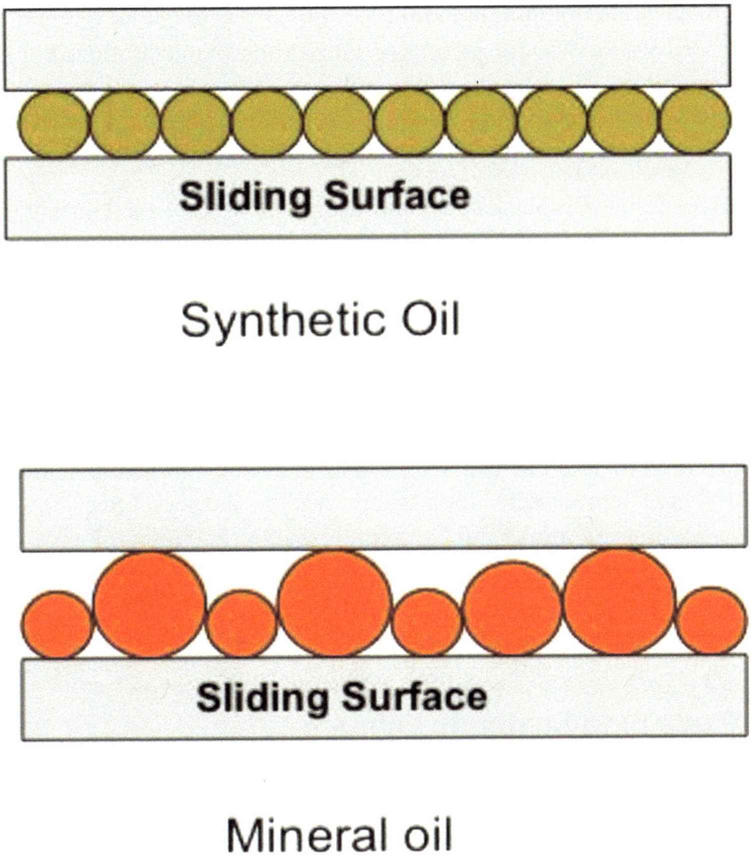

Figure 15.1 Synthetic oil versus mineral oil molecular structure.

2. **Poly-Alkylene Glycol (PAG)** – also referred to as Polyglycols are organic chemical fluids possessing excellent lubricity (friction reducing capability) and an inherent ability to volatilize (clean burn) any decomposed or oxidized products at high temperatures, leaving no sludge, acid build up, or damaging particles should oxidation take place. PAGs are not always oil soluble, but many grades are water-soluble. These lubricants are polar and excel when used for sliding wear protection. PAG's are polymers of alkylene oxides and find primary employment as industrial compressor oils, hydraulic oils (water glycol type) and severe duty gear oils, especially phosphor bronze worm gears . With a pour point of – 60F and a VI greater than 150 they have excellent biodegradability, but fall short in terms of mineral oil and PAO compatibility.

3. **Dibasic Acid Ester (Di-Ester)** – originally saw use at the end of the Second World War for jet engine lubrication thanks to their high shear VI stability of 150 and above. Formulated from a reaction between alcohol and acid laden oxygen, they can suffer from poor hydrolytic stability (reacts to water presence) and poor seal compatibility.
4. **Polyol-Ester** – replaced Di-Ester as the preferred jet aviation lubricant thanks to its increased thermal stability. With a low pour point of – 95F and a VI of 160 plus it is also the preferred lubricant for gas turbines and two cycle oil applications, and also serves as refrigerant oil. This type of synthetic is expensive and like its di-ester cousin suffers from poor hydrolytic stability, seal compatibility and corrosion stability.
5. **Silicone** – one of the most expensive lubricants on the market. Silicone's very high flash point and VI stability of over 250 make it attractive in very high temperature applications despite its poor lubricity. Typical applications are brake fluid and fire resistant hydraulic oils.
6. **Blends** - The first four base fluids account for more than 90% of the synthetic fluids used worldwide.

15.4 Examining Synthetic Lubes

Understanding the principal features and attributes of synthetic fluids will place the potential user in a position to prescreen applicable synthetics and question suppliers whose offer or proposal seems at odds with these performance stipulations.

15.4.1 Synthetic hydrocarbon fluids

Synthetic hydrocarbon fluids (SHF), such as those with a polyalphaolefin (PAO) base, provide many of the best lubricating properties of petroleum oils without their drawback; even the best petroleum oils contain waxes that gel at low temperatures, and constituents that vaporize or readily oxidize at high temperatures.

SHF base fluids are made by chemically combining various low molecular weight linear alpha-olefins to obtain a product with the desired physical properties. They are similar to cross-branched paraffinic petroleum oils because they consist of fully saturated carbon and hydrogen.

These man-made fluids have a controlled molecular structure with predictable properties. They are available in several viscosity grades and range from products for low temperature applications to those recommended for high temperature uses. They are favored for their hydrolytic stability, chemical stability and low toxicity.

15.4.2 Organic esters

Organic esters can be dibasic acid or polyol types. Dibasic acids have shear-stable viscosity over a wide temperature range (-90°F to 400°F), high film strength, good metal wetting properties and low vapor pressure at elevated temperatures. They easily accept additives, enhancing their use in many commercial applications and especially as compressor lubricants.

Polyol-esters have many of the performance advantages of dibasic acid esters and can be used at even higher temperatures. They are used principally in high-temperature chain lubricants, for industrial turbines and in some aviation applications.

Phosphate esters are organic esters that, when used with carefully selected additives, provide a group of synthetic fluids that can be used where fire resistance is required. Even when ignited, the phosphate esters will continue to burn only if severe conditions required for ignition are maintained. Some phosphate esters are less stable in the presence of moisture and heat. The products of the resulting degradation are corrosive and will attack paints and rubbers. The poor viscosity index (VI) limits the operating temperature range for any given phosphate ester product.

15.4.3 Polyglycols

PAG or Polyglycols can be manufactured from either ethylene oxide, propylene oxide or a mixture of both. The propylene oxide polymers tend to be hydrocarbon soluble and water insoluble, while the ethylene oxide tends to be water soluble and hydrocarbon insoluble.

In many applications, the physical properties of the finished product can be engineered by adjusting the ratio of ethylene oxide and propylene oxide in the final molecular structure.

Polyglycols have excellent viscosity and temperature properties and can be used in applications from -40°F to 400°F, and have low sludge-forming tendencies. A major application for polyglycol lubricants is in compressors that handle hydrocarbon gases. This is due to the non-hydrocarbon-diluting properties inherent in polyglycols. The polyglycols' affinity for water results in poor water separability

Table 15.1 Generalized properties of synthetic hydrocarbon lubes*.

	Mineral Oils	Polyisobutenes	Polyalphaolefins	Alkylated Aromatics	Polyalkylene Glycols
Viscosity Temperature Behavior (VI)	Moderate	Poor	Very Good	Moderate	Very Good
Low Temperature Behavior (Pourpoint)	Poor	Moderate	Excellent	Good	Good
Liquid Range	Moderate	Poor	Very Good	Good	Good
Oxidation Stability (Aging)	Moderate	Moderate	Very Good	Moderate	Good
Thermal Stability	Moderate	Moderate	Moderate	Moderate	Good
Evaporation Loss, Volatility	Moderate	Moderate	Very Good	Good	Good
Fire Resistance, Flash Temperature	Poor	Poor	Poor	Poor	Moderate
Hydrolytic Stability	Excellent	Excellent	Excellent	Excellent	Good
Corrosion Protection Properties	Excellent	Excellent	Excellent	Excellent	Good
Seal Material Compatibility	Good	Good	Very Good/	Good	Good
Paint and Lacquer Compatibility	Excellent	Excellent	Excellent	Excellent	Moderate
Miscibility with Mineral Oil	—	Excellent	Excellent	Excellent	Poor
Solubility of Additives	Excellent	Excellent	Very Good/	Excellent	Moderate

Perfluoroalkyl Ethers	Polyphenylethers	Dicarboxylics Acid Esters	Neopentyl phosphate Esters	Trialkyl phosphate Esters	Triakyl phosphate Esters	Silicone Oils
Moderate	Poor	Very Good	Very Good	Poor	Excellent	Excellent
Good	Poor	Excellent	Very Good	Moderate	Excellent	Excellent
Excellent	Poor	Very Good	Very Good	Very Good	Good	Excellent
Excellent	Very Good Good	Very Good/	Very Good	Very Good	Moderate	Very Good
Excellent	Excellent	Good	Very Good	Very Good	Good	Very Good
Excellent	Good	Excellent	Excellent	Very Good/	Very Good/	Very Good/
Excellent	Moderate	Moderate	Moderate Very Good/	Excellent/ Very Good/	Excellent/	Good
Excellent	Excellent	Moderate	Moderate	Moderate	Good	Good
Poor	Moderate	Moderate	Moderate	Moderate	Moderate	Good
Excellent	Good	Moderate	Moderate	Poor	Poor	Good
Very Good/	Moderate	Moderate	Moderate	Poor	Poor	Good
Poor	Good	Very Good/	Very Good/	Moderate	Moderate	Poor
Poor	Very Good/	Very Good/	Very Good/	Excellent	Excellent	Poor

(*continued*)

Table 15.1 (*Continued*)

	Mineral Oils	Polyisobutenes	Polyalphaolefins	Alkylated Aromatics	Polyalkylene Glycols
Lubricating Properties, Load Carrying Capacity	Good	Good	Good	Good	Very Good/
Toxicity	Good	Excellent	Excellent	Poor	Good
Biodegradability	Moderate	Poor	Good/	Poor	Excellent
			Moderate		Very Good/
Price Relation Against Mineral Oil	-	3*5	3-5	3-5	6-10
*Source: Technische Akademie Esslingen, Germany					

15.4.4 Silicones

Silicones have been in existence for many years and offer a number of advantages as lubricants. Silicones have good viscosity versus temperature performance, excellent heat resistance, oxidative stability and low volatility. Silicones are chemically inert and have good elastomer compatibility. Poor metal-to-metal lubricating properties and high cost limit their use to specialized applications where their unique properties and high performance can be justified.

15.4.5 Synthetic lubricant blends

Blends of the synthetic lubricants with each other, or with petroleum lubricants have significant synergistic results. In fact, many of today's synthetic lubricants being sold consist of a blend of two or more base materials to enhance the properties of the finished product.

Table 15.1 (*Continued*)

Perfluoroalkyl Ethers	Polyphenylethers	Dicarboxylics Acid Esters	Neopentyl phosphate Esters	Trialkyl phosphate Esters	Triakyl phosphate Esters	Silicone Oils
Excellent	Excellent	Very Good/	Very Good/	Excellent	Moderate	Poor
Excellent	Good	Good	Good	Very Good//	Moderate/	Excellent
			Good	Poor		
Poor	Poor	Excellent	Excellent	Very Good/	Very Good/	Poor
		Very Good/	Very Good/			
500	200-500	4-10	4-10	5-10	5-10	30-100

Synthetic lubricants have been steadily gaining industrial acceptance since the late 1950s. In many applications today, they are the specified lubricant of the compressor manufacturer. This is especially true in rotary screw and rotary vane air compressors.

While the greatest industrial acceptance has been with air compression, many other industrial applications can be economically justified. Synthetic lubricants are currently being used in compressors processing such diverse materials as ammonia, hydrogen, hydro-carbon gases, natural gas, hydrogen chloride, nitrogen and numerous others.

Synthetic lubricants are not limited to compressors but are used in gear boxes, vacuum pumps, valves, diaphragm pumps and hydraulic systems. Synthetic lubricants are being used in applications that require more efficient, safe lubrication or where the environmental conditions preclude the use of traditional petroleum products.

15.5 Properties and Advantages

Synthetic lubricant fluids provide many advantages when compared to mineral oils. These include:

- Improved thermal and oxidative stability
- More desirable viscosity-temperature characteristics
- Superior volatility characteristics
- Preferred frictional properties
- Better heat transfer properties
- Higher flash point and auto-ignition temperatures.

Experience clearly shows that these advantages result in the following economic benefits:

- Increased service life of the lubricant (typically four to eight times longer than petroleum lubricants)
- Less lubricant consumption due to its low volatility
- Reduced deposit formation as a result of good high-temperature oxidation stability
- Increased wear protection resulting in less frequent maintenance
- Reduced energy consumption because of increased lubricating efficiency
- Improved cold weather flow properties
- Reduced fire hazard resulting in lower insurance premiums
- Higher productivity, lower manufacturing costs and less downtime because machines run at higher speeds and loads with lower temperatures
- Longer machinery life because less wear results in more production during life of machine and tools.

Synthetic lubricant base stocks, while possessing many of the attributes needed for good lubrication, require fortification with additives relative to their intended use. An experienced formulator will take into consideration a range of requirements:

15.5.1 Contaminant dispersion

It is important to keep internally and externally generated oil-insoluble deposit-forming particles suspended in the oil. This mechanism reduces the tendency of deposits, which can lower operating efficiency, and form in critical areas of the machinery. Additives that impart dispersing characteristics are called "dispersants" and "detergents." A dispersant is distinguished from a detergent in that it is nonmetallic, does not leave an ash when the oil is burned and can keep larger quantities of contaminants in suspension.

15.5.2 Protecting the metal surface from rust and corrosion

Humidity (water) type rust and acid type corrosion must be inhibited for long surface life. An oil film itself is helpful but this film is easily replaced at the metal surface by water droplets and acidic constituents. Additives that have an affinity for a metal surface, more so than water or acids, are used in oils to prevent rust and corrosion and are generally referred to as simply "rust inhibitors."

Oxidative stability

Oils tend to thicken in use, especially under conditions where they are exposed to the atmosphere or where oxygen is present. This phenomenon is chemically termed "oxidation."

Oxygen reacts with the oil molecule initiating a chain reaction that makes the molecule larger, thereby decreasing fluidity. Conditions that assist the oxidation process are heat, oxidation catalyzing chemicals, aeration and other mechanisms that can allow the oxygen to easily attach itself. Additives that retard the oxidation process are termed "oxidation inhibitors."

Wear prevention

Inevitably the metal surfaces being lubricated come in contact. Whenever the speed of relative motion is low enough, the oil film does not stay in place. This can also happen if the loading on either or both surfaces is such that the oil film tends to be squeezed out. When moving metal surfaces come in contact, certain wear particles are dislodged and wear begins. Additives that form a protective film on the surfaces are called "anti-wear agents."

Viscosity index improvers
Viscosity index improvers function to improve viscosity/temperature relationships, that is, to reduce the effect of temperature on viscosity change.

Foam suppressants
Foam suppressants allow entrained air bubbles to collapse more readily when they reach the surface of the oil. They function by reducing surface tension of the oil film.

Oiliness additives
Oiliness additives are materials that reduce the oil friction coefficient.

Surfactants
Surfactants improve the ability of the oil to "wet" the metal surface.

Alkalinity agents
Alkalinity agents impart alkalinity or basicity to oils where this is a desirable feature.

Tackiness agents
Tackiness agents impart stringiness or tackiness to an oil. This is sometimes desirable to improve adhesive qualities.

Obviously then, the lubricant supplier or formulator has to choose from a number of options. There are technical considerations to weigh and compromises to make. Close cooperation between supplier and user is helpful; formulator experience and integrity is essential.

15.6 Case Histories

The following are highlights from many successful case histories when changing from a mineral based oil to a synthetic oil has made a difference.

15.6.1 Circulating oil system for furnace air preheaters

Several major refineries in the U.S. and Europe had experienced frequent bearing failures on their slow-rotating heat exchangers while lubricated with manufacturer-recommended mineral oil. With bearing housings typically reaching temperatures around 270°F, the cooled and filtered mineral oil would repeatedly overheat to the point of coking. Bearing failures after six months of operation were the norm. After changing to a properly formulated synthetic, a lubricant with superior high-temperature capabilities and low volatility, bearing lives were extended to several years. In one refinery alone, an annual savings of $250,000 pa in today's funds has been documented since the changeover.

15.6.2 Right-angle gear drives for fin fan coolers

A European facility achieved a disappointing mean-time-between-failures (MTBF) of only 36 months on 36 hypoid gear sets in a difficult to reach, elevated area.

Using mineral oil (ISO VG 160), a drain interval of six months was necessary to obtain this MTBF. Each oil change required 12 man-hours and temporary scaffolding at a cost of $1,000. Change-over to an appropriate synthetic, i.e., a synthetic with optimized temperature stabilizers, wear reducers and oxidation inhibitors, has allowed drain intervals to be increased to two years while obtaining a simultaneous increase in equipment MTBF.

Detailed calculations showed a net benefit of $1,950 per year per gear set. Combined yearly saving: approximately $150,000 in today's funds, with no credit taken for power reduction or avoided production curtailments.

15.6.3 Plant-wide oil mist systems

An oil mist lubrication system at a Southeast Texas chemical plant experienced an unscheduled shutdown as a result of cold weather. Twenty-seven mist re-classifiers in this system were affected. These re-classifiers provided lubrication to several fin fans, two electric motors and the rolling element bearings in 14 centrifugal pumps. Wax plugging of the mist re-classifiers brought on by the cold weather caused the unexpected shut-down. As a result, several bearings failed because of lubricant starvation. An ISO VG 68 grade conventional mineral oil was identified as the wax source.

The oil mist system had to be isolated and blown out to avoid further bearing failures. In addition to the downtime costs, significant labor and hardware costs were required to restore the unit to normal operation.

For this reason, a synthetic wax-free lubricant was identified as a replacement for the the mineral oil. Neither the oil feed rate nor the air-to-oil ratio required adjusting after switching to the synthetic.

Since converting, to a diester-based oil mist system, the following has resulted:

- No cold weather plugging of the mist re-classifiers has been experienced.

- No lubricant incompatibility has been detected with other components of the oil mist system.
- The synthetic lubricant is providing proper bearing wear protection as evidenced by no increase in required maintenance for pumps, fans or motors served by the oil mist system.
- Downtime, labor and hardware replacement costs attributed to cold weather operational problems have been eliminated.
- Savings in contractor and plant manpower used to clean the re-classifiers have averaged close to $130,000 per year.
- Two failures of pumps and motors were assumed to be prevented via use of wax-free lubricant. The savings equaled $25,000 per year for each of the two pumps.

Total net credit has been estimated at over $350,000 per year in 2023 dollars. This does not include any process losses associated with equipment outages.

15.6.4 Pulverizing mills in coal-fired generating plant

A large coal-fired power generating station in the southwestern U.S. was having lubrication problems with their coal pulverizing mill. The equipment, a bowl mill pulverizer, was experiencing the following problems with lubrication of gears driving the mill:

- The lubricant was losing viscosity and had to be changed every four to six months.
- Air entrainment in the lubricant was causing cavitation in the pumps that circulated the lubricant
- The gears were experiencing an unacceptable level of wear as measured by a metals analysis on the lubricant.
- On very cold mornings, the lubricant was so viscous it had to be heated before the unit could be put in service.
- The petroleum-based lubricant's initial viscosity varied significantly.

After evaluating options, it was decided to use a synthetic-based lubricant. In cooperation with a major lubricant manufacturer, a synthetic hydrocarbon base stock with the proper additive package was chosen. Additive package

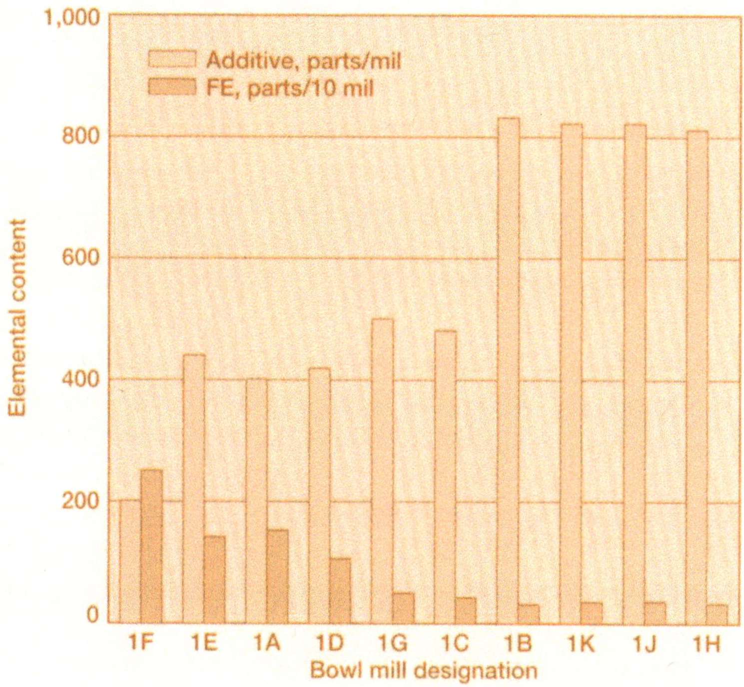

Figure 15.2 Bowl mill wear, 1,000 operating hours.

concentrations were evaluated in a number of bowl mills simultaneously to establish the optimum level and composition. Figure 15.2 shows the dramatic effect on metal gear wear accomplished over a 1,000-hour trial period.

The synthetic hydrocarbon base stock proved to be extremely shear resistant. One of the bowl mill was closely monitored during 54 months of operation (Figure 15.3) to establish viscosity stability. The data represents only operating hours, not total time elapsed, since the unit is not operated continuously. The performance was deemed excellent and lubricant life exceeded 60 months.

The synthetic hydrocarbon lubricant was compared to two petroleum-based lubricants supplied by major oil companies. The tests were run on three bowl mills that had recently been reworked and tested. All three bowl mills were fed the same amount of coal during the test period. All three gear oils were the same ISO 320 viscosity grade.

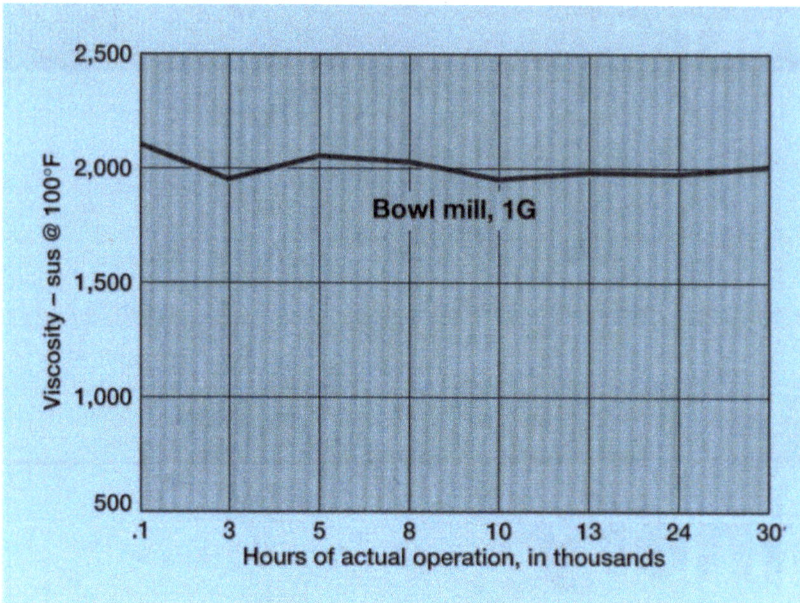

Figure 15.3 Viscosity stability of synthetic lubricant in bowl mill gear unit.

The average current draws were:

Product	Amps
Petroleum #1	70
Petroleum #2	75
Synthetic hydrocarbon gear oil	68

The lower amp difference shown by the synthetic hydrocarbon is the result of the lower coefficient of friction shown in Table 15.2.

Efficiency gains can be very sizeable and the resulting reduction in energy cost will often pay for the higher cost of synthetic lubricants within months. Table 15.3 shows a typical cost benefit analysis.

As demonstrated, the synthetic hydrocarbon gear oil has solved the original problems and provided additional benefits not anticipated. The switch to synthetic lubricants has clearly improved performance and achieved significant savings in operating costs, as calculated in Table 15.3.

The total annual savings for all of the above example categories amount to $13,545 p.a. In addition, savings in reduced wear resulting in fewer repairs and reduced downtime are certain to be realized.

A forward-looking process plant needs to explore the many opportunities for often substantial cost savings that can be achieved by judiciously

Table 15.2 Physical properties of ISO VG 320 gear oil.

	Petroleum	Synthetic hydrocarbon
Viscosity index	95	140
Thermal conductivity, Btu/hr/ft²	0.071	0.085
Coefficient of friction	0.101	0.086
Pour point, °F	+5	-4

applying properly formulated synthetic lubricants. To determine if synthetic lubricants can play a role in your plant start by answering three basic questions:

Q1. When should I consider use of a synthetic lubricant?"

- When it is cost effective (increased productivity, extended lubricant life, etc.)
- When a conventional lubricant has not worked (problem solver).
- When it enhances safety or environmental aspects of an operation (higher flash & fire points, reduction of used lubricant requiring disposal).
- When it reduces risk (failure to change out systems, reduced chance of misapplication through lubricant consolidation).

Q2. What type of synthetic lubricant should I use?
Key considerations to consider here are temperature extremes in operation, material compatibility, equipment requirements and methods of application. Discuss these with your lubricant supplier

Q3. What are the requirements for effective use of the selected synthetic lubricant?
In selecting a lubricant for demanding lubricant applications, there are several key imperatives that must be satisfied for things to work. Temperature extremes, lubricant service life, extreme loads, safety and environmental aspects are the usual key drivers. One or more of the demands will promote the selection of a synthetic for a specific application.

Table 15.3 Cost benefit analysis table.

1	**Annual Lubricant Consumption Cost**	
	Petroleum oil cost per gal	$20.00
	Petroleum oil changes per year	2
	Volume of gear box, gal	300
	Petroleum oil cost per yr	
	($20.00/gal)(2 changes/year)(300 gal/unit) =	**$12,000 p.a.**
	Synthetic oil cost per gal	$48.00
	Synthetic oil changes per year	0.2
	Volume of gear box, gal	300
	Synthetic oil cost per year	
	($48.00/gal)(0.2 changes/year)(300 gal/unit) =	**$2,880 p.a.**
	Annual savings on lubricant cost—	**$7,200 p.a.**
2	**Reduced Maintenance Cost Savings**	
	Petroleum oil changes per year	2
	Maintenance cost per change	$1,500
	Petroleum oil maintenance cost per year	
	(2 changes/year)($1,500/change) =	**$3,000 p.a.**
	Synthetic oil changes per year	0.2
	Maintenance cost per change	$1,500
	Synthetic oil maintenance cost per year	
	(0.2 changes/year)($1,500/change)	**$300 p.a.**
	Annual savings in scheduled maintenance cost	**$2,700 p.a.**

Table 15.3 *Continued*

3	**Lubricant Disposal Costs**	
	Petroleum oil used per year, gal	600
	Disposal cost per gal	$1.00
	Cost of disposal	$600
	Synthetic oil used per year, gal	60
	Disposal cost per gal	$1.00
	Cost of disposal	$60
	Annual savings in disposal cost per year—	**$540 p.a.**
4	**The reduction in energy consumption also provides significant savings:**	
	Average annual power cost using petroleum oil lubricant	$49,920
	Average annual power cost using synthetic lubricant	$46,815
	Annual savings in power consumption—	**$3,105 p.a.**
	TOTAL ANNUAL SAVINGS (1+2+3+4)	**$13,545 p.a.**

Synthetic lubricants should also be considered when looking for the following benefits:

- Increased productivity
- Enhanced Equipment Performance
- Cost Savings
- Enhanced Safety
- Enhanced Environmental Aspects

15.7 Synthetic Lubricants for Use in Extreme Pressure and Temperatures

A quality synthetic oil can both flow freely in frigid arctic temperatures that "freeze" conventional mineral oils stiff (Figure 15.4), while at the same time, keep its viscosity at steel-mill temperatures, (Figure 15.5), that can turn

Figure 15.4 The arctic demands free-flowing synthetic lubricants.

mineral oils to watery liquids. At the same time, it must also resist oxidation and demonstrate excellent volatility control at high temperatures—for long-term, reliable service.

Synthetic lubricants that are adapt and can excel in temperature extremes are those using high-quality polyalphaolefin (PAO) base stocks.

15.7.1 Polyalphaolefins (PAOs) make the difference

To understand polyalphaolefins, it helps to start with conventional paraffinic lubricants Long-chain paraffin molecules of 20 to 40 carbon atoms have many excellent properties such as oxidation stability and viscosity that does not change drastically as temperature goes up or down. Most important, they lubricate well because they cling to metal surfaces and slide past each other easily.

In fact, these molecules might be ideal lubricants except for one serious drawback: somewhere around room temperature (depending on the length of the chain), they can crystallize and pack together like sticks of dry spaghetti. The result is a solid matrix of wax.

In petroleum-based lubricants, paraffins work because they occur naturally attached to cyclic structures that crystallize at lower temperatures. For the wax crystals that do form, added pour-point depressants help keep them from growing large enough to cause trouble.

But petroleum-based lubricants still have problems at temperature extremes: cyclic components get too thin at high temperatures and don't

Figure 15.5 Steel mill temperature environment calls for synthetic lubricants.

resist oxidation well—and even the best pour point depressants lose effectiveness at extremely low temperatures.

The ideal solution would be a petroleum-based paraffin that couldn't crystallize into wax. It should be exceptionally pure and uniform, with a narrow boiling range and virtually no variation in batch-to-batch properties.

That solution exists in the polyalphaolefin (PAO) base-stocks used in superior synthetic EP industrial gear oils.

PAOs are specially synthesized branched paraffins with three to five 10-carbon chains united in a star-like structure. The shape virtually defies crystallization. The PAO molecules can resist freezing down to -40°C(-40°F) or lower, and they come close to the ideal lubricant in other ways too: they excel at maintaining viscosity, resisting oxidation, and controlling high-temperature volatility.

With the addition of a quality additive package, PAO EP lubricants can support a Timken OK load in excess of 100lbs, compared to 60lbs for a conventional petroleum-based EP gear oil. Figures 15.6 through 15.9 convey the performance advantages that can be obtained from the many grades of this PAO-based synthetic EP industrial gear oil.

Oxidation of an oil—breakdown due to heat and oxygen—causes viscosity to increase, and it creates soft sludges and hard deposits that can lead to equipment failure. Among conventional petroleum-based products, a premium EP gear oil offers outstanding oxidation performance at operating conditions up to 93°C(200°F). As shown in Figure 15.5, Exxon's Spartan Synthetic EP carries that performance to the extreme, staying clean up to 121°C(250°F).

Superior volatility control is illustrated in Figure 15.9, while pour point and viscosity characteristics can be compared in Figures 15.10 and 15.11. As can be seen, a quality EP PAO lubricant is ideal for arctic and other cold-weather environments, because it keeps flowing even at -30°C (-22°F) and colder. Refer to typical inspections Table 15.4. Most conventional petroleum-based gear oils become too thick to pour below -10°C (14°F). Machines start easier and gear boxes run more efficiently with high quality synthetic EPs.

15.8 Case Histories Involving PAO-based Synthetic EP Oils

Case History One: when a triple-race roller bearing on a crown roll fails, it can cost $25,000 or more to replace. One south-eastern paper mill was replacing each bearing at least every two years. In one instance, a bearing lasted only nine months.

The problem arose from the crown roll's unusually low rotational speed. The machine (Figure 15.6) did not generate enough centrifugal force to keep the rollers pressed against the bearing's outer raceway, so rollers leaving the load zone stopped rotating. As each roller reentered the load zone, it skidded like an air-plane tire first touching ground. Damage to the bearing raceways quickly led to bearing failure.

The competitive petroleum-based EP gear oil used by the mill couldn't stop the destruction, and use of a higher viscosity oil was not possible as the oil is shared by the hydraulic system, which would have suffered startup problems with a higher viscosity.

The switch to a high-quality Synthetic EP brought immediate benefits: its high viscosity index meant good fluidity at cold startup conditions. Also, its outstanding lubricity protected the expensive bearings. After two full years with the synthetic product, the bearings showed no evidence of unusual wear.

Case History Two: involves a 20-ton over-head crane at a major Midwestern steel company which had trouble every time the weather got cold. Located above an open railcar entryway, the gear boxes of the crane (Figure 15.7) were exposed to ambient temperatures, making the crane hard to start and difficult to keep running whenever the mercury dropped. When temperatures fell to -40°C (-40°F) one winter, the crane drew so much power trying to start that it blew the system's circuit breakers.

Table 15.4 Typical inspections for a synthetic EP oil. (Source: Exxon Company).

Grade	100	150	220	320	460	680	1000
AGMA number	3 EP	4 EP	5 EP	6 EP	7 EP	8 EP	8AEP
ISO viscosity grade	100	150	220	320	460	680	1000
Viscosity @ 40°C, cSt	106	151	220	327	427	680	989
@ 100°F, SUS	519	778	1149	1712	2241	3159	5154
@ 210°F, SUS	78	100	130	180	225	289	378
Viscosity index	151	150	151	162	167	160	154
API gravity	36.5	35.4	35.0	34.5	34.1	32.3	31.0
Demulsibility, ASTM D 2711							
Total free water, ml	86	85	85	84.5	84	85	8.4
% Water in oil	0.4	0.5	0.8	0.8	0.8	0.7	0.8
Emulsion, ml	0	0	0	0	0	0	0
Rust test, ASTM D 665A & B Four-ball EP test, ASTM D 2783	Pass	Pass	Pass	Pass	Pass	Pass	Pass
Weld point, kg	250	250	250	250	250	250	250
Load wear index	54	53	46	46	47	47	47
Flash point; °C (°F)	231 (448)	236 (457)	249 (480)	250 (482)	254 (490)	270 (518)	280(536)
Pour point, °C (°F)	-55 (-67)	-52 (-62)	-46 (-50)	-46 (-50)	-40 (-40)	-30 (-22)	-20 (-4)
FZG test, pass Oxidation tests	12 stages	12 stages	12 stages	12 stages	12 stages	12 stages	12 stages

Based upon manufacturer 's specifications, the crane's gear boxes were lubricated with a conventional ISO 320 EP gear oil. The pour point of this oil was -9°C (15°F), so the oil varied from stiff to solid in cold weather.

Working together, the technical services group of the oil manufacturer, steel-mill personnel, and the crane manufacturer determined that PAO Synthetic EP ISO 220 would be an acceptable substitute for the petroleum-based gear oil.

Synthetic EP solved the problem. Its low pour point made cold-weather startup easy, eliminating excess power drain. The high viscosity index provided good film thickness in summer temperatures, even at the lower viscosity

Figure 15.6 Paper machine rolls in severe duty service.

grade. The high Timken OK Load rating ensured outstanding extreme-pressure protection. In addition, the Synthetic EP appeared to provide energy savings.

Case History Three: to get more tonnage run the paper machine faster, and turn up the heat. To get less downtime, slow the machine, and lower the heat.

A Synthetic EP now helps a Southeastern mill run fast and hot *and* minimize downtime.

One major problem for this mill was presented by the drive gears in the press section. Because of high ambient temperatures plus heavy loads, these gears operated continuously at 200°F. The conventional EP gear oil used by the mill oxidized to a viscous black sludgy material so fast that even oil changes every six months were not frequent enough to ensure that gears and bearings would stay adequately lubricated.

After the mill switched to quality PAO Synthetic EP, two years later, when the original charge of the product was routinely tested, it still met the gear manufacturer's requirements.

15.9 Diesters: Another Synthetics Option

Earlier, we considered synthetic extreme-pressure industrial gear and bearing lubricants manufactured from synthesized hydrocarbons, predominantly

Figure 15.7 Overhead crane gear box exposed to severe ambient environment.

PAO's designed to provide outstanding performance under severe temperatures and loads. Applications include gear boxes, industrial differentials and high-loaded rolling contact bearings.

The thermal and oxidative stability of a quality Synthetic EP, superior to that of conventional gear oils, provides excellent resistance to sludging and helps ensure long lubricant life, when under the severe conditions encountered in small, high-temperature gear boxes. Because the synthesized hydrocarbons contain no wax, these lubricants can be used in mist lubricators and in low-temperature applications.

While there is some overlapping of applications for PAOs and diesters, diester-based lubricants are generally formulated to give outstanding

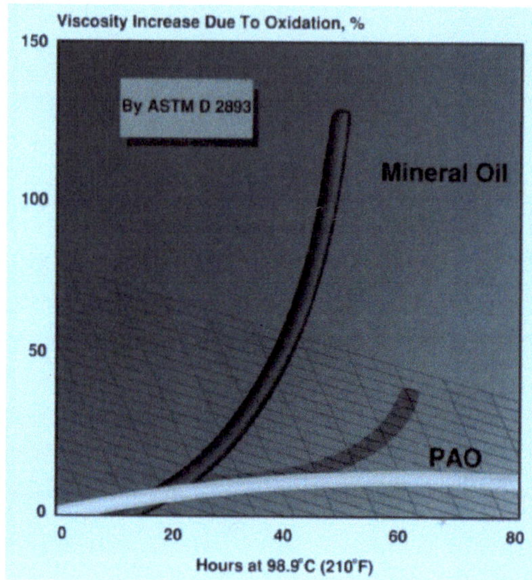

Figure 15.8 Viscosity increase due to oxidation, mineral oil vs. synthetic EP product. (Source: Exxon Company, USA, Houston, Texas.)

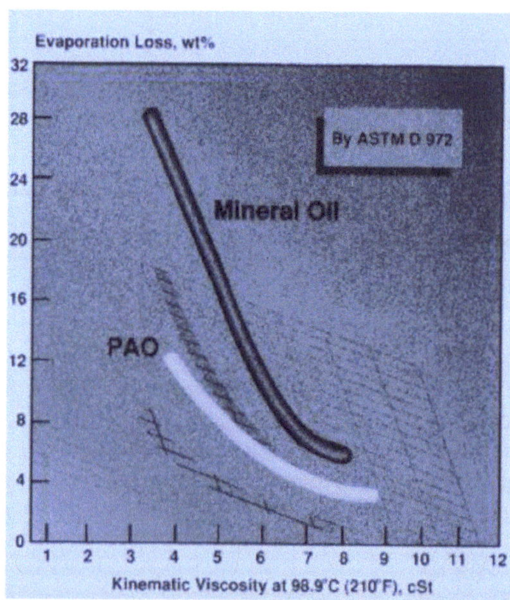

Figure 15.9 A superior synthetic EP oil will give superior volatility control to provide long-term lubrication effectiveness. Plus, the low volatility and high flash point compared to conventional gear oils give an added margin of safety at high operating temperatures. (Source: Exxon Company, USA, Houston, Texas.)

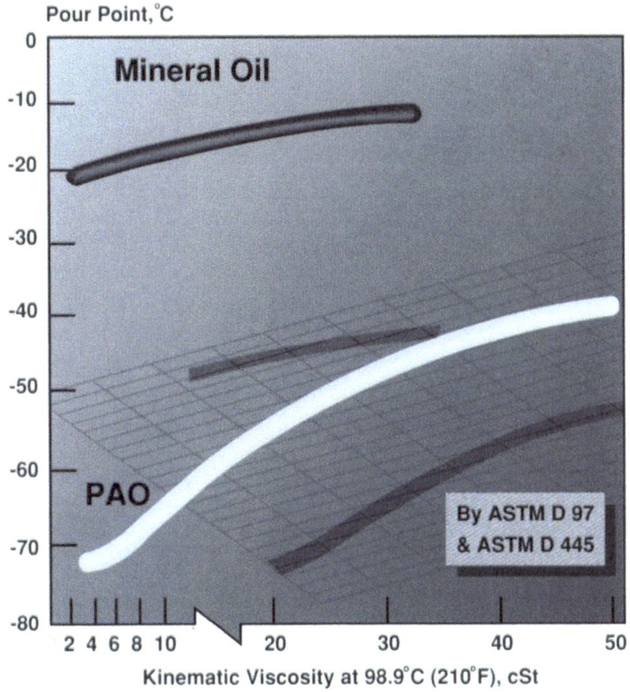

Figure 15.10 PAO-based synthetics have superior cold-weather performance. (Source: Exxon Company, USA, Houston, Texas.)

performance in air compressors, hydraulic systems, mist lubrication systems, air-cooled heat exchanger drives, and bearings in pumps and electric motors.

Changing from a conventional petroleum-based lubricant to a diester can minimize the frustrating and costly problems of hot-running equipment, premature bearing failures, damaging deposit build-up, cold-weather wax plugging and the necessity for frequent oil changes. The superior lubricity of diester lubricants vs. comparable petroleum oils permits bearings to run cooler, thus extending the life of the bearings and the lubricant. In addition, low volatility helps reduce lubricant consumption. Diester lubricants are rust-and-oxidation inhibited and have excellent anti-wear properties, with very low carbon-forming tendencies due to their diester base.

The application range of superior diester-based synthetics is best illustrated by briefly reviewing four case histories.

Case History One: in a petrochemical refinery, a diester-based synthetic lubricant eliminated bearing failures in a 4,000-HP electric motor and lowered oil temperatures by more than 50%.

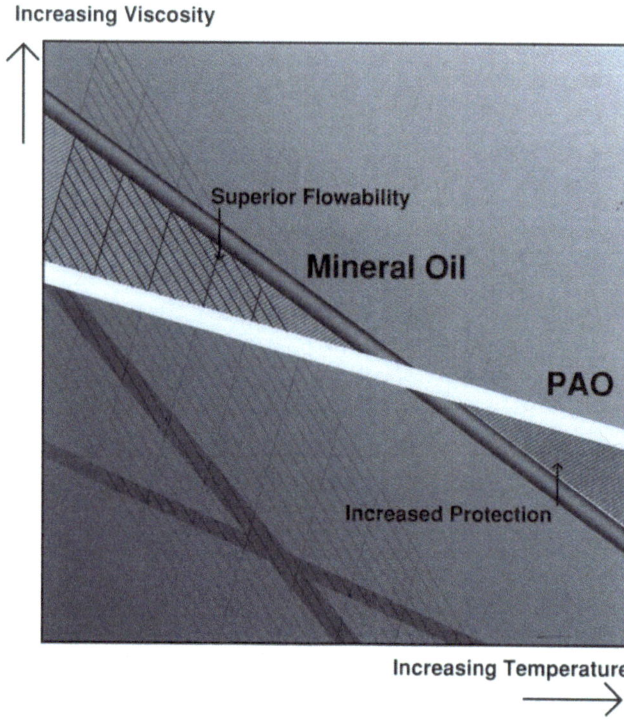

Figure 15.11 High-quality polyalphaolefin base stock helps keep viscosity stable over a wide range of temperatures. The viscosity index for Spartan Synthetic EP gear oils ranges from 150 to 167, compared to 90 to 100 for a conventional petroleum-based oil. Source: Exxon Corporation, USA, (PLIF 2).

Case History Two: In a chemical plant a VG100 diester-based synthetic ended a cold-weather wax-plugging problem in an oil mist system and saved $9,375 in the first year of operation.

Case History Three: In a British petrochemical operation, a VG32 diester-based synthetic reduced valve overhauls in a reciprocating compressor from 24 per year to one; saved $4,300 per year in energy costs; and extended drain intervals by a factor of eight.

Case History Four: In a major steel plant, a VG100 diester-based synthetic virtually eliminated coke deposits that had caused four reciprocating compressors to be shut down every 1,500 hours for cleaning. The compressors now run 6,000 hours and longer with no problems.

Case History Five: A pair of 50-HP reciprocating air compressors in a chemical plant were in alternate service (one week continuously on, then off),

Table 15.5 Typical inspections for a diester synthetic lubricant.

Grade		32	68	100	150	EP 220
Viscosity						
cSt @ 40°C	ASTM D 445	30.0	65.0	96.0	143	215
cSt @ 100°C		5.5	7.5	10.1	13.6	27
SUS @ 100°F	ASTM D 2161	153	339	502	740	1108
SUS @ 210°F		44.7	51.4	60.6	73.0	132
Viscosity index	ASTM D 2270	120	77	88	89	160
Pour point, °C	ASTM D 97	-57	-37	-40	-34	-43
°F		-71	-35	-40	-30	-45
Flash Point, °C	ASTM D 93	266	271	252	249	174
°F		510	520	485	480	345
Autogenous ignition temperature, °C		385	393	413	418	393
°F		725	740	775	785	740
Specific gravity @ 15.6°C (600F)		0.926	0 957	0.963	0.959	0.884
Timken OK load, lb	ASTM D 2782	—	—	—	—	75
FZG gear test, stages Demulsibility @ 54.4°C (130°F), ml oil/water/emulsion	ASTM D1401	40/40/0	40/40/0	40/40/0	40/40/0	12 40/40/0
Evaporative loss, 22 hr @ 98.9°C (210°F), mass %	ASTM D 972	1.7	0.20	0.30	0.30	
Copper strip corrosion, 3 hrs @ 100°C (212°F)	ASTM D 130	la	la	la	la	la
Rust prev., 24-hr dist. water	ASTM D 665	No rust	No rust	No rust	No rust	No rust
Panel coking, 6 hrs @ 316°C (600°F), mg deposit		9.8	9.6	6.0	9.9	
Conradson carbon residue, mass %	ASTM D 189	0.00	0.00	0.00	0.00	
Four-ball, 40 kg, 1800 rpm, 65.6°C (150°F), 30 min., scar diam., mm		0.30	0.40	0.40	0.40	—

Figure 15.12 Photo on left - No. 1 discharge valve after 4 months' service with ISO 150 mineral oil. Photo on right - No. 1 discharge valve after 6 months' service with Exxon SynESSTIC 100 diester synthetic oil. Source: Exxon Corporation, USA, (PLIF 2).

using an ISO 150 mineral oil. Carbon deposits on discharge valves caused such operating problems that the machines required maintenance every three months. In an operational test, one compressor was switched to VG 100 diester-based synthetic. After more than six months, discharge valves on this compressor were substantially cleaner than they were on the unit that used mineral oil for four months. The Diester synthetic allowed compressor maintenance intervals to be doubled from three months to six, a significant saving in labor and material. The comparison photos (Figure 15.12) tell the story.

15.9.1 High film strength for better wear protection

Diester synthetic lubricants outperform both mineral oils and competitive synthetics in film strength and lubricity. That means less wear, reduced maintenance and longer operating life for your machinery.

One key to film strength is the polar molecules of the diester base stock. These molecules line up on metallic surfaces like the nap of a carpet (see Figure 15.1), creating a strong lubricant film that helps prevent metal-to-metal contact.

As illustrated in Figure 15.13, many synthetic lubricants excel in wear protection tests over equivalent viscosity mineral oils. The bearing wear

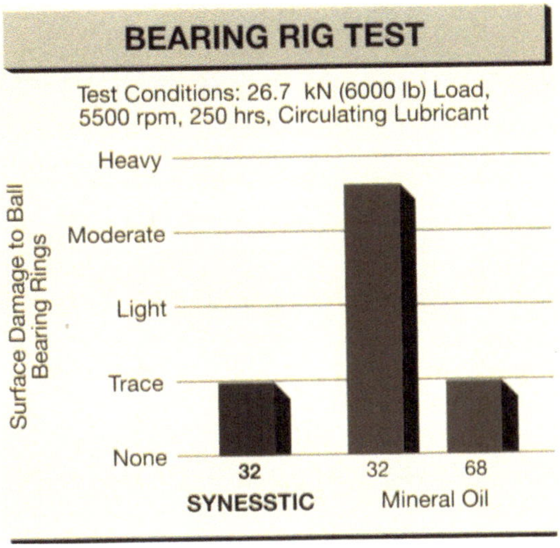

Figure 15.13 Laboratory studies with an instrumented bearing test rig demonstrate that a low viscosity SynESSTIC can provide the same protection as a higher viscosity mineral oil. In the test illustrated in this graph, the additive packages of all three lubricants were identical—only the base changed. Because a lower viscosity grade can achieve the same degree of protection, you may be able to save energy and reduce costs (Source, Exxon)

experience with an ISO grade 32 synthetic is typically similar to that of an ISO grade 68 mineral oil.

15.9.2 Long-term oxidation resistance

Lubricants must be able to resist degradation when exposed to oxidizing conditions for long periods. The test results below show the long-term stability of an Exxon synesstic diester fluid. In this severe laboratory (110°C copper analyst, warm air current, as illustrated in Figure 15.14), full of the oils showed initial control of oxidation. But once the oxidation inhibitor in the conventional mineral oil was consumed, that oil oxidized rapidly. Its degradation produced acidic by-products.

Unlike the mineral oil, all three "synesstic" diester grades resisted oxidation for 3,000 hours and more, evidence of the inherent chemical stability of this lubricant.

15.9.3 Negligible carbon deposits

When mineral oils are heated enough, they break own, leaving varnish, carbon and coke deposits that can be extremely damaging. Diester synthetics offer dramatic improvements in thermal stability.

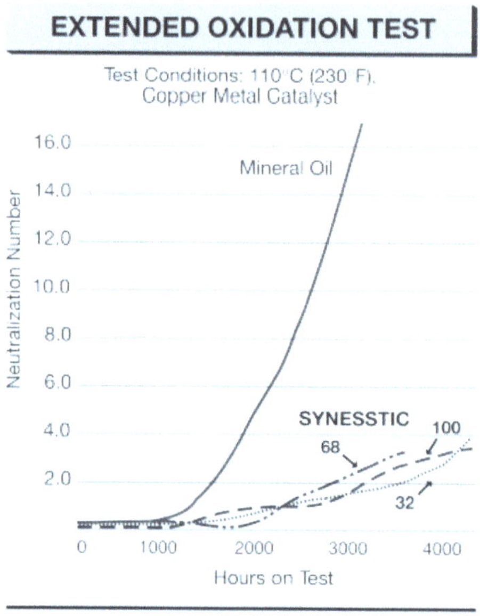

Figure 15.14 Oxidation tests show diester-base synthetics excel over mineral oils (Source: Exxon).

15.9.4 Low pour point advantage

Diester fluids flow easily at temperatures where conventional lubricants almost refuse to budge. This means easy startup for intermittent and cold-weather operations, and less startup wear. Unlike petroleum oils, which typically contain some wax, diester synthetics have no wax to hinder their flow.

Diester lubricants invariably show lower pour points than mineral oils of comparable grade. Although mineral oils generally undergo cold solvent treatment and filtration to remove most of the waxy hydrocarbon fractions, traces of wax that remain can freeze out at low temperatures. That translates to less-than-optimum lubrication for cold-weather and intermittent operations. Figure 15.15 demonstrates how a diester synthetic stays fluid in extreme cold conditions while its mineral oil counterpart solidifies.

15.9.5 Easy cold startup, low friction and energy savings

The microscope shows why diester synthetic lubricants help machinery start so easily in the cold. Magnified 360 times (Figure 15.16) VG 32 diester being tested is free of wax at -18°C (0°F), while a comparable weight mineral oil shows significant crystallization. Moreover, the diester-base lubricant reduces

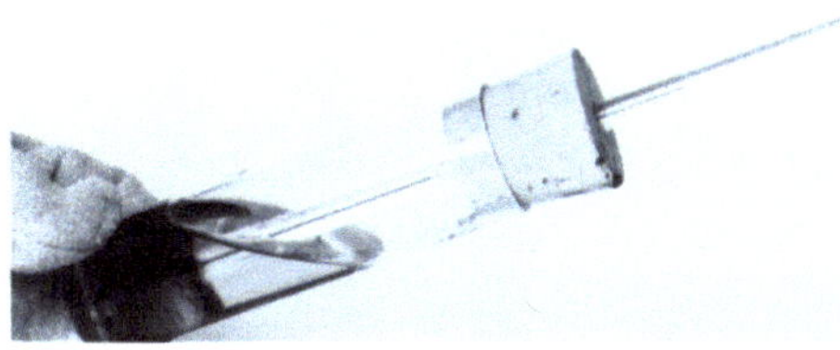

SYNESSTIC 32 @ -41°C (-42°F)

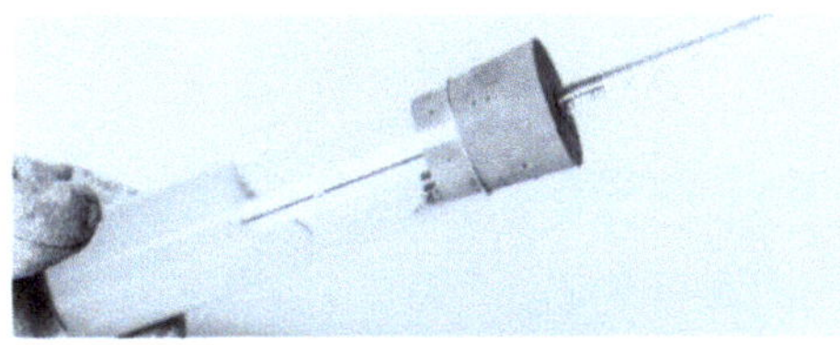

Mineral Oil ISO VG 32 @ -37°C (-34°F)

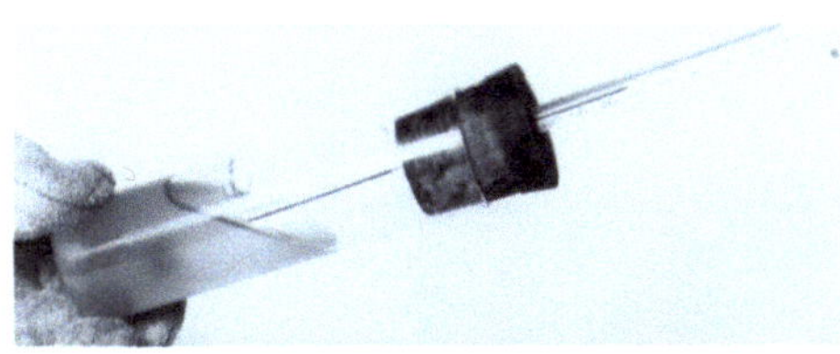

SYNESSTIC 100 @ -31°C (-24°F)

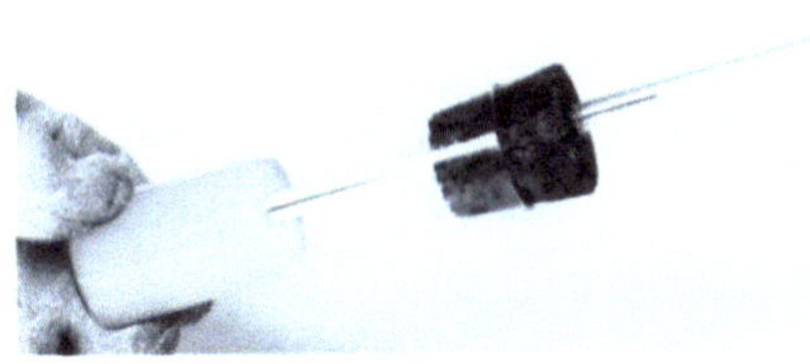

Mineral Oil ISO VG 100 @ -28°C (-18°F)

Figure 15.15 At comparative test temperatures, the mineral oils solidify while the Diester 32 and 100 grades remain free flowing. Source: Exxon Corporation, USA, (PLIF 2).

friction so effectively that the temperature of lubricated bearings remains much lower than it does in bearings lubricated with mineral oil. Lower temperature helps the lubricant last longer. This means that bearings last longer too, because a cooler lubricant has a relatively higher effective viscosity and maintains a more reliable film thickness. Lower bearing temperatures also provide a better margin of safety against thermal fatigue effects.

15.9.6 Reduced maintenance, fuel savings

At a Gulf Coast plant location, engineers ran a series of tests to measure the energy savings that could be achieved using synesstic synthetic lubricants.

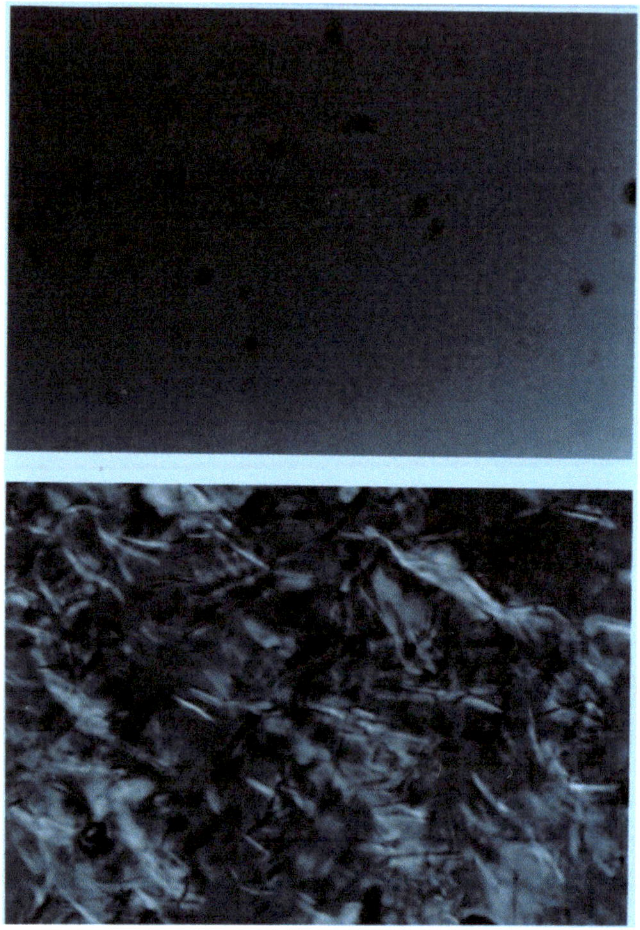

Figure 15.16 At –18°C, Exxon's Synesstic ISO 32 diester (top photo) has no wax crystals, while a comparable mineral oil (bottom photo) shows significant crystallization. The dark spots in the diester photo are air bubbles in the sample. Source: Exxon Corporation, USA, (PLIF 2).

The tests were conducted in two different compressor types, with the following results:

- Ingersoll-Rand TVR-21 Reciprocating Compressor, a 1,200-HP compressor with six double-acting compressor cylinders. In tests comparing synesstic 32 with a premium ISO 100 mineral oil, synesstic 32 achieved:
    - 2.9% less power consumption
    - $5,340 in natural gas fuel savings per year

Table 15.6 Diester-base synthetic lubricants extend oil drain intervals.

EQUIPMENT	METHOD	PREVIOUS	CURRENT
Rotary Screw Compressor	Circulation	2000 hours	8000 hours
Centrifuge Gear Case	Splash	3 months	24 months
Fin Fan Gear Unit	Splash	6 months	2 year plus
Parallel Shaft Gear Reducer	Splash	3 months	1 year plus
Sundyne® Gear Unit	Circulation	6 months	2 year plus
Reciprocating Pump	Splash	3 months	8 months plus

- $6,160 in maintenance savings per year
- $6,596 net operating savings,

- Ingersoll-Rand Centac 4-stage Centrifugal Compressor, a 900-HP (670 kW) electric motor drive operating between 22,400 and 47,900 rpm. In this compressor, a diester 32 was substituted for a premium ISO 32 mineral oil, and achieved:
    - 1.0% less power consumption
    - $3,624 savings in electricity
    - $1,936 net operating savings per year

15.10 Application Summary for Diester-base Synthetic Lubricants

Diester-base synthetics are cost justified whenever a plant requires a combination of exceptional oxidation resistance, outstanding high-temperature stability, low-temperature fluidity, deposit prevention and long-term cleanliness.

Diesters can extend drain intervals, as illustrated in Table 15.6. Moreover, these diester-base lubricants inevitably help prevent the high cost of machine servicing and replacement. They help avoid lubricant losses attributable to foaming and high-temperature evaporation; they can reduce energy costs and exhibit favorable viscosity-temperature relationships.

In reciprocating compressors, diesters resist high temperatures, avoid fouling of discharge valves, extend drain intervals, reduce friction and wear, and offer potential energy savings.

Figure 15.17 Diester film strength offers improved bearing protection for all types of compressors.

In rotary compressors, where there is continuous intermingling of lubricant and gas, the excellent oxidation performance of diester synthetics helps prevent deposits and reduces downstream oil carryover. Compressed gas is cleaner, and the compressor consumes less lubricant.

The good film strength and anti-wear protection of diester synthetic lubricants deliver excellent field performance in high-speed integrally geared pumps and other equipment under moderate loads. Field experience shows that diester 32 can be an effective and truly superior replacement for the automatic transmission fluids commonly found in high-speed gear boxes.

15.10.1 Oil change procedures

Diester synthetic lubricants have very good solvency characteristics that help keep machine components clean. However, this means that you should observe certain precautions when changing from a mineral oil to a diester synthetic. The following procedures are guidelines only; for help, contact your local supplier.

Once-through: Oil mist systems, reciprocating compressor cylinder lubrication, oil-injected screw compressors.

1. Drain mineral oil from reservoir and fill with the proper grade of diester synthetic.
2. If possible, flush diester through lube lines for five to ten minutes to remove old deposits.
3. If step 2 is not possible, operate lubrication system normally and observe lubrication points closely for any signs of plugging from displaced mineral-oil deposits.
4. When all deposits have been removed, return the machine to normal service.

Small Sumps: Pumps, steam turbines, small gear sets, other units with reservoirs of 10 gallons or less.

1. Drain mineral oil, inspect internal condition of the machine, and hand clean as much as practical.
2. Fill with proper grade of diester and run machine for one to two hours.
3. Drain diester synthetic into suitable container, refill and start up machine.
4. Take sample for oil analysis after two days.

Large Sumps: Gear units, reciprocating compressors, units with reservoirs of 10 gallons or more.

1. If possible, switch to diester synthetic after a machine overhaul. Ensure that most residual mineral oil has been removed, fill with proper grade of diester; and start up using normal procedures.
2. If step 1 is not possible, drain mineral oil; inspect internal condition; hand clean as much as practical.
3. Fill with proper grade of diester synthetic and start up the machine using normal procedures.
4. Take a sample for oil analysis after one and seven days.
5. If condition of lubricant deteriorates significantly between the one-day and seven-day sample, drain and refill with diester synthetic. Repeat step 4.

Circulating Oil: Centrifugal compressors, flooded rotary compressors, gear units, large pumps, etc.

1. Drain mineral oil from the system as completely as possible.
2. Clean the suction strainers of lube oil pumps and install new filter elements, if applicable.
3. Fill with the proper grade of diester synthetic and start up the system using normal procedures.
4. Monitor the filter condition carefully for any sign of plugging during the first week of operation.

15.11 Semi-synthetic Fluids

A more recent addition to the marketplace, semi-synthetic base oil blended with mineral base oil products are often mistakenly marketed, and purchased as "synthetic oils".

However, these blends are not controlled and will differ greatly from one manufacturer to another. Semi-synthetic blends generally contain up to 20% pure synthetic product and are less expensive than pure synthetic oils.

Because of the lack of standardization with this product, their engineered value continues to be debated.

Bibliography

[1] Halliday, Kenneth R.; "Why, When and How to Use Synthetic Lubricants." Selco, Fort Worth, Texas, 1977.

[2] Morrison, F.F., Zielinski, James R., "Effects of Synthetic Industrial Fluids on Ball Bearing Performance," ASME Paper 80-Pet-3, presented in New Orleans, Louisiana, Feb. 1980

[3] Zielinski, James, and Perrault, Cary E.; "Survey of Commercial Experience With Diester-Based Synthetic Lubricants in Refinery Equipment," NPRA Paper AM-83-20, presented at the 1983 Annual Meeting of the National Petroleum Refiners Asso., San Francisco, March 20-22, 1983.

[4] Douglas, Patrick J.; "An Environmental Case for Synthetic Lubricants," *Lubrication Engineering*, September, 1992, pp. 696–700.

16

Grease Lubrication

Many situations exist in which lubrication can be accomplished more advantageously with grease than with oil. Most lubricating greases consist of petroleum oils thickened with special soaps that give them an unusual ability to stay in place. Grease is often used, therefore, in applications for which it is not practical to provide a continuous supply of oil.

Though the retentive properties of grease—also resistance to heat, water, extreme loads, and other adverse conditions—depend primarily on the proportion and type of soap, frictional characteristics themselves are related almost entirely to the oil content.

Base-oil viscosity is a determining factor in the ability of the grease to provide a proper lubricating film.

The word grease is derived from the Latin word "Crassus" meaning "fat, dense, or thick," which adequately describes its consistency. Ancient Romans and Egyptians are both believed to have been amongst the first to create real grease by combining olive oil with lime (Calcium Carbonate) to make Calcium based grease.

Making grease is similar to making soap, both rely on a chemical reaction to take place between oil, fat or fatty acids (often present in the oil or added), and an alkali base material (referred to as the "thickener") to form soap like material, a reaction known as saponification.

Grease uses a variety of metal hydroxide alkaline to make and define the grease type. E.g., aluminum hydroxide makes aluminum grease, lithium hydroxide makes lithium grease, calcium hydroxide makes calcium grease, etc.

To give grease a wider temperature application range and enhanced properties a second thickener known as a complexing agent is added to the mix. This agent is a salt, usually of the same metal hydroxide used to originally thicken the grease. If lithium is used as the alkaline agent, lithium salts are then added to the mix to create lithium complex grease.

When applied, it is always the base oil, which constitutes 80% to 95% of the grease product that lubricates the bearing surfaces. The chemical soap fraction acts like a sponge to hold the oil and by default is designed to "wick" or release the oil into the bearing area when the bearing temperature rises. When the bearing ceases operation and/or the temperature cools, the grease will reverse its action and "wick" back the oil into suspension, acting as a semi-solid living reservoir for the oil.

16.1 Considering Grease as a Lubricant of Choice

Whenever rotating or interacting moving surfaces are encountered in a machine design, the designer must decide early on what lubricant is to be used.

Grease's unique capabilities give it both advantages and disadvantages over just oil as described in Table 16.1, Grease Versus Oil Comparison.

Note: Because oil is always the lubricating medium, all the decision factors involved in choosing the right viscosity and additives for the application will still apply. Before using specific grease, check with the grease vendor or manufacturer to ensure that the grease oil viscosity is suitable for your application.

16.1.1 Grease characteristics

Greases are classified and characterized in many ways. These characteristics are identifiers that allow us to rate and compare the ability of various greases with one another and can be found on the lubricant specification sheet provide by the manufacturer. The most common characteristics are:

16.1.2 NLGI consistency rating

Greases are manufactured in a variety of consistencies. These consistencies are distinguished into 9 specific classifications using the NLGI National Lubricating Grease Institute number rating system from a #000 grease representing a grease with a very fluid appearance at room temperature, all the way up to a #6 grease that appears block solid at room temperature. See Table 16.2. The most popular grease for use in grease guns is a NLGI #2 rated grease that appears soft looking at room temperature. The recommended grease consistency for use in automated centralized grease lubrication system is a NLGI #1 or lower.

NLGI consistency is determined in the laboratory using an ASTM (American Society of Testing Materials) D-217 cone penetration test (Figure 16.1) in which grease is placed in a cup and a cone is dropped from

Table 16.1 Grease versus oil comparison.

Grease	Oil
Advantages	
Stays where it is placed, easy to control	The most effective lubricant
Excellent Sealing capability - can be used as part of a labyrinth seal to keep out moisture and dirt	Excellent cleaning and flushing capability, especially in re-circulative systems
Can withstand heavy shock loads	Requires lower pump pressure and line size to move oil around in centralized systems
Requires less frequent applications than oil - better for remote locations and occasional use bearings	Can be used in re-circulative systems
	With correct application, no limit to machine speed
	Within centralizes systems, generally more stable than grease
Disadvantages	
Has a lower operating temperature than oil, can "cook" and harden in bearing in higher temperature applications	More difficult to control at bearing surface area
Introduced contamination from poor manual greasing process will not settle out and can harm bearing	Requires mechanical seals (e g radial lip seal) to control leakage
Dependent on frictional heat to allow oil to wick into contact area - bearings run hotter than oil- only lubricated bearings	
Grease lubricated bearings consume more energy to overcome grease fluid friction	
Requires greater pump pressure and line size to move grease around in centralized systems	
Total loss only lubrication	

a specified height at a room temperature of 77°F and allowed to penetrate the grease for five seconds. The depth of penetration is then measured carefully in tenths of a millimeter and rated according to an NLGI classification chart that assigns a rating number to penetration depth ranges. E.g., if the cone penetration is between 265 (26.5mm) and 295 (29.5mm) the grease is classified as a NLGI #2.

Note: Automatic grease lubricant delivery systems are designed for use with a NLGI 1 or less grease. NLGI 2 greases can be handled by most systems but will be difficult to pump in low temperature conditions

Table 16.2 NLGI grease classification # system.

NLGI #	ASTM D217 Penetration*	Appearance
000	445-475	Very Fluid
00	400-430	Fluid
0	335-385	Semi-Fluid
1	310-340	Very Soft
2	265-295	Soft
3	220-250	Medium Hard
4	175-205	Hard
6	80-115	Block

* 1/10th mm @ 77° F

If grease is classified with an EP letter rating prior to the NLGI number, e.g., EP2 grease, this will signify that the grease is formulated with an Extreme Pressure additive such as Molybdenum Disulfide for use in low-speed, high-load applications.

16.1.3 Appearance

As we have seen in the NLGI rating system grease is classified by its appearance at room temperature and ranges from very fluid to soft or smooth to a block like appearance. Descriptions may vary slightly from manufacturer to manufacturer, e.g. soft versus smooth.

16.1.4 Color

Manufacturers often color grease with dyes for identification purposes. Grease can be black, green, blue, red, brown, or white—usually reserved for food grade products.

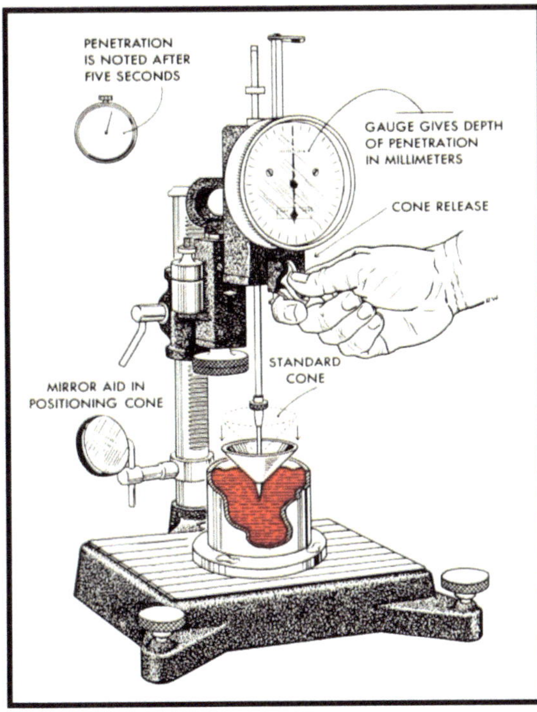

Figure 16.1 ASTM D217 Cone penetration test stand. Source: Gulf Oil, (PLIF 3).

16.1.5 Pumpability

Details the grease's ability to flow easily under pressure through a distribution system over a given temperature range

16.1.6 Slumpability

Sometimes referred to as "feedability," slumpability refers to grease's ability to relax in its reservoir under gravity and be fed into the pump. If a grease has poor slumpability, a reservoir follower plate will be required to ensure the grease level stays constant for dispensing purposes.

16.1.7 Dropping point

This is the temperature at which grease becomes liquid or soft enough to drip; important information when greasing bearings near heat sources. Note: This is not the maximum operating temperature

Table 16.3 Grease thickener types.

Grease Thickener(s)	Operating Temperature Range
Lithium	-20 to 350 F
Lithium Complex	-50 to 375F
Calcium (Lime)	-50 to 230F
Calcium Complex	-20 to 350F
Sodium	0 to 250F
Barium,	Up to 380F
Aluminum Complex	-50 to 350F
Bentonite (Clay)	-20 to 500F

Note: temperature ranges shown are approximate; always check the manufacturer's specification sheet for actual temperature range

16.1.8 Operating temperature

The recommended optimum grease operating temperature range.

16.1.9 Water resistance or washout

Grease's ability to withstand the effects of water spray before it affects its ability to lubricate and prevent rust formation.

16.1.10 Shear stability

As grease is "worked" or sheared in operating conditions its consistency may change. Greases better able to retain their consistency in working conditions have a better shear stability and are preferred. Greases that harden as they are worked are known as rheopectic grease, whereas greases that soften are known as thixotropic grease.

16.1.11 Grease thickeners

The thickener tells us what alkali base (soap base) or non-soap base (see Figure 16.2) was used to manufacture the grease. Most popular thickeners are identified in Table 1.6. Grease Thickener Types.

17

Pastes, Waxes, and Tribo-systems

17.1 Lubricating Pastes

Lubricating pastes are cohesive lubricants made up of base oil (mineral and/or synthetic oil), additives, and solid lubricant particles. They are predominantly designed to be applied under extreme conditions to prevent fretting corrosion, stick-slip and adhesive wear. Depending on their composition, lubricating pastes can be resistant to water and water vapor, with good anti-corrosion characteristics. Metal-containing pastes may be suitable for service temperatures up to 1200°C.

Lubricating pastes are classified in terms of:

- solid lubricant type (MoS_2, graphite, metals, PTFE, other plastics)
- base oil (synthetic oil, mineral oil, and mixed)
- application range (lubricating and assembly paste, high-temperature paste, conductive paste, etc.)
- special characteristics (color, EP properties, etc.)

The base oil and the solid lubricant particles have different tasks, depending on the type of paste. For example:

Lubricating and Assembly Paste - The solid lubricant improves the base oil's lubricity.

High-temperature Paste - The oil must distribute the solid lubricant particles over the friction surface. At temperatures of about 160 to 200°C all the base oil evaporates and leaves a coherent lubricant film on the friction surface.

Conductive Paste - The solid lubricant particles contained in thermally and electrically conductive pastes compensate the insulating effect of the base oil. Conductive pastes must contain a certain percentage of solid lubricant powder.

Screw Paste - These pastes are used to ensure precise assembly (tightening torque).

High-temperature Screw Compounds - The dry residue left after the base oil has evaporated must be "crumbly" to avoid sticking of the thread in the bore.

17.2 Lubricating Waxes

Lubricating waxes are formulated using a combination of synthetic hydrocarbons of high molecular weight and additives. Wax emulsions will contain an emulsifier and water.

Upon reaching a specific design temperature, lubricating waxes are designed to change state from a coherent to a fluid state. Similar to a candle, the melting point will depend on the waxes' ingredients, and their structure is reversible. If the tribological requirements are mainly about corrosion protection, a coherent structure is of advantage.

Lubricating waxes and wax emulsions have an advantage over traditional lubricants due to their excellent inherent lubricity, and special anticorrosion properties. In addition, they provide a non-tacky protective film when applied below their melting point.

Their main disadvantage is the lack of heat dissipation until the incorporated water has evaporated. Lubricating waxes do not flow below their melting point, which is of special importance for relubrication.

The exhibited positive behavior effects of waxes include:

- Adhesiveness
- Ability to stick to metals
- Polar properties
- Corrosion Protection
- Good lubricity
- Wear protection
- Ability to produce a dry film.

Their individual characteristics permit an application in both boundary and mixed friction regimes. In this context, the non-tacky wax film has an additional advantage of attracting less dust or dirt. This film ensures quasi-dry lubrication. When the friction point is heated up, the wax melts and is redistributed, whereas the perimeter areas remain below the melting point.

Table 17.1 Typical machine elements and components lubricated by wax and wax emulsions.

• Bolts	Springs
• Seals	• Sliding points
• Dowels	• Nails
• Switches	• Ropes
• Plug-in contacts	• Pins
• Screws	• Chains

17.2.1 Wax application

Lubricating waxes and wax emulsions can be successfully applied to many assorted machine elements and components similar to those listed in Table 17.1.

Waxes are suitable for most metallic material pairings that include Al alloys/ferrous metals, and Cu alloys/ferrous metals. In addition, they can be used for pairings with elastomers, plastics or wood. This ability permits them to be used in the automatic assembly of mass-produced parts, their wax film ensuring clean and easy operations. Table 17.2 shows selection criteria that can be used to aid in the selection of a suitable lubricating wax. Table 17.3 illustrates typical inspections for refined waxes.

17.2.2 Lubricating release agents

Some branches of industry require petrochemical products for mold release purposes. These release agents consist of liquid hydrocarbons and/or solid lubricants, a solvent to carry the solid lubricants, and an emulsifier to ensure miscibility with water.

Water is either used for cooling or for obtaining a concentration suiting the individual application. The performance of a product depends on the adequate combination of its ingredients. In addition to the releasing effect, a product may also be required to ensure good lubricity or to protect mold or tool surfaces.

Petrochemical release agents can be applied in the following industrial examples:

- Industrial baking tins (see Figure 17.1)
- pouring ladles

Table 17.2 Selection criteria for lubricating pastes.

Lubricating paste for	Base oil / solid lubricant	Service Temp' range (°C)	Density at 20.C (g/cm³) DIN 51757	Worked penetration DIN ISO 2137 (0.1 mm)	Base oil viscosity DIN 51561 (mm^2/s) 40°C	Notes
High load, wide service temp' range, good resistance to water and humidity, not suitable for dry lubrication	Synth. oil/ inorganic solid lubricant, soap thickener	-45-150	1.23	300 - 340	380	Threaded spindles, adjusting nuts, spline shafts, bolt and socket joints, bolt supports, toothed wheels, shaft seals, O-ring seals, V-ring seals, rubber cups, rolling bearings—very low sliding speed, assembly aid
High load, wide service temp' range, good resistance to water, not suitable for dry lubrication	Min. oil/ Ba complex soap, inorganic solid lubricant	-15-150	1.40	250 - 270	42-50	Clamping chucks
Low friction coefficient at low and high load high temperatures, suitable for dry lubrication	Min. oil/MoS_2, graphite, inorganic solid lubricant	-10-450, dry lube from approx. 160	1.60	250 - 270	70	Pressing-in of pins, bolts, bushings, , slideways, cams profile guides, sliding guides
High humidity, high temperature suitable for dry lubrication	Synth. oil/ graphite inorganic and lubricant	-30-200, above that dry lubrication up to 1200	1.01	270 – 310	120	Machine screws, plain bearings for low to medium sliding speeds
Low temperatures, high humidity, suitable for dry lubrication	Synth. hydrocarbon oil, graphite, inorganic solid lubricant	-25 to 200, above that dry lubrication up to 1000	1.30	300 - 330	800	Chains, assembly aid

- tire molds
- Stamping dies

The structure and ingredients of the various release agents depend on the requirements they have to meet, these include:

- temperature resistance
- corrosion protection

Table 17.3 Typical inspections for a balanced line of highly refined waxes (Exxon's PARVAN wax products).

Typical Inspections Grade	127	129	131	137	138	142	145	147	152	154	158	161
Melting point, ASTM D 87, °F	126	130	131	136	137	142	144	147				
Congealing point ASTM D 938, °F									151	153	157	159
Oil content, ASTM D 721 mass %	0.1	0.1	0.1	0.1	0.1	0.1	0.1	0.2	0.3	0.3	0.2	0.2
Saybolt color, ASTM D 156	+30	+30	+30	+30	+30	+30	+30	+30	+30	+30	+30	+30
Pounds per gallon @174°F	6.41	6.42	6.42	6.44	6.44	6.48	6.49	6.50	6.53	6.54	6.56	6.60
Blocking point, °F ASTM D 1465						118	121					
Flash point, °F ASTM D 92	420	420	425	440	446	450	460	470	490	500	500	500
Kinematic viscosity, ASTM D 445,												
cSt@210°F	3.4	3.5	3.6	3.9	4.1	4.7	5.1	5.4	6.2	6.6	7.3	7.4
Needle penetration @100°F												
ASTM D 1321, mm/10	105	95	90	35	30	25	25	25	25	20	20	20
UV absorbance*							Pass					

*FDA 21 CFR 172.886 and 21 CFR 178.3710

- reduction of friction
- suitability for use in contact with food products
- neutrality towards rubber and plastics.

It may also be required to apply a product that ensures a separating effect and is at the same time neutral towards plastics.

17.3 Tribo-system Materials

Tribo-system materials are a combination of a lubricant and a base material to form a self-lubricating design element. By adding a lubricant, high-strength

Figure 17.1 Baking pan mold release lubrication rail (Courtesy ENGTECH Industries Inc.)

plastics are imparted better tribological characteristics (e.g., thermoplastics with incorporated lubrication).

Plastics with good friction behavior can also be improved with strength-enhancing additives (e.g., PTFE compounds). Providing the following advantages, tribo-system materials are a good alternative to traditional lubricants:

- simplified structural design because no extra lubrication is required
- lifetime lubrication without relubrication
- no contamination caused by the lubricant
- no corrosion
- excellent resistance to chemicals

PTFE based tribo-system lubricants are suitable for vacuum applications. They can be applied at low and high temperatures, prevent stick slip, protect against wear and ensure a low friction coefficient. When selecting a proper material and in the design phase it is important to take into account the peculiarities of plastic materials.

Figure 17.2 Tribo-system materials (Klüberplast).

Tribo-system materials are used in many applications, for example:

- Machine tools (e.g., coating of slideways)
- Packaging machinery (e.g., plain bearings or sliding films in conveyors)
- Compressors, pumps (e.g., piston rings, guide rings)
- Medical equipment (e.g., plain bearings in sterilizers)
- Textile industry (e.g., sliding guides in loom grippers)
- Conveyor systems (e.g., guide roller bearings)
- Valves (e.g., sliding rings and seals, also in drinking water valves)

Typical tribo-system materials are illustrated in Figure 17.2; for selection criteria, refer to Table 9.7.

17.3.1 Tribo-system coatings

Tribo-system coatings are procedures for the application of dry lubricants for tribo-systems. The service life of tribo-system coatings depends on four main factors (see Figure 17.3): the dry lubricant used, the component's design, the load and stress factors and the manufacturing conditions.

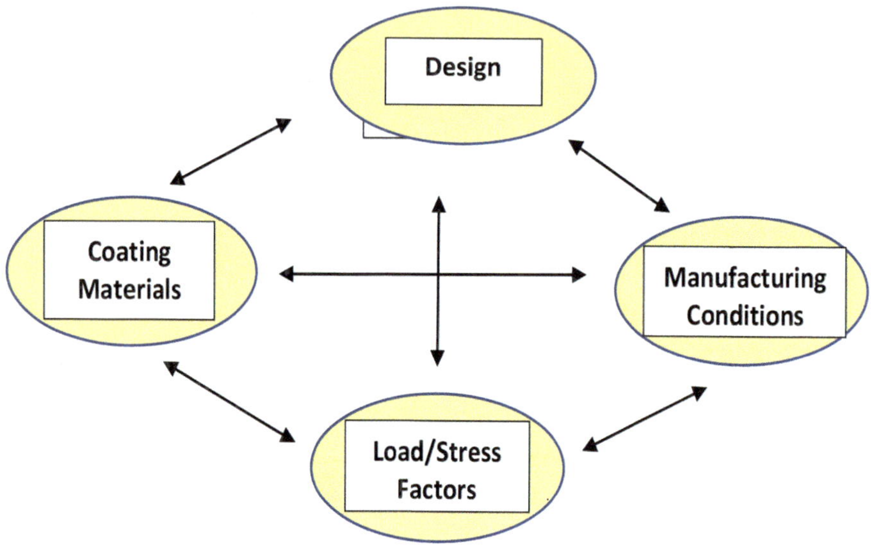

Figure 17.3 Factors influencing service life of tribo-system coatings.

Coating costs are mainly determined by the coating technique. In most unfavorable cases, the coating may be five times as expensive as the material to be coated.

Dry lubricants for tribo-systems can be applied with most of the standard methods used for lacquers: by brush, spraying, immersion, centrifugation or tumble processing.

For mass-produced parts the tumbling process is suitable provided that it is possible to entirely coat a component. This automated method provides a high degree of film thickness constancy. Tumble processing is a very cost-effective method for the coating of small to medium-size mass-produced parts.

17.3.2 Dry lubricants for tribo-systems

Dry lubricants consist of solid lubricants, a binding agent and a solvent. Their tribological behavior is determined by the type and quantity of solid lubricants.

Wear resistance mainly depends on the binder. The solvent distributes the dry lubricant over the component and evaporates during the hardening process, not having any direct impact on the friction and wear behavior of the lubricating film. Dry lubricants for tribo-systems ensure a mostly coherent film between 3 and 15 μm thickness, depending on the applied product.

They are characterized by an extremely wide service temperature range (between -180°C and 450°C) and excellent resistance to chemicals.

Their lubricating effect may be optimized by incorporating various solid lubricants. For example, products containing graphite have very good tribological behavior, and those containing MoS_2 are also suitable for vacuum applications. Dry lubricants with PTFE have a very low friction coefficient. In addition, adhesive friction is lower than sliding friction, which means that there is no stick-slip. Dry lubricants for tribo-systems provide advantages wherever:

- traditional lubricants cause contamination
- penetration may cause malfunctions
- service temperature limits of oils and greases are exceeded
- aggressive media, humidity and dust have an impact on the friction points
- lubricating oils and greases would impede the operational process
- uniform tribological conditions are required in a very wide temperature range
- the entire component requires protection against corrosion

The dry lubricant's effect is based on "transfer lubrication," a type of "erosion" of the top layers. If the lubricant layer is used up, the friction point will fail.

17.3.3 Application

Dry lubricants for tribo-systems can be applied in various ways, such as immersion, spraying, tumbling or electrostatic coating. The surface to be coated must be treated by cleaning all components to remove any grease residue prior to applying the lubricant. Sand blasting or grinding will ensure better adhesion. Phosphating improves the protection against corrosion. The selected binder system determines the hardening process.

Today's dry lubricants for tribo-systems are used in many applications, for example (Table 17.4) rolling bearings, bolts, screws, nuts, washers,

Table 17.4 Application methods for dry lubricants.

Process	Equipment requirements	Production costs	Reproducibility of film thickness
Application by brush	low	low	poor
Manual spraying	low	medium	poor-medium
Automatic spraying	high	medium-high	very good
Electrostatic spraying	high	medium	very good
Immersion by hand	low	low	medium
Immersion and centrifugation	high	low-medium	good
Immersion and tumbling	medium	low	good
Tumbling	medium	low-medium	good

Figure 17.4 Nuts coated with Klübertop.

springs, ropes, slideways, toothed gears and racks, O-ring seals, rotary shaft lip seals, threaded spindles, and more. Some of these coated components are shown in Figure 17.4.

SECTION 3

Lubricant Delivery

18

How Much and How Often? Calculating Bearing Requirements

If you were asked to choose which bearing is most likely to achieve its intended design L10 in the following service scenarios, which would you choose?

Scenario 1: a pillow-block bearing is placed in service in a HEPA-filtered clean-room manufacturing environment. The bearing runs under light-load conditions for eight hours a day; is set up using a laser-aligned and balanced drive shaft; is continually lubricated using an engineered automatic lubrication-delivery system.

Scenario 2: an identical pillow-block bearing is placed in service in a hot and dirty foundry, operating two full shifts (16 hours per day). The machinery is set up using a manual "eye-ball" alignment technique, and is manually lubricated with a grease gun on a PM schedule with a job task that states "lubricate when required."

If you are like most, you will likely have voted scenario #1 as the prospective winner. In reality, both bearings are candidates for premature failure if maintenance has not understood how failure can occur and planned accordingly to prevent it. In over 40 years of investigating real-world bearing failures, I have assembled a top-ten list of root causes listed below (not presented in any specific order).

- Lack of lubrication training
- Lack of lubrication-application engineering
- Poor housekeeping (lack of order and cleanliness)
- Over-lubrication of bearings
- Under-lubrication of bearings
- Use of dirty or contaminated new lubricants

- Infrequent oil/filter changes
- Bearing lubricant contaminated with an incompatible lubricant
- Bearing lubricated with the incorrect lubricant
- Bearing mounted out of square or misaligned when set up

Note that nine out of 10 items on this list are due directly or indirectly to ineffective lubrication practices. Many failures will result from multiple causes on the list, but by far the most prevalent failures result from **Over** and **Under** lubrication of bearings.

18.1 The Need for Precise Lubricant Delivery

The majority of rolling element bearings are lubricated with grease direct from the factory to a ready to use fill level between 25% to 50% depending on the bearing style, intended service (speed and load), and if shields/seals are used. **Note:** if your plant has already gone through a lubricant consolidation program, bearings should be ordered in pre-greased with your program required grease to avoid any lubricant cross-contamination issues

Under lubrication can result in a bearing operating in a boundary condition or dry running condition that may result in a series of micro seizes or full seize.

Over lubrication is every bit as detrimental as under lubrication. When a bearing cavity is filled there is no place left for the lubricant to go but "out" of the bearing. Pumped under pressure far greater than any seal can hold in check the excess lubricant will compromise the seal and allow contamination to find its way into the bearing surfaces creating rapid wear. In addition, the churning effect of excess grease in the bearing creates fluid friction resulting in overheated bearings, rapid lubricant loss, and increased energy consumption.

Under and over lubrication are caused by ineffective lubrication practices. This occurs when there is no planned lubrication program in place, or when a PM work order is issued with a non-engineered instruction that simply states "lubricate as necessary." Such instruction can result both extended, missed or forgotten manual lube cycles, or accelerated cycles; allow auto lube reservoirs to run dry, or incorrectly set the lube timers. Figure 18.1 shows the result of an incorrectly set lubricator that is overlubricating the bearing with no one checking on if the lubrication system is operating correctly.

In addition, when using multiple styles/types of grease guns, each can deliver a different "shot" amount. When the PM dictates "X" shots of grease,

Figure 18.1 Incorrectly set up pitman arm style auto lubricator creating a very messy and detrimental over-lubrication condition. Courtesy: ENGTECH Industries Inc.

bearings may be under or over lubricated depending on the shot size and number of strokes demanded on the work order. Figure 18.2 illustrates a manually overlubricated pillow block with a blown bearing seal allowing the grease to become a dirt collection device.

Because of the premature failure consequence due to the imprecise "over" and "under" lubrication practice, it is imperative that lubricant be delivered to the bearing point in a concise measured amount. Figure 18.3 illustrates that lubricant must be delivered in a regulated amount on a regular basis to ensure the bearing is operating in a relative frictionless full-film

Figure 18.2 Manually over-lubricated pillow block bearing. Courtesy: ENGTECH Industries Inc.

condition (represented by the black line in the green zone.) If insufficient lubricant is present, or the bearing is under lubricated as represented by the black line in the lower red zone, the bearing can be seen to operate in a boundary condition. Conversely, if the too much lubricant is present, fluid friction will cause the bearing to overheat as represented when the black line enters in to the in the upper red zone.

Figure 18.3 also shows that to achieve continued optimum full film lubrication, the right lubricant must be delivered in the right amount at the right time. When this occurs, precision lubrication is in effect

To achieve precision lubrication, the user must calculate how much lubricant is required, and how often that amount should be applied.

18.2 How Much and How Often?

There are two primary methods of calculating bearing lubrication requirements for both oil and grease. Method one is the primary method of choice for lubrication delivery system manufacturers for both oil and grease lubrication.

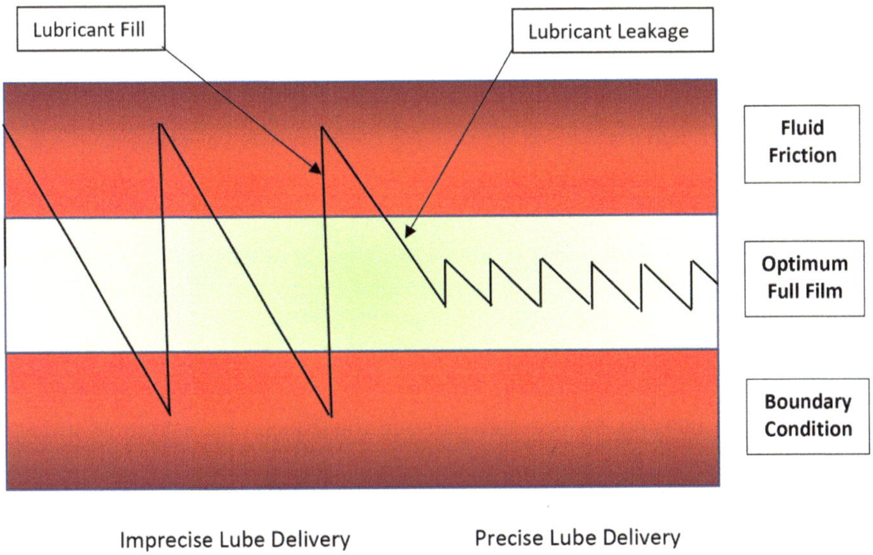

Figure 18.3 Lubricant delivery pattern.

This method is volumetric (cc/c inch displacement) based and application amount varies based on bearing type, size and a corresponding rate of replenishment calculation dependent on the lubricant and delivery method.

Method two is a weight requirement calculation method most often used by bearing manufacturers for grease applications only. This method is weight based (grams/ounces) and is calculated based on bearing type, bearing speed, and size.

It should be understood that both calculation methods provide a starting point from which to tune a bearing's actual needs that might change according to its condition and operating environment. Commonly referred to as "service factors", these change factors can include machine condition (age and wear), speed variations, temperature excess/change (hot, cold), process contamination (water, dirt, etc.), and shock or variable loading conditions.

18.2.1 Method one

When lubricating a bearing, the object is not to fill the void, but rather to provide full film lubrication as illustrated in Figure 18.1. The replenishment rate will vary depending the lubricant type (oil/grease) and its application method (manual/automated delivery). Table 18.1 shows the recommended lubricant replenishment rate under normal operating conditions.

Table 18.1 Lubricant replenishment rate (R) for normal operating conditions.

Lubricant and Delivery Method	Film Thickness	Time
Manual Application – Terminating Grease	.002" / .050mm Film	4 Hours
Automatic Application – Terminating Grease	.001" / .025mm Film	4 Hours
Automatic Application – Terminating Oil	.001" / .025mm Film	1 Hour
Automatic Application – Circulating Oil	.001" / .025mm Film	1 Minute

The volume of oil/grease required per replenishment cycle (V), is determined by multiplying the bearing surface area (A) by the lubricant replenishment rate (R) shown in table 18.1.

$$\mathbf{V = A \times R}$$

There are four bearing surface area calculations based on the type of surface being lubricated.

Area One: **Cylindrical plain bearings**, bushings, sleeves, etc.

$$\mathbf{Area\ (A) = \pi DL}$$

where: $\pi = 3.14$

D = Shaft Diameter (in/mm)

L = Bearing length (in/mm)

Area Two: **Linear/Flat bearing surface**s, gibs, ways, slides, etc.

$$\mathbf{Area = Total\ contact\ surface\ area}$$

where: Area (in/mm)

Area Three: Anti-Friction Bearings, ball, roller, needle, etc.

$$\mathbf{Area\ (A) = D^2R}$$

where: D = Shaft Diameter (in/mm)

R = Number of Rows

Area Four: Gears (each single gear), spur, herringbone, etc.

$$\mathbf{Area\ (A) = \pi DW}$$

where: $\pi = 3.14$

D = Pitch Diameter of Gear

W = Gear Face Width

18.2.2 Method Two

While method one is favored by the lubrication delivery system manufacturers, method two is the preferred method adopted by most bearing manufacturers.

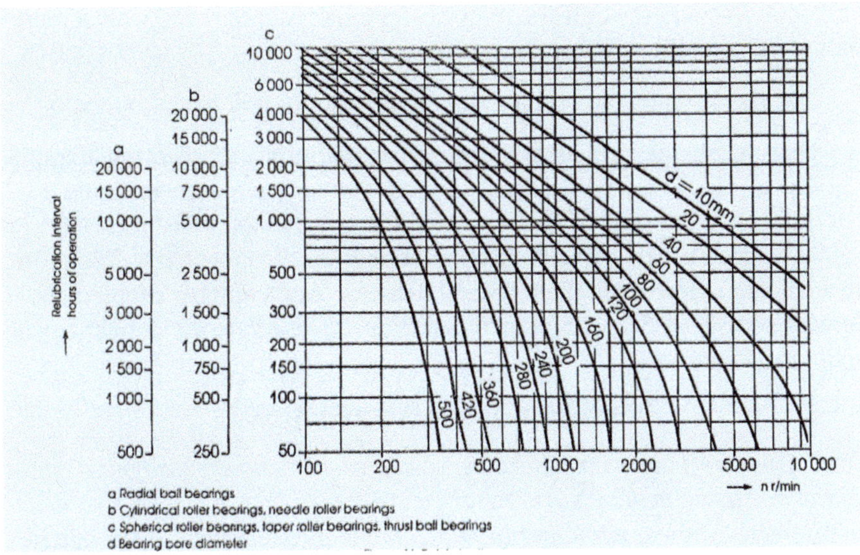

Figure 18.4 Typical bearing company relubrication chart. Source: McGuire Bearing.

Method two is for grease lubrication only and the amount of grease required for each relubrication application uses the following simple formulae.

For Imperial bearings:

$$\textbf{Amount in Fluid Ounces} = \textbf{Bearing OD (in)} \times \textbf{Bearing Width (in)} \times \mathbf{0.114}$$

For Metric bearings:

$$\textbf{Amount in Grams} = \textbf{Bearing OD (mm)} \times \textbf{Bearing Width (mm)} \times \mathbf{0.005}$$

The second method uses a nomograph style chart to establish the relubrication frequency interval as depicted in Figure 18.4.

Example: we are looking to establish the amount and lubrication frequency for a standard ball bearing with an OD of 85mm, ID of 45mm, and a width of 19mm. Its operational speed is 1750 rpm for two shifts (16hrs) per day, 7 days per week, 52 weeks per year.

Lube amount in grams = 85 × 19 × 0.005 = **8 grams**

Using the chart in Figure 18.4, find 1750 on the Y axis (horizontal) and draw a vertical line until it intersects the bore diameter line (45mm). Then, draw a horizontal line and read off the relubrication interval hours for scale a representing radial ball bearing requirements.

In this example, the relubrication interval is **10,000 hrs** of operation

The bearing operates 5824 hours per year, therefore the bearing requires 8 grams of grease every 1.72 year period. As we have seen in figure 18.3 it is more advantageous to deliver a small, precise amount of lubricant on a regular basis rather than one large amount on a less frequent basis. Therefore, we are better to space out the lubricant delivery over a number of intervals. If we were to use a 1.5cc (1.5 grams) per stroke grease gun, the delivery cycle would round out to:

1 stroke of grease every 15 weeks

Note: *when working with fluid ounces, the reader should recognize that a fluid ounce is a common North American measurement of a fluid's volumetric property, whereas an imperial ounce is the measurement of the mass or weight of a solid material. Grease gun and lubrication pump displacement is usually measured in Cubic inches or grams. To convert a fluid ounce into cubic inches simply multiply by 1.805. In SI units, 1 gram is equal to 1ml (fluid) or 1cc (solid).*

References

Lubrication for Industry, Second Edition, Industrial Press. Bannister Kenneth E 2007 How Often Should I grease my Bearings? McGuire Bearing,

19

Manual Lubrication Delivery Systems for Oil & Grease

19.1 Manual Oiling

Although not as prevalent as manual greasing, when a bearing surface is contained, it makes more sense to lubricate with pure oil. This is especially the case when lubricating small mechanical devices such as a watch, clock, or industrial timing mechanisms. Applications of this kind usually only require a single drop of two of lubricant.

Oil is applied using an oil can with a straight tapered spout, or a plunger style pump that allows the lubricator to control the oil flow directly onto the lubricated surface(s), or into an oil cup that allows oil to be externally applied and drain. Figure 19.1 shows a typical plunger manual oil pump, while Figure 19.2 shows an old lubrication poster promoting oil lubrication to conserve the machine health that shows an oil cup in the lower right of the picture.

Although manual oiling is still in use, manual greasing is much more prevalent in the industrial environment.

19.2 Manual Greasing

The lowliest and most prolific device found within the world of lubrication recently celebrated its centennial anniversary. Designed as a simple, yet efficient mechanical gateway device, the grease nipple serves only to connect a manual grease delivery gun and provide a protected one-way access to the bearing cavity for filling purposes.

Prior to Arthur Gulborg's invention of the grease gun in 1918, bearings were greased individually using a grease cup device. The grease cup, aptly named, resembles a large thimble sized reservoir sitting above the bearing entry point. The reservoir, (grease cup), has a screw-in lid that upon removal allows grease to be manually paddled in place. Once filled, the lid is partially

Figure 19.1 Plunger style oil can. Images Source Unknown.

Figure 19.2 Lube poster showing oil cup. Images Source Unknown.

screwed back in place. In doing so, the screw-in action mechanically forces (pushes) new grease into the bearing through an orifice in the cup's base screwed into the bearing. As the machine continues to operate, at regular intervals, the operator, or designated lubricator takes time to screw in each cup lid a partial or number of turns to force lubricant into the bearing This effort is repeated over a period of time until the cup is empty and requires manual refilling. Figure 19.3 shows a grease cup cutaway.

While working for the Alemite Die Casting and Manufacturing Company in Chicago, Illinois, Arthur Gulborg was responsible for ensuring all machine lubrication cups were operated and filled regularly.

Recognizing an opportunity for efficiency, Gulborg set about and designed the World's first pressurized grease gun that could be directly coupled to the grease fitting, thereby making the grease cup obsolete. He named his design the "Alemite High Pressure Grease System, which, when adopted in 1918 as the greasing standard for the US Army, made the company's fortune. See Figure 19.4.

By 1922, Gulborg's grease gun design had evolved into a Push/Pump style gun that could be placed up against his newly designed button-head compression grease fitting that opened under the grease gun pressure

Figure 19.3 Grease cup cutaway. Source: Unknown

Figure 19.4 Arthur Gulborg's grease gun lubrication system invention. Source: ENGTECH Industries Inc.

allowing the lubricant to flow through to the bearing point. As this was not a positive "lock on" connection, the grease gun operator had to use both hands to assure the grease gun end was precisely positioned square to the grease fitting's face to ensure no grease leakage occurred while operating the grease gun.

At the same time, not far away in Kenosha, Wisconsin, an eccentric Hungarian immigrant named Oscar Ulysses Zerk was busy at work inventing a simple lubricant receiving nipple shaped spring activated, high-pressure grease fitting, destined to take the lubrication world by storm.

The Alemite Company, continued development of its high-pressure grease system and under its new ownership with the introduction of the hand trigger operated grease gun that now connected to the first positive lock grease nipple they called the pin-type grease fitting. The fitting style, still in use today, is used wherever a guaranteed high-pressure leak proof seal is required. By 1922, this new system was successfully being promoted for, and used on, mass production automobiles.

In 1923, Oscar Zerk received his patent for the Zerk grease fitting and began marketing his invention through the Allyn-Zerk Company of Cleveland, Ohio. His smaller, less expensive, more compact snap on-off ball lock design proved a worthy competitor to the more cumbersome Alemite pin lock design. So much so that in 1924, the Alemite Company purchased the Allyn-Zerk Company outright and standardized the Zerk grease fitting for use with its new high-pressure grease gun systems we know and still use today.

The mid 1920's then witnessed a buy out by its present owners Stewart-Warner corporation, who continued development of the grease gun and grease nipple under the engineering guidance of Joe Bystricky. In 1933, Bystricky reinvented and patented (US Patent # 2,016.809) the positive sealing slip/lock grease gun nozzle and improved grease nipple design still commonly, and incorrectly, referred to as the zerk fitting system. See Figure 19.5.

A testament to the Bystricky design, his grease nozzle/nipple design has changed little over the past century. Ask anyone to describe a manual lubrication system and most will immediately visualize a trigger or lever action grease gun connected to a "zerk" grease nipple. However, not all of today's grease fittings are the same and employing or replacing grease fittings will require some investigation.

19.2.1 Choosing the right grease fitting for your needs

Grease nipples are truly simple devices made up of a shaped and threaded housing, ball and retaining spring; rarely do they fail. Should it prove difficult or impossible to pump grease into the fitting it is usually indicative of a hydraulic lock caused by a problem with the bearing itself. That being said, grease nipples can corrode or be physically damaged when struck and may need replacement from time to time. When this is the case, a maintainer or planner should be aware of the multiple choices available.

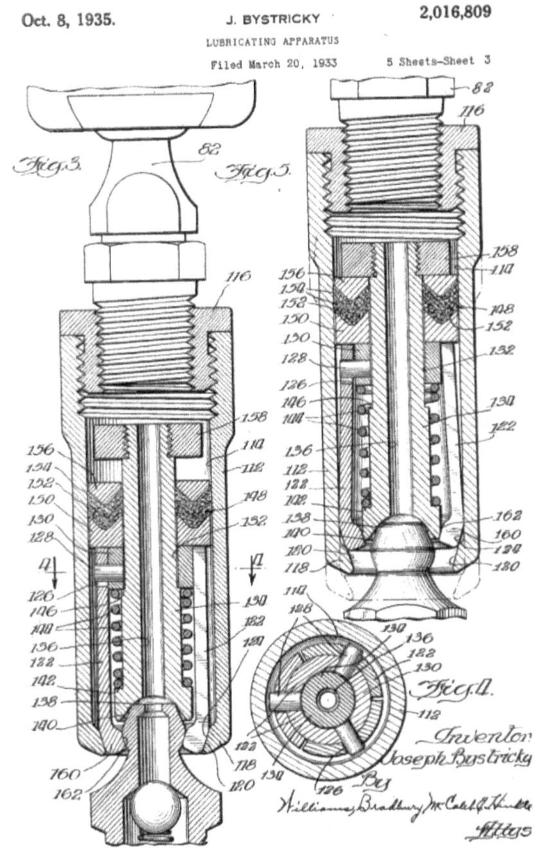

Figure 19.5 Joseph Bystricky grease gun nozzle patent. Source: US patent # 2,016,809

19.2.2 Material choice

The most popular and least expensive grease nipple material is mild steel. In wet, humid, and/or corrosive environments, choosing a nipple made from Zinc plated steel, aluminum, stainless steel or brass may make for a better decision. Fittings made from monel are recommended in highly corrosive or caustic environments.

19.2.3 Thread choice

Depending on where the machinery was made or specified, grease nipple threads may be imperial or metric. Depending on the size and application, the thread may be a straight (parallel) or taper thread; British Pipe, National

Figure 19.6 Assorted grease fitting styles.

Pipe or Unified thread. When in doubt, use a grease fitting thread gauge or a set of thread pitch gauges to determine the correct thread size. Fitting an incorrect threaded nipple can cross thread the bearing and cause grease leakage to occur.

Note: for light duty applications, non-threaded drive fittings are available

19.2.4 Style choice

Depending on access to the grease nipple the fitting style may need to be considered. Most grease nipples are straight where the nipple is directly lined with the bearing entrance. For more poor access areas, an angled grease nipple may prove advantageous. Grease nipples can be purchased in 90, 60, 45degree angles as seen in Figure 19.6. Extended barrel nipples can be purchased when reach is a problem. If protrusion of any kind is a problem, a non-protruding or flush grease nipple can be employed.

19.2.5 Specialty fittings

For heavy, high-powered machinery, military cannons, large automobiles, etc., a large button head fitting (sometimes referred to as a DIN fitting) is

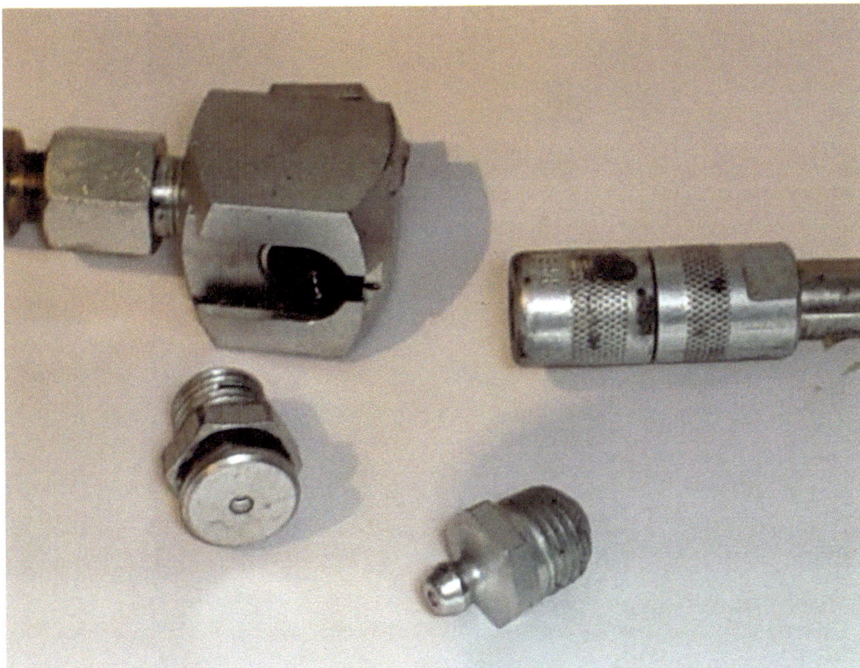

Figure 19.7 Button Head (DIN) fittings alongside regular zerk grease fitting and adaptor. Photo: Courtesy ENGTECH Industries Inc.

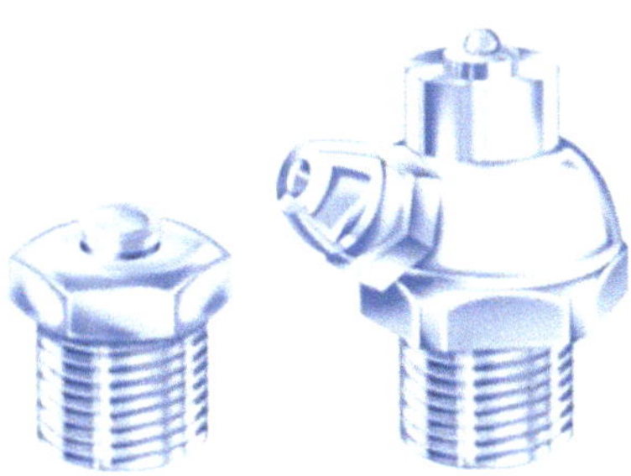

Figure 19.8 Pressure relief and Hydraulic shut-off fittings. Source: Mobil Oil.

available. This is a positive lock fitting and will require its own grease gun slide-on applicator device. This is demonstrated in Figure 19.7 below.

Where absolutely no leakage is allowed, the original Alemite pin-lock fitting is still available for use, and like the button head fitting, will require a specialized grease gun applicator device.

Pressure relief grease fittings are compact safety valves available in pressures as low as 0.25 psi up to 1,600 psi, (Figure 19.8, LH fitting) used to relieve internal bearing pressure where continued positive internal pressure can cause problems.

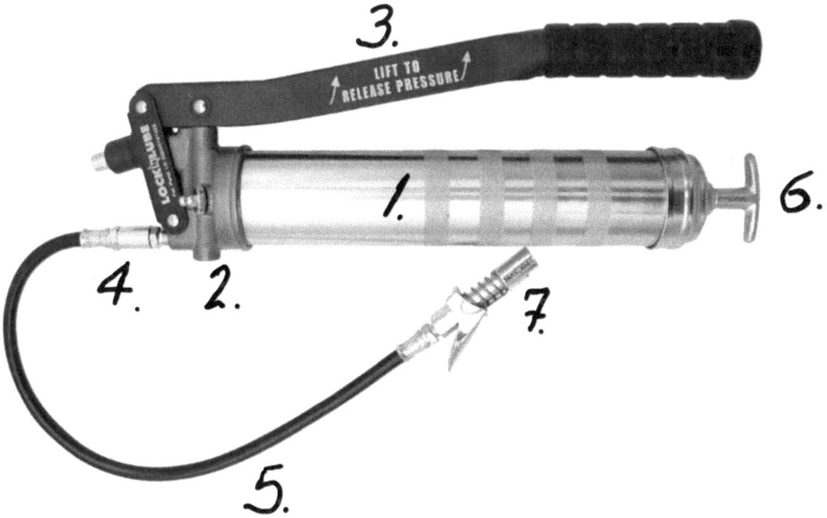

Legend

1. Grease Barrel
2. Grease gun head with pressure relief nipple shown
3. Lever (trigger)
4. Hose Swivel
5. Flexible Hose
6. Tee Handle
7. Positive Lock Grease Coupler

Figure 19.9 The gold standard professional lever action grease gun design. Image Courtesy: LocknLube Corporation.

Similarly, hydraulic shut off grease fittings (Figure 19.8, RH fitting) are designed to positively close at a pre-determined back pressure to prevent over-lubrication and the rupture of bearing seals during the filling operation. This combination of fittings is commonly used to protect electric motor bearings from over-lubrication and the forced release of grease into the motor windings. See manually greasing an electric motor later in this chapter.

19.3 Anatomy of a Grease Gun

Grease guns are manufactured in many different styles, capacities, delivery, and pressure ratings.

Figure 19.10 illustrates a similar, lighter duty grease gun. This gun is fitted with a transparent grease barrel that allows the lubricator to see and determine, at a glance, the exact grease lubricant being pumped. This is a handy feature that helps eliminate cross contamination of greases.

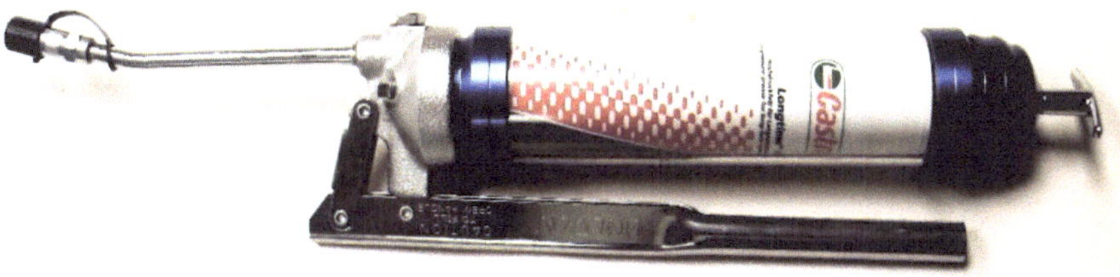

Figure 19.10 A lever action grease gun with see through barrel. Note: barrel ends are also color-coded blue. Image Courtesy: ENGTECH Industries Inc.

19.3.1 Output and delivery matters

A grease gun is an inherently simple device designed to operates on basic hydraulic principles. Depending on the devices internal design, a few strokes of the trigger, or lever, can produce an astounding level of delivery pressure ranging from 2,500 psi up to 15,000 psi. With bearing seals designed to withstand only a fraction of that type of pressure, it is easy to understand how an untrained grease gun operator with a few strokes of a grease gun can overfill the bearing cavity, reach enough pressure to rupture the bearing seal and over-lubricate the bearing. The result is invariably premature bearing failure and equipment downtime. This is sometimes referred to as "killing a bearing with kindness." A grease gun in the hands of an untrained operator is literally a lethal weapon.

Sadly, this is not all the grease operator's fault as the grease gun pressure, or delivery displacement is rarely, if ever, stamped on the device. If a grease gun's maximum pressure is unknown it can be determined using a simple pressure test rig comprised of a fixed 20,000 psi gauge connected to a grease fitting. The grease gun is connected to the nipple using a positive lock on grease coupler and pumped until the grease gun hydraulic locks up or "dead heads". The pressure is then read on the gauge and tagged, stamped or marked on the grease gun in permanent marker.

Similarly, a grease gun's delivery output, or displacement is not marked on the grease gun. Grease guns aren't standardized or built to the same design specifications. Therefore, their displacement output volume, or "shot" size will likely be different between grease gun make, style and model. This poses enormous problems for a plant when a PM task calls for 2 shots of grease and the grease guns in the plant are not standardized. Keep in mind that 2 shots of a 3cc displacement gun will deliver six times more lubricant than 2 shots form a similar looking 0.5cc displacement grease gun. Table 19.1

Table 19.1 Grease Gun comparison chart. Courtesy ENGTECH Industries Inc. and Alemite Corporation.

Grease Gun	ID #	Capacity	Delivery	Pressure
	F104	3oz bulk	1oz-25 stroke	**2500psi**
	525	14oz cartridge	1oz-33 stroke 1oz-60 stroke	**4500psi 10,000psi**
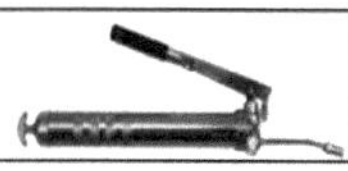	4015	14oz cartridge	1oz-7 stroke	**1700psi**
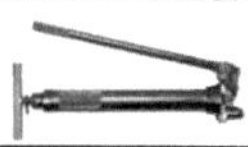	6268	12oz bulk	1oz-24 stroke	**15,000psi**

compares different styles of grease guns and illustrates perfectly the differences between them and the need to know your grease gun's capability.

Some grease gun manufacturers further confuse the delivery output issue by marketing grease gun reservoir capacity and shot displacement amount in fluid weight (grams and fluid ounces). Be aware that fluid weight is based on the specific gravity of water. Grease manufacturers formulate their greases using different ingredients of varying specific gravity and weight that can result in one grease tube weighing in at 300 grams versus another weighing in at 400 grams. Both require the same displacement into the bearing. To avoid confusion in this matter, always strive to use volume displacement as your grease unit of measure.

To measure the grease pump delivery output, a test tube or syringe marked with cubic centimeters or cubic inches (Figure 19.11) and pump in 10 shots of grease. Note: one complete lever or trigger cycle equals one shot.

Divide the total mount shown on the measure by 10 to arrive at the actual shot size. For example, if ten grease shots measured 20cc on the syringe or test tube, the actual shot amount is equal to 20 divided by 10, which equals 2cc per shot. Pneumatic, or electric operated grease guns can also be calibrated for each pulse, or can be measured using a calibrated flow meter attachment.

19.4 How to Load a Grease Gun

Although manual greasing appears intuitive and easy to accomplish, the reality is many maintenance practitioners have never received any formal

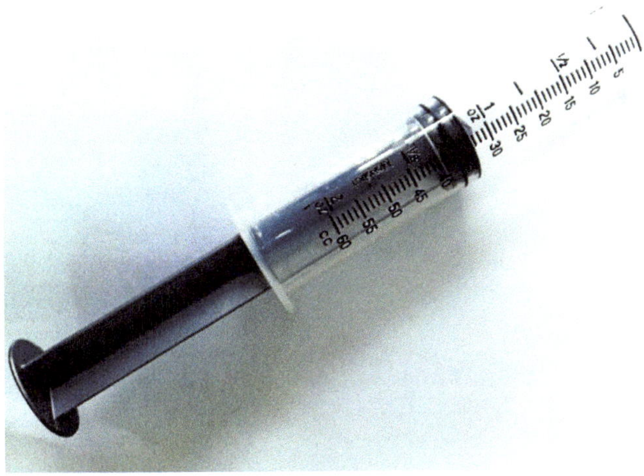

Figure 19.11 Marked syringe used to determine grease gun displacement. Image Courtesy: ENGTECH Industries Inc.

training in assessing the capability of a grease gun and the actions required to ensure its proper use.

How well do you know your grease gun? Do you know how to correctly cartridge and/or bulk load your gun? Do you know how to expel trapped air from it? Do you know it's shot size in grams, fluid ounces, cubic centimeters or cubic inches? Do you know how much pressure your gun develops? If you don't, you're not alone.

19.4.1 Grease cartridge loading

The majority of grease guns purchased today look similar to those shown in Figures 19.9 and 19.10. This style of grease gun can usually be cartridge loaded or bulk loaded. Cartridges sizes are standardized across the industry and filled with either 400 or 300 grams of grease fill depending on the grease's specific gravity. To correctly load a new cartridge:

1. Wipe the grease gun barrel (1) clean with a lint-free rag.
2. Unscrew the grease gun head (2) from the barrel (1), and place on a clean surface or clean paper towel.
3. Firmly hold the grease barrel (1) in one hand and with the other hand, firmly pull out the center rod using the rod tee handle (6) at the end of the barrel until it can go no further. Once out as far as it will go, lock the rod in the extended position. Depending on the rod style, it may have

Figure 19.12 End of grease barrel showing Tee handle and lock slot. Courtesy ENGTECH Industries Inc. and LocknLube Corporation.

a friction-lever lock built into the end of the barrel that automatically holds the rod in place when extended (see Figure 19.12.) Or, the barrel end may have a slotted hole that requires the extended rod to be positioned across the slot into the locked position.

4. If replacing a used cartridge already in the barrel, carefully remove the spent cartridge from the open end of the barrel, taking care not to cut a finger with the sharp open edge of the spent cartridge. Discard the empty cartridge in accordance with the local authority's contaminated waste lubricant requirements (See chapter xx)

5. Wipe the barrel (1) clean with a lint free cloth and place beside the grease gun head (2) on the clean surface.

6. Ensure the new grease cartridge is filled with the same grease as the old cartridge. If not, the grease gun head (2), hose (5), and end connector (7) will need to be thoroughly degreased and cleaned to ensure lubricated bearings are not cross-contaminated with two different greases in the grease gun.

7. Pull the plastic end cap off the new grease cartridge and immediately insert the cartridge into the grease-gun barrel open end first. Once fully inserted, remove the pull-tab foil end from the cartridge and discard both cartridge endcaps in accordance with local authority's contaminated waste lubricant requirements.

8. Fully screw the grease gun barrel (1) back on to grease gun head (2) and back off (loosen) one turn.

9. Release the rod tee handle (6) by push releasing the friction-lock lever, or returning the lever rod back across the slot to its center position, then slowly push the rod back in place in the barrel as far as it will go.

10. Pull the trigger, or pull/push the lever (3) until grease securely tighten the gun head (2) to the barrel (1).

11. If no grease flows, an air lock is likely to blame. To release trapped air, pump the grease gun a couple times then, if fitted on the grease head, push the air-release valve and re-pump the grease gun. If grease still doesn't appear, repeat the process. On smaller grease guns with no air-release valves, back off (loosen) the barrel a couple of turns, and pump until grease appears, then retighten the barrel. *Note: Some grease guns offer a combination air bleed nipple/bulk filler port mounted in the grease gun head (shown in Figure 19.9) This can also be used to attach the grease coupler when the grease gun is being stored.*

12. Once again, wipe the grease gun clean with a lint-free cloth, and then be sure to transport all spent materials in a contaminated-waste container for removal.

13. If no see-thru barrel is employed (see Figure 19.10), indicate on the grease barrel by color coding or writing the grease type on a barrel sleeve.

19.4.2 Bulk grease loading

Many grease guns have a dual-fill feature that allows the barrel to accommodate a standard grease cartridge, or to be bulk loaded. For such guns, and those designed specifically for bulk loading:

1. Clean the grease gun with a lint-free rag,

2. Unscrew the grease-gun head (2) from the barrel (1), and place on a clean surface or clean paper towel.

3. Immerse the open end of the grease barrel into the bulk-grease container and slowly plunge it into the grease while drawing back the rod tee handle (6) until fully extended and the grease barrel is full.

4. Alternatively, the barrel (1) can be hand-packed from the open end with the tee rod (6) extended and locked. NOTE: This type of fill can be messy and prone to dirt and air inclusion.

5. Fully screw the grease-gun head back (2) on the barrel (1) and back off (loosen) one turn.
6. Release the rod tee handle (6) by push releasing the friction-lock lever or returning the lever rod back across the slot to its center position, then push the tee rod back in place in the barrel as far as it will go.
7. Pull the trigger, or push/pull the lever (3) until grease begins to dispense, then securely tighten the barrel to the gun head.
8. If no grease flows, an air lock is likely to blame. To release trapped air, pump the grease gun a couple times then, if fitted on the grease head, push the air-release valve and re-pump the grease gun. If grease still doesn't appear, repeat the process. On smaller grease guns with no air-release valves, back off (loosen) the barrel a couple of turns and pump until grease appears, then securely retighten the barrel.

   Note: Some grease guns offer a combination air bleed nipple/bulk filler port mounted in the grease gun head (shown in Figure 19.9) This can also be used to attach the grease coupler when the grease gun is being stored.
9. If no grease flows, an air lock is likely to blame. To release trapped air, pump the grease gun a couple times then, if fitted on the grease head, push the air-release valve and re-pump the grease gun. If grease still doesn't appear, repeat the process. On smaller grease guns with no air-release valves, back off (loosen) the barrel a couple of turns, and pump until grease appears, then retighten the barrel,
10. Once again, wipe the grease gun clean with a lint-free cloth, and then be sure to place all spent materials in a contaminated-waste container for removal.

19.4.3 Cleaning & storage practices

After use, the grease gun should be wiped clean with a lint free cloth and made ready for its next assignment by placing the lever in the closed or collapsed position for storage.

Ideally, the grease gun should be stored singularly in a grease caddy as shown in Figure 19.13. Grease storage caddies can be magnetized to a toolbox or metal cupboard, or screwed to a wall.

This type of storage device will ensure your grease gun remains damage free, dirt free, and, very important, cross-contamination free from contact with other grease guns. Figure 19.14 shows how NOT to store a grease gun.

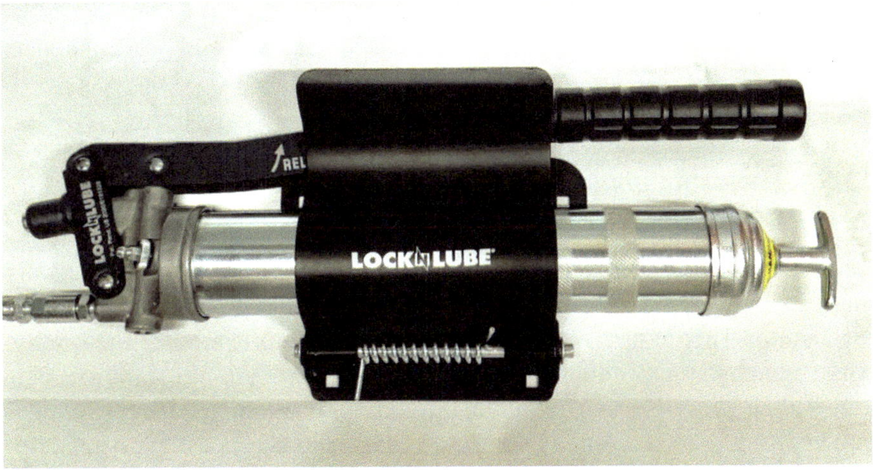

Figure 19.13 Horizontal grease gun caddy. Source: ENGTECH industries Inc. and LocknLube Industries.

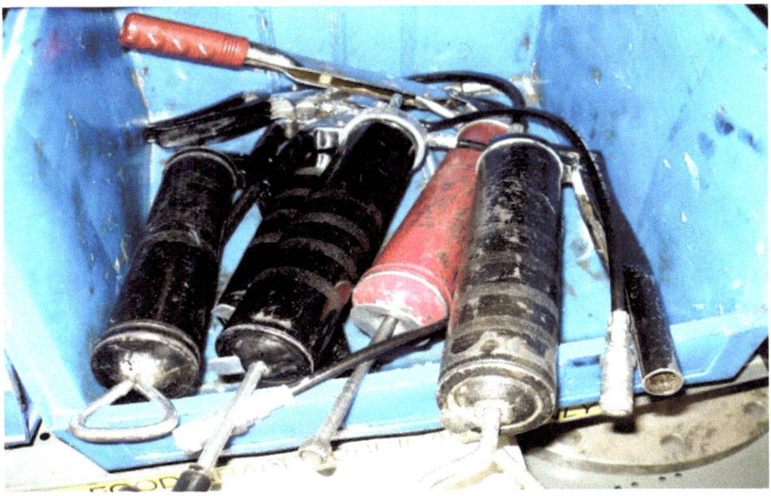

Figure 19.14 Typical grease gun storage method particularly prone to dirt and lubricant cross contamination, Courtesy: ENGTECH Industries Inc.

19.5 How to Grease a Bearing - In Seven Steps

The grease gun's "connect and pump" theory is overwhelmingly simplified when compared to the necessary actions required to grease effectively. The effects of poor manual greasing control are exacerbated when:

- lack of lubricator training is coupled with the mythical belief that if a little lubrication is good, then a lot of lubrication must be better.

- Use of simplistic job plan instructions such as "lubricate as necessary"—perpetuated by equipment O&M manuals and work planners alike.

19.5.1 Perform simple act of greasing in a consistent and controlled manner

Ideally, bearing cavities require between 25% to 50% grease fill to effectively lubricate the bearing. Over filling creates friction through the grease "churning" action. This in turn requires additional energy to overcome the friction while creating an unwanted spike in bearing temperature.

When every bearing requirement and application frequency is calculated according to size, speed, load, and usage and all grease guns standardized to a single design, the grease gun "shots per bearing" can be calculated and a formal program be instituted.

In the lack of a formal grease program or training, maintainers can exercise control over their greasing actions and promote extended bearing life by consistently following the seven simple steps listed below

Step one – Action check

Check Work Order or machine plate to determine the following:

- Grease point location (if numbered or colored)
- Amount of grease required per grease point (only if some grease point engineering has taken place)
- Grease specification matches the grease found in the grease gun
- If this information is missing, compile a data page in a lubrication notebook for each machine and stay with a consistent standard manual grease choice until a product consolidation program has taken place

Step two – Pre-clean

- Remove protective grease nipple cap, if fitted. See Figure 19.13.
- Clean grease fitting(s) to be greased with a clean lint-free cloth

Note: A lint-free cloth is usually made from a cotton fabric with sewn and overlapped edges; the fabric is close woven and does not readily 'shed' or release fabric fibers. Purchase or rent lint-free mechanic's cloths from a cleaning or safety supply house. Mechanic's cloths can be cleaned and

Figure 19.13 Grease-point colored collar ID and dust protector tag. Courtesy: Fluid Defense Systems.

reused over and over again. Disposable lint-free paper-based products are also available.

Rag style cloths made from discarded garments are NOT *recommended for wiping grease nozzles or fittings. Rags are cut or torn into size and can be made from any fabric type, including wool (wool fibers are easily shed and can roll themselves into pill balls). The torn and cut fabric edges easily release errant fibers that can adhere to the grease fitting or grease nozzle and then be injected into the grease fitting. Fibers can be abrasive and promote premature bearing wear.*

Step three - Prepare gun for greasing

Ensure the grease gun is charged with grease, primed and cleaned ready for grease delivery:

Trigger Style Grease Gun

- Clean grease fitting(s) to be greased with a clean lint-free cloth.
- Slowly squeeze trigger toward grease gun barrel until the grease discharges approximately ¼ inch or 6mm of grease from the end of the grease gun nozzle

Figure 19.14 Connecting a manual grease gun using a grease coupler and lever action grease gun (note gun lever in closed position.) Courtesy LocknLube Corporation.

- Wipe the grease gun nozzle/coupler clean with a clean lint-free cloth

Lever Style Grease Gun

- Collapse lever to the barrel causing grease to discharge from the nozzle
- If lever is already collapsed, open lever and collapse discharging a shot of grease from the grease nozzle.
- Wipe the grease gun nozzle/coupler clean with a clean lint-free cloth

The gun is now ready – continue to Step 4 immediately, if gun is dropped or placed down, contamination of the grease gun nozzle/coupler is likely and step 3 must be repeated

Step four: Connect nozzle to the grease nipple

Zerk style grease fitting

- Method one: Align the grease gun nozzle with the zerk grease fitting and push the nozzle firmly on to the fitting. Ensure nozzle is securely connected by slowly rocking the nozzle side to side and trying to gently pull the grease gun away, if resistance is felt, the grease nozzle is connected. Figure 19.7 shows the regular zerk grease fitting and connector.

- Method two: if a positive lock grease coupler is employed, connect coupler directly onto the grease nipple and pull back slightly on the coupler to ensure it is locked in place, then begin the greasing procedure. See Figure 19.14.

DIN – Button Head style fitting

- Slide the grease gun button head nozzle's location slot across the button shaped head of the grease fitting until it comes to a firm stop. Figure 19.7 shows the DIN fitting and connector,

Step five - Deliver lubricant

Note: if the bearing grease calculation or grease gun-shot size output is unknown, it is difficult to impossible to know exactly how much grease is required to fill the cavity to a 30% to 50% fill. If this is the case, the best tactic is to "feel" the grease into the bearing cavity until slight resistance or back pressure is felt against the trigger, or lever, signaling the bearing cavity is now full. **Warning!** Pumping beyond this point will almost certainly blow the bearing seal and create an open channel for contamination to get into the bearing cavity and cause premature failure.

Trigger Style Grease Gun

- Gently Squeeze trigger (use only finger tips) towards grease gun barrel and "feel" grease into bearing until resistance or back pressure is felt.
- Relax hand and allow the trigger return spring to reset the trigger back to the loaded position
- Repeat if more lubricant is required

Lever Style Grease Gun

- Gently push lever towards grease gun barrel using palm of hand and "feel" grease into the bearing until resistance or back pressure is felt
- If continuing to grease, pull lever back to the fully extended position and repeat if more lubricant is required

Note: if no resistance is felt, check to see if grease gun reservoir/cartridge is empty. If grease is present, check to ensure bearing seal is not compromised.

Note: One squeeze action of a trigger style grease gun and one lever action of a lever style grease gun = one shot of grease.

Step six – Disconnect grease gun

Zerk style grease fitting

- Method one: Rock the standard grease gun nozzle side to side and firmly pull the grease gun away from the grease fitting.
- Method two: if a grease coupler is used, squeeze the coupler's spring cover to open the coupler jaws and extract the coupler from the grease fitting

DIN – Button Head style fitting

- Slide the grease gun nozzle's location slot across the button shaped head of the grease fitting in a reverse manner until disengaged

Step seven - Post clean

- Wipe residual grease from the grease nozzle and grease fitting using a lint-free cloth
- Refit protective grease fitting cover

Note: in very dirty locations, when no protective grease fitting is utilized, it is often preferable to leave a grease smear on the grease nipple to ensure dirt does not build up on the nipple surface and can be wiped away successfully at the next manual grease cycle.

Good grease gun hygiene automatically makes you part of the solution, and remember that greasing is a deliberate act requiring a gentle touch and active restraint.

19.6 Greasing Distribution Devices/Systems

Machines often require manual grease lubrication to multiple points positioned in multiple locations across the machine. Often these grease points can be difficult to access and may require special connector/devices to help apply the grease. The following demonstrates some of the specialized manual lubrication adaptors/tools available for use.

Grease Injector Needle is for use with hand operated grease guns and is employed to introduce fresh lubricant in to a sealed bearing. The needle is inserted under, or through, the seal's rubber cover.

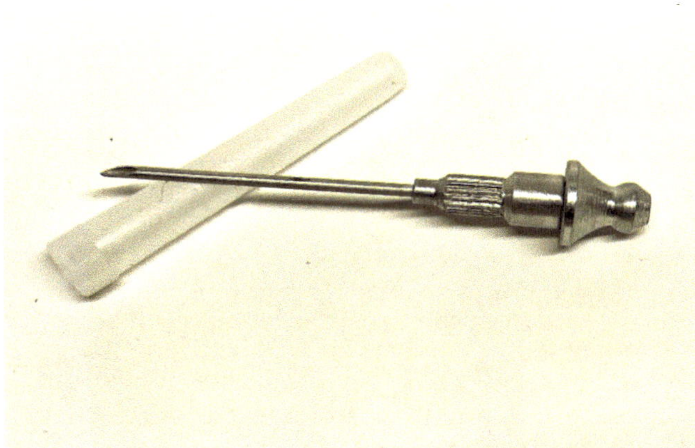

Figure 19.15 Grease injector needle. Courtesy: ENGTECH industries Inc. and LocknLube Industries.

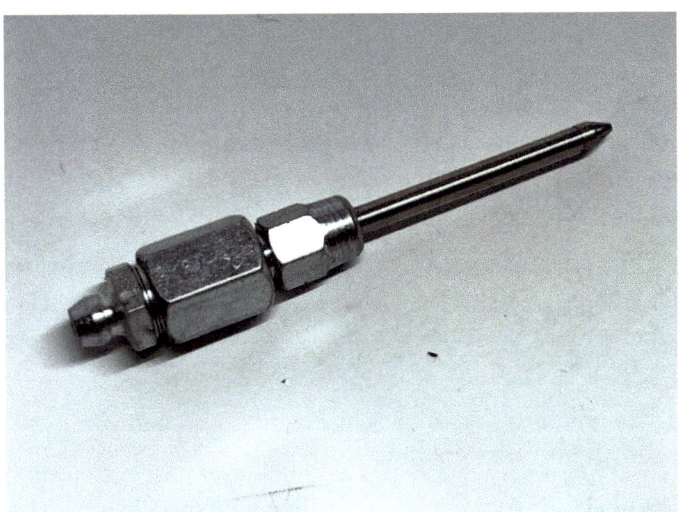

Figure 19.16 Narrow needle nose dispenser Courtesy: ENGTECH industries Inc. and LocknLube Industries.

Narrow needle nose dispenser is used to dispense lubricant in a fine line for shafts and linear applications in hard-to-reach area

The 90° Adaptor is for use with hard-to-reach grease fittings where the flexible hose cannot radius enough to attach to the grease fitting. Two versions of the right-angle adaptors are shown

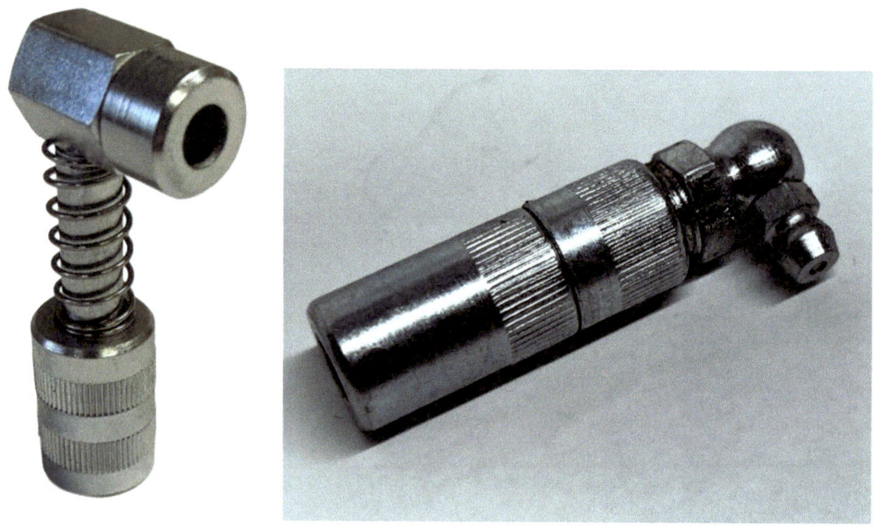

Figure 19.17 Two styles of 90° - grease fitting adaptors. Courtesy; ENGTECH industries Inc. and LocknLube Industries.

Seal-off Grease Adaptor is used for grease points that do not have Zerk fittings in place. The rubberized end is pushed into the open lubrication hole to make a seal and allow the lubricant to enter the bearing surface under pressure with minimal mess.

Recessed Fitting Extension is a rigid, narrow extension for the grease coupler enabling the grease gun to reach fittings in confined areas. The tip has a concave cup shape to allow angular flexibility when positioning the adaptor onto the zerk fitting.

Adaptor Kits similar to that shown in Figure 19.20 below are an excellent addition to any lubrication technician's toolbox.

Grease Fitting Multi-tool is used to facilitate the installation and extraction of grease fittings. This multi tool is also able to re-thread the grease fitting hole

19.7 Ultrasonic Greasing

As referenced in Chapter 18 – How Much, How Often? Calculating Bearing requirements, manual lubrication is based on a preventive basis that is time and fixed displacement based. Ultrasonic greasing differs in that lubricant is displaced based on the bearing condition requirement that is sensed using ultrasonic (U/T) wave detection.

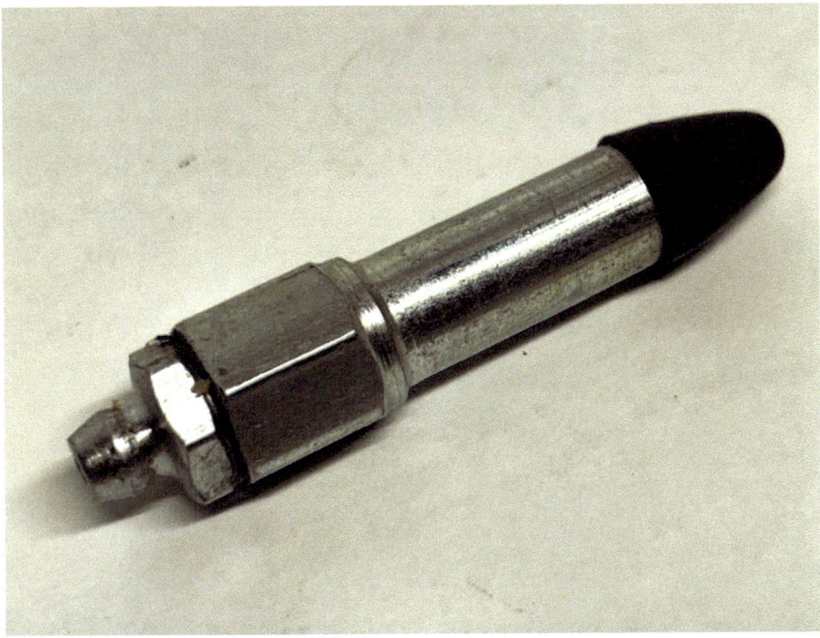

Figure 19.18 Seal-off grease adaptor. Courtesy: ENGTECH industries Inc. and LocknLube Industries.

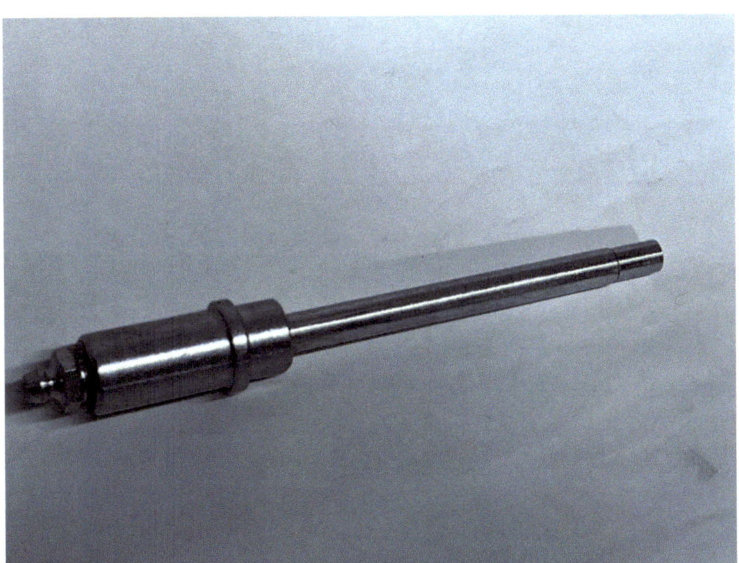

Figure 19.19 Recessed fitting grease extension adaptor Courtesy: ENGTECH industries Inc. and LocknLube.

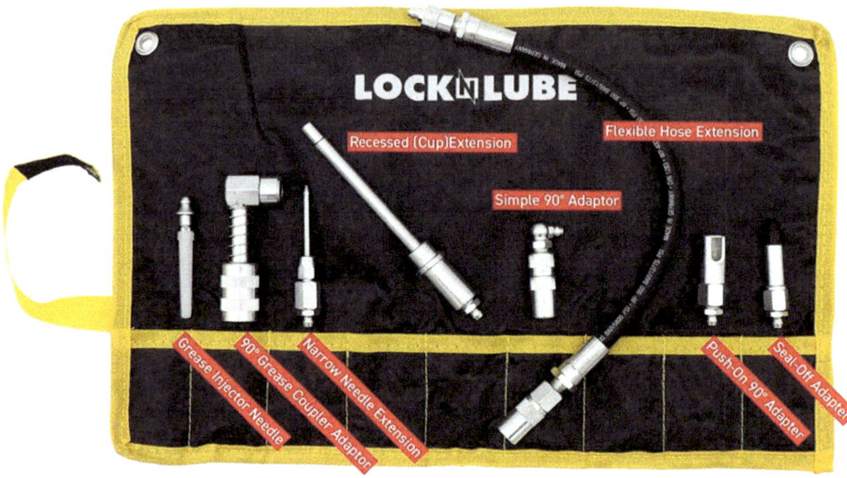

Figure 19.20 Grease gun adaptor kit. Source: LocknLube.

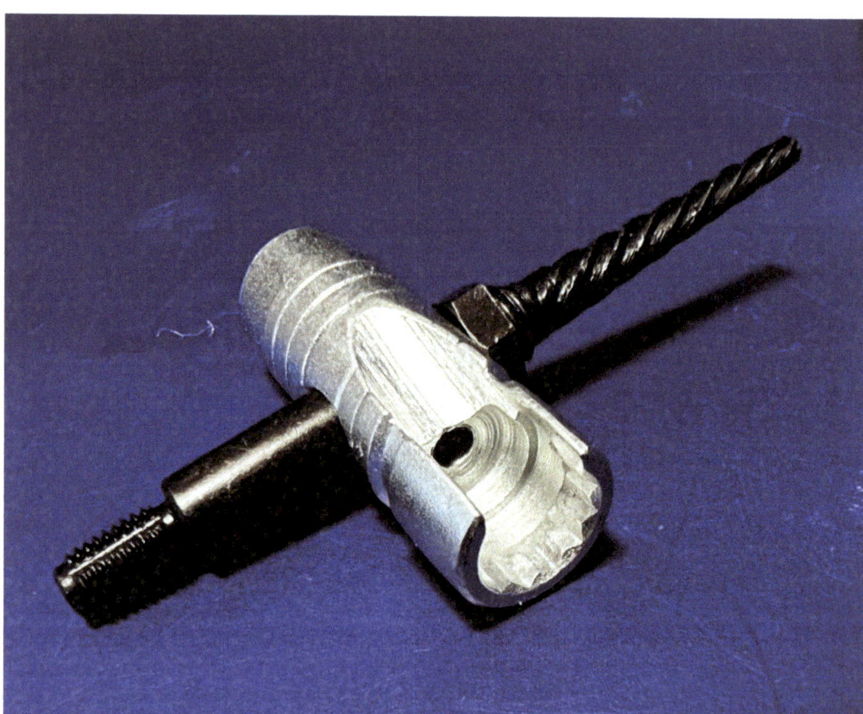

Figure 19.21 Grease fitting multi-tool. Courtesy: ENGTECH Industries Inc.

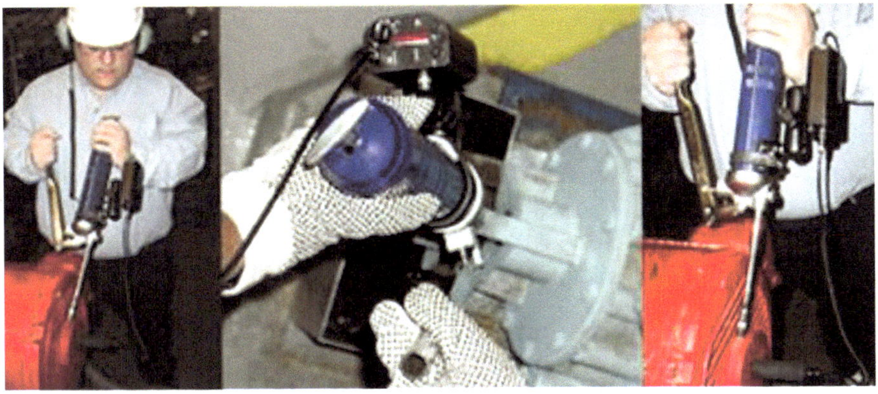

Figure 19.22 Operator performing manual greasing using a UE systems Ultraprobe 201 grease gun system. Source: UE Systems Inc. Elmsford, NY, USA.

Ultrasonic greasing is a unique way of manually greasing a bearing using a specially adapted grease gun that is connected to an ultrasonic sensor, amplifier and operator headphones.

The premise is quite simple, as the lubricant diminishes in the bearing from full-film condition to a boundary level condition (see Figure 18.3), friction levels begin to rise. This friction manifests into localized ultrasonic sound waves in the 20-100 kHz range that can be detected by magnetically mounting a piezoelectric ultrasonic sensor close to the bearing.

When the U/T sensor and adapted grease gun is locked on the bearing, a trained grease gun operator can differentiate the high frequency U/T sounds transmitted to the sensor that have been converted using a heterodyning process into the human audible sound range of less than 16.5 kHz. This audible decibel reference allows the operator to "sense" the bearing condition and amount of grease in the bearing cavity. If the audible noise level dictates the bearing is in boundary condition the operator slowly applies grease into the bearing cavity until the decibel noise signature changes to an acceptable level indicating the bearing is now operating in a full-film condition. Optimum decibel conditions can be used to set baseline requirements for each individual bearing making it easy to test for and replicate on a regular basis. Figure 19.22 shows an operator using a UE systems ultrasonic adapted grease gun system.

Although this is a condition based instrument, like all other condition based testing equipment such as vibration and oil analysis, this method still requires regular attendance checking of the bearing to set up a time based lubrication schedule.

19.8 Implementing a Manual Grease Program

Manual greasing has a place in a lubrication program. If it is to be successful in extending bearing life as intended, it will require an engineered and disciplined approach to its use. In the hands of properly trained and managed lubrication technicians, manual grease lubrication systems provide a very inexpensive and effective method of sustaining and increasing bearing life.

To ensure the effectiveness of a manual greasing program implement an engineered manual greasing program based on the following six steps:

Step 1: Consolidation. Perform a lubricant consolidation program to minimize the number of greases products in use within the plant (Refer to Chapter 35)

Step 2: Standardization. Standardize all Grease Guns within the plant. To facilitate this change management process, allow the maintenance staff to take home and keep the old grease guns. Replace old guns with new, identical specification grease guns, these can be cartridge or bulk fill style so long as there is only one style chosen. Once chosen, perform a volume displacement check and mark each gun with the pressure and volume displacement rating.

Step 3: Determine Bearing Requirements. To reflect the new grease gun volume displacement per shot, all PM tasks will need to be updated. An engineering study may be required to determine if the current bearing requirements and schedule application periods are correct. (Refer to Chapter 18)

Step 4: Introduce Engineered Distribution Devices. To facilitate the lubrication process and assure consistent lubrication metering every time, investigate replacing the zerk fittings or non-engineered "ganged" fittings (Figure 19.22) with an engineered progressive divider lubrication distribution block. In a progressive divider style distribution block, shown in Figures 19.23 and 19.24, multiple points can be lubricated with an exact engineered amount of lubricant in a matter of seconds through one zerk fitting on the block. Grease is pumped into the block until the "tell-tale" pin moves in and out fully to indicate a complete cycle of the block has taken place, and all attached lubrication points having received an engineered lubricant amount. As part of the sales process, the engineered system vendors will usually perform the bearing calculation requirement analysis for the divide block sizing

An added advantage to these blocks is the ability to hook up directly to an automated lubrication pump at a later date.

In an engineered block system, the system operating sequence follows three steps:

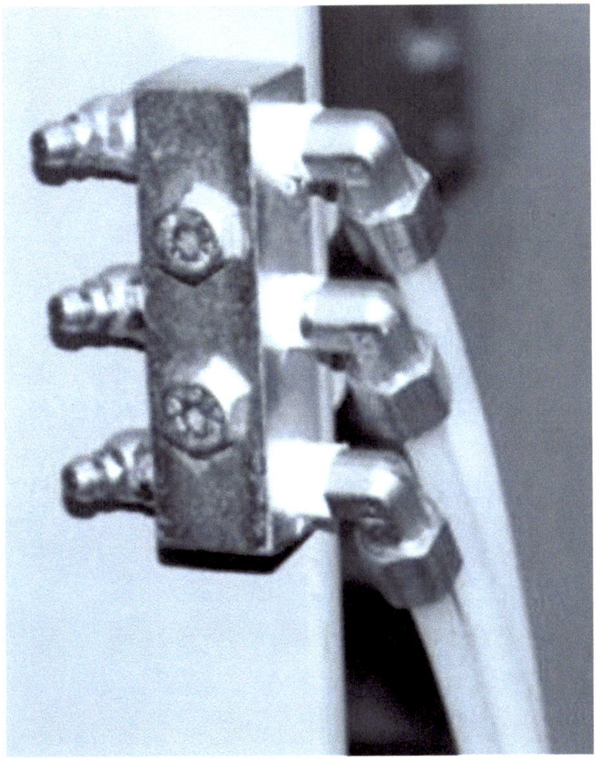

Figure 19.22 Non-engineered triple gang grease block. Courtesy: ENGTECH Industries Inc.

1. The lubricant is delivered to the divider valve block through a machine mounted or hand operated grease gun
2. The positive displacement divider valve dispenses grease in measured amounts directly to each bearing through attached feed lines
3. The cycle pin indicator visually signals the completion of lubricant flow to the bearings

Step 5: Color Coding. Managing with color is excellent for ensuring lubricants are not mixed. Different greases are assigned their own color and dedicated grease guns are marked with the assigned color, or purchased in that color. If grease is to be applied singularly at the bearing point the grease point can then be marked with its appropriate grease color using a simple grease point collar protective tag shown in Figure 19.13, further ensuring that lubricants are not mixed.(Refer to Chapter 35)

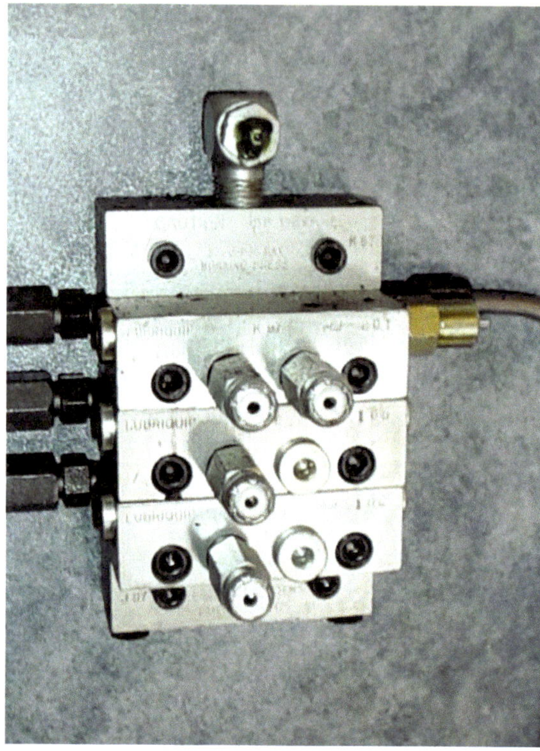

Figure 19.23 Lubriquip/Trabon manual progressive divider valve block with cycle pin and over pressure indicators. Courtesy: ENGTECH Industries Inc.

Step 6: Perform Operator Training. In house training can be provided from both vendors and lubrication management experts who must provide training tailored to your plant environment and the specific needs of the individuals to be trained.

Manual lubrication systems can be very effective but can take considerable effort to set up and manage correctly. Constant monitoring is required to ensure grease guns are only replaced with like specification guns, lubrication frequency and grease type remain consistent. Any updates to the program must be documented and trained to all lubrication/operations staff.

Bibliiography

Bannister Kenneth E., Lubrication for Industry, Second Edition, Industrial Press, NY, NY. 2007.

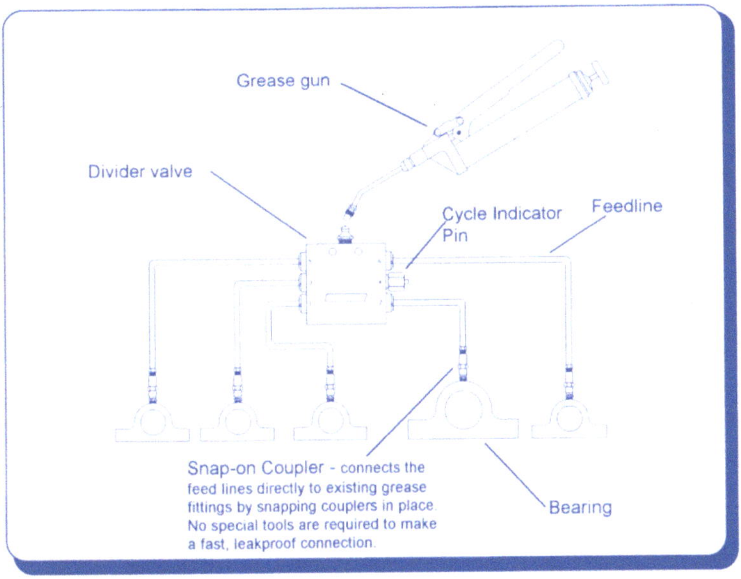

Figure 19.24 Simple manual centralized grease lubrication system. (Source: Lincoln Company, St. Louis, Missouri.)

Lubrication Guide: Ultrasonic Condition Based Lubrication White Paper, UE Systems Inc. Elmsford, NY

Bannister Kenneth E., "Good Greasing is a Feeling", *Efficient Plant Magazine*, 2017

Bannister Kenneth E., "Four Cornerstones of Effective Manual Greasing", *Efficient Plant Magazine*, 2016

Bannister Kenneth E., "How Well Do You Know Your Grease Gun?", *The RAM Review*, 2021

Bannister Kenneth E., "The Grease Nipple: 100Years and Counting", *The RAM Review*, 2021

20

Automated Lubrication Delivery Systems for Oil & Grease

20.1 Automated Single Point Lubrication (SPL) Devices and Systems

Anyone who has visited the engine room of a steam ship, steam train engine, or has witnessed an old steam engine at work will have seen and appreciated the beautiful brass and glass chambers sitting full of oil directly atop each major bearing point. These pioneering single point automatic lubricating devices were elegantly simple; the majority operated by incrementally opening a tapered valve to a determined point to allow oil to flow by gravity into the bearing cavity, or onto an intermediate transfer wick, or brush, in contact with the bearing surface.

In 1872, while working as an oiler for the Michigan Central railroad, Elijah McCoy changed the game considerably when he invented the world's first automatically pressurized (non-gravity activated) lubricator that used steam from the engine to activate and force feed lubricant from the device to the bearing surface. So successful was McCoy's patented lubricator that the railroad companies shunned all other designs in favor of McCoy's, coining the phrase by wanting only "the real McCoy!"

Using a similar design, grease can also be successfully dispensed using gravity and the aid of spring tensioned follower plate inserted in the lubricant reservoir chamber. Grease is filled from the bottom of the reservoir via a nipple that allows grease to hydraulically push against the follower plate and spring to load the chamber. Once loaded, the grease is expelled by the loaded spring pushing the grease against the hydraulic back pressure set up by the bearing clearance; the bigger the bearing, the bigger the clearance, the more grease is expelled into the bearing area—see Figure 20.1. At higher altitudes where the atmospheric pressure is lower, the grease can be expected to expel at a faster rate.

Figure 20.1 Single-point spring loaded grease unit, Courtesy Lubesite by Bijur-Delimon.

Although expressly designed as a one bearing, single point system, the gravity lubricator could, and was, often coupled to a single inlet, multiple outlet manifold to lubricate multiple points simultaneously, albeit in a non-metered manner.

As a testament to their original design these devices are still available for sale, with many original lubricators still in use today largely due to their simplicity, quality of manufacture, and ability to be refilled easily by the user as seen in Figure 20.2. Their modern predecessors are in a different league, and are sophisticated devices that use chemical, electro-chemical, and electro-mechanical pumps controlled by built in electronics to move oil and grease at pressures great enough to use progressive styled metered divider block delivery systems.

20.1.1 Constant level oilers (Bottle oilers)

Many small machines, and especially centrifugal pumps employ constant level oilers, more commonly referred to as bottle oilers." Unfortunately, some of the most popular brands of constant level oilers will maintain their

Figure 20.2 Filling a late 1800s original Single Point Lubricator with oil—still in use in 2016. Courtesy ENGTECH Industries Inc.

level setting only if the dynamic, or operating pressure inside the bearing housing remains the same as the static, or non-operating equilibrium pressure that existed when the bottle oiler was being set up on the non-running machine. Since these pressures may be different due to changing pressure drop conditions across bearing housing vents, breathers, or bearing housing seals, it would be prudent to use only pressure-balanced constant level oilers, Figure 20.3.

When the liquid in the bearing recedes because of liquid consumption, the liquid seal on the inside of the lubricator is temporarily broken. This allows air from the air intake port (smaller threaded connection in Figure 20.3) to

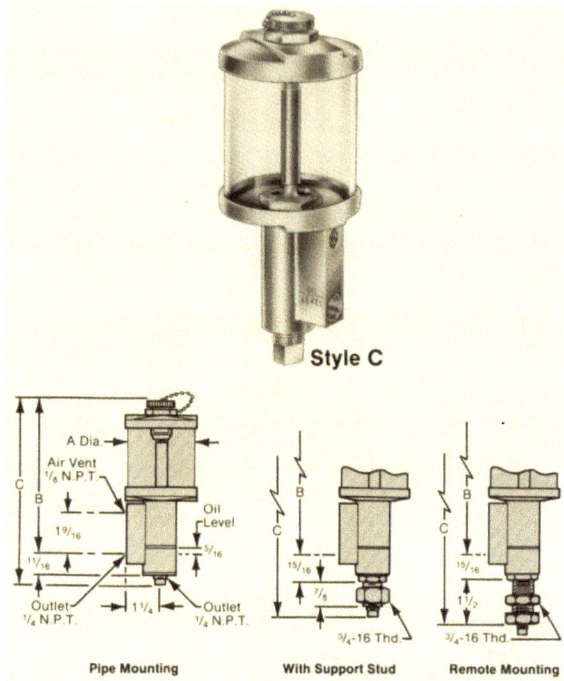

Figure 20.3 Piping "Air Vent" back into the equipment bearing housing produces a reliable pressure—balanced lubricator. (Source: Oil-Rite Corporation, Manitowoc, Wisconsin.)

enter the lubricator reservoir, releasing the liquid until a seal and proper level are again established. For reference, a liquid level line is scribed on the base. The unit is being refilled through a top filler cap. It should be noted that the reservoir need not be removed for refilling. A shut-off valve holds the liquid in the reservoir when the filler cap is removed. Once the cap is tightly screwed down again, the lubricator resumes normal functioning.

This particular oiler becomes a pressure balanced device when the air vent is piped back to bearing housing, thereby equalizing any existing pressure or vacuum.

Numerous variants of these highly reliable lubricators are available. Figure 20.4 shows the device fitted with a low-level safety switch; Figure 20.5 depicts the same constant level oiler with a large sight glass for viewing the liquid level and condition of the liquid. Used in conjunction with properly selected face-type bearing housing seals, pressure-balanced constant-level lubricators with integral sight glasses allow for the hermetic sealing of virtually any bearing environment. Pressure balance is very important. Sadly, it is often overlooked by both machine manufacturers and owner-users.

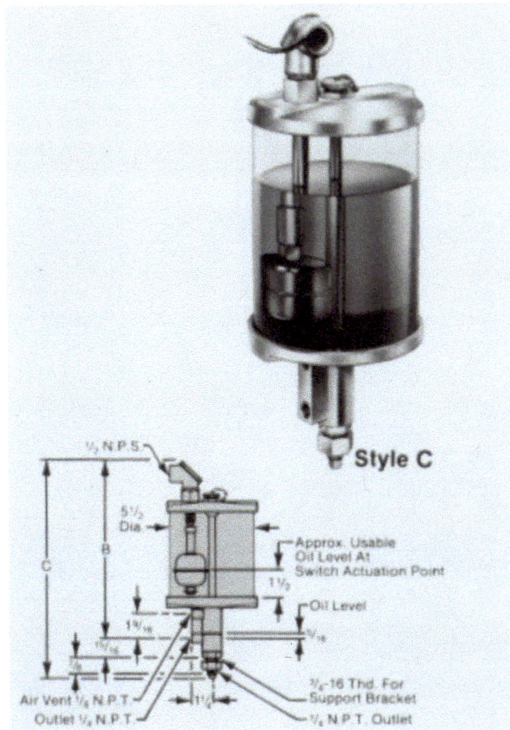

Figure 20.4 Pressure-balanced lubricator with low level safety (warning or shut-down) switch. (Source: Oil-Rite Corporation, Manitowoc, Wisconsin.)

Lack of pressure balance often causes insufficient oil levels in bearing housings. Hermetic sealing refers to the exclusion of atmospheric contaminants, including of course dirt, dust and water vapor. It is these contaminants which are largely responsible for bearing degradation and damage. Hermetically sealed bearing housings no longer incorporate open vents, breathers, expansion chambers, desiccant cartridges, check valve-type vents, filter inserts, or similar components offered to the average maintenance community.

20.2 How a Modern SPL Functions

Altogether, there are five basic styles of single point lubricators (SPL's). These include the gravity lubricator seen above, that can be refilled over and over again requiring a preventive strategy to ensure the unit never runs out of lubricant; the chemical activated SPL designed to continue to exhaustion once activated, and the electro-chemical SPL that introduced an on/off

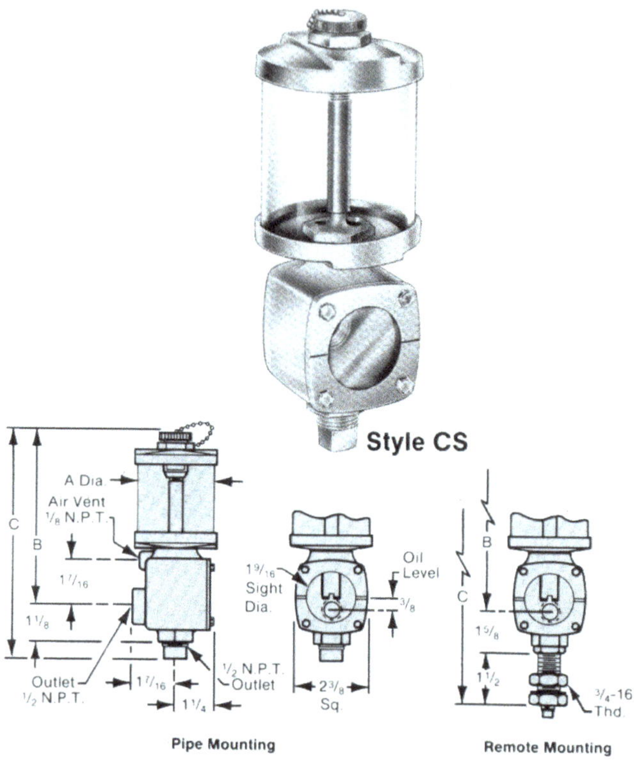

Figure 20.5 Pressure-balanced constant level oiler with integral sight glass. (Source: Oil-Rite Corporation, Manitowoc, Wisconsin.)

switch allowing the user more control over lubricant delivery term and rate. Continual improvement introduced maintenance to the electro-mechanical unit that allows for refills of lubricant, flow and rate control, and enough output pressure to allow hook up to a progressive divider style distribution block making it into a multi-point lubrication system

20.2.1 Chemical activated SPLs

One of the earliest innovators and pioneers of the modern styled SPL's is the German company Perma , who developed an inexpensive disposable chemical activated SPL in the early 1960s—still available and in use today. As seen in the cutaway Figure 20.6, this simple "permalube" unit uses a chemical reaction to develop a gas contained within a sealed expandable bellows unit.

The maintainer activates the unit by releasing a fixed chemical charge pellet into the bellows that is forced to react with an electrolyte to produce

Figure 20.6 Cutaway of Perma classic chemical activated SPL. (Source:Perma USA)

an expandable gas. As the gas slowly expands within the bellows it pushes the lubricant out of the unit into the bearing area. Different chemical charge amounts are used to vary the dispensing time from days to months, depending on the bearing's needs. Once the pellet is released, the unit cannot be deactivated.

With a 4-oz (120ml) reservoir, this style of unit, in its current updated version, continues to be one of the most popular SPL units offered for sale for the dispensing of grease. As the gas generated in this type of unit may be flammable, check with the distributor and /or manufacturer for suitability when an intrinsically safe unit is required (E.g. for use in mines, flour and rice mills where dust can be ignited.)

20.2.2 Electro-chemical activated SPLs

Moving forward to the early 1980s, the first electronic controllable SPL is designed by ATS Electrolube. Using an electro-chemical reactor cell, the user

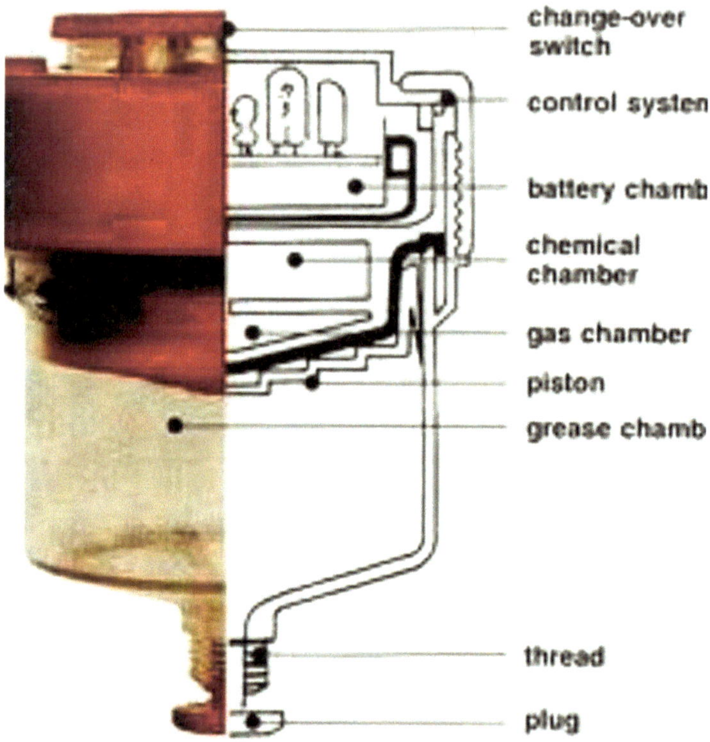

Figure 20.7 Programmable electro-chemical SPL (Source: ATS Electrolube)

activates the unit via a series of time selector switches connected to a battery-operated electronic circuit board (See Figure 20.7.)

Once activated, a pulsed electrical current is sent through a contained electrolyte-soaked sponge to create an electro-chemical reaction. This reaction generates an inert nitrogen gas to form inside a hermetically sealed bellow gas chamber that expands and pushes against the oil or grease charge contained in the lube chamber section of the unit. Unlike the chemical style unit, the discharge can be controlled or turned off completely by the circuit board selector switches. Like it's chemical predecessor, this is also a single use disposable unit, as the bellows cannot be collapsed. Modern modified variations of this unit style now allow for a refillable reservoir with only the power unit and bellows requiring replacement when exhausted. Offered in many differing reservoir sizes, the most popular size unit offered continues to be the 4oz (120ml) unit.

Figure 20.8 Electro mechanical SPL. (Source: Howard Marten Co. for Memolube)

20.2.3 Electro-mechanical activated SPLs

One of the major drawbacks of all units addressed thus far is the lack capability to deliver controlled lubrication to multiple points. The original gravity lubricators only ever had a single point of control and residual line pressure, whereas the chemical and electro-chemical lubricators only developed between 50 and 60 psi. The latest generation of SPLs shown in Figure 20.8 address this issue with its battery-operated rotary mechanism driving a positive displacement pump that can deliver output pressures from 350 psi to 900 psi—more than enough to move a small multiple outlet series progressive divider valve built into the pump, or piped remotely to the pump.See Unlike previous designs, the core unit is reusable with refreshed lubricant and batteries. Still very affordable, these units offer a viable centralized lubrication system alternative to the bigger systems it now competes with.

20.2.4 SPL Pros and cons

Initially, the early gravity units were designed to relieve the continuous attention required from the lubricator whose job was now to check on the unit's reservoir level and fill as necessary, while the bearings reaped the benefit of continued lubrication in small amounts. The advent of the newer auto style units allowed out of sight bearings such as those found on overhead cranes or roof top HVAC fan units to receive continued lubrication for weeks or months, again relieving the burden on the lubricator whose role now changed to checking and marking the lubricator reservoir and performing a unit changeout only when required. For no capital outlay, these low-cost units have successfully continued to extend the life of many bearings.

Because individual units are inexpensive and convenient, many maintenance departments fall into the trap of using them on every bearing with a grease nipple. Performance notwithstanding, widespread indiscriminate use can become very expensive, very quickly. Their use should be monitored against the cost of putting in place the more expensive electro-mechanical SPL with hard piped divider delivery systems, or one of the more robust standard types of centralized lubrication delivery systems designed to deliver lubricant to hundreds of points simultaneously.

20.2.5 Tips for set up and use of SPLs

Tip #1: When using disposable units and disposable cores their disposal must follow municipal and state guidelines for hazardous waste (lubricant, chemicals and batteries).

Tip #2: When using manual, chemical and electro-chemical units the user must understand that both styles of units rely on atmospheric backpressure to control the flow from the lubricator. Using the manufacturer's recommended settings is fine for units used below 1000 feet elevation. Above 1000 feet elevation the settings change approximately 5% for every 1000 ft. As the air thins at higher elevations, the backpressure is reduced and the unit flows at a faster rate, which can mean over-lubrication of the bearing and an empty reservoir sooner than anticipated. Know your elevation and follow the manufacturer's charts for settings at different elevations.

Tip #3: Working in Northern climates means hot summer weather and cold winter weather that will affect the lubricant viscosity in all styles of units. Using a #2 grade grease on a cold day in an outside location, e.g. rooftop and fan units, will stall the unit and starve the bearing of lubricant. Enter a seasonal PM in the computerized maintenance management (CMMS) or work order system to change out #2 grease units to a #1 or #0 grease unit in the late fall, and another PM to change back to a #2 grease in the spring.

Figure 20.9 Electro-chemical lubricator with date in service clearly indicated by the installer (Courtesy ENGTECH Industries Inc.)

Tip #4: Always pre-prime lubricant delivery lines with a grease gun and fresh lubricant to the bearing on any new installation before screwing the lubricator in place.

Tip #5: To ensure continuity of lubricant to the bearing, all units require switching on to allow the unit to prime to the outlet point prior to installation. This may mean switching on the unit 12-24 hours prior to intended use. Check the manufacturer's recommendations for the correct startup procedure for the unit.

Tip #6: Always clearly mark the date of installation on the unit in large visible letters (See Figure 20.9). Startup dates allow the maintainer to check the actual delivery time versus the setting time and adjust accordingly.

Tip #7: At each PM check of the unit mark the reservoir level to ensure the unit has actually delivered lubricant since the last check.

SPLs aren't for every application, but they are an important part of every plant's lubrication management program and are here to stay.

20.3 Automated Centralized Lubrication Systems

Whenever multiple lubrication points are required, the most accurate, efficient and cost-effective way to lubricate is to employ an automated centralized lubrication system. Centralized systems can even be fully automated to

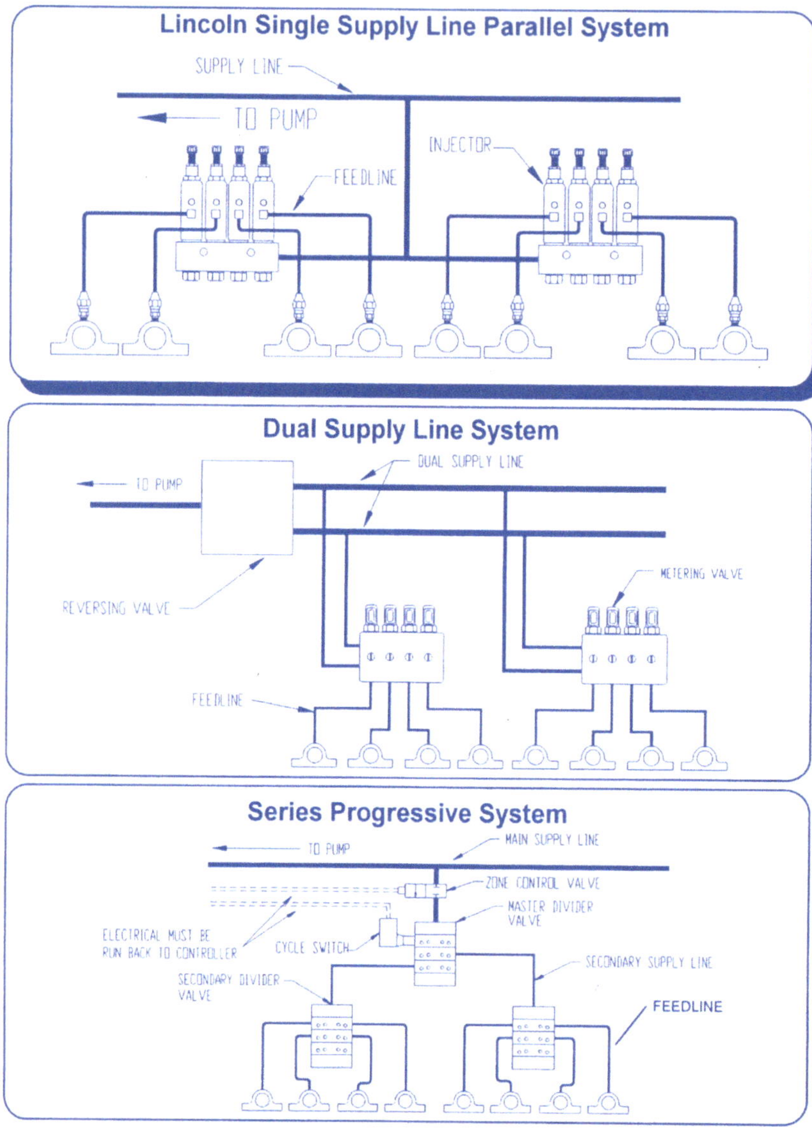

Figure 20.10 Different versions of automated centralized grease/oil systems. (Source: Lincoln Company, St. Louis, Missouri.)

serve entire plants with a choice that includes single line resistance, pump-to-point, series progressive, single and dual line parallel (Figure 20.10), and mist systems all available to the user.

Properly engineered, centralized greasing systems are ideal for a wide range of lubrication requirements. Modular in design and easily expandable,

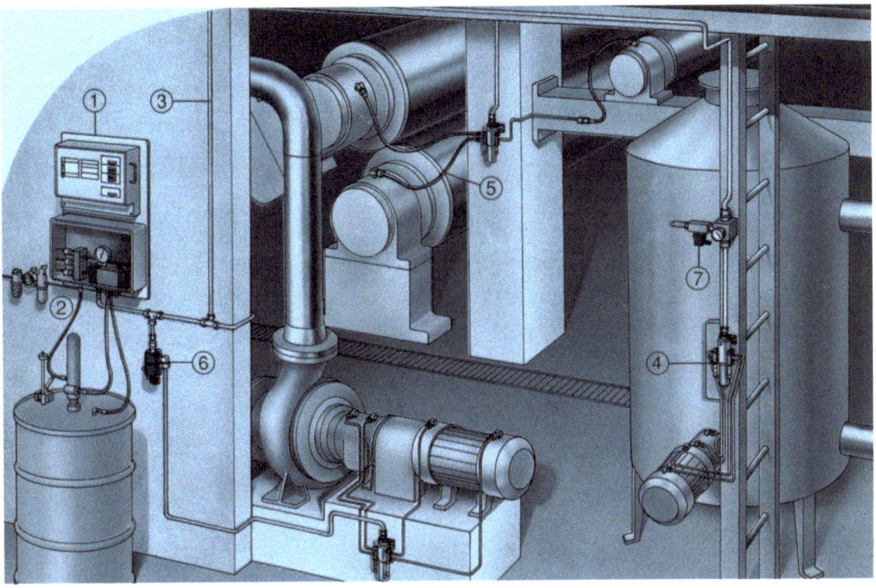

Legend
1. Control center
2. Pumping center
3. Main supply line
4. Metering module
5. Lubrication line
6. Remote-controlled shut-off valve
7. Pressure switch

Figure 20.11 Safematic SG1-grease lubrication system (Source: Safematic, Inc., Alpharetta, Georgia, and Muurame, Finland.)

they are suitable for machinery with just a few lubrication points as well as installations covering complete production plants and involving thousands of points. Systems of the type shown in Figure 20.11 are designed for the periodic lubrication of anti-friction rollers and sleeve bearings, guides, open gears and joints.

Depending on plant and equipment configuration, engineered automatic lubrication systems consist of a single-or multi-channel control center (Figure 20.22, Item 1), one or more pumping centers or pumping stations (Item 2), appropriate supply lines (3), Metering module (4), supply tubing (5) and a remote controlled shut-off valve (6).

Different size dosing modules are used to optimally serve bearings of varying configurations and dimensions. The dosing modules themselves are individually adjustable to provide an exact amount of lubricant and to thus avoid over-lubrication. A pressure sensing switch (7) completes the system.

The control center starts up a pump, which feeds lubricant from the barrel through the main supply line to the dosing modules. When pressure in the system rises to a preset level the pressure switch near the end of the line transmits an impulse to the control center, which then stops the pumps and depressurizes the pipeline. The control center now begins measuring the new pumping interval. If for some reason the pressure during pumping does not rise to the preset level at the pressure switch, an alarm is activated and the lubrication center will not operate until the problem has been rectified and the alarm subsequently reset.

Special multi-channel controllers are available with state-of-art automatic lubrication systems. These have the ability to provide lubrication to installations requiring a variety of lube types, or consistencies. Even different timing intervals can be controlled from single multi-channel controller locations. These systems have proven their functional and mechanical dependability in operating environments ranging from –35°C to 150°C. One Finnish manufacturer tests every type of grease supplied by user/client companies under these temperature extremes and leaves no reliability-related issues open for questioning.

20.3.1 Cost studies prove favorable economics of automated lubrication systems

Direct contact with user companies in Finland in 1996 proved revealing and educational.

In one mill alone, 3798 lubrication points were covered by a dual (two) line automatic grease systems. More recent installations have opted for equally reliable, flexible, but less expensive one-line systems. Washers, agitators, pumps, electric motors, soot blowers, barking drums, chippers, screens, presses, conveyors and other equipment are automatically lubricated at this facility.

A machine, which prior to lube automation had five to ten lubrication-related bearing failures per year, now experiences none, this translates into 30-60 hours of additional machine time, and profit gains of $90 000-$180 000 annually. The plant reports a decrease in total maintenance downtime from 470 hours before lube automation, to 148 hours per year after the implementation of automatic lubrication on just one major machine. Plant-wide maintenance costs have been reduced by 23% over a period of six years. Grease consumption is now only 85% of the amount used previously with manual "hit-and-miss" lubrication. Over-greasing has been eliminated and the ranks of lubrication/preventive-maintenance workers have been reduced from 20 to now only 12 technicians. This explains why, since the early 1990s, this mill-wide reliability and availability upgrade approach has been employed in many retrofit as well as grass-root installations.

Figure 20.12 Graco/Trabon 250-point Centralized grease system using a drum pump with control station and progressive divider distribution installed on a large walking beam furnace. Courtesy: ENGTECH Industries Inc.

Interesting to note that in Scandinavia alone, there are over 200 paper machines equipped with automated wet-end lubrication systems. Payback for these systems, both originally supplied as well as retrofitted, typically ranges in the six months to three-year time frame. This could explain why Scandinavian paper producers, whose workers have higher incomes that most of their American counterparts, are profitable and able to compete in world markets. Automated lubrication has consistently yielded increased plant uptimes ranging from 0.1% to 0.5%.

20.3.2 Elements of a quality dual-header lubrication system

A modern two-header system is characterized by its versatility. Modular in design, reliable in operation and capable of accepting a wide range of dosing modules it is suited to just about all industrial requirements—from lubrication of the smallest joint to the largest of roller bearings.

20.3.3 Key features of single-header lubrication systems

Modern single-header systems utilize an improved spool technology in combination with traditional technology now in use in dual-line systems. One such design leaves the spool ports open during the pressurization, thus eliminating any grease separation risk.

20.3.4 Comparing manual and automatic grease lubrication provisions

Three principal disadvantages of manual lubrication are generally cited:

- Long relubrication intervals allow dirt and moisture to penetrate the bearing seals. Well over 50% of all bearings experience significantly reduced service life as a result of contamination.
- Over-lubrication occurrence during grease replenishment, which causes excessive friction and short-term excessive temperatures. These temperature excursions cause oxidation of the oil portion of the grease.
- Under-lubrication occurrence as the previously applied lube charge is being depleted, and prior to the next regreasing event.

In contrast, automated lubrication has significant technical advantages. Time and again, statistics compiled by major bearing manufacturers have shown lubrication-related distress responsible for at least 50%, and perhaps as much

Figure 20.13 Modern steel mills use automatic grease lubrication systems.

as 70% of all bearing failure events world-wide. Well-engineered automatic lubrication systems, applying either oil or grease, are now available to forward-looking, bottom line-oriented user companies. In paper, pulp, refining and steel plants (Figure 20.13), these systems are ensuring that:

- The time elapse between relubrication events is optimized.
- Accurately predetermined, metered amounts of lubricant enter the bearing "on time" and displace contaminants.
- The integrity of bearing seals is safeguarded.
- Supervisory instrumentation and associated means of monitoring are available at the point of lubrication for critical bearings.

20.4 Circulating Lubrication Systems

Circulating (liquid oil) lubrication systems are typically utilized in equipment where oil performs cooling or heating duties in addition to the lubrication of parts. Applications include paper machine drying sections and steel industry rolling mills.

A typical circulating system used in the paper and steel industries is shown in Figures 20.14 and 20.15. It consists of a circulation lubrication center (1), comprised of a stainless-steel reservoir, twin filters and pumps, and one

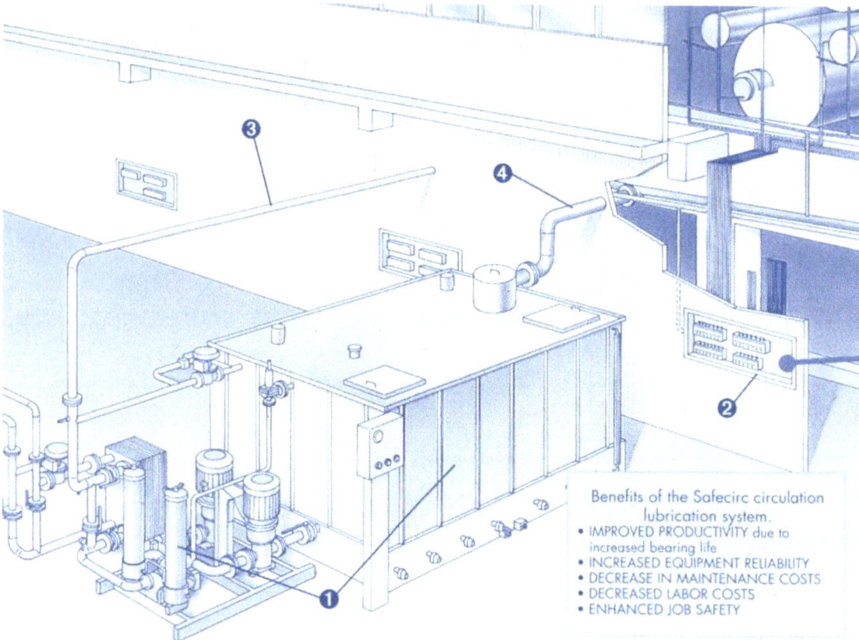

Figure 20.14 Circulating lubrication system comprised of filters/cooler/pumps and reservoir (1), oil flow metering modules (2), pressure piping (3), and return piping (4). (Source: Safematic, Inc., Alpharetta, Georgia, and Muurame, Finland.)

Figure 20.15 Large recirculating oil—Circoil™ pumping unit used of large paper mill applications. Courtesy: Howard Marten Company ON, Canada

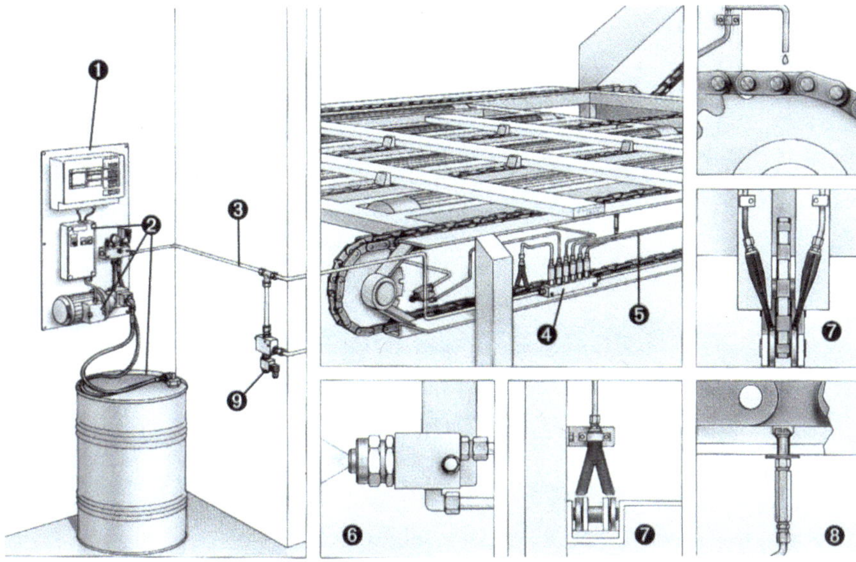

Figure 20.16 Open, centralized oil lubrication system. (Source: Safematic, Inc., Alpharetta, Georgia, and Muurame, Finland.)

or two cooler or heater sets. Each of these elements would normally be furnished with supervisory instrumentation. A well-planned system would further include oil flow meter groups (2), pressure piping (3) and return piping (4).

A modern, closed circulating system utilizes the lubrication center (see also Figure 20.11) so that each lubrication point receives the correct amount of high quality, clean oil at the required temperature. Such a system would be sized to accommodate the exact requirements of the equipment it serves. Moreover, each flow meter would again be equipped with appropriate alarms or similar annunciation devices. It should be noted that a good flow meter offers easy calibration and readability. Flow calibration in accordance with the viscosity characteristics of the oil should be possible over a fairly wide range without requiring meter replacement.

20.5 Open, Centralized Oil Lubrication Systems

Open, centralized oil lubrication systems are designed for the cyclic lubrication of industrial conveyors, guides, and other heavy-duty machinery. Figure 20.16 illustrates the principal components of one such system, comprising a control center (1), pumping station (2), main supply line (3), metering module (4), branch tubing (5), appropriately configured spray nozzles (6, 7 & 8), and a remote-controlled shut-off valve (9).

Figure 20.17 Forest product processing machinery using open, centralized lubrication. (Source: Safematic, Inc., Alpharetta, Georgia, and Muurame, Finland.

Experience shows that open, centralized systems reduce energy consumption and unscheduled downtime. These systems can operate in widely varying temperature environments.

Many different lubricants can be accommodated and delivered at the point of usage as a clean, metered quantity. Greatly reduced component wear and a three-fold overall increase in machine life are not unusual on forest product machinery (Figure 20.17), and other equipment.

20.6 Air-oil Lubrication Systems

In high-speed rotating equipment such as machine spindles, turbines, blowers, etc., bearing and gearing speeds can approach DN surface speed values approaching 2.2 million DN (DN value is calculated by multiplying the bearing diameter in mm (D) by the rotational speed of the spindle in RPM (N)).

Traditional wet sump oil and grease lubrication methods cope poorly in these high-speed environments as they struggle to dissipate the additional heat load created by speed and fluid friction. This will often result in significant reductions in lubricant life, energy loss, and machine speed capability. Traditional mist lubrication systems allow for higher rotating speeds over traditional oil lubrication systems but are not able to provide the exact metering requirements needed for extended bearing service life, and as mist is a micro droplet form, it is susceptible to becoming airborne in the plant environment and viewed a health and safety problem by some.

The air-oil system was developed as a total loss oil system to meet the specific needs of the high-speed bearing environment and has been refined to

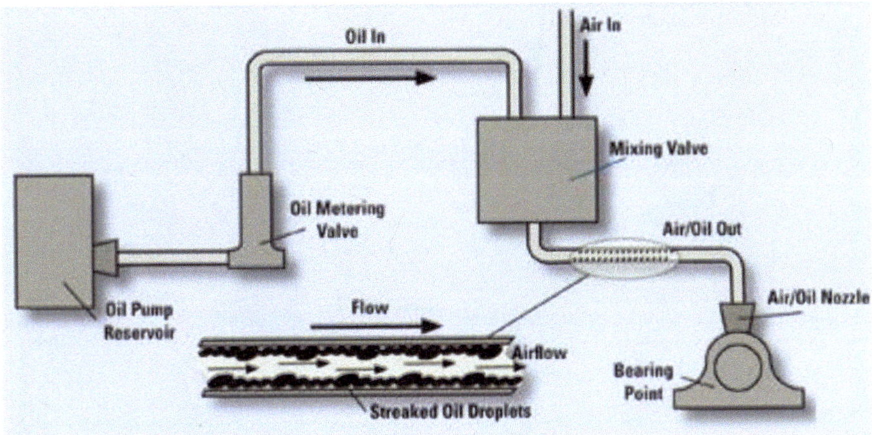

Figure 20.18 Air Oil lubrication schematic. (Source: Ken Bannister for Lubrication Management & Technology magazine's "Anatomy of a Centralized Lubrications System" article series

the point where it is a viable and effective lubrication system for any rotary or linear machine application.

A relatively new lubrication system design, the air-oil lubrication system can be regarded as a hybrid system that utilizes the metering capability of existing single line resistance, positive displacement injector, and progressive divider delivery systems. Oil is metered in the traditional manner in minute quantities to a mixing block connected to a clean and dry compressed air supply. The individual oil drops are dispensed into a small diameter (usually 4mm–3/16inch diameter) delivery tube where the oil is "streaked" by the air into macro droplets and transported along the inner walls of the delivery tube to a dispensing nozzle located at each individual lubrication point. The small nozzle diameter creates a venturi effect allowing the air and the droplet to envelop around the bearing surface in an almost oil-free manner. The air forces the lubricant film across the bearing surfaces creating additional cooling and a positive pressure in the bearing area that aids in sealing out external contaminants such as coolant and dirt. (Figure 20.18)

Air-oil lubrication systems, Figures 20.19 and 20.20, are designed to provide continuous metered flow in minimum, exact quantities to points of application. There are many options and configurations; they typically include pressure switches, pressure gauges, check valves, and controllers ranging from elementary to the most advanced electronic models. Air flows can be fixed or adjustable to each point of application.

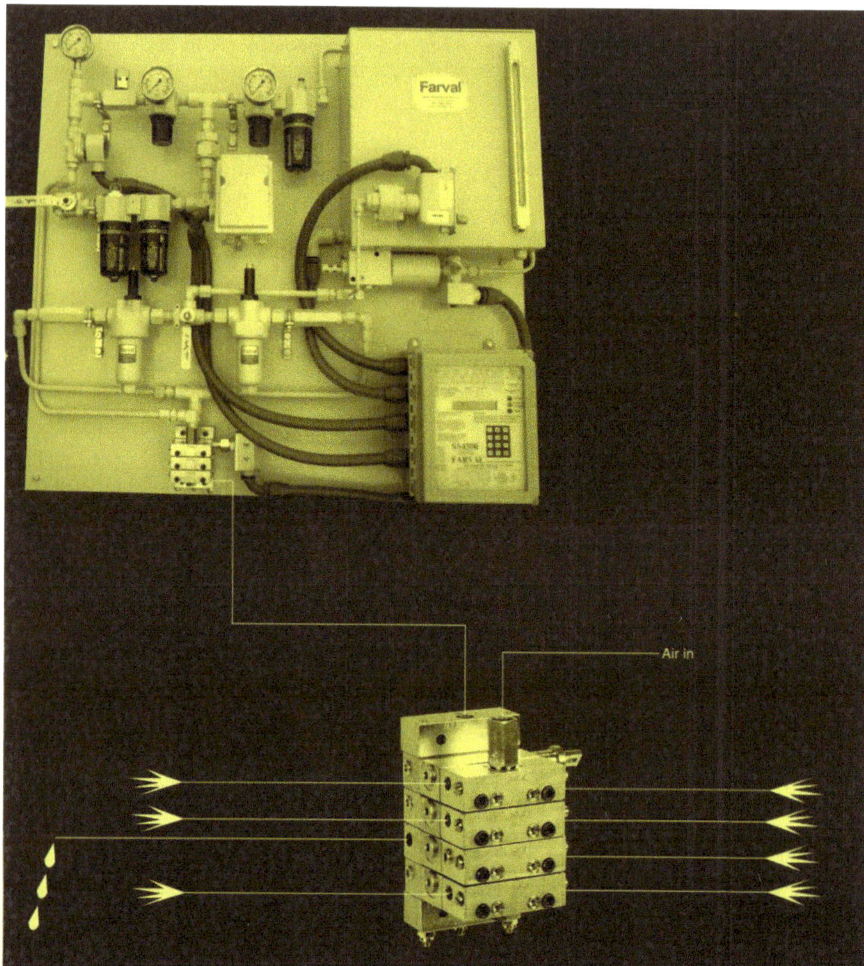

Figure 20.19 Typical air-oil lubrication system. (Source: Farval Lubrication Systems.)

Depending on selection criteria, air-oil lubricator devices use mixing chambers remote or integral with the metering valve. The four-section air-oil valve assembly illustrated in Figure 20.21 features integral air-oil mixing valves. The associated pumping stations can be either electrically or pneumatically powered.

The units depicted in Figures 20.22 and 20.23 operate on the principle of filtered air entering the main lubricator. Separate lines carry air and oil to the oil delivery control unit inside the manifold block assembly. Coaxial distribution delivery tubes separately carry oil and air near the points of

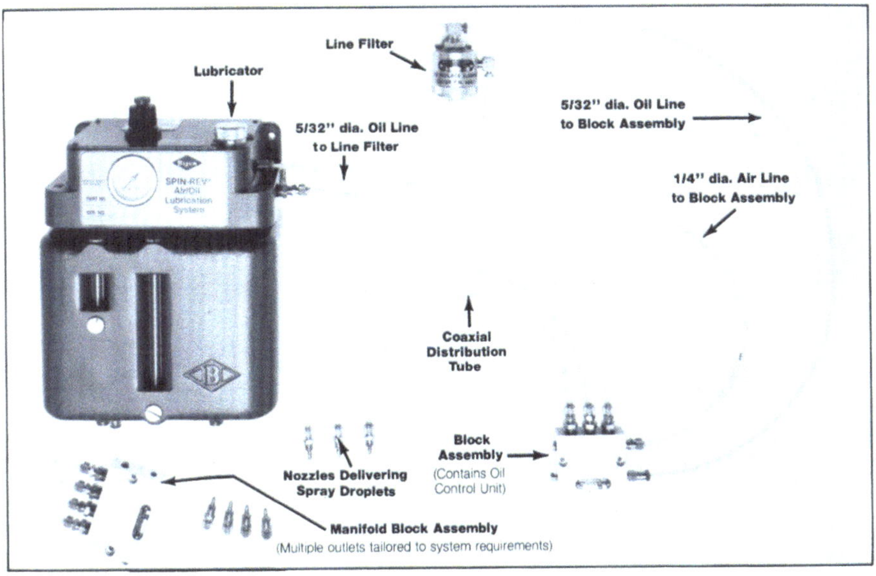

Figure 20.20 Small air-oil lubrication system. Courtesy: Bijur Delimon International.

application. Air and oil are then mixed at the nozzle assembly and controlled spray droplets (not mist) generated at the ends of the distribution line nozzles. The system is monitored for low oil level and low oil and air operating pressure.

20.6.1 Pros and cons

When set up correctly the air-oil system boasts an impressive resume of benefits that include:

- Highest possible bearing surface speeds,
- Bearing temperature rise over ambient of < half that of traditional lubrication methods,
- Energy consumption reduction,
- Health and safety approved system
- Continuous fresh oil delivery,
- Up to 90% lubricant use reduction compared to mist lubrication systems,
- Up to 99% lubricant use reduction compared to grease systems,

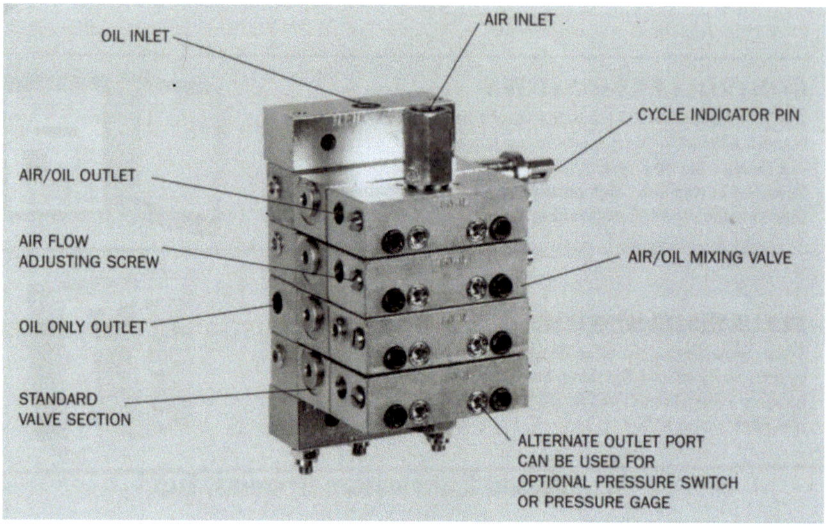

Figure 20.21 Four-section air-oil valve assembly. (Source: Farval Lubrication Systems, Inc.)

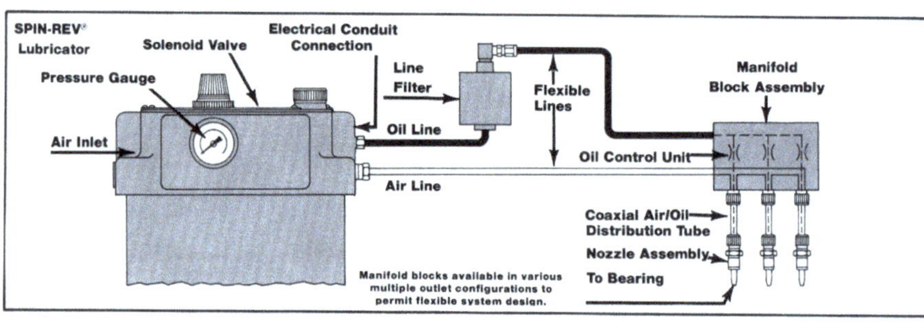

Figure 20.22 Air-oil lubrication systems with coaxial delivery tubing. Courtesy: Bijur Delimon International.

- Extended bearing and seal life expectancy,
- Small environmental footprint in comparison to other traditional lubrication cleanup/disposals methods.

Because the air-oil system is a hybrid system, it requires the purchase of a traditional lubrication system and the additional costs of adding mixing control valves and a clean and dry air supply network to each mix valve assembly. Due to the small aperture of the delivery nozzle, solids additives such as moly and PTFE are discouraged from use as they can "bridge" across the nozzle, form deposits and cause an oil starvation situation. The air-oil system

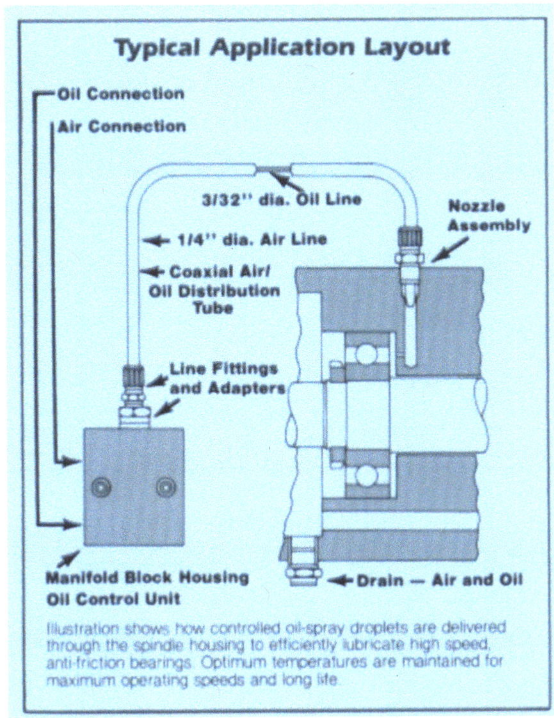

Figure 20.23 Typical application layout for air-oil lubrication. Courtesy: Bijur-Delimon International.

is easier to incorporate in the machine design stage and much more difficult to design as an add-on system later on. single bearing point when the pump is operated, and can be designed to accommodate up to 200 delivery points in a single pump system design.

20.7 Single Line Resistance (SLR) Systems

The single line resistance system is a fully engineered central system designed to apportion pumped oil manually in a single shot (total loss) method, in an automated cyclical (total loss) manner, or in a continuous circulative manner. The system is a low-pressure system engineered to deliver an apportioned amount of lubricant to every single bearing point when the pump is activated. This system type can accommodate up to 200 delivery points in a single pumpo system design.

Today, this system is arguably the most copied lubrication system on the market; was originally designed for automobile lubrication, and later on

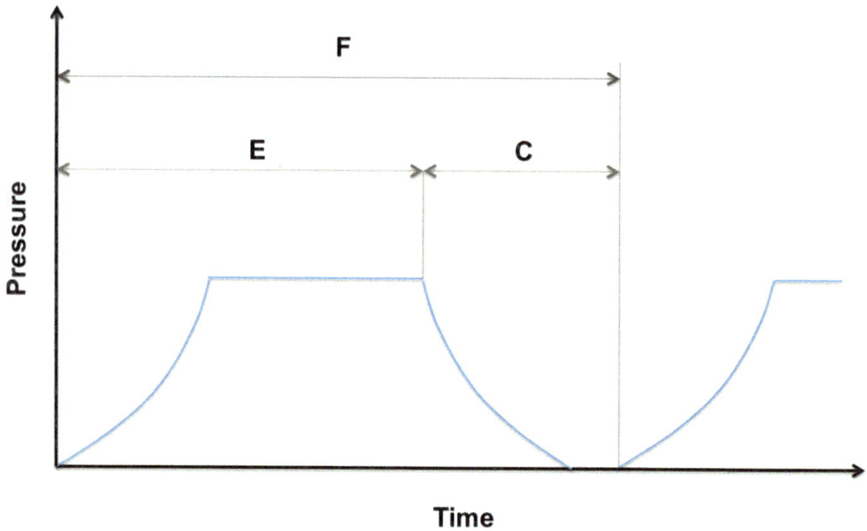

C = Interval Time
D = Working cycle time
E = Lubrication time
C + D = F

Figure 20.24 Single line resistance (SLR) metering valve system lubrication cycle.

adapted for small to medium machine tools and manufacturing equipment. Introduced to market in 1923, the system was a US design product by Joseph Bijur of the then Bijur Lubricating Corporation—who today operates as Bijur Delimon International and still remains a leader in the manufacture of lubrication delivery systems

20.7.1 How the system works

In a typical total loss one-shot or cyclic system shown in Figure 20.25, a piston pump delivers lubricant through a 5/32" (3mm) or 3/16" (4mm) diameter line at a pressure of approximately 60psi. When pump operates, all points are being continually lubricated, the amount delivered is a proportion of the total pumped amount based on the apportioning size of the metering or control unit in comparison to the sizing of all other points, the number of total points in the system, and the length of time the pump operates.

Figure 20.24 depicts a lubrication cycle diagram for a single line resistance (SLR) metering valve system. The operating time of the pump is referred to as the lubrication time (E). Working cycle time (F) is calculated by adding the lubrication time (E) and the interval time (C). When the pump shuts off – usually from a timer, the internal spring-loaded metering valves

Figure 20.25 SLR metering units. Courtesy: ENGTECH Industries Inc.

slowly close as the residual line pressure normalizes to between 2-5psi. Once the timer reactivates the pump, cycle (D) recommences. Note: if the system is designed as a continuous lubrication system, the pump runs continuously until the machine is turned off, therefore the working cycle time is equal to the machine runtime. In this system type the metering valves require no spring return shut off mechanism.

In a continuous circulating system, a gear pump delivers continuous oil flow through a flow proportioning device known as a control unit. Both meter and control units pictured in Figure 20.25 appear identical from the outside but differ considerably in how they are constructed. A meter unit contains a metering pin of a controlled diameter that floats in an accurately reamed cylindrical passage producing an annular orifice of known flow rate. The clearance between the pin and the cylinder wall determines the meter unit flow rate designation. A control unit used in continuous systems has no check valve and uses a helical screw to meter the unhampered lubricant flow.

Figure 20.26 demonstrates a typical line diagram representation for a SLR system. This representation shows a manual hand pump. The system layout is the identical for either hand pump or automatic pumps.

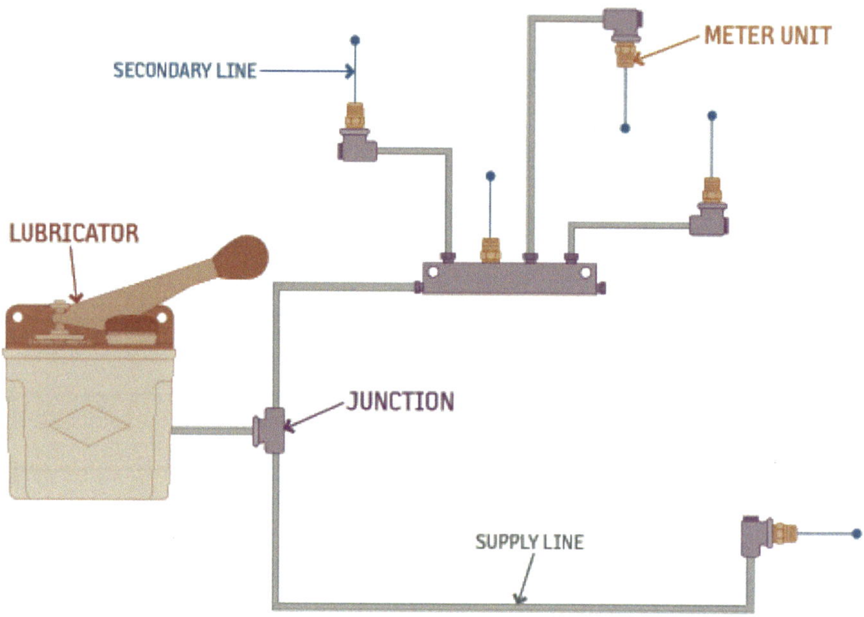

Figure 20.26 Single line resistance system. Courtesy: Bijur Delimon International.

20.7.2 Pros and cons

The single line resistance system is a simple, inexpensive, engineered lubrication system designed for small to medium sized machinery.

Unfortunately, the system can only be used with oil, and does not produce a control signal.

Because the system metering units are piped in series (hence the single line designation), care must be taken to ensure all fittings are leak proof and that meter or control units are never allowed to be "drilled" out to increase rate of flow. Both instances have the effect of a broken line that allows oil to take the path of least resistance, and effectively "starve" all bearings simultaneously.

With basic care and understanding, the SLR system will provide excellent bearing life cycle management at a minimal cost.

20.8 Single Supply Line Injector System (Positive Displacement Injector)

Also known also as a single line parallel system, the single line Positive Displacement Injector (PDI) system was first developed and introduced to

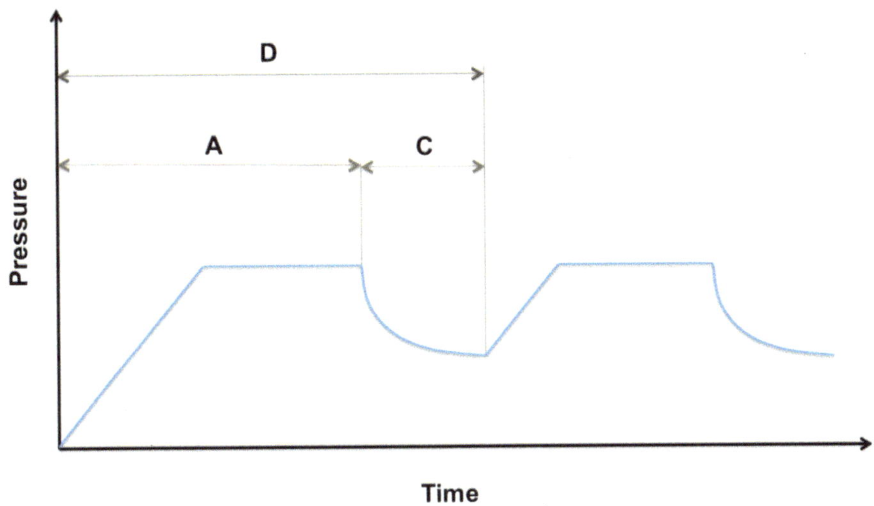

A = Lubrication Cycle Time
C = Interval Time
D = Operating Cycle Time
A + C = D

Figure 20.27 Single line positive displacement injector lubrication cycle.

industry by the Lincoln Industrial Corporation in 1937 to accurately displace metered quantities of oil or grease in a cyclical manner to small/medium sized industrial equipment.

In contrast to Single Line Resistance and Progressive divider type systems, each metering valve, or point, can be set independently, adjusted, or easily changed, without affecting the system design. This enables additional injectors (lube points) to be added into the system at a later date, without the need to re-engineer the entire system.

20.8.1 How the system works

All PDI systems utilize either a pull handle manual or automated pump to force oil or grease into the main line and injectors (all connected together in a single line) to a pressure greater than 800 psi or 55 Bar.

In fully automated systems, a pressure switch located at the very end of the main line is set up to shut off the pump once line pressure is achieved. In manual pump systems, an optional pressure gauge is often employed to enable the operator to see the built-up line pressure and discontinue pumping once suitable line pressure is achieved. Figure 20.27 depicts a lubrication cycle diagram for a single line PDI system.

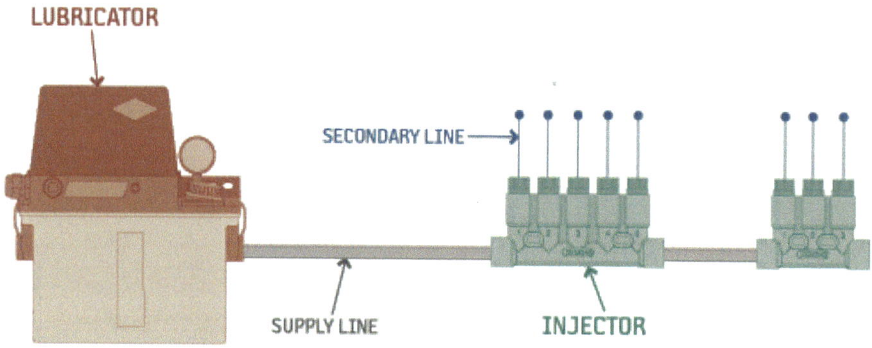

Figure 20.28 Single line parallel (PDI system). Courtesy: Bijur Delimon International.

Once the pump is signaled to operate, line pressure begins to build causing a metered quantity of lubricant in the injector "firing" chamber to be directly forced into the bearing point. Main line pressure is regulated by an end of line pressure regulating valve that once a pre-determined line pressure has been reached will signal the pump to turn off. During this pressure relief time (B), pressure is released through the pressure relief valve and line pressure will relieve until it reaches residual line pressure. This action causes a compression spring within the injector to allow a metered amount of lubricant to enter into the firing chamber ready for the next lubrication cycle. Interval (C) time is controlled by the pump timer.

Each lubrication point requires its own injector and is connected directly to the lubrication point via a secondary delivery line as shown in Figure 20.28.

20.8.2 Pros and cons

Because it can be used with oil and grease, does not require much system engineering, and can easily have additional points added to the system, the PDI system has long enjoyed a reputation as both a versatile and universal system.

PDI systems that use fixed injector displacement caps are preferred over injector types that allow the user/ operator to readily adjust the piston output via an external adjustment wheel or lever on the side of the injector. (Figure 20.29). User/operator adjustable injectors are easily tampered with and can easily lead to an over-or-under lubrication condition unless they are access controlled.

Although a main open line failure can easily be detected through a time out switch located at the end, no secondary line failure device is

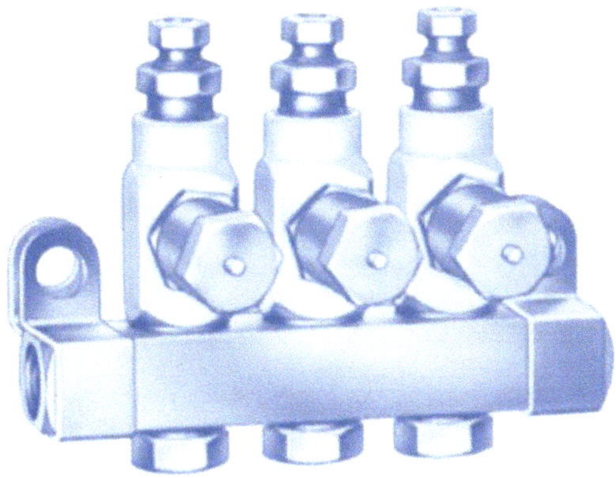

Figure 20.29 Lincoln style positive displacement injectors. Source: Lincoln Industrial.

available. Users must perform system line integrity checks as part of their PM program.

20.9 Dual Line Parallel System

First introduced to market by the Farval Lubrication company, now part of the Bijur Delimon International, the dual line lubrication system—also known also as the twin line parallel system—was designed to accurately displace and deliver oil or grease to as little as 20 lubrication points, up to many hundreds of points, over great distances, from a single pumping station. The system's heavy-duty construction, and its use of small-bore piping and tubing made it an ideal choice for automated lubrication in medium/large sized industrial equipment typically employed by the steel, cement, mining, pulp and paper, power generation and Petro-chemical industries.

The dual line system bears many similarities to the PDI single line parallel or positive displacement injector system, in that each metering valve, or point, can be set independently, or easily adjusted while in operation. This unique feature also enables additional injectors (lube points) to be added into the system at a later date, without the need to re-engineer the entire system.

20.9.1 How the system works

As suggested by its name, the system employs two main lubrication lines that run in parallel from the pump to the last lubrication point through a series of lubrication delivery valves as shown in Figure 20.31. Once the pump is activated, line pressure is built up on the pressure or delivery supply line to fire

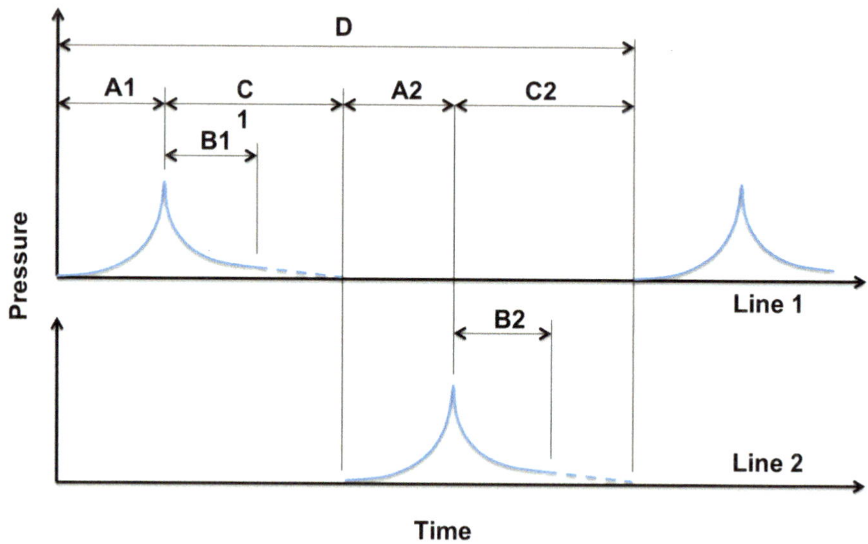

A1 / A2 = Lubrication Cycle Time
B1 / B2 = Partial Pressure Relief Time
C1 / C2 = Partial Interval Time
D = Operating Cycle Time

Figure 20.30 Dual line parallel injector system lubrication cycle.

the lubrication point injectors while simultaneously venting the second return line back through a reversing valve to the reservoir.

Dual line injectors differ to single line injectors in that they do not use a spring arrangement to fire and load the injector, but rather employs a dual acting hydraulic spool valve set up to feed two separate lubrication points (one per each pressure cycle). Once an end of line pressure switch signals that a preset line pressure has been reached and all injectors have fired, the system has completed one pressure cycle, or one half-lubrication cycle. The reversing valve is now actuated to its changeover position to allow the previous venting line to become the new primary pressure line and full-lubrication cycle.

The system can operate in both a manual mode with a pull handle pump or automatic mode as shown in Figure 20.32.

A dual line system operates in a similar manner to a single line PDI system except there are two main lines in parallel fed by a single pump. The first half of the operation begins on line 1. The pump is activated and lubricant is routed to line 1 through a diverter valve allowing all injectors to discharge lubricant into their connected bearings. Once a predetermined system pressure is reached a pressure switch turns the pump off, and pressure begins to relieve (B1) allowing the injectors to reset. As this occurs a reversing valve

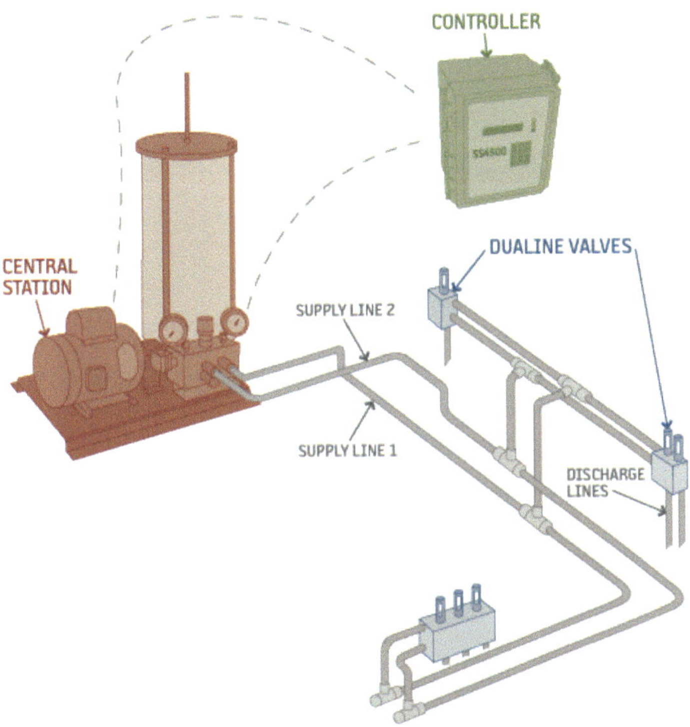

Figure 20.31 Dual supply line parallel system. Source: Bijur Delimon International.

switches the pump flow to line 2 and the process repeats itself on the second main delivery line. Once both lines have lubricated their bearings the pump can be controlled by a timer to continue, or wait for a designated cycle time (D) before the cycle repeats again. A dual line lubrication cycle is represented in figure 20.30.

20.9.2 Pros and cons

Because it can be used with oil and grease, the system engineering is not demanding and will easily accommodate the addition or reduction of system points post installation.

As with single-line systems, the adjustable injectors are easily tampered with and can lead to an over or under lubrication condition unless they are access controlled.

A pressure line failure is easily detected through a time out switch located at the end of line; no secondary line failure device is available. Users must perform system line integrity checks as part of their PM program.

Figure 20.32 Dual line farval pump system in use at a small hydro generation facility. Courtesy: ENGTECH Industries Inc.

20.10 Series Progressive Divider Systems

The most complex engineered of all lubrication systems is the series progressive system. This system is designed to pump oil or grease in either a cyclical (total loss) or continuous circulative manner. Its divider block design is engineered to positively deliver an exact displaced amount of lubricant to every single bearing point, to operate in severe environments and accommodate upwards of 200 delivery points in a single pump system design.

Today, most lubrication OEM's offer a version of this most popular system type, with the original system designed and developed in the US by the Lubriquip organization in the earlier part of the 20th century and marketed under the Trabon name, which continues to be sold today.

20.10.1 How the system works

A lubricant pump is connected to an engineered network of series progressive divider blocks, and via a controller is allowed to pump lubricant in a

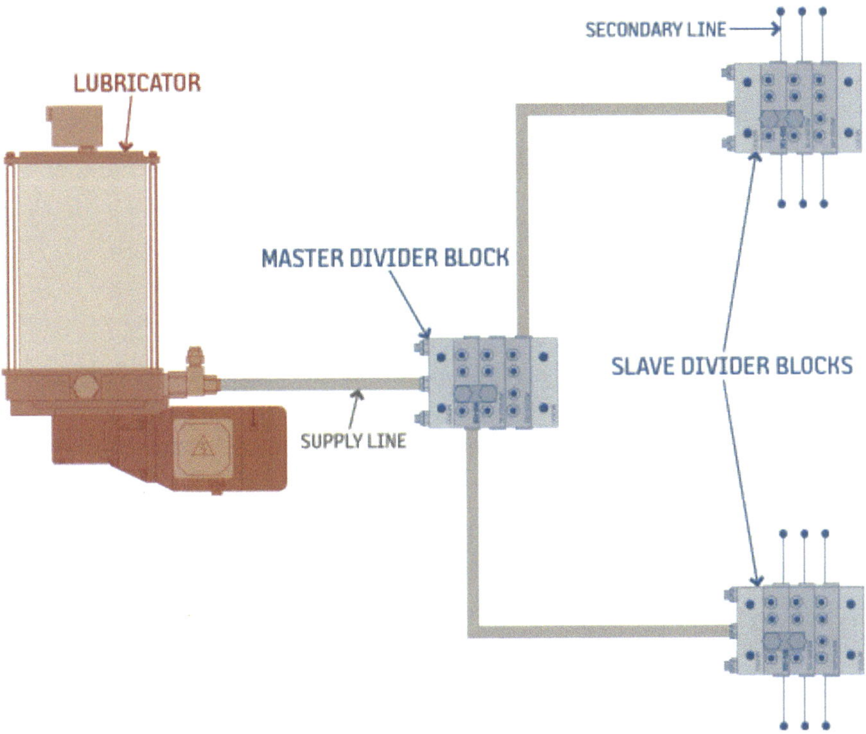

Figure 20.33 Series progressive system. Courtesy: Bijur Delimon International.

continuous or controlled cyclical systematic manner to each divider block as depicted in Figure 20.33.

Divider blocks are built in a modular manner and contain a series of lapped hydraulically actuated spool valves sized for varying displacement. The ability to "cross-port" a valve results in "doubling" the delivery of lubricant on one side of the valve only. The valves are progressively linked together in series causing each valve to "shuttle" over to one side of the block in a progressive pattern, then to "shuttle" back to their original position as the lubricant continues to be pumped through the block.

In a Progressive divider system, the lubrication cycle time (A) as shown in the lubrication cycle diagram 20.34 is determined by the pump operating time which in most cases shuts off once all divider blocks have cycled a minimum of one time ensuring all points in the system receive an engineered amount of lubricant. As depicted in the diagram, during interval time (C), the lube system is partially pressure relieved. Once the pre-determined interval time (C) is complete, the pump timer re-activates the pump and the operating cycle begins again Because of its hydraulic nature, as the valve is shuttled

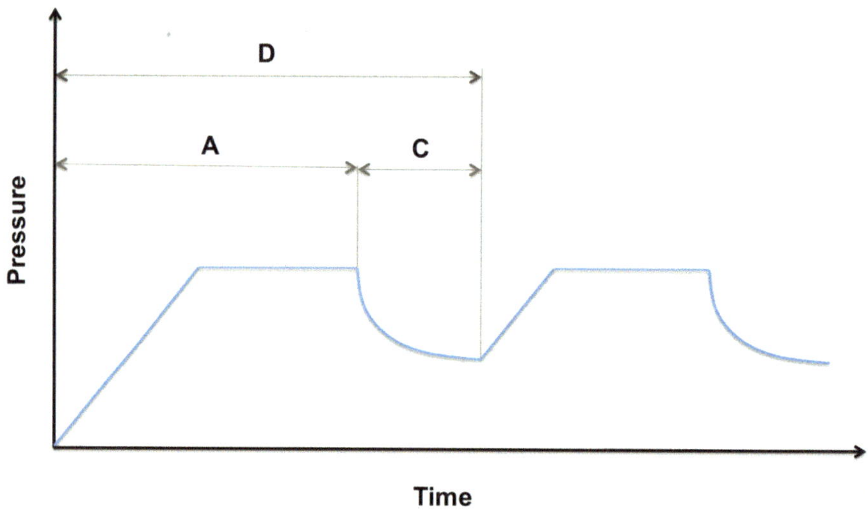

A = Lubrication Cycle Time
C = Interval Time
D = Operating Cycle Time
A + C = D

Figure 20.34 Series progressive divider system lubrication cycle

back and forth, it both displaces a lubricant charge on one end of the valve to the bearing point while simultaneously filling the void on the other side of the valve in preparation for displacement once the valves "shuttle" back to the other side. Figures 20.35 to 20.38 demonstrate one complete cycle of the metering block.

The center inlet passage in white is connected to all piston chambers, at all times. Only one piston is free to move at any one time while the remainder hydraulically lock in place until the moving piston charge has been released into the bearing, momentarily relieving system pressure, allowing the next piston to move. This repeats until all pistons in the block have released their charge on either side of the block having completed one full cycle of the block.

Figure 20.35 picks up the cycle when all pistons A, B and C are positioned far right. Application of lubricant in the inlet passage locks in place pistons B and C, while piston A moves to the left. This action discharges a measured quantity of lubricant through the pink colored passageway to outlet number one and on to the bearing as depicted in Figure 20.31.

Now piston A has discharged its lubricant, the center passage is now connected to through piston A to the right end of piston B via the center pale blue passage. Piston B now moves over to the left while pistons A and C

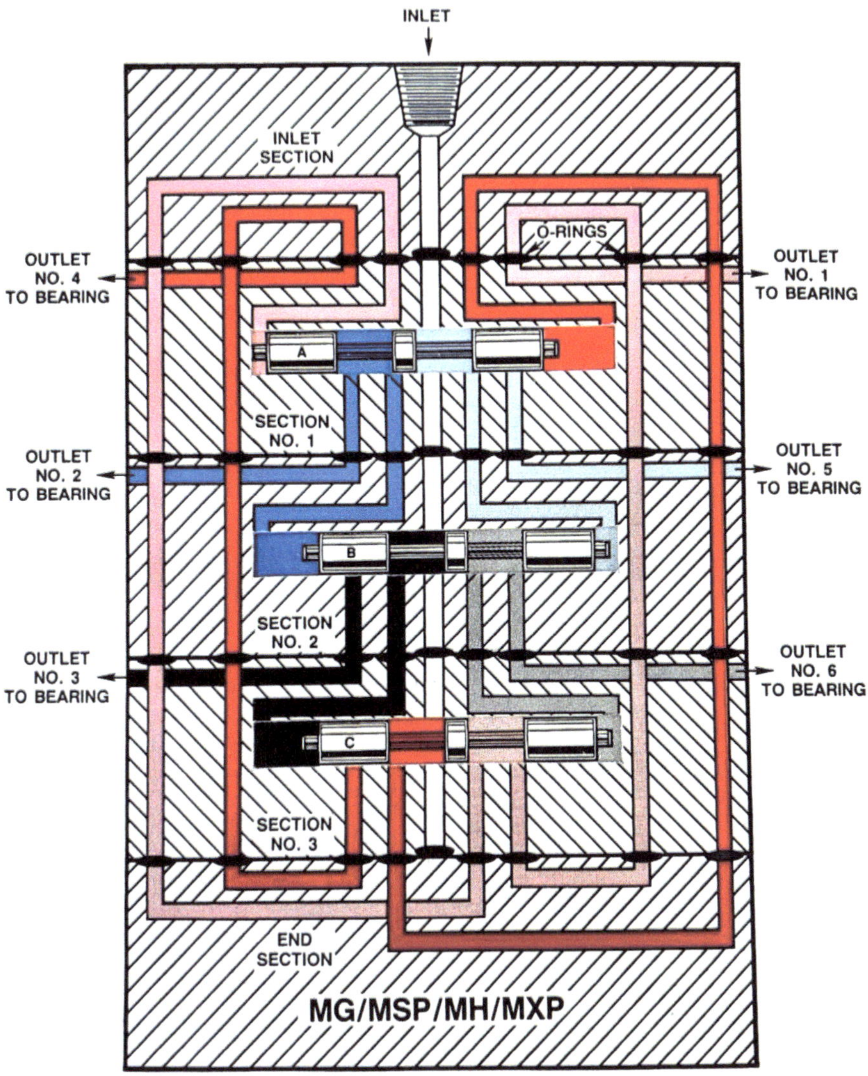

Figure 20.35 Piston A – Outlet one discharge (lubriquip slide card 412-C).

remain locked in place discharging a measured amount of lubricant through the dark blue passageway to outlet number two and on to the bearing as shown in Figure 20.36.

Once Piston B completes its stroke to the left, the gray center passage opens up to allow lubricant through to the right-hand end of piston C forcing it to move to the left while piston A sn B remain locked in place. Lubricant

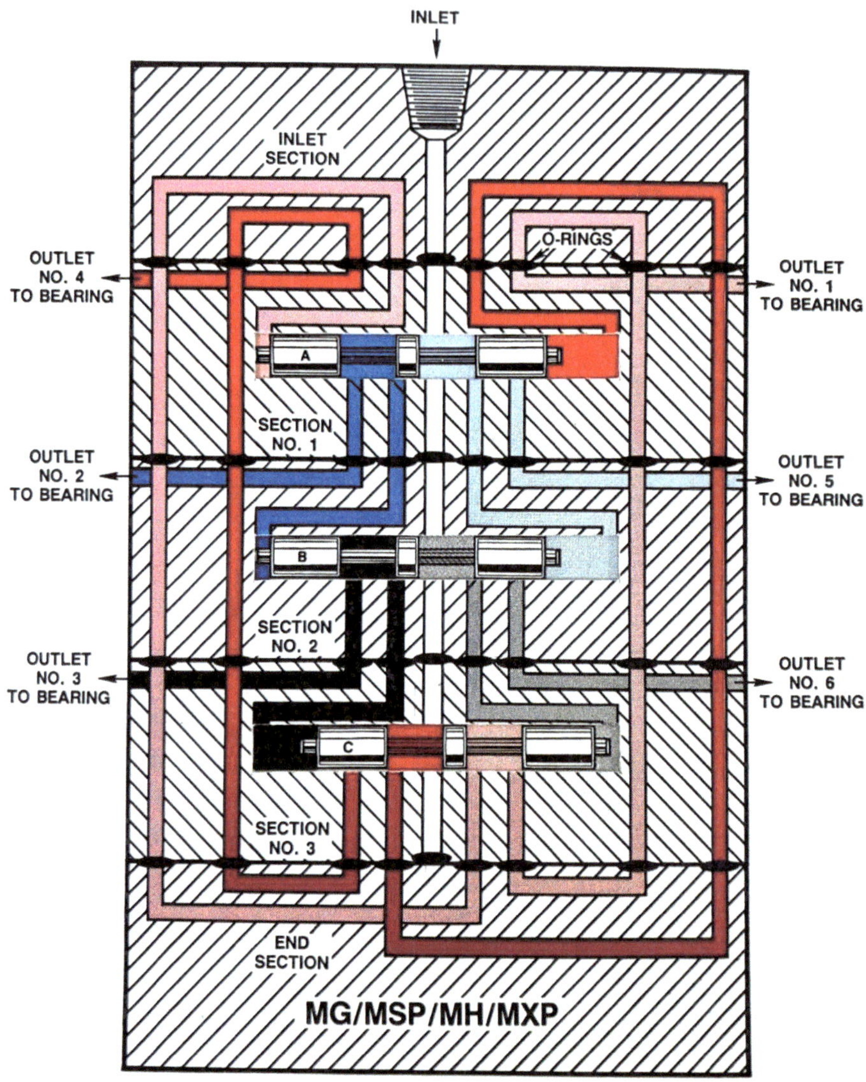

Figure 20.36 Piston B – Outlet two discharge (Lubriquip slide card 412-C).

now discharges through the black passageway to outlet number three, on to the bearing point as shown in Figure 20.37

With Piston three locked to the left lubricant is now allowed through the center of piston three and through the left-hand pink passageways up to the left-hand side of Piston A. Piston A now moves to the right while piston B and C lock in place, and allow lubricant to make its way through the red

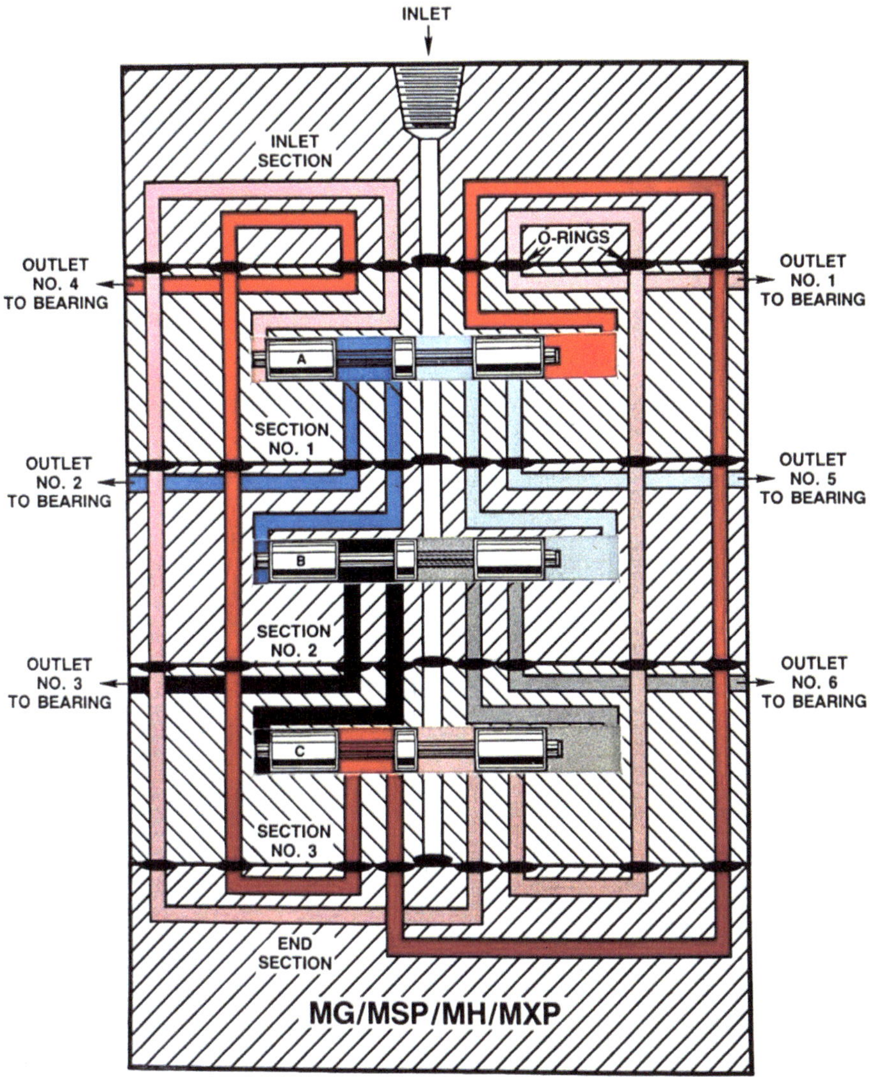

Figure 20.37 Piston C – Outlet three discharge (Lubriquip slide card 412-C).

passageways to outlet number four, and on to the bearing point as shown in Figure 20.38.

The block cycle continues with Piston B moving its charge through the pale blue passageway to outlet five bearing, and on to piston C that now moves its charge through the gray passage on to outlet number six bearing.

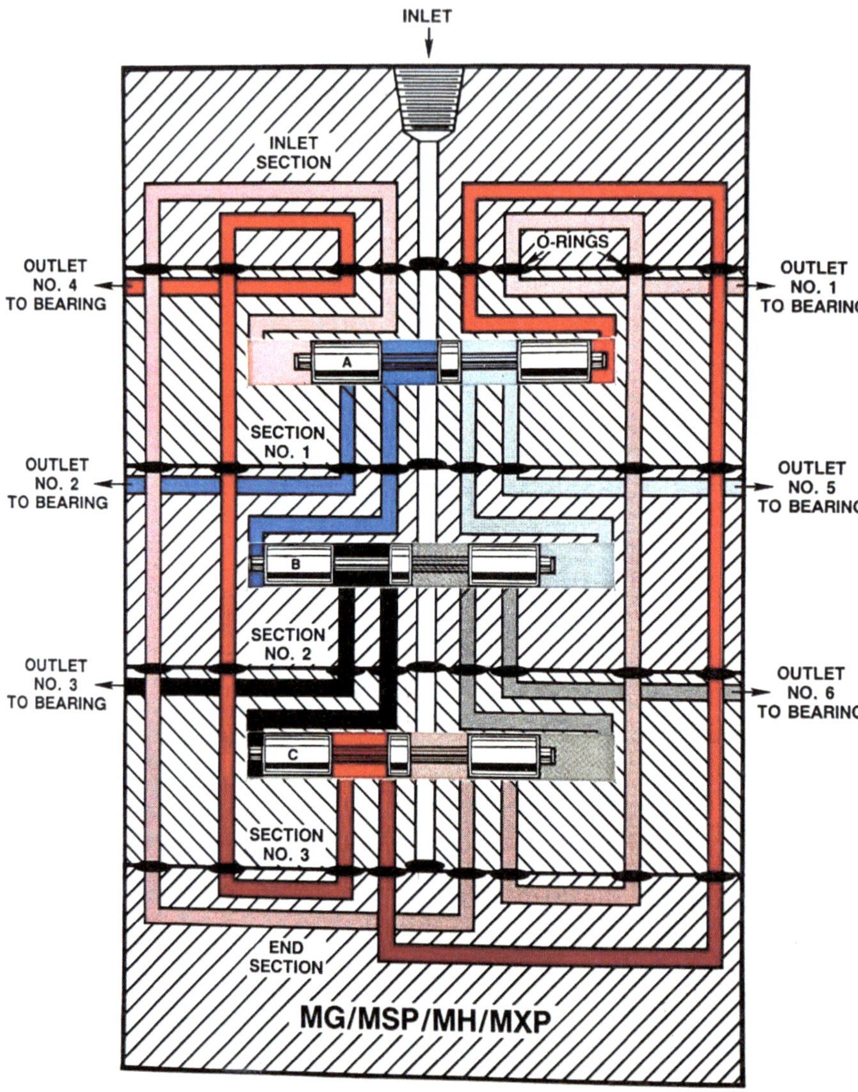

Figure 20.38 Piston A – Outlet four discharge (Lubriquip slide card 412-C).

This completes one full cycle of the block where every outlet on the block has delivered an engineered amount of lubricant to the bearing.

20.10.2 Monitoring the System

A cycle pin can be connected to the right-hand end of piston A that visually show the operator or lubricator that the block is cycling. Attaching a counter/

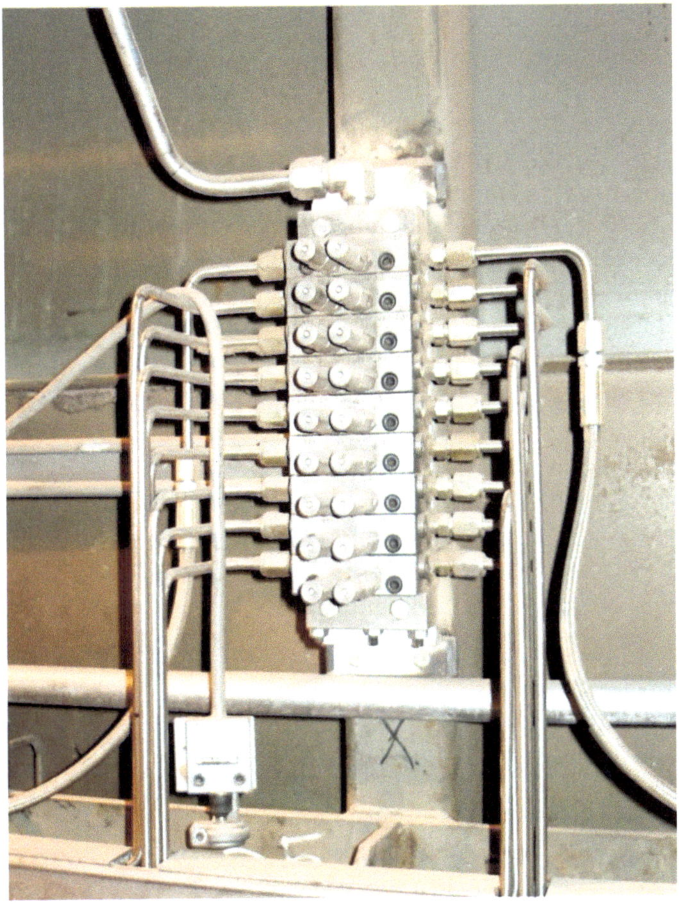

Figure 20.39 Trabon progressive divider block with indicator pins. Courtesy: ENGTECH Industries Inc.

timer control to the cycle pin will indicate if no delivery has taken place within a given time period and signify a broken main delivery line.

Because all pistons are hydraulically interconnected, a blocked bearing or pinched line can produce a hydraulic lock up in the system. This can also be detected by lack of movement on the cycle pin when connected to a controller time-out sequence.

Blocked and crimped lines can also be visually detected via simple mechanical overpressure indicators connected at the secondary delivery line block outlet as shown in Figure 20.39. Whenever a "restriction" caused hydraulic backpressure is sensed, a visual indicator pin "pops" up to indicate the exact line/bearing point requiring maintenance. If no electronic alarm sensors are used, operations and/or maintenance must perform regular visual checks for visual alarm indications.

20.10.3 Pros and cons

With basic TLC this system type will provide trouble free service for many years due to its engineering and tamper proof design. Blocks are modular, and replacement parts are relatively inexpensive to purchase, stock, and replace. As depicted in the manual greasing section's Figure 19.23 divider valve photo, this system can even be utilized with a manual grease gun to deliver an engineered amount to each bearing point and automated at a later date. (Photo clearly shows block cycle pin and four overpressure indicators in "run" position).

The down side of the system is the difficulty to add points once the initial system is installed (although achievable); in comparison to other single line systems, series progressive systems demand more system engineering.

However, the plus side to this is every system is engineered by the vendor on your behalf, and provided with schematics and a Bill of Material to include in your CMMS and/or maintenance files, usually all included in the cost of the system

20.11 Pump-to-point Lubrication System

Pump-to-point systems, sometimes referred to as multipoint systems, owe their pedigree to the industrial revolution when a cam shaft set in an oil reservoir was connected for the first time via a pitman arm to a rotary or rocking motion shaft as part of a steam engine bearing lubrication system. A cam, attached to a series of in line rocker arms, in turn was attached to a series of individual pistons tasked to draw oil from the reservoir to lubricate the camshaft and rockers and pump oil to individual bearing points on the machine via copper lines; this now meant the lubrication individual need only watch and fill one reservoir feeding 6-8 lubrication points. Figure 20.40 shows a box cam multipoint oil lubricator still employed today on a 19th century Victorian textile mill steam engine. Simple in concept, cam box lubricators were expensive to manufacture, limited to oil only, and could only accommodate a small number of bearing points.

The next evolution of the pump-to-point design provided an independent air driven pump with a reservoir mounted atop of the pump, with the ability to dispense oil or grease through a series of ported outlets mounted around the periphery of the pump chamber. As the pump piston is actuated--usually from

Figure 20.40 Victorian steam engine box cam pump to point lubricator (still in use today) Coutesy: ENGTECH Industries Inc.

an electronic timer or counter--it passes by each "metered by restriction" outlet, positively displacing lubricant into each line piped to the individual bearing points. The air valve turns off and the spring-loaded pump returns drawing lubricant into the firing chamber as it resets for the next lubrication cycle. (See Figure 20.41).

These new pump-to-point systems are inexpensive to purchase and install, and are most popular for small to medium size machinery with less than 40 points. They are also very popular for use as chassis lubrication systems for trucks and tractors with on board compressors, used to lubricate the fifth wheel, shackles, and steering components while the vehicle is in motion.

20.11.1 Pros and cons

Early cam box units were oil only devices constrained to their designed number of points that could easily be individually adjusted for flow to bearing points. No line breakage or blockage protection devices are available.

Later pump-to-point systems can be used with oil and grease and system engineering is not demanding. Within the confines of the pump size, it will accommodate the addition or reduction of system points post installation. Flexibility of lubricant delivery can be adjusted by the frequency of the

Figure 20.41 Modern multipoint pump-to-point lubricator. Courtesy: Interlube Corporation.

pump operation. A failure top pump incident can be detected via the controller, but secondary line protection is difficult to implement.

20.12 Injector Pump Systems

Injector pump systems (Figures 20.42 and 20.43) are used for dispensing adjustable quantities of oil, grease (or, occasionally, other liquids) for such varied duties as

- point lubrication
- application of cutting and cooling oils
- air-line lubrication
- oils for punching and stamping operations

Other Liquids would include food additives, application od adhesives, inks, printing fluid, dyes, chemicals, and more.

Properly engineered, injector pump systems do not exhibit "after-drip", can operate in any position, eliminate vapor lock and cavitation, and require no system priming. Most systems are available in single-feed or multi-feed outlets. In general, injectors are adjustable and can be operated using a controller that can cycle the injector by the second, or by the day as the operation requires.

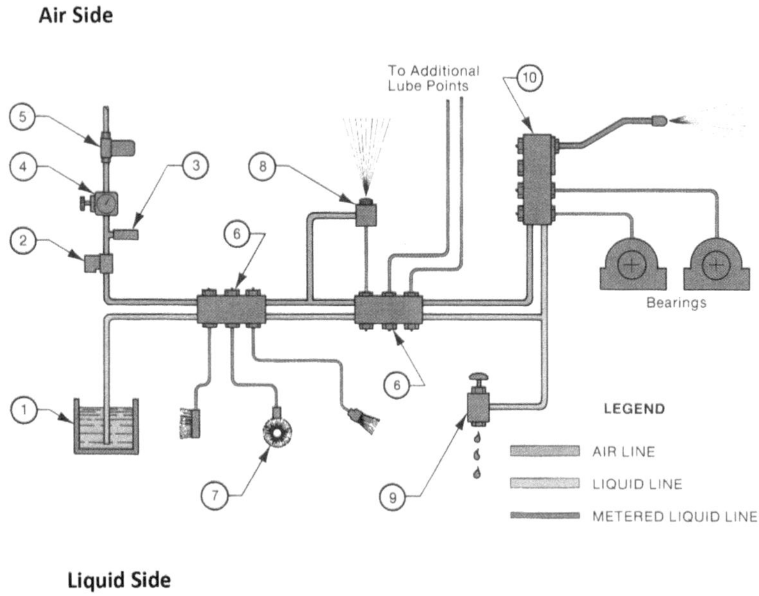

1. Reservoir	2. 3-Way Solenoid Valve
3. Pressure Switch	4. Air Regulator
5. Air Filter	6. 3-Feed Air Operated Injector Pump
7. Liquid Application Brushes	8. Remote Spray Valve
9. Manually Operated Injector Pump	10. 2-Feed Air Operated Injector Pump with Single Feed Spray Valve

Figure 20.42 Air-operated injector pump system. (Source: Oil-Rite Corporation, Manitowoc, Wisconsin.

20.13 Oil Mist Lubrication Technology and Applications

Oil mist systems have come a long way since their introduction to the European textile industry in the mid 1900's. Initially designed to resolve high-speed spindle bearing failures that would not respond to conventional oil and grease lubrication tactics, oil mist first saw favor in the North American market in the 1960s/70s with the use of oil mist lubrication in the refining and petrochemical industries when companies such as Exxon and Chevron began to apply oil mist to pump bearings [1]. By the early 1970's oil mist lubrication was being applied to rolling element bearings of electric motors in the

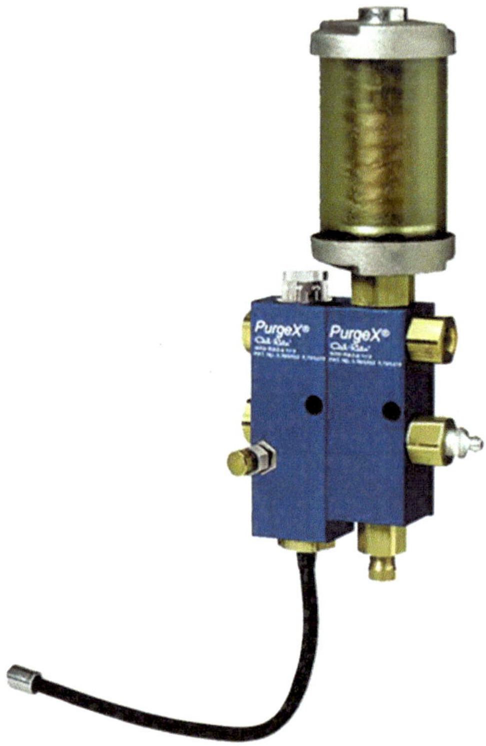

Figure 20.43 Air-powered "purgex" grease injector pump. (Source: Oil-Rite Corporation).

refining industry [2]. In the mid-1990s, 77% of the major, multi-location US refining companies had at least one large-scale oil mist system in at least one of their refineries [3].

Oil mist technology has kept pace with advances made by process industries and the mist systems being designed and installed today are far superior and more efficient than those installed in the 1980s. Some of the advances and new applications that are utilized in modern day systems are:

- Microprocessor controlled central oil mist generators compatible with central distributive control systems
- More efficient and effective distribution system design practices
- Improved oil mist manifolds
- Environmentally clean mist collection containers

- Drain leg designs and components, which eliminate waste and venting
- Efficient, environmentally clean, closed-loop oil mist systems
- Demisting system for the textile industry
- Miniaturized, closed-loop lubricators
- Portable mist density monitors
- New applications for oil mist:
  - Rotary lobe blowers
  - Defibrator press

20.13.1 Benefits and description of an oil mist lubrication system

As oil mist became more mainstream and moved into the large manufacturing industries, its open vent system design—combined with the ability for operators and maintainers to easily adjust the air flow pressure at will—often created an oily environment for workers. Unfortunately, an ever-increasing awareness of workplace environment safety forced a ban on both well-designed and poor designed oil mist systems in many plants.

With the introduction of micro process control, vortex style mist generation and closed loop system design, coupled with airborne mist detection systems, the oil mist system witnessed a dramatic resurrection.

20.13.2 How the system works

In the original oil mist system design a pressure controlled, filtered compressed air supply is passed through a venturi. Oil is then siphoned from a reservoir by the quickened airflow directly following the venturi and is directed at a baffle plate causing the oil to atomize into very fine droplets known as "dry mist." Anything less than a 1part oil to 200,000 part oil ratio (approximately 1.5 microns or 0.00006" diameter) falls back to the reservoir as heavy oil particles; mist then is then piped into a 2" diameter scheduled pipe header at between 5-40 inch water pressure creating a velocity of up to 24 feet per second. As it approaches the bearing point, the mist is then passed through a series of mist metering fittings that allow an engineered amount of oil to enter the bearings.

Depending on the fitting type used, the "dry mist" or partially reclassified mist then "envelops" and "wets" the entire bearing surface area with a thin lubricant film while imparting a partial positive pressure within the bearing housing that works to prevent any contamination influx.

In this original design, the mist was allowed to vent to atmosphere. New system designs employ sloped lines to carry reclassified oil within the lines directly back to the reservoir, and more importantly, employ a closed loop design with drain legs and components designed to capture coalescent waste and eliminate open venting. An improved Vortex air chamber design to replace the old venturi design has resulted in a more efficient mist generation that can now carry over distances of 600 feet—three times that of the original system design—translating into the ability to lubricate substantially more points from a single mist generation unit. The coalesced oil return improvements have turned a once total loss system design into a much more effective partial recovery system design using less lubricant.

Oil mist is a centralized lubrication system that continuously produces, conveys and delivers mist lubrication to bearings and metal surfaces. Oil mist lubrication has been shown to significantly reduce the number of lubrication related bearing failures when compared to oil splash and grease lubrication.

Example: in one large petrochemical plant, bearing failures in two similar units with a population of about 200 pumps each were compared. One had mist, while the other had conventional [oil splash] lubrication. The unit on oil mist realized approximately 85 percent fewer bearing failures. Figure 20.44 is a graphic illustration of typical findings.

In a 1994 comprehensive research study performed Texas A&M University, oil mist lubricated bearings were found to run cooler by about 10°C compared to oil sump lubricated bearings. The oil mist lubricated bearings also ran with about 25% less friction than oil sump lubricated bearings [5]. Since the air under pressure in the housing escapes through the housing enclosures or vents, the entrance of moisture and grit is retarded. In addition, oil mist lubrication continuously supplies only clean, fresh oil to the bearings. The two factors combine toward a full bearing life expectancy. Because rolling element bearings require very little lubricant, oil consumption is comparatively small [6]. Recall, again, that with "wet sump" (purge mist) the mist floats in the space above the liquid oil. In the "dry sump" (pure mist) application method, oil is applied in the form of tiny droplets carried by air. These tiny droplets coalesce and "plate out" on the bearing's rolling elements.

By the mid-1970s, sufficient experience had accrued to single out dry sump oil mist methods as best suited for plant-wide petrochemical complexes.

In addition to technical and failure statistics, plant owners also turn to cost reduction and return on investment to justify use of oil mist systems.

Figure 20.44 Repair cost comparison for 2 identical petrochemical facilities, with vs. without oil mist lubrication.

One of the largest cost savings attributed to the use of oil mist lubrication is reduced equipment repair and lower maintenance cost.

Other factors that add to the justification for use of oil mist include:

- Reduced lubricant consumption
- Greater manpower flexibility
- Reduced spare parts inventory
- Low mist system maintenance requirements
- Higher equipment availability

Oil mist systems are extremely reliable and have a fifteen-to-twenty-year useful life without major overhaul. Oil mist systems can be installed with new projects or retrofitted to existing facilities. When savings are compared to total installation costs, the payback period normally calculates to between one and two years [8]. Given the 20-year life of the systems, the discounted rate of return (DCF) on the investment in oil mist is typically 50% to 100%, meaning it represents a very attractive investment project.

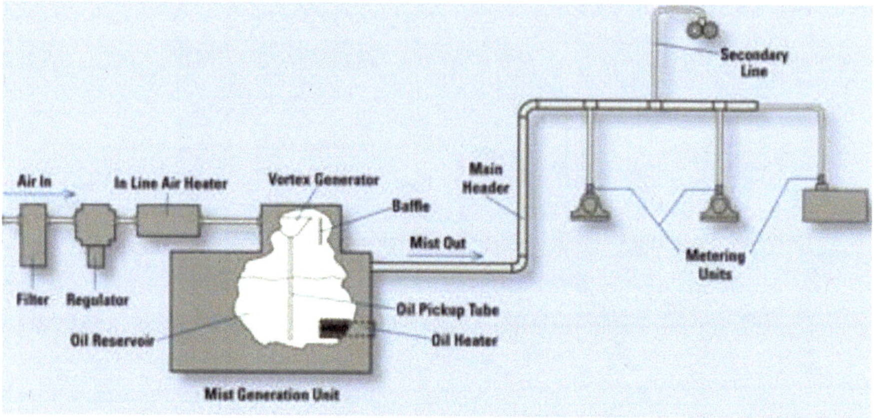

Figure 20.45 Typical mist system design. Source: Ken Bannister for Lubrication Management & Technology magazine's "Anatomy of a Centralized Lubrication System" article series.

20.13.3 Conventional oil mist system

The key components of a conventional, "one-way" oil mist system (Figure 20.45) include:

1. Central oil mist generator and oil supply tank.
2. Distribution piping to convey the mist.
3. Collecting containers at the equipment receiving the oil mist.
4. Mist manifolds to divide and direct the mist.
5. Stainless steel tubing to direct mist to each application point.
6. Re-classifiers to measure and apply the mist.
7. Drain lines to an oily water sewer or some type of collection container.

Each of these component types existed in oil mist systems that proved successful in 1975, but significant design improvements to each have made the modern systems even more effective, efficient, reliable and environmentally friendly.

20.13.4 New central mist generator design

The heart of the system is the generator (Figure 20.46) which utilizes the energy of compressed air, typically from the instrument air system, to atomize oil into micron-size particles. In modern, large-scale systems the mist generator is fully monitored and microprocessor controlled. Solid-state

Figure 20.46 Central oil mist generator and supply tank. (Source: Lubrication Systems Company, Houston, Texas.)

pressure and temperature transducers and level sensing devices have replaced the old-style electro-mechanical switches. Rather than using gauges, all monitored variables are displayed on demand by an alpha-numeric panel that not only shows typical gauge values but also provides messages describing the operating condition.

The control panel is password-protected, meaning only those operators trained and authorized have the capability to set and adjust operating and alarm conditions

Therefore, the possibility for making well intended but incorrect adjustments is minimized.

This new generation central mist generator has the following properties:

- Meets Class 1, Division 2, Group B/C/D, standards.
- Continuously monitors at least eight (8) operating variables including mist density.
- Factory plus user set control ranges allow for establishing sequential and system specific alarm limits.
- Alarm save and recall function allows for efficient and effective troubleshooting.

- Independently set and monitored mist and regulated air pressure control ranges to safeguard against improper troubleshooting and alarm elimination.
- Independent 4 to 20 milli-amp current signal which allow for external monitoring of each operating variable.
- Large capacity, nine-gallon (35 liter), internal reservoir, constructed of stainless steel, aluminum, or painted carbon steel.

In addition to improved reliability, the microprocessor control of the new mist generator units provides for customizing operating set points and alarm limits to exact user requirements. High temperature cutout controls are factory set. The user sets operating conditions and alarm limits for all monitored variables, meaning the system can be optimized for that particular application. For example, users can tailor the unit to mist very heavy viscosity lubricants.

20.13.5 Internal reservoir design

The internal reservoir of the new oil mist generators is compartmentalized and equipped with baffles. This design allows for efficient heating of the oil and elimination for the possibility of coking. The design also allows for settling of any contaminants that may be present in the lube oil. These reservoirs have a bottom that slopes to a low point where a bulls-eye sight glass allows for inspection of the oil. There is also a low point, valved drain port to draw off contaminants. The internal reservoirs of the new oil mist generators are much more than a simple rectangular shaped container as used on older units. A United States Patent [9] describes these features in greater detail.

20.13.6 Distribution header system design

The oil mist produced in the central oil mist generator is transported throughout the process unit through header pipe. Typically, this is 2-inch schedule 40 galvanized, threaded and coupled piping. In the hydrocarbon processing industry, the header pipe is normally installed in overhead pipe racks. The header pipe must be installed without traps or sags as pressure in the header is only .050 bar (20 inches of water column). The latest design specifications state that the oil mist can be transported using standard installation practice up to 180 meters (600 feet) horizontally from the central generator. Older design standards limited this run length to 60 meters (200 feet).

Today's designs call for the header to be sloped back to the central generator; none of the header should be sloped away from the generator [10].

Figure 20.47 Oil mist distribution piping. (Source: Lubrication Systems Company, Houston, Texas.)

This design promotes oil usage efficiency since the oil mist that coalesces in the header is returned to the mist generator for reuse (Figure 20.47, and 20.48). Older technology allowed sloping the header away from the mist generator towards drain legs[11].

Where elevation changes do not allow for sloping back to the generator, drain legs are installed. The drain leg prevents accumulation of oil in the main header because such accumulation of oil would block the flow of mist downstream of the trap. In drain legs utilizing the prior art, the collected coalesced lubricant has been manually drained, either to a sewer or a container which needed to be manually emptied, while the drain leg continuously vented oil mist to atmosphere. Today's distribution systems utilize automated drain leg assemblies that do not require manual operation and can be fully integrated into closed-loop systems.

These assemblies are equipped with an air-activated level switch and pump. They collect the coalesced oil and automatically pump that oil overhead to a point in the distribution header that does slope back to the central generator [12].

These automated drain leg assemblies can also be retrofitted to older, once-through mist systems, thus enhancing oil recovery and minimizing oil flow to sewers.

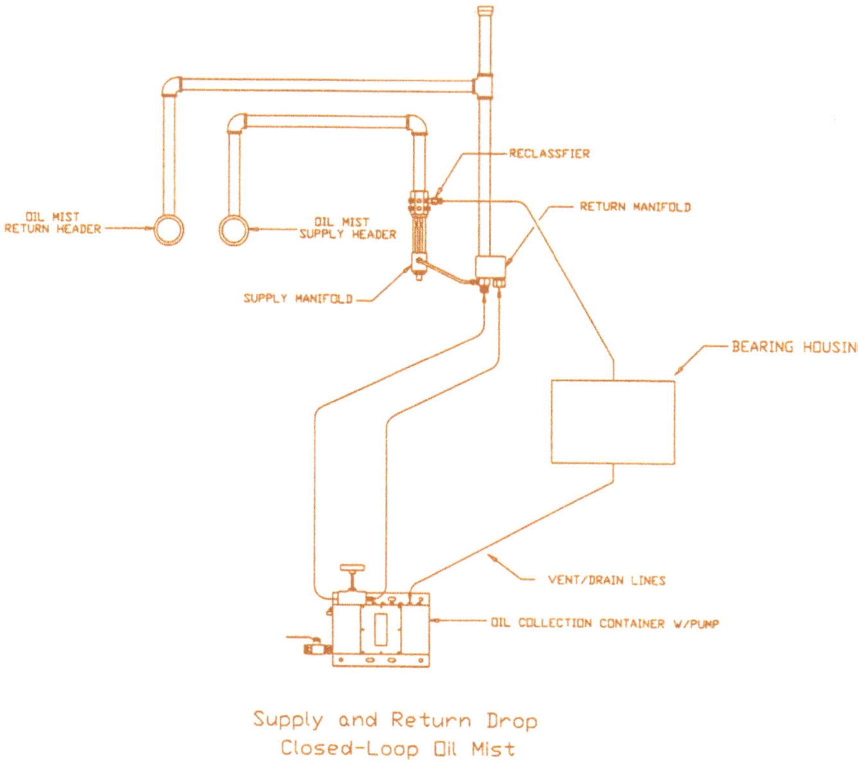

Figure 20.48 Closed-loop oil mist system schematic. (Source: Lubrication Systems Company, Houston, Texas.)

20.13.7 Mist manifold

Above the equipment that is to receive oil mist, a "drop" is installed. Drop piping is normally 3/4-inch galvanized steel. The drop rises from the top of the 2-inch header so that liquid oil and any particulate contamination are not carried to the lubricated equipment. The drop terminates in a manifold assembly that divides the mist flow to the individual lube points that receive oil mist. Often the reclassifier or application fitting, an orifice metering device, is mounted in the manifold block. From the manifold block, stainless steel tubing is used to direct the mist to the application point.

In older mist systems the manifold was typically a rectangular metal block with ports drilled for mist flow. Most of these blocks were equipped with a snap acting valve for draining of the collected coalesced oil but the level of the collected oil was not visible. Modern mist manifolds (Figure 10.57) contain a high temperature glass viewing chamber that allows for visual monitoring of the level of collected, coalesced oil [13]. Operators can see when oil needs to be drained. When the manifold is drained, the flow of mist can be seen in the viewing chamber. Thus, mist flow is inspected without venting to atmosphere.

20.14 Lubrication Delivery System Pumps—Mechanical and Pneumatic Activated Types

Thus far, in this chapter we have addressed all variations of lubricant distribution delivery systems. With exception to the mist, air/oil, and single point lubricating devices—that all utilize a unique integral lubricant pumping unit—centralized lubrication system designers have a variety of lubricant pumping options to choose from.

With the distribution delivery system design finalized, the next step for the system designer is to review the lubricated machine design and customer cost constraints to select a suitable lubricant pump. To narrow the selection process, the designer must first answer some basic questions:

1. Does the design budget allow for the cost of a fully automated pumping system, or is the design restricted to much less expensive manual actuated pump that be upgraded or automated at a later date?
2. Does the machine have an integral lubricant reservoir the pump can be mounted in or on, or does the pump require its own lubricant reservoir?
3. Does the equipment have a mechanical power takeoff point, hydraulic or pneumatic power source available?
4. Is an electrical power source available?

20.14.1 Mechanical powered pump units

If no electrical power source is available, the designer has no choice but to use a mechanically actuated pump. The pump design will usually employ a positive displacement piston whose output delivery can be adjusted by restricting the length of the piston stroke. For most manual operated pumps, a lever arm is mechanically connected to a cam that moves a single acting piston pump back and forth (some designs use a spring returned pump) with lubricant fed from an attached reservoir. The pump is actuated by manually moving the lever arm in a back-and-forth arc motion drawing lubricant into the piston chamber that is in turn pumped into the distribution system through an internal check valve.

Figure 20.49 Mechanically actuated grease pump. Courtesy: ENGTECH Industries Inc.

If reciprocating or rotary machine motion is available, the lever arm of the manual pump can be replaced with a power takeoff pitman arm linkage attached to the motion device. Figure 20.49 shows a series progressive distribution system with a mechanical pump attached to a pitman arm arrangement attached to the end of the large diameter rotating machine shaft. The shaft attachment point is offset from the center to produce a reciprocating (up and down) motion of the arm that produces a rocking motion at the pump shaft, emulating the back-and -forth rocking motion of the manual lever arm. By changing the length relationship of the pitman arm attachment point and arm length, the degree of arc will change and speed up, or slow down, the amount of pump strokes per hour. In the example photo you can see that the pump setting is incorrect, evident by the excessive grease being pumped out of the bearing seal.

In the smaller single line resistance "oil only" type systems, a spring return piston is employed. A single push of the lever pushes lubricant out through the meter valves to the lube points. As the lubricant is apportioned, the line pressure dissipates and the spring return piston draws in the next lubricant charge.

20.14.2 Pneumatic powered pumps

Pumping lubricant to many points, over large distances, through large diameter lines is typical of large progressive and dual line type systems in heavy industries. These systems will typically employ a high-pressure pneumatic pump as shown in Figure 20.12.

Pneumatic barrel or drum pumps are unique in that they are designed to sit straight on top of a standard grease or oil drum, eliminating the need for a reservoir. The pump can deliver an output pump pressure of up to 70:1 airline input pressure. Pump design is again piston style, and is usually controlled by a stroke counter or by line pressure depending on the distribution system requirements.

In the advent years of the centralized lubrication systems the automotive industry utilized "on board" vacuum operated lubricating oil pumps to automatically lubricate the suspension and steering components of luxury cars while they moved.

20.15 Maintaining Your Centralized Lubrication Delivery System

A centralized lubrication system is an integral and critical part of the host machine it lubricates and as such, must be maintained accordingly.

20.15.1 Cleanliness and contamination control

The old adage "cleanliness is godliness" is a mantra to live by when dealing with lubricants and lubrication systems. Induced system contamination is a major causal factor for premature bearing and lubrication system wear. Great care must be taken when transferring lubricants into lubrication system reservoirs so as peripheral dirt is not introduced into the system and passed through to the bearing points. By their very design and nature, lubrication system components are not dirt tolerant as many of the systems employ fine tolerance pistons and spool valves in their pumps and delivery blocks and injectors similar to those found in the bearings they are to lubricate. Utilizing the following simple maintenance and set up tips will alleviate most contamination problems:

20.15.2 New installations

1. To avoid cross lubricant contamination, ensure the lubricant reservoir tag identifying the correct lubricant is the same lubricant about to be dispensed into the reservoir
2. Clean the reservoir fill area and fill the pump reservoir with clean lubricant using a filter cart with a cleaned dispense nozzle and a clean dedicated transfer funnel in the case of oil, or a fully cleaned positive coupled air powered grease barrel pump when filling a grease reservoir
3. Start pump and purge lubricant through the pump before connecting the main delivery lines
4. Clean all lube lines of swarf and debris before connecting the lines. Use a air powered "wad" cleaning system to shoot wadding through the lines to ensure no dirt is in the lines prior to start up
5. Connect lines to cleaned dispensing blocks and purge with lubricant before connecting and purging the secondary lines prior to connecting top the bearing points.
6. Check for system leaks and repair any leak immediately cleaning up all traces of leaked lubricant
7. After system had run for a number of hours, perform a second leak check
8. After system has run for a number of days, perform a lubrication check at each bearing point to ensure no lubricant has purged through the bearing. If lubricant is evident the system will require further calibration.

20.15.3 Existing Installation

1. Set up a PM task to clean the lubrication pump and reservoir regularly
2. To avoid cross lubricant contamination, ensure the lubricant reservoir tag identifying the correct lubricant is the same lubricant about to be dispensed into the reservoir
3. Clean the reservoir fill area and fill the pump reservoir with clean lubricant using a filter cart with a cleaned dispense nozzle and a clean

dedicated transfer funnel in the case of oil, or a fully cleaned positive coupled air powered grease barrel pump when filling a grease reservoir

4. Perform a system leak check

More detailed information on contamination exclusion and control refer to Section 5: Chapter 33

20.15.4 The power of adjustability

Lubrication programs, systems, and methods once implemented and operational, are rarely scrutinized or updated. Such behavior presents golden opportunity for improvement that can pat huge dividends in extended bearing life, reduced lubricant consumption, reduced energy and carbon footprint and lower maintenance and operational costs.

When initially purchased, most rotating equipment will be delivered with some form of lubrication system or approach in place. These can be as complex as a fully integrated and computer controlled automated lubrication system with each bearing receiving a measured amount of lubricant based on time, cycle, or condition trigger, e.g., temperature, energy draw. Alternately, the machine can be delivered with a modest lubrication approach utilizing grease nipples and oiling points left to the end user to cope with a "lubricate as necessary approach to bearing care. From one extreme to another, the maintenance department has the power to and ability to tune their lubrication set up to improve and optimize their lubrication approach.

When lubrication systems are designed for use they inherently allow for adjustability to sallow for system tuning in one or more of three major areas that include: The metering devices; pump output; pump cycle control.

20.15.5 Regular PM/Operator maintenance

Daily Lubrication system checks are essential for ensuring the lubrication system is operating as designed and that there is lubricant in the system. This is often best performed on a daily basis by the equipment operator who visually checks the entire system in a quick system walk-around and only notifies the maintenance department when an exception is found. Check functions can include:

1. Check reservoir fill level—is the level between the Lo and Hi mark on the reservoirs? See Figure 20.50.

2. Check for and notify immediately of any system leaks

Figure 20.50 Gearbox reservoir Bulls Eye sight gauge. Copyright Des-Case Corporation, 2022.

3. Check for apparent system damage including line crush and any over-pressure indicator signal denoting back pressure in the system caused by a damaged or blocked bearing or lin
4. Check for controller warning signals/lights and report immediately
5. Check any pressure filter (used in recirculating oil and hydraulic systems) is not showing a red flag signal indicating the filter is full and in bypass mode

Of course, the type of lubrication system and lubricant will dictate the level of checking required. For example, circulating oil systems are prime candidates for oil condition analysis that indicates lubricant condition allowing the lubricant to be changed only when needed, based on its condition.

References

[1] Bloch, H.P., "Large Scale Application of Pure Oil Mist Lubrication in Petrochemical Plants," ASME Paper 80-C2/Lub-25 (1980).
[2] Miannay, C.R., "Improve Bearing Life with Oil Mist Lubrication," *Hydrocarbon Processing Magazine*, (May 1974), 113–115.
[3] Ward, T.K., "1995 Refinery Oil Mist Usage Survey," Lubrication Systems Company Internal Memo, (February 1996).
[4] Towne, C.A., "Oil Mist Lubrication for the Petrochemical Industry," Proceedings of the 11th International Pump Users Symposium, The Turbomachinery Laboratory, Texas A & M University, College Station, Texas, (1994), p. 107.
[5] Shamim A., Kettleborough, C.F., "Tribological Performance Evaluation of Oil Mist Lubrication," Texas A&M University Research Paper, (December, 1994).

[6] SKF USA Inc., Bearing and Installation Guide; SKF Publication 140170, (February, 1992), p. 35.
[7] Bloch, "Oil Mist Lubrication Handbook," Gulf Publishing Company, Houston, Texas, USA (1987).
[8] Bloch, H.P., and Shamim, A., *Oil Mist Lubrication, Practical Applications*, The Fairmont Press, Lilburn, Georgia, USA (1998).
[9] Ehlert, C.W., Inventor, United States Patent 5,125,480, (Issued June 30, 1992).
[10] Lubrication Systems Company of Texas, Inc., Design and Installation of LubriMist® Oil Mist Lubrication Systems, (September 1996).
[11] Stewart-Warner Corporation, Oil Mist Lubrication Systems for the Hydrocarbon Processing Industry, (1982)
[12] Lubrication Systems Company of Texas, Inc., LubriMist® Accessories Brochure, (March 1996)
[13] Ibid.
[14] Ehlert, C. W., Inventor, United States Patent 5.318.152, (Issued June 7, 1994)
[15] Lubrication Systems Company of Texas, Inc., LubriMist® Accessories Brochure.
[16] Ibid.
[17] Ibid.
[18] United States Patent 5.318.152.
[19] Ibid.
[20] Ehlert, C.W., Inventor, Patent Application, (January 30, 1996)
[21] Arnold, D.R., "Improving Rotary Lobe Blower Reliability," Blower Reliability Conference, (April 26, 1995).
[22] Ibid.
[23] Smook, G.A., Handbook for Pulp & Paper Technologists, Joint Textbook Committee of the Paper Industry, Atlanta, Georgia, (1987).
[24] Bradshaw, Simon; "Investigations into the Contamination of Lubricating Oil in Rolling Element Pump Bearing Assemblies," *Proceedings of the Texas A&M Pump Users Symposium* (2000)
[25] Lubriquip Series Progressive Divider Valve Slide Card 412-C

Bibliography

Bannister, Kenneth E., Manual Lubrication Delivery Devices, *Lubrication and Fluid Power* magazine, Feb 2006
Bannister, Kenneth E., Anatomy of a Centralized Lubrication System article series, *Lubrication Management Technology* magazine, 2011–2013

Bannister, Kenneth E., Industrial Lubrication Fundamentals Article series, *Lubrication Management & Technology* magazine, 2013–2015

Bannister, Kenneth E., Liquid Gold—Lubrication Principles Workshop Training Manual, *ENGTECH Industries Inc.*, 2008–2011

Bannister, Kenneth E., Industrial Lubrication Fundamentals—Certification Preparatory Training—Level 1 Workshop Manual, *ENGTECH Industries Inc.*, 2012–2016

Bannister, Kenneth E., "Understand and Improve your Lubrication System Delivery Setup", Maintenance Technology Magazine, 2016

Bloch, Heinz P., "Dry Sump Oil Mist Lubrication for Electric Motors," *Hydrocarbon Processing*, March 1977

Bloch, Heinz P., "Oil Mist Lubrication Cuts Bearing Maintenance" (*Plant Services Magazine*, November 1983)

Bloch, Heinz P.; "Benefits of Oil Mist Lubrication for Electric Motor Bearings" *(Plant Services Magazine*, April 1986)

Bloch, Heinz P.; "Preservation by Oil Mist Application," (*Plant Services Magazine*, November 1987)

Bloch, Heinz P.; "Oil Mist Lubrication: Is it Justified and How Should it be Executed in the 90s?" *Hydrocarbon Processing*, October 1990, pp. 25

Bloch, Heinz P.; "Best-of-Class" Lubrication for Pumps and Drivers," *Pumps & Systems,* April 1997

Bloch, Heinz P.; "Oil Mist Lubrication for Electric Motors," *Hydrocarbon Processing*, August 2005

Bloch, Heinz P.; "Applying Oil Mist," *Lubrication & Fluid Power*, February 2005

Bloch, Heinz P. and Fred Geitner; "Machinery Uptime Improvement," (2006) Elsevier-Butterworth-Heinemann, Stoneham, MA (ISBN 0-7506-7725-2)

Bloch, Heinz P.; *Improving Machinery Reliability*, (1982 and all later Editions), Gulf Publishing Company, Houston, TX, ISBN 0-87201-376-6; ISBN 0-87201-455-X; ISBN 0-88415-661-3

Bloch, Heinz P. and Allan Burris; *Pump User's Handbook—Life Extension*, (2006) 2nd Edition, The Fairmont Press, Lilburn, GA 30047, ISBN 0-88173-517-5 SKF USA Inc, "Bearings in Centrifugal Pumps, Application Handbook" *Publication 100-955* (1995), pp. 20

Wilcock, Donald F., and E.R. Booser, 1957, *Bearing Design and Application*, McGraw-Hill Book Company, New York, NY 1012

21

Lubrication Delivery System Design Components

21.1 Reservoirs

Webster's dictionary describes a reservoir as "a place where something is kept in storage: such as a part of an apparatus in which liquid is held". For lubricants there are two classes of reservoirs, 1) Storage reservoirs, e.g., totes, pails, drums, etc., and 2) Production, or working reservoirs.

21.1.1 Lubricant storage reservoirs

Lubricant storage reservoirs are simple storage containers used to transport and hold lubricants for later transfer into smaller working reservoirs. Practical examples include tanker trucks that deliver lubricants to site, which are then in turn often transferred into interim on-site storage reservoirs. Typical on-site storage reservoirs are shown in Figure 21.1.

Polyethylene totes are popular storage reservoirs as they never rust, and their light color makes it easy to determine oil fill level by sight. Due to their light wall construction, poly totes are often contained in steel frames to prevent their walls bulging under load; this also allows them to be easily stacked in storage locations.

Grease storage containers differ in that they are commonly sold and delivered in one-time use drums and pails, whose contents can then be transferred manually or by external pump at a later time to a machine or working grease lube system reservoir. Alternatively, a drum or pail is employable as a replacement working reservoir by simply mounting the machine lubrication pump to the container.

Typically, storage style reservoirs are simple box or tube style vessels that contain a lid or large capped opening for bulk fill purposes. Oil storage reservoirs will often be equipped with a drain port for draining off moisture. Depending on reservoir design, oil is transferred through the fill port using an

Figure 21.1 Typical on-site interim lubricant storage reservoirs (Courtesy ENGTECH Industries Inc.)

auxiliary pump, or employs gravity to dispense lubricant through a tap valve mounted in the lower region of the reservoir.

Storage reservoirs are rarely ever filled 100% and as the fluid level falls, the resulting "head space" volume between the top of the fluid and its roof becomes larger. This space will contain water and solids contamination caused either by ambient heating and cooling cycles, and/or open filling of the reservoir. To alleviate contamination a desiccant style breather installed in the reservoir roof is recommended. This style of breather allows air to flow from inside the reservoir to outside the reservoir and vice-versa, thereby neutralizing any tank pressure build up while capturing moisture and solid contamination in the breather filter media.

21.1.2 Lubricant working reservoirs

Working reservoirs are often integral to the machine, system or lubrication pump they are employed to hold lubricant for; consequently, their design is more critical and complex than a regular storage reservoir due to their footprint having to concede to the space made available by the machine, system or pump design. In addition, their design must accommodate a variety of system control demands noted below. Figure 21.2 shows a closed loop oil system with pumps connected outside of the reservoir.

When a closed loop circulating oil system is employed, a design balance must be struck between the bearing lubrication requirements, the size

Figure 21.2 Closed loop oil system with multiple pumps connected to single reservoir Courtesy ENGTECH Industries Inc.

of the lubricating pump, and the size and shape of the lubricant reservoir. Typical key design features can include:

- Engineered reservoir capacity and size so that:
  - the reservoir is sized large enough to hold enough lubricant for the entire delivery system at full load and have enough room to accommodate all returned oil when the system is at rest without overflowing.
  - the reservoir is large enough to accommodate a baffled cooling circuit. In a circulating oil system, the oil is "washed" over the bearing to both lubricate and cool the bearing. Cooling is performed through the transfer of heat from the bearing to the oil, which in turn is returned to the reservoir. Part of a working reservoir's job is to allow the oil to cool considerably before being pumped back to the bearing for it's next cycle. This is achieved through the use of internal tank baffles forcing the oil to travel slowly across the reservoir from the return to the delivery side using the reservoir and baffles as a heat

sink to transfer and dissipate oil heat as it travel through the reservoir. Rapid cooling is achieved through the use of built in radiator fin style exchange cooling circuits.

- ○ the reservoir is sized large enough if it has an external built-in deep-cleansing filter loop (can require up to an additional 20% reservoir capacity)
- ○ the reservoir has does not have a large headspace (distance between the lubricant level and the inside top of the reservoir) when the pump is at rest

- For circulative oil systems, the pump will usually reside on top of the reservoir or sometimes inside the reservoir if it also serves as a machine system housing, e.g., gear pump, vane pump
- For large circulative oil systems, the reservoir can be designed with an in-service decontamination system to filter out machine process contaminants (see below for detailed description)
- For circulative oil systems, the reservoir can have "quick-lock" style connection for connecting an external filter cart to perform deep filter cleansing if the oil whilst in operations
- Note: for grease systems the pump can be an attachment assembly that fits directly on to a grease drum or pail container that is discarded once empty
- For total loss oil systems and grease, the reservoir must be ideally sized so that it only requires refilling weekly (minimum) to monthly (maximum). Sizing will depend on number of points and machine operational time. This fill time requirement ensures that the system is looked at and assessed regularly

For total loss oil and grease systems, the pump can fit on top of the reservoir as shown in Figure 21.3, or visa-versa. Here the reservoir is also constructed in metal and see through plastic

A working reservoir is likely to have an internal screen type suction filter plumbed to the pump delivery line to protect the pump. Most reservoirs will indicate lubricant level using a plumbed in external site gauge, which could also be sensor connected to an alarm or smart signal-receiving device. Total loss oil systems can employ low and hi limit reservoir fill sensors that can be set up to auto activate a solenoid fill valve connected to a storage

Figure 21.3 Total loss reservoir over pump style lubrication system (Courtesy ENGTECH Industries Inc.

reservoir allowing the unit to fill automatically as long as oil is present in the storage reservoir.

Similar to storage reservoirs, one would expect to see a desiccant style breather in use when the reservoir is atmospherically pressured. A closed reservoir system is employed when a reservoir is to be positively pressurized as in a spray type lube system.

In its basic form a working reservoir is filled through a simple capped fill port that is easily prone to contamination ingress during the filling process itself, or through forgetting to refit the cap in place after filling is complete! Newer sophisticated systems now employ direct connect/disconnect fittings that when connected to compatible filter carts allow the reservoir to be filled with the cleanest oil possible. The same system can also be used with the filter cart to perform an "off-line" bypass cleaning through the cart filter, often while the machine is running, returning it directly into the reservoir saving money and downtime on regular oil changes. (Note: see reservoir filter setup in the filter section later in this chapter.)

When employed outdoors in cold climates or unheated winter conditions, many reservoirs are fitted with thermostatically controlled electric heat plugs, similar to those used in automotive engine oil reservoirs, used to ensure the oil always operates in its optimum viscous range. Other heating devices can be magnetically attached to the reservoir exterior; can be a trace wire attached mechanically or glued to the reservoir casing; can be a thermal blanket encasing the reservoir.

21.2 Lubrication Lines and Hoses

The lubrication delivery line is an important and integral component within any lubrication system, whether a centralized automated system or simple centralized manual delivery system; making the wrong choice of lubricant delivery line can easily compromise machine reliability!

The function of the lubrication delivery line is simple, connect each bearing point to a lubricant source (indirectly from a meter or gang block, or directly from the pump,) and allow the lubricant to be contained within the line to flow without constriction. As lubricant delivery systems are hydraulic in nature, the line must also be capable of withstanding pressures ranging from hundreds to in some cases, many thousands of psi pressure.

21.2.1 Line size and material

Correct choice of line size and material is essential if a lubrication line is to prove reliable in service. For the most part, lubrication delivery lines play a passive role within the system and are fixed to the side of the machine, the exception being in the case of a lubricated part that moves independently of the fixed machine in which case, the line is used to provide the flexible connection.

Before a delivery line can be specified a number of basic questions regarding the system design must be addressed, these include:

- ***Is this system automated or manual?*** A crucial question for assessing line material, line diameter and wall thickness, which relate specifically to the line material burst pressure rating.
    - Manual systems designed to "gang" grease nipples in a central block can be lubricated by grease guns capable of developing up to 15,000 psi pressure

  - Manual hand pump and automated pump systems operate at much lower pressures between 100 psi up to 5000psi

- ***What type of automated/engineered delivery system is specified?*** **Some** system designs require a single line size throughout, whereas other designs require a main and secondary line of different diameter and flow rate. For example:
  - Single line resistance (SLR) oil systems and pump-to-point (P2P) oil/grease systems are low pressure systems designed to deliver the total amount of lubricant in one pump cycle (total loss system) in which a single size diameter delivery line(s) is sufficient. In most cases the line size is standardized at 5/32" or 4mm diameter for SLR systems and 1/4" or 6mm for P2P systems.
  - Single line positive displacement injector, dual line injector and progressive divider systems require multiple cycles of the pump connected to a larger diameter main line used to rapid fill the injectors/ main distribution block, and smaller diameter secondary line to connect the metering outlets to the lubrication points,
  - Recirculating oil systems usually require a single size diameter delivery line and a larger diameter return line system
- ***How many lubrication points are included in the system and where are they located on the machine?*** This is required to map out a central pump location and injector or delivery block locations so the line distances can be measured for material take off amounts, and in the case of long line lengths, to calculate pressure drop so the correct line diameter(s) can be calculated
- ***What lubricant type and grade/viscosity are you planning on using?*** Grease requires higher pressure than oil to move the fluid through the blocks and lines, which will affect line material type and line diameter choice
- ***In what type of working environment will the system be used?*** Ambient and working temperatures can affect line integrity, and if unprotected, copper, brass and plastic lines to be easily damaged in high traffic areas—especially where lift trucks are used regularly
- ***What is your budget?*** Although this should not be a factor in line choice, it often is. Figure 21.4 depicts progressive divider blocks, one piped in correctly rated plastic tubing and the other in correctly rated steel tube. Although steel tubing takes considerably longer to install, the

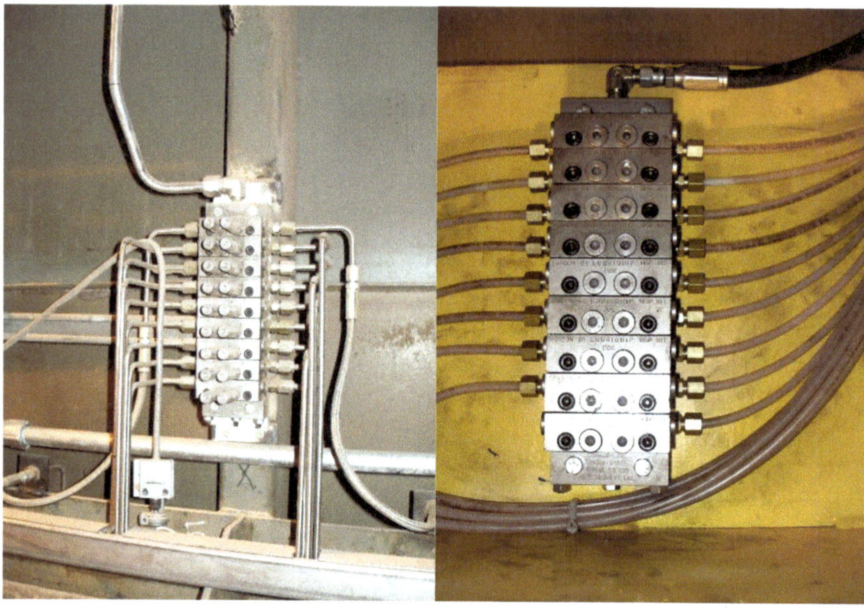

Figure 21.4 Compare steel line routing versus nylon line routing. Images Courtesy of ENGTECH Industries Inc.

additional minor up-front cost can pay long-term dividends, especially when a problem occurs such as a leak or a blocked line. The plastic tubing is bundled together and is difficult to individually trace a line from the pump to the lube block. In addition, lines are difficult to physically attach to the machine frame and as such are more vulnerable to "kink" and cut damage. The steel lines, all have line ID tags making them very easy to trace and trouble shoot. The system looks engineered and permanent in comparison to the bundled plastic example.

Once this information has been determined, present it to your lubrication system designer, or the system manufacturer /supplier who will now be able to work with you to determine the correct line choice for your application.

21.2.2 Main causes of line problems

The majority of line problems manifest themselves as leaks or blockages. A leaking line will starve lubricant from one or many bearing points and seriously compromise machine reliability. Leaks are invariably found at the connection points and line bend areas. Some considerations to be looked at are:

- Copper lines are very soft and can easily work harden at the bend points if there is excessive bending during fitting and/or machine vibration when in use. In addition, copper can attack the antioxidant packages found in both industrial and automotive oils. Note: copper is the primary material used in laboratory tests to accelerate oxidation when testing lubricants! Not all lines that appear to be copper are actually copper; many manufacturers offer copper coated steel lines to prevent rusting. Steel lines are much more difficult to bend by hand, so always perform the hand bend test to know what material you are dealing with and take action accordingly
- Nylon lines can be easily over tightened or not cut square at the connection points causing a leak at the compression fitting
- The single chamfered compression fitting designed for nylon/plastic lines can be mistakenly used on steel (steel requires double chamfered compression fittings – see Figure 21.5) and compromise the seal, causing a leak at the compression fitting
- Steel lines are easily tagged and identified when troubleshooting the system. Nylon or plastic lines are often "bunched" or tie wrapped together making it less expensive to install, but much more difficult to maintain, line trace and troubleshoot later
- To reduce cost, nylon or plastic lines are used to substitute flexible hose line with hard end fittings typically recommended for moving bearing points found on machine slides, rams, etc. Plastic lines are in most cases not rated for cyclic repetitive movement duty and will eventually harden and crack. If used outdoors, nylon line is susceptible to accelerated UV aging from sunlight.

Blockages in lubrication lines usually occur when they are pinch damaged through being hit by a foreign object that crimps or flattens the line shut. This then causes a line backpressure that can blow the fitting or eventually stall an entire progressive divider system starving many bearings in the process. Steel lines offer the best defense against pinched lines.

To ensure no bearing is starved after lube system implementation or a line replacement **ALWAYS** pre-fill the lubricant line with the correct grease lubricant before final fastening to the bearing, or in the case of oil, operate the lube system and open each bearing point to ensure oil is flowing at the point before final tightening at the bearing point.

Line choice is a choice, lubrication system reliability depends on making the correct lubricant line choice

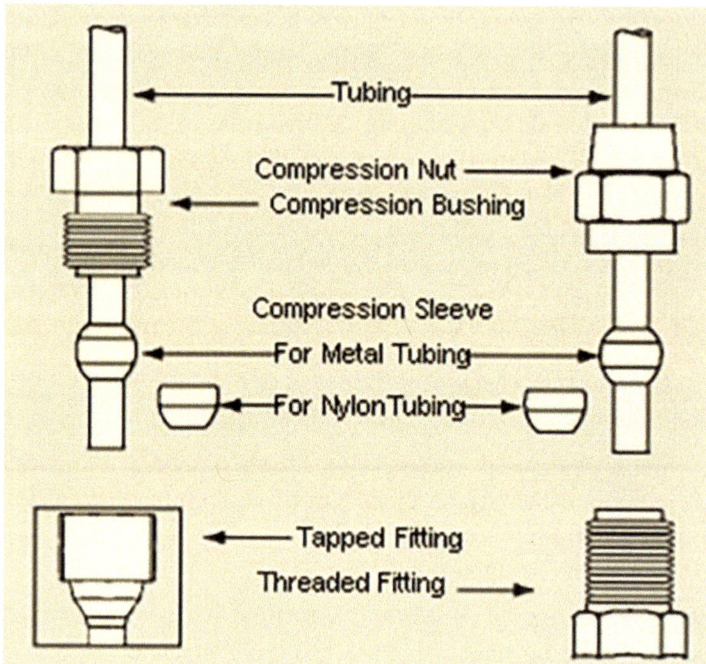

Figure 21.5 How a compression nut system is put together showing the difference between the metal sleeve and nylon tubing sleeve. Image courtesy of Bijur-Delimon International.

21.3 Lubrication Seals

If your equipment employs lubricants to protect its moving bearing surfaces, chances are it will require some form of mechanical sealing device to contain the lubricant in the bearing surface area.

Sealing devices must not only contain lubricant leakage within its reservoir storage or bearing surface area, it must simultaneously prevent any outside contaminants (water and dirt) from entering into the bearing surface area. In short, seals are designed machine elements, or "gatekeeper" devices that separate spaces containing different fluids or substances that may or may not be subject to pressure differentiation.

When asked "what is a seal", a maintainer is most likely to describe one of two types of shaft seals. Shaft seals come in many different shapes, sizes and materials and must be matched accordingly to the application, temperature, and lubricant properties.

Choosing a suitable lubricant for your application will require knowledge of the operational conditions (shaft speed, bearing load, hours of operation), the ambient working conditions (hot, cold, dirty, wet), and the type of shaft seals employed.

To meet the shaft seal requirements and deliver long term reliability, the lubricant must meet specific design criteria that:

- Ensures damage fee installation of the seal
- Dissipates frictional heat
- Increases sealing effect
- Prevents adhesion of the seal even after a long standstill
- Permits easy assembly/disassembly of the seal
- Compatibility with the sealing material and resistant to ambient media

Two distinct shaft seal designs dominate industry today, these being the radial lip seal, and the labyrinth seal. They differ considerably in that the inexpensive radial lip seal is referred to as a contact seal and is primarily used for isolating oil systems, while the more expensive labyrinth seal is referred to as a non-contact seal that can isolate lubricants as well as gases, typical requirements in steam turbines.

21.3.1 Labyrinth seals

In their 2009 Journal of Mechanical Science and Technology paper titled *"Comparative analysis of the influence of labyrinth seal configuration on leakage behavior"*, authors Tong Seop Kim and Kyu Sang Cha eloquently describe a labyrinth seal as "a non-contacting sealing device that consists of a series of cavities connected by small clearances [in which] flow loses its total pressure while it sequentially experiences acceleration into the clearance due to contraction, friction through the clearance, and dissipation of kinetic energy at the cavity". The cavities and small clearances create a torturous pathway that result in turbulence acting as a curtain or barrier to restrict outward flow (egress) of lubricant or gas, and inward flow (ingress) of contaminants.

Figure 21.6 depicts a typical steam turbine labyrinth steam seal designed to prevent steam from mixing with turbine generator lube system. The diagram clearly shows the multiple clearances and cavities along the shaft.

Because a labyrinth seal is non-contacting, it technically will not wear out and should never need replacing

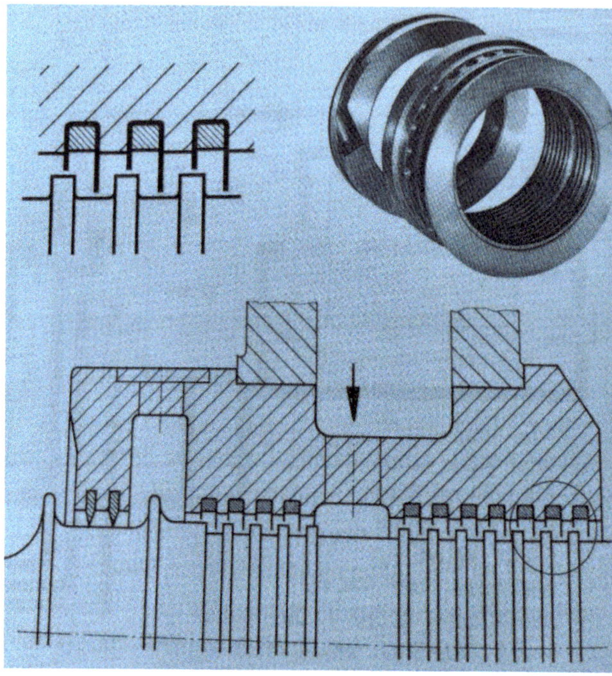

Figure 21.6 Steam turbine labyrinth seal.

21.3.2 Radial lip seal

Radial seals are not new. There first uses employed cloth woven string, rope, and leather that were oil soaked and "packed", or trapped in the bearing clearance area. The swelling of the material ensured a pretty good seal that stopped a lot of lubricant loss. As speeds started to get faster, the advent of rubber and plastic (flouro-plastic and flouro-elastomer) materials surpassed the original packings cementing the radial lip seal as a proven and reliable mechanical seal, still in use today.

A modern radial lip seal is a point contact seal made up from a circular metal outer band that captively fits into the stationary housing bore while its bonded elastomer sealing lip is designed to sit and seal dynamically and statically against the rotating shaft. Most radial lip seals also employ a garter spring that ensures the single (or double) point contact is always engaged on the shaft. See Figure 21.7.

When the shaft is rotating at speed the shaft seal runs on a thin layer of lubricant between the elastomer contact lip and shaft. Lubricant is also

Figure 21.7 Typical radial lip seal.

pumped into the areas hydrodynamically by the centrifugal pumping action of the rotating shaft. This lubrication effect is beneficial to the life of the seal but is only as good as the oil condition! If the oil is in poor condition, sliding contact surface will cause the seal to heat up and prematurely wear and lose its sealing qualities.

Radial lip seals are classed and purchased according to surface speed ranges. Once known, a suitable elastomer type can be chosen according to the operating temperature and lubricant choice. For example, a natural Buna-N rubber elastomer is good from -40° F to 225° F and is excellent for petroleum-based lubricants. In more demanding temperature conditions, a Viton seal offers a much larger temperature range capability from 0° F to 400° F and is excellent for all synthetic fluids that can "soften" other elastomer type seals.

For truly demanding conditions, PTFE (PolyTetraFlouroEthylene) is a hard plastic that requires greater care on installation, but can operate in conditions from -100° F to 400° F and is compatible with most fluids.

Although inexpensive to purchase, radial lip seals are full contact designs and as such require the seal to come into contact with the lubricant, making lubricant choice a decisive factor when purchasing the seal.

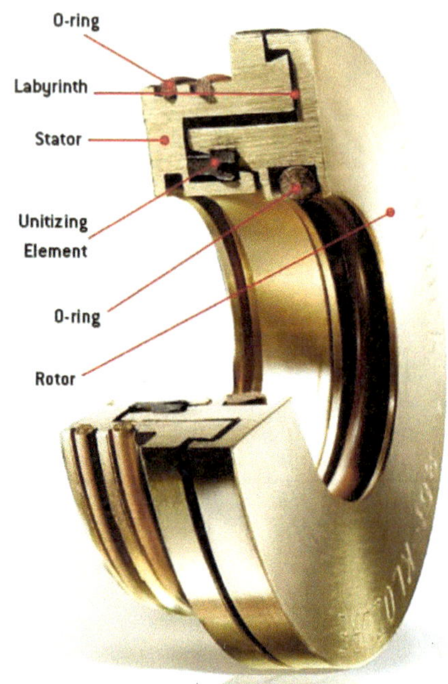

Figure 21.8 Modern bearing isolator. Image courtesy Impro-Seal.

21.3.3 Bearing isolators

A more recent seal design gaining popularity is the bearing isolator – Figure 21.8. This hybrid compact unit design is a dynamic non-contact two-piece unit consisting of a fixed piece known as the stator, which interconnects and marries up to the moveable rotor piece attached to the shaft. The bearing isolator is an easy to install split hybrid design that incorporates O-rings and an elastomer connecting ring known as the unitizing element.

As always, know your operating conditions and consult with your lubrication and bearing seal providers to make the best seal decision for your assets.

21.4 Lubrication System Controllers and Signal Devices

To complete a centralized lubrication system design, the designer must tie the pump and delivery system together, and synchronize its operation with a combination of control and signal devices. The type of controller and signal devices used will depend on budget, level of system protection required, the type of pump and distribution system employed, and the ability of the host machine to interpret and act upon the control signals.

21.4.1 Controllers

A lubrication system controller is a device often described as the lubrication system's "brain." Most controllers are multi-function stand-alone devices housed in a control panel. The exceptions to this can be found in the single point lubricators (SPLs) and some of the smaller electro mechanical oil delivery pumps that have built-in circuit board style controllers that can be programmed through an LCD touch screen or a series of mechanical switches. These are usually simple control devices that turn the lubricating pump on or off, and control its rate of operation to speed up or slow down the rate of lubricant flow being pumped.

A controller's primary function is to turn a pump's power source on or off. In the case of a pneumatically powered pump the controller opens an electrically operated air solenoid valve to allow air into the piston and fire the pump. With electrically driven piston and gear pumps, the motor is electrically energized allowing the pumping action to commence. Once the controller receives a signal informing it that the pump has fired, a given time has elapsed, a lubricant line pressure has been achieved, or a distribution block cycle has been completed, the pump's power source is shut down until the next lubrication cycle commences. (The only exception to this is a recirculating oil system that is powered up on machine startup and runs continually until it is turned off when the machine is idle or shut down.)

Lubrication cycles can be controlled by a counter that counts the number of machine or production operations, a programmed or set timer, or by a condition signal. E.g., amperage draw meter indicating an energy draw increase on the machine system motor(s) due to a mechanical friction rise caused by lack of lubrication. (This is a popular control mechanism that measures the amperage of the conveyor drive and take-up motors to activate and deactivate the "power and free" conveyor chain and pin lubricators used in automotive assembly plants.).

A controller 's secondary function is to take an emergency signal, shutdown the system and activate an alarm. This alarm can be a simple light or buzzer wired directly to a solenoid in the control panel, to the activation of an alarm email or work order in the CMMS/ EAM maintenance management software—or both.

A controller 's level of sophistication can range from a manual on/off device, to a simple count driven on/off device, all the way up to a very sophisticated programmable PLC/computerized PC device. The controller sophistication is usually underwritten by the lubrication system's consequence of failure where public safety is a concern. E.g. nuclear industry, chemical industry, etc., or where production losses are a major concern when the lubricated machinery is a production constraint, or machinery failure results in high down-time costs.

21.4.2 Signal devices

Signal devices are used for both control and system protection and can be mechanical, hydraulic, pneumatic or electrical in design, and active or passive in operation.

Different delivery system designs will use controls differently. E.g., in most single or dual line system designs, the pump must continue to pump until the line pressure has reached an end-of-line line pressure of at least 800 psi allowing the injectors to fire. Once attained, a pressure signal switch informs the controller to shut off the pump, which in turn also reverses a flow valve allowing the lubricant to relieve back to the reservoir and allow the injectors to reset. A timer then counts operations or elapsed clock time and tells the controller to start the process all over again. In this system type, the pressure switch can also be coupled to a time out switch set to signal an alarm state if the system doesn't achieve its line pressure (due to a broken line or no lubricant) in a set time period.

Progressive divider systems can employ simple counters attached to the top piston cycle pin indicator in the primary delivery block. Once every outlet in the block has fired lubricant the top pin will have moved in and out the block once signaling one complete operation of the block. (For block operation see Figure 20.35 to 20.38.) A mechanical or electronic signaling device can be mounted to the block taking its signal from the cycle pin. This can then be linked to the controller that can now actuate the pump based on the number of block cycles called for.

Progressive blocks will also utilize passive hydraulic blocked line indicators that actuate when hydraulic lockup occurs in a line blockage situation, IN this case the lubricant back pressure causes the blocked line indicator pin in the front of the block to "pop" out visually indicating which line is blocked. This simple device can speed up the troubleshooting process. Note: if left ignored too long, eventually the entire system can hydraulically lock up producing a situation in which the pump will then stall and no bearing points will receive lubrication. (See Figure 21.9.)

These are many control and protection devices available to the customer, question your suppliers to ensure you have the right system and right level of control you need.

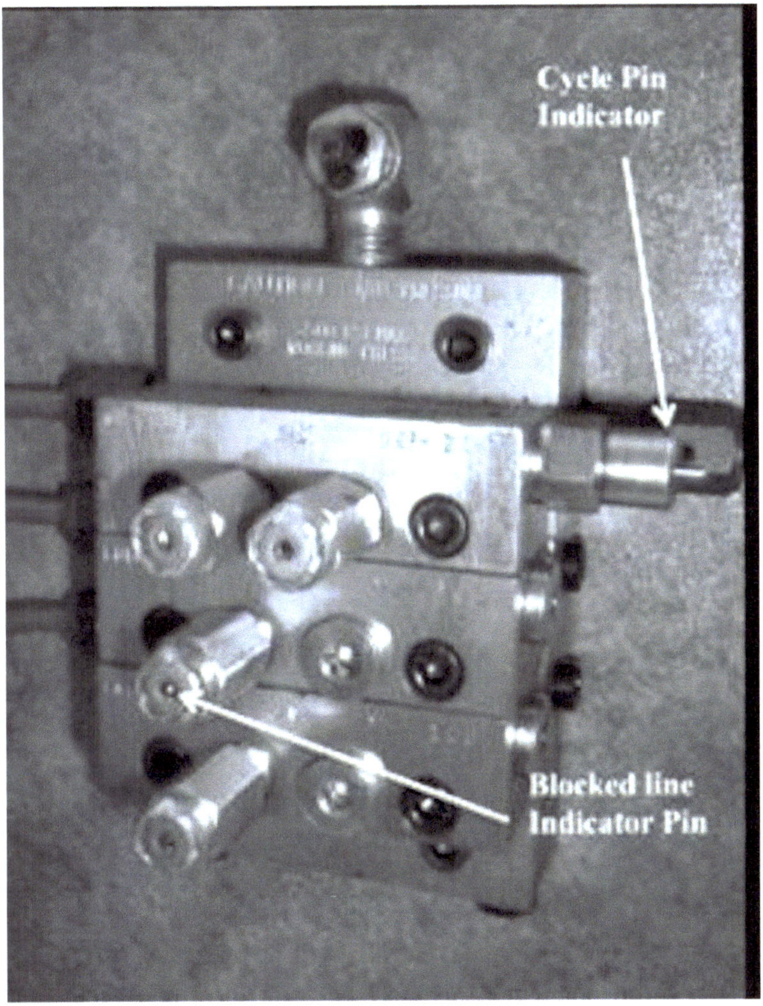

Figure 21.9 Progressive divider block signal indicators (Courtesy ENGTECH Industries Inc.)

21.4.3 Broken line, or pressure loss protection

Blocked lines are simple to detect in virtually any system with the aid of pressure switches and count cycle devices, however, a loss of system pressure is not always available on every system.

Loss of lubricant due to an empty reservoir can be detected simply by installing a reservoir level switch that signals low lubricant and can perform a system (pump) shutdown from the same signal source.

A broken line however, can be easily detected and signaled on a single/dual line injector system through and end of line pressure switch. In this case used the pressure switch serves double duty to sense when to shut off the pump at peak pressure; if the pump fails to make pressure after a prescribed time limit due to loss or lack of lubricant, the switch again is used to shut down the pump and signal an alarm state.

In a single line resistance (SLR) or progressive divider system, a line break is much more difficult to detect. The SLR system provides not open line protection, only reservoir level protection. A series progressive can have pressure switches mounted individually at each bearing point. Because of the expense, this detection method is usually reserved for placement at the secondary block inlet to indicate which arm of the system is affected, and at bearing points deemed critical to the operation.

21.4.4 Reservoir fluid level gauge devices

Reservoirs are designed to hold a specific range of lubricant to ensure the pump does not stall or cavitate. In cases where the reservoir houses lubricant for a gear assembly and not a lubricant delivery system, both an upper and lower-level limit must be managed/observed. Overfilling the gearbox can cause the lubricant to "churn" (aeration through excessive turbulence) and foam, causing premature depletion of the lubricant's antifoam additive. In addition, changes in reservoir levels will affect the available reservoir "headspace" (space between the level of lubricant and the top of the reservoir) that can create problems with reservoir breathing and internal pressure stabilization.

Webster's dictionary describes a gauge simply as "a [telltale] device attached to a container to show the height of its contents". Google expands on this description by stating a gauge is "an instrument or device for measuring the magnitude, amount, or contents of something, typically with a visual display of such information."

In any typical industrial plant, there are many oil reservoirs in use for lubrication and hydraulic systems where numerous types and configurations of level gauges are employed to facilitate recognition of the internal lubricant level, and depending on the gauge, the lubricant condition. Sight level gauges are inexpensive, "passive", human/machine interface (HMI) visual indicators that are inadequately exploited in most PM programs.

The most efficient type of level gauge is one that does not require the maintainer to physically open the gearbox to atmosphere, and possible contamination, to check lubricant level and lubricant condition such as that typically required when using a dipstick style gauge.

The majority of passive sight level gauges on offer today fall under three specific design categories, these being 1) the planar sight glass, 2) the column sight gauge, and most recent to market, 3) the 3-dimensional sight glass. Each category has different attributes to offer and all can be upgraded to make them multi-functional telltale devices.

21.4.5 Planar sight glass

The planar sight glass is a two-dimensional glass porthole screwed or press fitted into the reservoir wall with its centerline positioned approximately at the ideal recommended reservoir fill level. (see Figure 21.14 depicting a very large planar style sight gauge in a hydraulic reservoir with both upper and lower limit level markers clearly defined)

Pros

- Very inexpensive
- Available in a variety of materials
- Available in a variety of diameter sizes

Cons

- Minimalistic
- If used in a recirculating lubrication or hydraulic system, the maintainer needs to know if the center level line is correct for when the machine is in operation, or when the machine is at idle and all lubricant has returned to reservoir. Will require notification on the reservoir, stating running or idling level.
- Because the sight glass is the inner wall of the reservoir and is in direct contact with the lubricant, it can be susceptible to staining and varnishing if the lubricant is not regularly changed. This can cause a false level indication or make it very difficult to see the actual lubricant level.
- Because there is no backlighting, it is very difficult does not show the color or oil condition

Upgrades

- Larger diameter sight glass is preferable
- An upper and lower fill level line can be printed on the glass if the porthole is large enough

21.4.6 Columnar sight level gauge

Columnar site level gauges are arguably the most common sight gauges in use and are available in two point or single point entry versions. Dual point entry gauges are solidly affixed to the reservoir allowing for very long column tubes (up to and over 60 inches in length in most commercially available gauges) and as the top entry is open through to the reservoir any pressure build up in the tube column is automatically vented into the reservoir. Single point entry gauges are usually much shorter as they are susceptible to movement and damage, and as they can allow a hydraulic pressure build up in the column tube that can result in a false level reading, they have a built in 2-5 Micron vent cap to equalize tube pressure. To accommodate a wide range of fluids, the gauge is offered in a variety of materials such as steel, brass, aluminum, etc. for both the body and plastic or glass for the tube column.

The tube can be customized with calibration marks depicting gallons or liter fill levels and can be fitted with fixed or movable upper and lower limit level indicators. In addition, the lower entry point, which is most often fitted into the reservoir drain port, can be fitted with an external drain port that can double as a sample port. This is only useful as a sample port if the lower entry port is higher than the potential sludge level—usually one quarter to one third of the reservoir height.

In circulating oil and fixed level reservoirs, the columnar sight gauge is most often favored. Figure 21.10 demonstrates a system that employs a dual point entry gearbox sight level gauge. A plate clearly indicates the operational Hi-Lo level location positions that allow the operator or maintainer to inspect at a glance if the level is correct. Also note, the plate indicates the correct oil to use, in this case an Omala 680 gear oil. A gearbox/motor identification plate is also attached to the gearbox reservoir casing. The reader may also note the piping on the right, which is connected to a remote fill point for ease of both oil fill and drain.

Pros

Inexpensive

- Available in a variety of materials
- Can be fitted anywhere on the reservoir
- Can accommodate very large reservoirs
- Delivers an almost 360-degree sightline to the level limit
- Can be calibrated to show reservoir volume

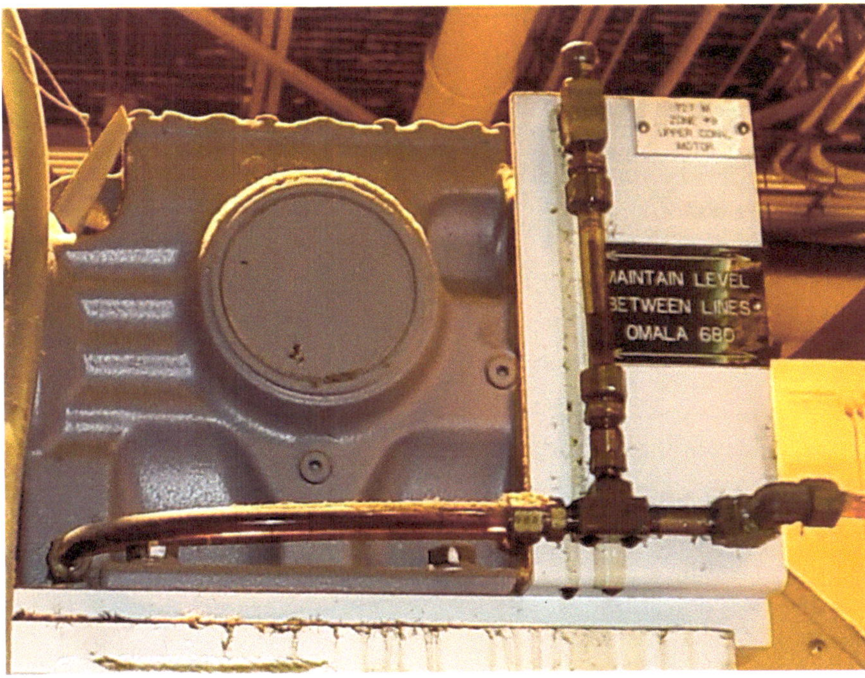

Figure 21.10 Gearbox level gauge system with Hi-Lo markers (Courtesy ENGTECH Industries Inc.)

- Can be calibrate to show running level and idle level
- Best for showing oil condition/color
- Can show emulsified water
- Can show aerated oil
- Can double as a reservoir drain
- Certain single- entry designs can double as a filler port
- Can be set up with a live oil analysis sampling port
- Less susceptible to glass staining

Cons

- Requires more fittings and gaskets that can leak if not installed correctly
- Does not show
- Single point entry units are more easily damaged

- Single point entry unit will require breather maintenance

Upgrades
- Multiple level and zone markers
- Calibrated columns
- Drain ports
- Sampling ports
- Filler port
- Color coded column backdrops

21.4.7 3-Dimensional sight glass

The 3-D sight glass is positioned in the reservoir in the same manner as the planar sight glass. The major difference is the site glass protrudes out from the reservoir wall (like a glass jam-jar) allowing the maintainer to see exactly what the oil is doing inside the reservoir.

Pros
- This is a true physical extension of the reservoir
- Uninterrupted sightline to the lube level
- Can be set up with a live oil analysis sampling port
- Can show oil condition
- Best for showing oil foaming
- Best for showing oil aeration
- Best for showing emulsified water

Cons
- Does not show running or idling level very well

Upgrades
- Probe style sampling port

The popular 3-Dimensional sight glass attached to the reservoir at the fluid level as seen in Figure 21.11. Although this style does not come with Hi-Lo markings it is easy to use a marker on to establish hi-Lo indicator levels as seen in Figure 20.50. This type of indicator is excellent for noticing foaming in the oil.

Figure 21.11 Modern 3-dimensional level sight gauge (Copyright Des-Case Corporation, 2022).

Other types of reservoir level indicators along with their features and benefits are shown in the Figure 21.12 Sight Level indicator comparison chart. A favorite of designers is the oil sight glass level monitor that can be installed on the drain port and used to drain off water or take an oil sample if needed. It comes with an adjustable upper-level indicator band (red) and a lower-level indicator band, again adjustable (Green). Note: anything freely adjustable can be accidently or deliberately moved. Once the levels are set in place always mark their place using an indelible ink marker to indicate the engineered levels). With only one entry point to the reservoir this type of indicator is more vulnerable to damage than the dual entry indicator.

In total loss reservoir over pump oil/grease system design, the reservoir is most often fabricated from a see through plastic. As depicted in Figure 21.13, this can also be made more user friendly by introducing a RAG – Red/Amber/Green indicator system that indicates the full level (Green), the time to fill level (Amber/yellow) level, and the emergency must fill NOW level, before pump cavitation takes place requiring a system bleeding and bearing starvation. This very quick and easy indicator system allows an operator. Maintainer, or lubrication technician to quickly establish the lubrication system requirements at a glance.

	360° View	Installs on Drain Port	Collect and Drain Water (Optional Magnet Drain Valve Available)	Stainless Steel Fittings Available	Monitor Oil Level	View Clarity and Color	Additional Benefits
3-D BullsEye®	✓				✓	✓	• See foaming in oil
Oil Sight Glass	✓	✓	✓	✓		✓	• Various volumes available • Can be installed horizontally for applications with limited space
Oil Level Indicator	✓	✓			✓	✓	• Filtered or close-looped installation • Optional ND-2 breather
Oil Sight Glass Level Monitor	✓	✓	✓	✓	✓	✓	• Dual port available for sampling oil • Filtered or close-looped installation

Figure 21.12 Sight level indicator comparison chart (Copyright Des-Case Corporation, 2022).

21.5 Reservoir Filters and Breathers Filters

The primary contamination control device used in an oil lubrication system is the fluid filter. Filters can be separated into two categories, 1) Surface type filter and, 2) Depth type filter. For semi-fluid systems using grease as a lubricant only surface type filters, known as strainers and mechanical filters are used.

21.5.1 Surface fluid filter—oil

Surface filters are the most common style of oil fluid filter in use and can be found in many design configurations. They are primarily designed to work in the direct flow path of the lubricant to capture any dirt particles (contaminants)

Figure 21.13 RAG sight level indication system on a total loss grease reservoir (Courtesy ENGTECH Industries Inc.)

held in colloidal suspension in the oil as it flows through/across the filter media, see Figure 21.14. Typically, this filter element is very porous so as minimize any differential fluid pressure loss across the filter media. Because of this, the filter exhibits low capture efficiency, dirt holding capacity, and a moderate life expectancy requiring that it be inspected, cleaned or replaced on a regular basis.

The actual design and capture capability of the filter will depend on its location within the lubrication system and the desired functional effect of the filter. For example, in a typical circulative hydraulic or lubrication system, a pump suctions oil from a lubricant reservoir and pumps it under pressure to a moving device such as a valve, cylinder or bearing. Once the lubricant has performed its job it is allowed to return, under gravity, to the reservoir where it can cool and be cycled again. In this typical oil delivery system, we can expect to find six to seven types of surface filter designs and one depth filter.

Filter #1—Pump protection: utilizes a suction filter positioned low in the reservoir and connected to the pump suction inlet via a suction tube. Filter

Figure 21.14 Surface filter media. Source: Engtech Industries Inc.

media is typically a wire mesh gauze, paper, or felt designed to capture and stop any large debris or metallic wear particles greater than 70 Microns in size from entering the pump.

Filter #2—Primary system protection: is a pressure filter positioned directly in the pump output delivery line between the pump and the first moving device in the lubrication system. This filter is the primary system protection filter that utilizes pleated paper, cellulose or fine porous metal media designed to withstand the pump system pressure and capture small micron particulate that have managed to move through the suction filter and pump gear set. A typical example of this filter style is a spin-on or cartridge style automotive filter

Filter #3—Wear metal and colloidal debris collection: positioned on the gravity return piping side of the lubrication system just before the reservoir

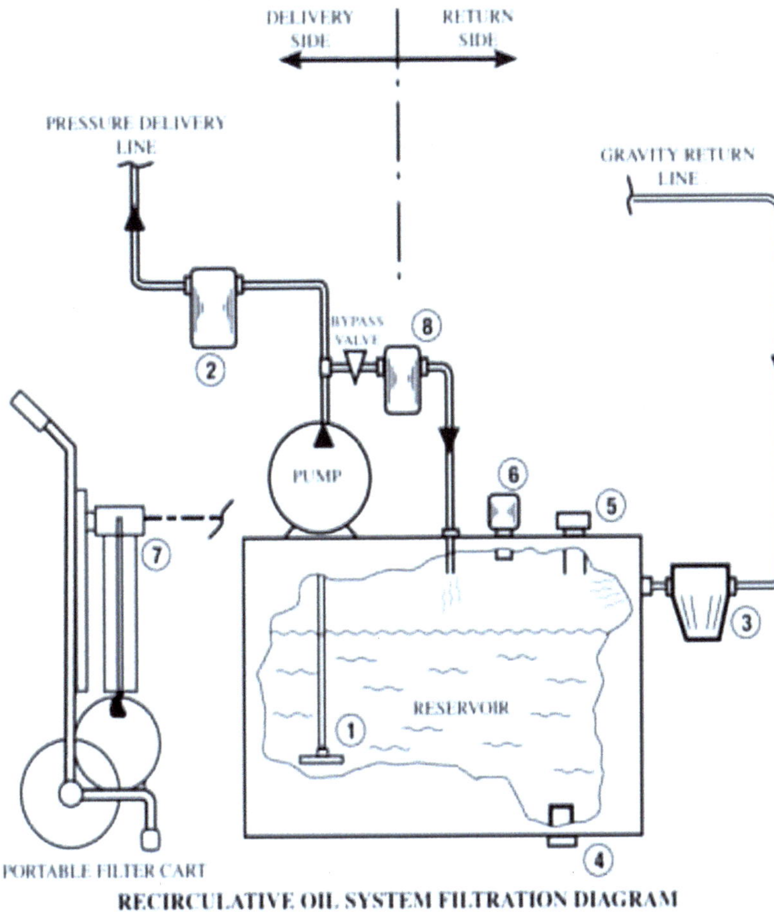

Source: Industrial Lubrication Fundamentals – Lubricant Lifecycle Management. *Lubrication Management & Technology* magazine - Ken Bannister

Figure 21.14 Different filter types found in a typical oil circulating lubrication system.

lubricant return inlet. Known as a gravity return filter, this filter generally employs a low-pressure paper medium deigned to capture wear metal and debris washed from the moving parts of the machine by the lubricant.

Filter #4—Ferritic Metals and Debris: is a passive filter designed to attract and hold any ferritic (magnetic) debris and wear metals that might of by-passed the return filter or have been left in the reservoir from start up. The filter is magnetic and often serves second duty as a reservoir drain plug.

Filter #5—Solids Ingress: is a large pore metal mesh strainer sock positioned in the inlet mouth of the reservoir fill port designed to capture any errant large particulate from making its way into the reservoir when filling is taking place, or if the fill cap has been left off the reservoir.

Filter #6—Airborne Contaminants and Moisture: is a breather style filter designed to equalize pressure in the reservoir and in its simplest form utilizes a wire wool media to keep out any particulate of 40 micron and above in the air from entering the reservoir. In a more sophisticated design, a breather style filter (see breather section below) employs a desiccant silica gel hydrophilic material designed to allow the reservoir to breathe and prevent outside airborne particulate 3 micron and above from entering the reservoir. This style also has the added advantage of being able to wick and capture moisture from inside the reservoir and prevent outside moisture from entering the reservoir. Once saturated/exhausted, the gel turns from blue to pink, visually indicating its need to be replaced, see Figure 21-

Filter #7—All Contaminants: is an optional filter utilized in larger systems employing a filter cart hooked up to the reservoir and used to 1) extract oil from the reservoir; 2) filter the oil using filters similar to the #2 pressure filter, or through a "bag" filter that employs a thick felt type sock in the form of a bag to provide fine filtration of the lubricant; 3) return the oil to the reservoir clean for further use, thereby extending the oil change frequency. This is often referred to as a kidney loop auxiliary system and although depicted as a portable filter cart, it can be hard piped as part of the reservoir piping system.

21.5.2 Depth fluid filter—oil

A depth filter differs from a surface filter in that it takes the lubricant through an indirect maze- like flow path designed to capture and accommodate large amounts of dirt. The filter is highly efficient and capable of withstanding high differential pressure. If the surface filter were to use a sheet of toilet paper as its media, the depth filter equivalent would be the entire toilet roll of paper as seen in Figure 21.15

Filter #8—Depth Cleaning and Polishing: this is a depth filter typically placed within a bypass circuit on the output pressure delivery side of the pump prior to the pressure filter. The filter medium can be made from cellulose, fiberglass, felt, or diatomaceous earth, designed to deep clean and polish the lubricant.

Figure 21.15 Typical depth filter media. Source: Engtech Industries Inc.

When choosing an oil filter medium and size, always consider the low temperature conditions as fluid viscosity increase may cause an increase in pressure differential through the filter

21.5.3 Grease filtration

Because grease does not typically flow the same as oil and is pumped at higher pressures than oil, metal strainers or wedge wire filters are fitted to the pressure side of the pump delivery system and are used to trap large debris usually introduced into the system during filling. The mesh or wedge wire (looks similar to a tightly coiled spring – see Figure 21.16) mechanically traps contaminants down to 145 microns and must be cleaned regularly.

Physical filter location and accessibility is crucial. Typical location and design parameters should include the following:

- Locate outside of a lock-out zone on the machine
- Locate outside of a designated confined space area
- Part numbers for housing and elements are prominently displayed/ tagged in a prominent location on or near the filter assembly
- Identify filtered lubricant information (type, viscosity and manufacturer) via tag or label in a prominent location on or near the filter assembly
- Provision of adequate space to remove and replace filter element and the filter housing when appropriate

Figure 21.16 Wedge wire grease contaminant trap. (Courtesy ENGTECH Industries Inc – ICML: MLT/MLA1 Certification Training Materials.)

- Condition indicators (Gauge) positioned for operator/maintainer viewing ease and kept clean
- When practical, always set up as a parallel duplex installation with accessible shut off valves located before and after each filter assembly so filter change out can occur while the machine is running, or with minimal down time
- Units must be protected from debris, and environmental contamination

If these conditions are not met with current lubricant deliver system design, the lubrication team must seriously consider upgrading the piping arrangement to accommodate and facilitate filter access as depicted in the above bullet list.

21.5.4 Hydraulic system bypass filter

Hydraulic lubricant systems differ slightly to circulating oil systems. When a hydraulic system is in operational use, it cannot be shut down to make an oil filter change as it may be in use applying hydraulic force. This can constitute great safety risk to people and machines alike if that force is rapidly relieved under load. Therefore, most hydraulic filters are designed with an auto bypass circuit on the filter if it becomes too contaminated. Whilst

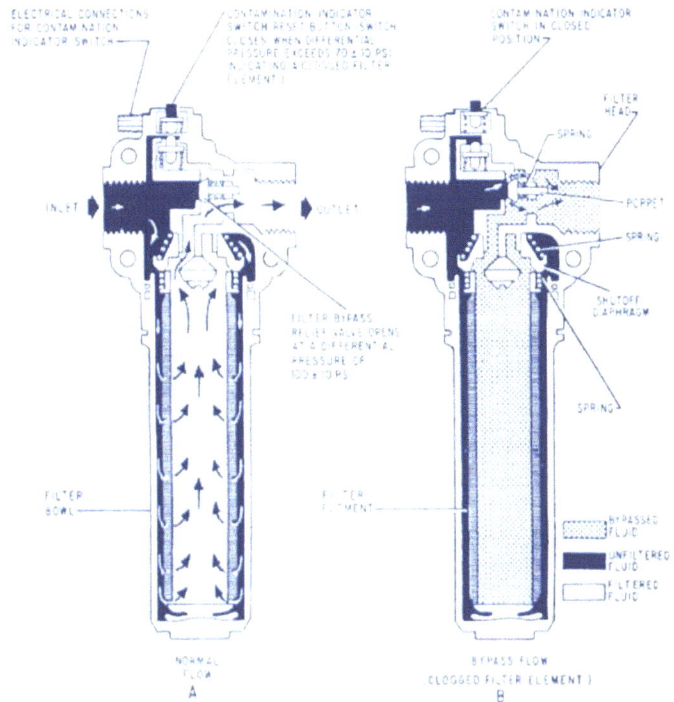

Figure 21.17 Hydraulic filter bypass circuit (Courtesy: UFI Hydraulic Filter).

this ensures a safe system, if bypass does occur, the entire lubricant must be replaced and the system cleaned properly as contaminated oil working in a tight tolerance environment can cause rapid component wear and system leakage making the system inefficient and often unsafe to use. Figure 21.17 shows a cutaway section of a hydraulic filter.

21.5.5 Measuring filter efficiency and beta ratio

Filter media are rated based on their ability to trap and capture particles down to a defined minimum micron size. Their efficiency is measured in how well the media captures those particles. Under laboratory conditions, the volume of **upstream** particles (of a defined micron size) is counted before entering the filter and counted again upon exiting the filter on the **down-stream** side. The difference is then calculated as a percentage to arrive at the filter efficiency rating. (See Figure 21.18).

To determine a filter's efficiency for a given particulate size, the ISO 4572 Multi-pass Test Procedure is employed. A filter's Beta ratio, or

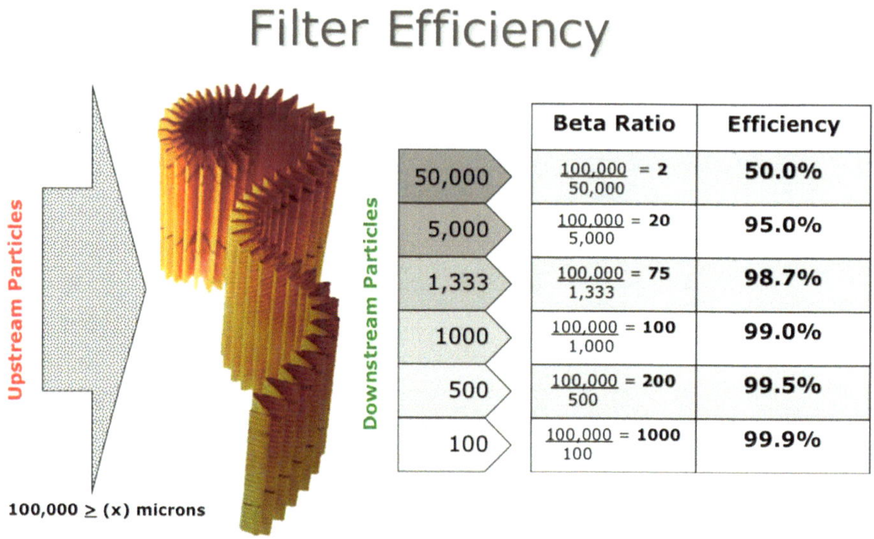

	Beta Ratio	Efficiency
50,000	100,000 / 50,000 = 2	50.0%
5,000	100,000 / 5,000 = 20	95.0%
1,333	100,000 / 1,333 = 75	98.7%
1000	100,000 / 1,000 = 100	99.0%
500	100,000 / 500 = 200	99.5%
100	100,000 / 100 = 1000	99.9%

Figure 21.18 Filter efficiency measured as a Beta Ratio. (Courtesy ENGTECH Industries Inc. ICML: MLT/MLA1Certification Training Materials).

Filtration ratio as it is sometimes referred to, is calculated dividin the number of **upstream** particles by the number of **downstream** particulate once they have passed through the filter.

Demonstrated in the Figure 21.18 example above, when 100,000 particles of a stated minimum micron size are flowed through a filter that only manages to arrest 50,000 particles, its measure of performance Bx is calculated using the formula:

Bx = # Upstream Particles / # Downstream Particles

Where x = > Specified particle size in Microns

I we look at Figure 21.18 and take the example shown for the first pass where 50,000 particles passed through the filter

Bx = 100,000 Particles / 50,000 Particles = 2

therefore,

Beta Ratio Bx = 2

with an efficiency rating of: **Efficiency x = (1 – 1/Beta) 100**

therefore,

Efficiency x = (1 – 1/2) 100 = 50%

Similarly, if the filter is successful in stopping 99,500 of 100,000 upstream particles of a minimum size

Bx = 100,000 Particles / 500 Particles = 200

therefore,

Beta Ratio Bx = 200

Efficiency x = (1 – 1 / 200) 100 = 99.5%

A higher Beta ratio number equates to a more efficient and higher quality filter.

21.5.6 Portable filter carts

The ability to control contamination is an important aspect of any lubrication management program, especially where lubricant cleanliness is concerned. A constant supply of clean oil is essential to lubricant life, and more importantly, bearing life.

One of the most efficient and practical tools available to ensure lubricant cleanliness is the portable filter cart, its primary function being to filter fluids. In a typical industrial environment, portable filter carts used to perform three specific operations within a lubrication program, these being:

1. To transfer oil from its original container into a machine reservoir
2. Pre-filter and clean virgin stock (new) oil in preparation for machine use
3. Recondition and clean-up of oil currently in service. In addition, use of specialized filters on the outlet side can also be used to extract any free and emulsified water present in the oil.

A typical filter cart design, as shown in Figure 21.19, employs a two-stage filtration approach where a gear pump is connected to both filters. The inlet, or suction side is the first stage low-pressure side (approximately 5psid) designed to capture larger contaminant particles above 150 microns in size. Oil is pumped through the inlet filter to the second stage high-pressure (approximately 25psid) outlet, or delivery side filter designed to capture much smaller particulate matter that can be filtered to 2 Microns in size, depending on the filter rating used.

The quality and cost of a filter cart is determined by the quality of the cart itself, the size of the pump, the efficiency rating of the filters, the type of couplings installed, and the quality of the workmanship.

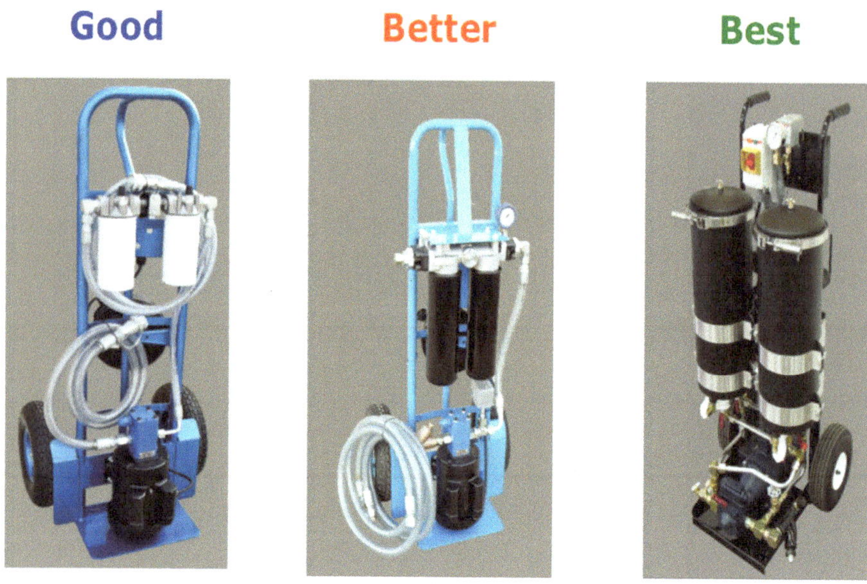

Figure 21.19 Three different but similar filter cart designs (Courtesy Fluid Defense Systems Inc.)

Filter carts are best purchased after a lubricant consolidation program has taken place as each lubricant requires a dedicated filter cart to minimize any cross contamination. In addition, the level of oil cleanliness must be determined and stated in order to choose the correct filter for the cart.

The success of any tell-tale device or warnings sent to management, such as the Red Amber Green (RAG) indicator on the side of the RH filter housing in Figure 21.20, is incumbent upon the user base adopting them into their work plans and abiding by their notices. In this case the lets the operator know when the filter condition mandates it be changed out. A similar type of sensing device is now being offered on "smart" filter carts that can send an electronic service notice for a filter change. Lack of action can disable use of the cart requiring a new filter and an electronic reset by a foreperson or reliability password owner. See IIOT Sensors later in this chapter for more information.

How clean should your oil be?

Oil cleanliness is universally measured using the ISO 4406 cleanliness code rating system. (Note: for more in-depth information regarding contamination control and the ISO 4406 cleanliness code rating system refer to Section five, Chapter 33.) This rating system quantifies the number of

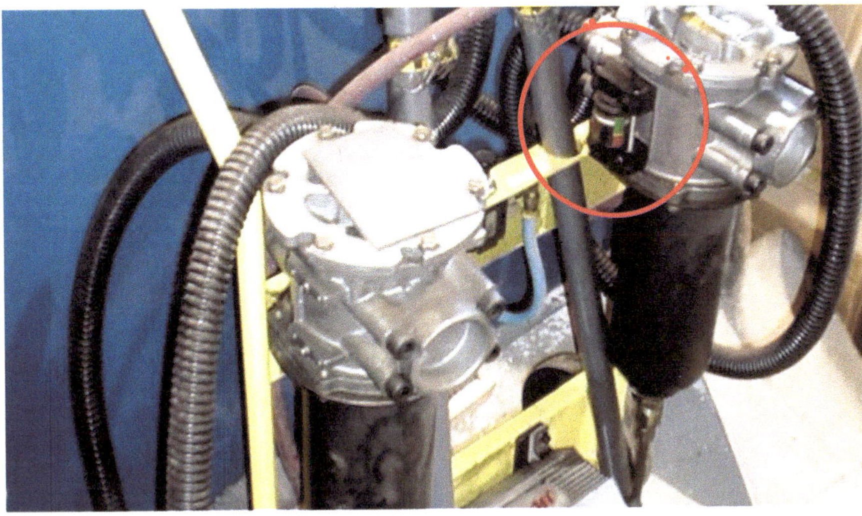

Figure 21.20 Filter cart with RAG indicators showing status of Filter cleanliness. (Courtesy ENGTECH Industries Inc.)

contaminant particles 4, 6, and 14 Microns in size present in a 1 ml lubricant sample, and compares them to a specifies particle concentration range to come up with the ISO range number value. For example, a 19/17/14 lubricant sample value (typical of new oil) translates to the presence of 2500 to 5000 particles >4 microns in size, 640 to 1300 particles >6 microns in size, and 80 to 160 particles >14 microns in size present in the oil sample. The lower the particle numeric values, the cleaner the oil as shown in Table 21.1 below.

When new or virgin stock oil is received from a supplier, many maintainers believe they are receiving a "ready to use" product. This is not always the case as depicted in Table 21.1 that shows a new oil being typically received around a 19/17/14 ISO cleanliness level that may only be suitable for non-critical gear systems. All other applications will require the oil to be cleaned and polished by passing it through a filtration system prior to use in service. We also see in Table 21.1 one that "In service" oil dirtier than 19/17/14 is unsuitable for any lubrication or hydraulic system and will require a complete replacement or clean up using a kidney loop set up with a portable filter cart and a full filter change.

The number of passes through the filter cart to achieve the appropriate cleanliness level will depend on the start and finish cleanliness level and the filter types and rating in use. Oil analysis will be required to establish cleanliness levels. Choosing a suitable combination of pump and filter size/type

Table 21.1 ISO 4406:1999 solid contamination code suitability matrix. (Courtesy ENGTECH Industries Inc.)

ISO Range	Description	Application	Life Extension	Dirt p.a.*
22/20/17	**Very Dirty**	• Unsuitable for any lubrication system (H,L,G)	.5x (H) .75x (G)	600 kg
19/17/14	**New Oil**	• Medium/low pressure systems (H,L) • Non-critical gear systems (G)	1	140kg
17/15/12	**Light Contamination**	• Standard hydraulic and lube oil systems (H,L) • Standard Gear system (G)	1.6x (H) 1.3x (G)	34kg
16/14/11	**Clean**	• Servos and high pressure hydraulics (H,L) • Critical gear systems (G)	3x (H) 1.7x (G)	17kg
<14/12/10	**Very Clean**	• All oil systems (H,L,G)	4(H)–2x(G)	8.5kg

H-Hydraulic Oil, L – Lube Oil, G – Gear Oil
* - Dirt thru pump at 200ltr/min, 230 work day, single shift in 1 year

will require consultation with the filter cart manufacturer who will need to understand your working environment and type/viscosity of oil(s) you use.

The rate of clean up (speed) will depend on the reservoir size, pump flow rate, and the cleanliness-rating delta. What can be measured immediately is the time to perform one complete filter pass through the filter cart that is calculated using the formula:

(Reservoir size x7) / filter cart flow rate in gpm = time for a single pass filtration

60-gallon x 7 / 10gpm = 42 minutes for a single pass filtration

Note: a single pass = filtration of 1 x the reservoir capacity

If the plant lubricants are consolidated and cleanliness levels are known, a matrix can be developed to determine how many passes are required to filter to an acceptable cleanliness level.

Best practices for use of a filter cart

As in all other facets of maintenance, there are best practices associated with the use of portable filter carts, the following are recommended practice:

- Work with the filter cart supplier to determine the right pump and filter choice for your plant requirements

- To eliminate cross-contamination of lubricants each filter cart must be dedicated to a single lubricant use for both transfer and cleaning of lubricants. Pilot the filter cart program with the most critical and/or or most utilized plant lubricant type
- Always clean the unit after each successful transfer operation, paying particular attention to the wand/coupling ends and open drip tray under the filters and pump area; Open oil is a dirt attractant and can be transferred unwittingly if the cart and its components are not kept scrupulously clean
- Unless specified, most filter carts are sold with open-end transfer wands fitted to the delivery and suction hose ends designed to slide easily into the reservoir openings of the donor and recipient reservoirs. In a program designed to filter contaminants from the oil, this type of delivery fitting can allow moisture and dirt contamination into the respective reservoirs during the transfer process. To combat this, and ensure a contamination free transfer process, fit the filter cart delivery /return hose ends and reservoir fill / drain ports with "quick-lock" style couplings. As the reservoir is now airtight it will also require a quality desiccant style breather to be fitted, and in the case of larger capacity reservoir, a closed loop expansion tank
- Specify "kink-resistant" flexible suction and delivery hose to prevent pump cavitation. Clear hoses allow a visual reference of the oil flowing through the lines;
- The electric motor will require access to electricity. Ensure an electrical outlet is within easy reach of the unit's electrical cord. If the cord is short in length, consider mounting a retractable electrical cord caddy on the unit with enough cord length to reach the nearest electrical outlet;
- Paint a lined box similar to a lay down area as close as possible to the oil reservoir being serviced so the cart can be positioned and used quickly without obstruction and within reach of its hose and wand assemblies
- Place the cart on a PM check program prior to every use to ensure the filters do not go into bypass mode from being too dirty.
- Place clean filter cart in a controlled access area to ensure they are only utilized for the purpose and lubricant they were purchased for. (Eliminates cross contamination)
- Do not use to transfer waste or contaminated oil

- Standardize all couplings size for machine lubrication systems and lubrication carts

21.5.7 Breathers

To breathe is to allow air to freely enter and exit a fully enclosed space unencumbered. From birth we as human beings are accustomed to breathing automatically without thought for the breathing process. When it comes to machines, the design engineer must be cognizant of enclosed spaces and mechanisms that can cause an internal air pressure build up, and design in the ability to relieve, or ventilate the excess air pressure build up within that space at a controlled rate that so that the space can be either neutral or slightly positively pressured.

The ability of internal mechanisms to breathe and equalize pressure has a profound effect on a machine's ability to perform work efficiently, and its component(s) lifecycle. A perfect example of this is found in early combustion engine design in which crude piston and ring technology inevitably allowed combustion gases to leak past the piston rings into the crankcase, a process known as "blow-by". With no engineered provision for venting the enclosed crankcase, pressure would be allowed to build enough to compromise seals and gaskets and allow crankcase oil to diffuse to atmosphere. This would result in engine power losses, oil contamination and constant leaks. This was eventually solved through improved piston and ring design and the introduction of a crankcase ventilation system that introduced fresh air in to the crankcase through the filler cap and allowed it to mix with combustion gases and draft out through an open tube connected to the crankcase known as a "road draft "tube. The system was further refined into the pressurized crankcase ventilation (PCV) system in use today.

An automotive crankcase is of course a reservoir similar to any gearbox or hydraulic oil reservoir. In all cases, the reservoir is an enclosed container used to house lubricant that is pumped through the machine to various bearings surfaces and allowed to return to the reservoir within the closed loop system. When designed correctly, the reservoir will always have an air space, referred to as "headspace", above the oil level designed to permit thermal expansion of the oil and allow the fluid to de-aerate (aerated fluids cause pump cavitation). To equalize any unwanted pressure build up created through the resulting changes in the oil level as the machine moves from rest to full operation, and vice versa, air must be allowed to enter and exit the reservoir through a device known simply as a breather.

Now that air is allowed to freely exchange in and out of the reservoir through the breather device, so is everything else contained within that air exchange. This can include free borne contaminants and moisture, both detrimental to the oil and the very bearings surfaces oil is designed to protect. This now requires the reliability engineer and/or maintainer to recognize the ambient working conditions and choose the appropriate breather style and type for the conditions and furthermore, exercise diligence through preventive maintenance checks to ensure the breather is always in place in the reservoir, is clean and unencumbered allowing it to work as designed.

21.5.8 The anatomy of a breather

Breathers come in all configurations, shapes and sizes and different styles accommodate different airflow, particulate size and working conditions. Most breathers are consumable devices. As such, they must be changed on a regular basis through the lubrication system preventive maintenance program. Change out schedule is based on application and ambient conditions. If the breather is a less sophisticated design that does not display its condition to the operator or maintainer, the rule of thumb is change out every three months in dirty environments such as a foundry to every six months in cleaner environments.

21.5.9 Filter/Breather combination unit

The most common breather in service is the combination filler /breather design that allows a single reservoir opening to serve two purposes. The device looks like a typical fill port with its screw on cap and tubular mesh basket designed to catch large debris from falling into the reservoir during the filling process. The difference from a regular fill port is found in the combination unit cap design, which contains a filter element to keep out airborne contaminants. Filters can be made from different media based on the filter size and airflow restriction requirements. For example, a polyurethane filter media is good for >10 Micron particulate and allows an airflow exchange of 550 litre/min (140gpm or 19cfm), whereas an impregnated paper media will capture >3 Micron particulate and allow an air exchange of 440 litre/min (110gpm or 15cfm). The cap shown in Figure 21.22 demonstrates a damaged breather that is open to air serving little protection against contamination breaching the reservoir.

21.5.10 Standard breather unit

A standard breather unit looks similar to a filler breather cap and is usually screwed onto a threaded pipe that provides air exchange through the top of the reservoir. Other style standard breather units can look similar to an automotive "spin on" oil filter.

In environments that experience large shifts in machine ambient temperatures and working conditions or high humidity, breather caps designed with a filter and a pressure valve can be specified. The pressure relief and vacuum breaker capability is designed to limit air exchange and provide a positive suction head at the pump inlet.

21.5.11 Desiccant breather unit

Desiccant breathers are recent additions to the breather family and differ in that they offer superior air exchange and condition control and are designed to visually indicate to the operator when replacement is required. Figure 21.21 demonstrates how the breather allows clean air in and similarly exhaust air out from the reservoir whilst filtering out solid contamination. The unit is designed with a see-through polycarbonate body filled with a silica gel absorbent designed to hold up to 40% of its weight in absorbed moisture that changes from a blue to light pink color when saturated, indicating a need for change. The unit also contains regular polyester filters designed to capture particulate down to 2 Micron in size.

Figure 21.23, clearly demonstrates the breathing action of an adsorbent media style breather.

Caution! A breather will only work when it is in place; breathers taken off to fill reservoirs, or for checking purposes must be replaced or refitted immediately if the reservoir environment is to stay protected from outside contamination. Breathers are an important and integral part of any reservoir-based lubrication system that require their own maintenance schedule.

21.5.12 IIOT sensing

IIOT, or the Industrial Internet of Things has for a while now made inroads on the lubrication asset management area. Simply put, the industrial internet of things is about electronic devices that can include sensors, instruments, and smart enabled devices on a machine, or within a lubrication system to be networked together with computerised application software and hardware. This networked communication is always in real time searching for flow, pressure,

Figure 21.21 Desiccant style silica gel breather (Copyright Des-Case Corporation, 2022).

cycle, time, energization, alarm triggers. Once an anomaly is detected either through loss of a signal, or through the detection of a signal, its connection to smart software is initiated.

In the case of lubrication devices and systems the smart software is likely to be linked into a SCADA (Supervisory, Control and Data Acquisition system designed to gather and analyze real time data and interact with equipment), the lubrication work management system, or the corporate asset

Figure 21.22 Hydraulic gearbox with damaged breather (Upper LH corner circled in red) c/w a planar style level window with upper (max) and lower level (min) markings. (Courtesy ENGTECH Industries Inc.)

management system. These systems are usually linked to the work management (work order) system that kicks out an automatic work order to a lubrication technician's phone and/or tablet, as well as other interested parties set up in the system to receive event notices.

In the case of lubrication systems, IIOT lends itself well to temperature sensors, level sensors, pressure sensors (end of line), timers, cycle control sensors (activate lubrication cycle(s) by number of cycles/operations/time, heat – cold or hot, loss of pressure, pressure build up, etc.).

Figure 21.24 demonstrates a breather with on board sensor technology that no longer employs color-changing gel but rather electronically senses when the unit is saturated and can no longer dry the air. The unit then sends an electronic message through a WiFi connection requesting change. As with any unit, they are only effective if changed in a timely manner once they spent.

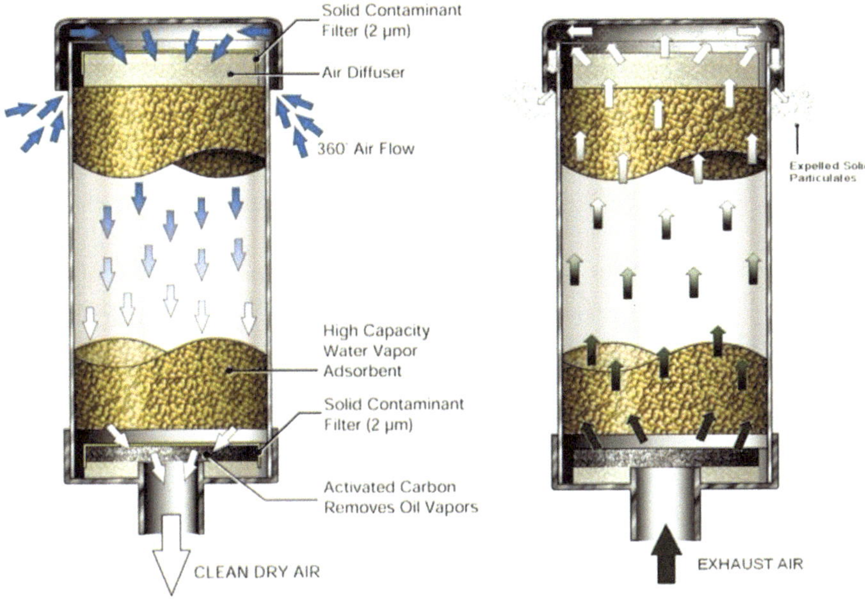

Figure 21.23 Breathing action of an adsorbent media style breather (Courtesy Howard Marten Company for Air Sentry Breathers)

Figure 21.24 Sensor activated IIOT breather (Copyright Des-Case Corporation, 2022).

Bibliography

Bannister, Kenneth E., "Lubrication Reservoirs Control Devices and Systems", *Maintenance Technology Magazine,* November 2018

Bannister, Kenneth E., "Lubrication Delivery Lines", *Maintenance Technology Magazine,* March 2017

Bannister, Kenneth E., "In Service Oil Decontamination", *Efficient Plant Magazine*, December 2018

Tong Seop Kim and Kyu Sang Cha, "Comparative Analysis of the Influence of Labyrinth Seal Configuration on Leakage Behavior", *Journal of Mechanical Science and Technology,* 2009

Bannister, Kenneth E., "Specify the Right Lube-Delivery Line", *Efficient Plant magazine*, April 2017

Bannister, Kenneth E., "Choose from Three Lubricant Seals", *Efficient Plant magazine*, October 2018

Bannister, Kenneth E., "All Sight-Level Gauges Aren't Created Equal", *Efficient Plant Magazine*, March, 2017

Bannister, Kenneth E., "Oil Systems Need to Breathe", *Efficient Plant Magazine*, February, 2018

Bannister, Kenneth E., "Industrial Lubrication Fundamentals - Lubricant Lifecycle Management", *Lubricant Management and Technology Magazine*, 2016

Parker Hannifin Corporation, "The Handbook of Hydraulic Filtration" ENGTECH Industries Inc., ICML MLT/MLA Level Training Presentation, 2015–2023

Bannister, Kenneth E., "Put Portable Filter Carts to Work", *Efficient Plant Magazine*, May, 2015

SECTION 4

Applied Lubrication

22

Bearing Lubrication

If rolling bearings are to operate reliably, they must be adequately and correctly lubricated to prevent direct metal-to-metal contact between the rolling elements, raceways and cages; to prevent wear, and to protect bearing surfaces against corrosion.

The following lubrication information and recommendations relate to bearings without integral seals or shields, used to keep lubricant in place in the bearing cavity. Bearings and bearing units supplied with integral seals or (shields) are generally pre-greased by the bearing manufacturer. Standard greases used by competent bearing manufacturers for these products have operating temperature ranges and working properties designed to suit the intended design application. As the service life of the grease can often exceed the lubricated bearing life, with some exceptions, no provision is made for re-lubrication.

A wide selection of grease and oil is available for the lubrication of rolling bearings. Lubricant choice will depend primarily on the bearing's operating conditions, i.e., temperature range, load, speed, working and ambient conditions. The most favorable operating temperature is obtained when the bearing is supplied with the minimum quantity of lubricant needed to provide reliable lubrication. However, when the lubricant has additional tasks, such as sealing or the removal of heat, larger quantities are often required.

Eventually, the lubricant in a bearing will gradually lose its lubricating properties during operation as a result of mechanical work, aging and contamination build-up. Therefore, it is necessary for grease to be replenished or renewed from time to time, and for oil to be filtered and/or changed at regular intervals (see "Re-lubrication" and "Oil Change," later in this segment).

Because of the large number of lubricants available and, particularly where greases are concerned, because there may be differences in the lubricating properties of seemingly identical greases produced at different locations, a bearing manufacturer cannot accept liability for the lubricant or its performance. *The user is therefore advised to specify the required lubricant properties in detail and to obtain a guarantee from the lubricant supplier that the particular lubricant will satisfy these identified demands.*

22.1 Grease Lubrication

Grease is used to lubricate rolling bearings under normal operating conditions in the majority of applications. For grease lubrication of spherical roller thrust bearings, refer to a later page in this chapter.

Grease has the advantage over oil in that it is more easily retained in the bearing arrangement, particularly where shafts are inclined or vertical, and it also contributes to sealing the arrangement against external solid contaminants, moisture or water.

An excess of lubricant will create a "churning" effect internally within the bearing as it's cage and rolling elements try to move through the wall of grease (not dissimilar to a person trying to move efficiently through 2-3 feet depth of water). This churning effect, particularly when running at high speeds, causes the bearing's internal operating temperature to rise rapidly and spike, while all the while drawing excess energy to overcome the frictional forces. As a general rule, therefore, only the bearing should be completely filled, while the free space in the housing should be partly (between 30 and 50 %) filled with grease. Refer to Chapter 18 "How Much and How Often" for more detailed information. Recommended grease quantities for the initial fill of bearing housings can often be found in the bearing manufacturers' housing tables.

Where bearings are to operate at very low speeds and need to be well protected against corrosion, it is advisable to completely fill the housing with grease.

Speed ratings for grease lubrication can be most often found in the bearing manufacturers' literature. Suffice to say that grease values are lower than corresponding speed ratings for oil lubrication to take account of the initial temperature peak which occurs when starting up a bearing which has been filled with grease during mounting, or just after relubrication. The operating temperature will sink to a much lower level once the grease has been worked and distributed throughout the bearing arrangement. The pumping action inherent in certain bearing designs, similar to that found in angular contact ball bearings and taper roller bearings, will become more accentuated as speeds increase, or due to the pronounced working of the grease which occurs, for example, in full complement cylindrical roller bearings. This can also make it necessary for the speed ratings for grease lubrication to be lower than those for oil lubrication.

22.1.1 Lubricating greases

Lubricating greases are thickened mineral or synthetic oils, the thickeners usually being metallic soaps. Additives are also included to enhance certain grease properties. The consistency of the grease depends largely on the type and concentration of the thickener used. When selecting a grease, the viscosity of the base oil, the consistency, operating temperature range, rust inhibiting properties and the load carrying ability are the most important factors to be considered.

The ***base oil viscosity*** of grease normally used for rolling bearings lies between 15 and 500 mm^2/s at 40°C. Greases based on oils having viscosities in excess of this range will bleed oil too slowly resulting in a bearing that will not be adequately lubricated. Therefore, if a very high viscosity is required because of low speeds, oil lubrication will generally prove more reliable.

The base oil viscosity also governs the maximum permissible speed at which a given grease can be used for bearing lubrication. For applications operating at very high speeds, the most suitable greases are those incorporating diester oils of low viscosity. The permissible operating speed for a grease is also influenced by the shear strength of the grease, which is determined by the thickener. A speed factor nd_m is often quoted by grease manufacturers to indicate the speed capability; n is the operating speed and d_m the mean diameter (mm) of the bearing,

$$d_m = 0.5 (d + D).$$

Greases are divided into various *consistency classes* according to the National Lubricating Grease Institute (NLGI) Scale.

The consistency of greases used for bearing lubrication should not change unduly with temperature within the operating temperature range or with mechanical working. Greases which soften at elevated temperatures may leak from the bearing arrangement; those that stiffen at low temperatures may restrict rotation of the bearing.

Metallic soap thickened greases of consistency 1, 2 or 3 are those normally used for rolling bearings. The consistency 3 greases are usually recommended for bearing arrangements with vertical shaft, where a baffle plate should be arranged beneath the bearing to prevent the grease from leaving the bearing.

In applications subjected to vibration, the grease is heavily worked as it is continuously thrown back into the bearing by vibration. Stiffness alone does not guarantee adequate lubrication; mechanically stable greases should be used for such applications.

Greases thickened with polyurea can soften and harden reversibly depending on the shear rate in the application, i.e., they are relatively stiff at low speeds and are soft or semifluid above a given speed. In applications with vertical shafts there is consequently a danger that a polyurea grease will leak when it is in the semi-fluid state.

The ***temperature range*** over which a grease can be used depends largely on the type of base oil and thickener and its additives. The lower temperature limit, i.e., the lowest temperature at which the grease will allow the bearing to be started up without difficulty, is largely determined by the base oil type and its viscosity. The upper temperature limit is governed by the type of thickener and indicates the maximum temperature at which the grease will provide lubrication for a bearing. It should be remembered that grease will age and oxidize with increasing rapidity as the temperature increases, and that the oxidation products have a detrimental effect on lubrication. The upper temperature limit should not be confused with the "dropping point" which is quoted by lubricant manufacturers. The dropping point only indicates the temperature at which the grease loses its consistency and becomes fluid.

Table 22.1 gives the operating temperature ranges for the types of grease normally used for rolling bearings. These values are based on extensive testing carried out by SKF laboratories and may differ from those quoted by other lubricant manufacturers.

Values are valid for commonly available greases having a mineral oil base and with no EP additives. Of the grease types listed, lithium and more particularly lithium 12-hydroxystearate base greases are those most often used for bearing lubrication.

Greases based on synthetic oils, e.g., ester oils, synthetic hydrocarbons or silicone oils, may be used at temperatures above and below the operating temperature range of mineral oil-based greases.

If bearings are to operate at temperatures above or below the ranges quoted in the table and are to be grease lubricated, the bearing manufacturer should be contacted for advice.

Bearings operated in extreme low temperature conditions can see general use grease solidify and stall the bearing. Similarly, in high extreme temperatures, the grease can separate the oil out quickly and cause the thickener to "cook" and solidify. Figure 22.1 shows and rotary casting head needle bearing that has seized in operation due to use of an incorrect grease that has "cooked" and solidified.

The ***rust inhibiting properties*** of a grease are mainly determined by the rust inhibitors which are added to the grease and its thickener.

Table 22.1 Operating temperature ranges for greases used in rolling element bearings.

Grease type (thickener)	Recommended operating temperature range			
	from	to	from	to
—	°C		°F	
Lithium base	-30	+110	- 22	+ 230
Lithium complex	-20	+140	-4	+284
Sodium base	-30	+80	-22	+176
Sodium complex	-20	+140	- 4	+284
Calcium (lime) base	-10	+ 60	+14	+140
Calcium complex	-20	+130	- 4	+266
Barium complex	-20	+130	- 4	+266
Aluminum complex	-30	+110	- 22	+230
Inorganic thickeners (bentonite, silica gel etc.)	- 30	+130	-22	+266
Polyurea	-30	+140	-22	+284

A grease should provide protection to the bearing against corrosion and should not be washed out of the bearing in case of water penetration. Ordinary sodium base greases emulsify in the presence of water and can be washed out of a bearing. Whereas, lithium and calcium base greases containing lead-base additives offer very good resistance to water and protection against corrosion. However, because of environmental and health reasons such additives are being replaced by other combinations of additives which do not always offer the same protection.

For ***heavily loaded bearings***, e.g., rolling mill bearings, it has been customary to recommend the use of greases containing EP additives, since these additives increase the load carrying ability of the lubricant film.

Originally, most EP additives were lead-based compounds and there was evidence to suggest that these were beneficial in extending bearing life where lubrication was otherwise poor, e.g., when K (calculated as explained in conjunction with Figures 22.7 and 22.8), is less than 1. However, for the reasons cited above, many lubricant manufacturers have replaced the lead-based additives by other compounds, some of which have been found to be

Figure 22.1 Solidified grease thickener that has destroyed the bearing. Courtesy ENGTECH Industries Inc.

aggressive to bearing steels. Drastic reductions in bearing life have been recorded in some instances.

The utmost care should therefore be taken when selecting an EP grease and assurances should be obtained from the lubricant manufacturer that the EP additives incorporated are not of the damaging type, or in cases where the grease is known to perform well a check should be made to see that its formulation has not been changed.

Note: It is important to consider the *miscibility* of grease when, for whatever reason, it is necessary to change from one grease to another. If greases which are incompatible are mixed, the consistency can change dramatically and the maximum operating temperature of the grease mix be so low, compared with that of the original grease, that bearing damage cannot be ruled out.

Greases having the same thickener and similar base oils can generally be mixed without any detrimental consequences, e.g., a sodium base grease can be mixed with another sodium base grease. Calcium and lithium base greases are generally miscible with each other but not with sodium base greases. (Refer to Table 16.4.) However, mixtures of compatible greases may have a consistency which is less than either of the component greases, although the lubricating properties are not necessarily impaired.

In bearing arrangements where a low consistency might lead to grease escaping from the arrangement, the next relubrication should involve complete replacement of the grease rather than replenishment (see segment "Relubrication").

22.1.2 Relubrication

Rolling bearings have to be relubricated if the service life of the grease used is shorter than the expected service life of the bearing.

The time at which relubrication should be under-taken depends on many factors which are related in a complex manner. These include bearing type and size, speed, operating temperature, grease type, space around the bearing and the bearing environment. It is only possible to base recommendations on statistical rules. For example, the SKF bearing company defines it's relubrication intervals as the time period at the end of which 99% of the bearings are still reliably lubricated, and represent L_1 grease lives. The L_{10} grease lives are approximately twice the L1 lives.

The information given in the following is based on long-term tests in various applications but does not pertain in applications where water and/or solid contaminants can penetrate the bearing arrangement. In such cases it is recommended that the grease is frequently renewed in order to remove contaminants from the bearing.

The *relubrication intervals* tf for normal operating conditions can be read off as a function of bearing speed n and bore diameter d of a certain bearing type from Figure 22.2. The diagram is valid for bearings on horizontal shafts in stationary machines under normal loads. It applies to good quality lithium base greases at a temperature not exceeding 70°C. To take account of the accelerated aging of the grease with increasing temperature it is recommended that the intervals obtained from the diagram are halved for every 15° increase in bearing temperature above 70°C, remembering that the maximum operating temperature for the grease given in Table 22.1 should not be exceeded. The intervals may be extended at temperatures lower than 70°C but as operating temperatures decrease the grease will bleed oil less readily and at low temperatures an extension of the intervals by more than two times is not recommended. It is not advisable to use relubrication intervals in excess of 30,000 hours. For bearings on vertical shafts the intervals obtained from the diagram should be halved.

For large roller bearings having a bore diameter of 300 mm and above, the high specific loads in the bearing mean that adequate lubrication will be obtained only if the bearing is more frequently relubricated than indicated by

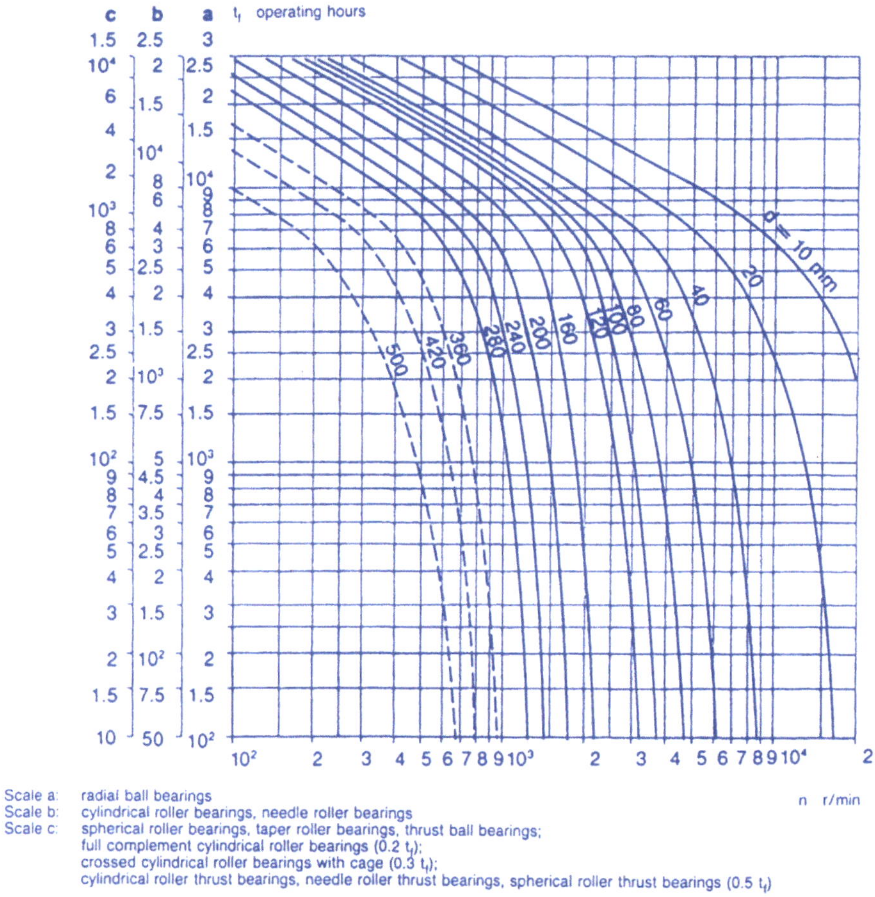

Figure 22.2 Grease relubrication intervals as a function of bearing type, size, and speed. (Source: SKF America, Kulpsville, Pennsylvania.)

the diagram, and the lines are therefore broken. It is recommended in such cases that continuous lubrication is practiced for technical and economic reasons. The grease quantity to be supplied can be obtained from the following equation for applications where conditions are otherwise normal, i.e., where external heat is not applied (recommendations for grease quantities for periodic relubrication are given in the following section).

$$G_k = (0.3 \ldots 0.5)\, D\, B \times 10^{-4}$$

Where:

G_k = grease quantity to be continuously supplied, g/h
D = bearing outside diameter, mm
B = total bearing width (for thrust bearings use total height H), mm

One of the two *relubrication procedures* described below should be used, depending on the relubrication interval t_f obtained:

- if the relubrication interval is shorter than 6 months, then it is recommended that the grease fill in the bearing arrangement be replenished (topped up) at intervals corresponding to $0.5t_f$; the complete grease fill should be replaced after three replenishments, at the latest;
- when relubrication intervals are longer than 6 months it is recommended that all used grease be removed from the bearing arrangement and replaced by fresh grease.

The six-month limit represents a very rough guideline recommendation and may be adapted to fall in line with lubrication and maintenance recommendations applying to the particular machine or plant.

By adding small quantities of fresh *replenishment grease* at regular intervals the used grease in the bearing arrangement will only be partially replaced. Suitable quantities to be added can be obtained from:

$$G_p = 0.005\ D\ B$$

Where:

G_p = grease quantity to be added when replenishing, grams
D = bearing outside diameter, mm
B = total bearing width mm (for thrust bearings use total height H)

To facilitate the supply of grease using a grease gun, a grease nipple should be provided on the housing. It is also necessary to provide an exit hole for the grease so that excessive amounts will not collect in the space surrounding the bearing. This might otherwise cause a permanent increase in bearing temperature. However, as soon as the equilibrium temperature has been reached following relubrication, the exit hole should be plugged or covered so that the oil bled by the grease will remain at the bearing position. The danger of excess grease collecting in the space surrounding the bearing and causing temperature peaking, with its detrimental effect on the grease as well as the bearing, is most pronounced when bearings operate at high speeds. In such cases it is advisable to use a grease escape valve rather than an exit hole. This prevents over-lubrication and allows relubrication to be carried out, without the machine having to be stopped.

A grease escape valve, Figure 22.3, consists basically of a disc which rotates with the shaft and which forms a narrow gap together with the housing end cover. Excess and used grease is thrown by the disc into an annular

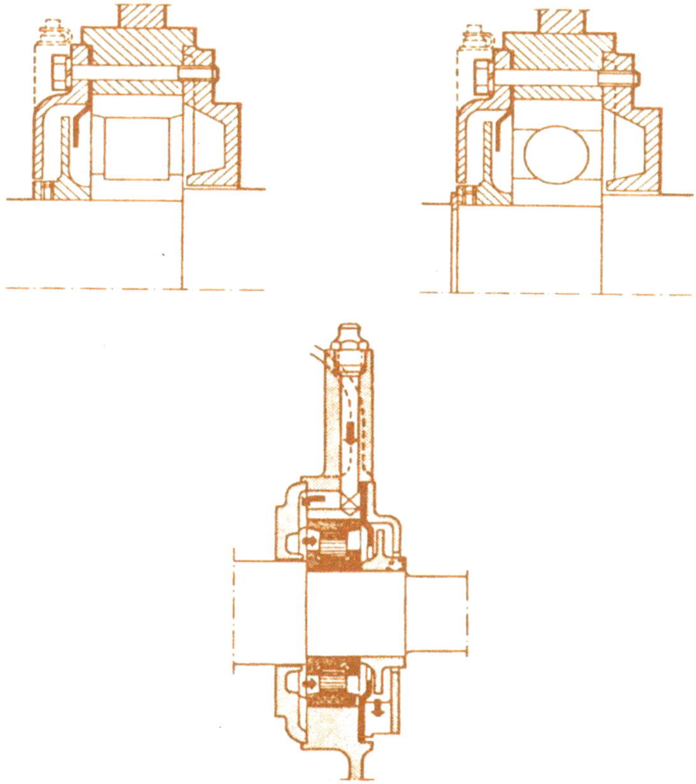

Figure 22.3 Advantageous, simple grease escape "valve." (Arrangements found in ASEA Electric Motors.)

cavity and leaves the housing through an opening on the underside of the end cover. The use of small check valves that are expected to perform as grease escape valves has proven problematic and should be discouraged.

To ensure fresh grease actually reaches the bearing and replaces the old grease, the lubrication duct in the housing should either feed the grease adjacent to the outer ring side face or, better still, into the bearing which is possible, for example, with spherical roller bearings and double row full complement cylindrical roller bearings.

Where centralized lubrication equipment is used, care must be taken to see that the grease has adequate pumpability over the range of ambient temperatures.

If, for some reason, it is necessary to change from one grease to another, a check should be made to see that the new and old greases are compatible (see under "Miscibility," earlier in this segment.

When the end of the relubrication interval t_f has been reached, the used grease in the bearing arrangement should be completely removed and replaced by fresh grease. As stated earlier, under normal conditions, the free space in the bearing should be completely filled and the free space in the housing filled to between 30 and 50% with fresh grease. The requisite quantities of grease to be used for a particular housing are usually stipulated in the manufacturer's literature.

In order to be able to *renew the grease fill* it is essential that the bearing housing is easily accessible and easily opened. The cap of split housings and the cover of one-piece housings can usually be taken off to expose the bearing. After removing the used grease, fresh grease should first be packed between the rolling elements. Great care should be taken to see that contaminants are not introduced into the bearing or housing when relubricating, and the grease itself should be protected. Where the housings are less accessible but are provided with grease nipples and exit holes or grease valves it is possible to completely renew the grease fill by relubricating several times in close succession until it can be assumed that all old grease has been pressed out of the housing. This procedure requires much more grease than is needed for manual renewal of the grease fill.

22.2 Oil Lubrication

Oil is generally used for rolling bearing lubrication only when high speeds or operating temperatures preclude the use of grease, when frictional or applied heat has to be removed from the bearing position, or when adjacent components (gears, etc.) are lubricated with oil.

22.2.1 Methods of oil lubrication

The simplest method of oil lubrication is the oil bath, Figure 22.4. The oil, which is picked up by the rotating components of the bearing or by a flinger ring, is distributed within the bearing and then flows back to the oil bath. The oil level should be such that it almost reaches the center of the lowest rolling element when the bearing is stationary. Speed ratings for oil lubrication given in manufacturers' bearing tables normally apply to oil bath lubrication.

Operating at higher speeds will cause the operating temperature to increase and will accelerate aging of the oil. To avoid frequent oil changes, circulating oil lubrication, Figure 22.5, is generally preferred; the circulation is usually produced with the aid of a pump. After the oil has passed through the bearing it is filtered and, if required, cooled before being returned to the

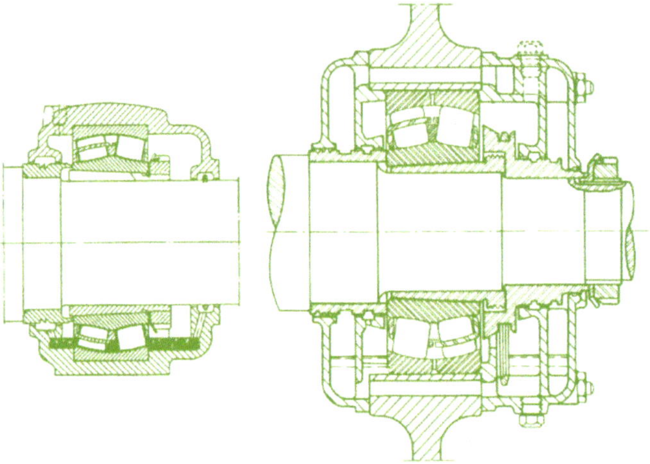

Figure 22.4 Oil bath/oil spray on double row spherical roller bearings. In the oil bath configuration (left), the oil level reaches the center of the rollers at the bottom. In the spray configuration (right), the oil is conveyed to the tapered flinger by an oil ring which dips into the oil bath. An important feature of this application is that the air pressure on both sides of the bearing and enclosures is equalized by connecting ducts. This prevents leakage of the lubricant when the housings are located in an air stream.

bearing. Cooling of the oil enables the operating temperature of the bearing to be kept at low level.

For very high-speed operation it is necessary that a sufficient but not excessive quantity of oil penetrates the bearing to provide adequate lubrication without increasing the operating temperature more than necessary. One particularly efficient method of achieving this is the oil jet method, Figure 22.6, where a jet of oil under high pressure is directed at the side of the bearing. The velocity of the oil jet must be high enough (at least 15 m/s), so that at least some of the oil will penetrate the turbulence surrounding the rotating bearing.

With the air-oil method, Figure 22.7, very small, accurately metered quantities of oil are directed at each individual bearing by compressed air. This minimum quantity enables bearings to operate at lower temperatures or at higher speeds than any other method of lubrication.

The oil is supplied to the points of application by a metering unit at given intervals. The oil is transported by compressed air; it coats the inside of applicator tubing or wires, and "creeps" along them. It is injected to the bearing via a nozzle. The compressed air serves to cool the bearing and also produces an excess pressure in the bearing arrangement which prevents contaminants from entering.

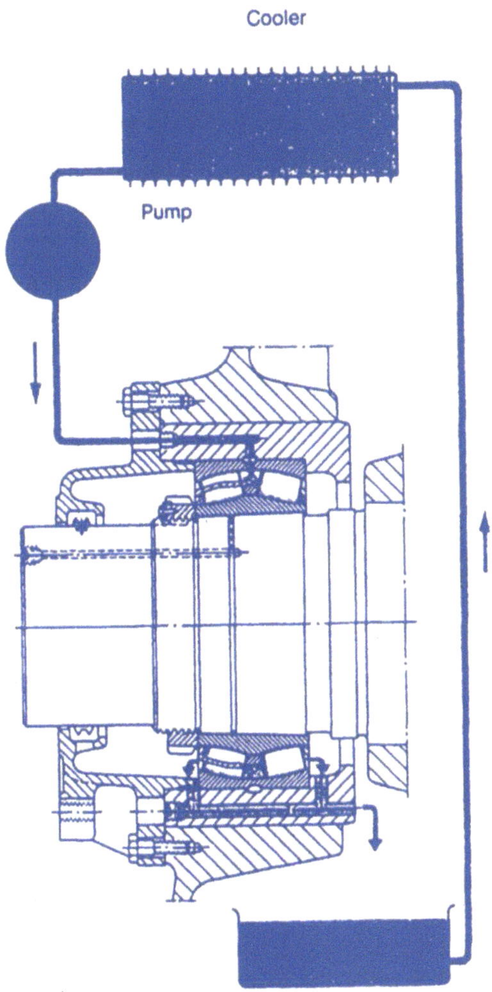

Figure 22.5 Circulating oil lubrication.

When using the circulating oil, oil jet and air-oil methods, it is necessary to ensure that the oil flowing from the bearing can leave the arrangement by adequately dimensioned ducts.

Straight mineral oils without additives are generally favored for rolling bearing lubrication. Oils containing additives for the improvement of certain lubricant properties such as extreme pressure behavior, aging resistance etc. are generally only used in special cases. *Synthetic oils* are generally only considered for bearing lubrication in extreme cases, e.g., at high loads, and very low or very high operating temperatures. It should be remembered that

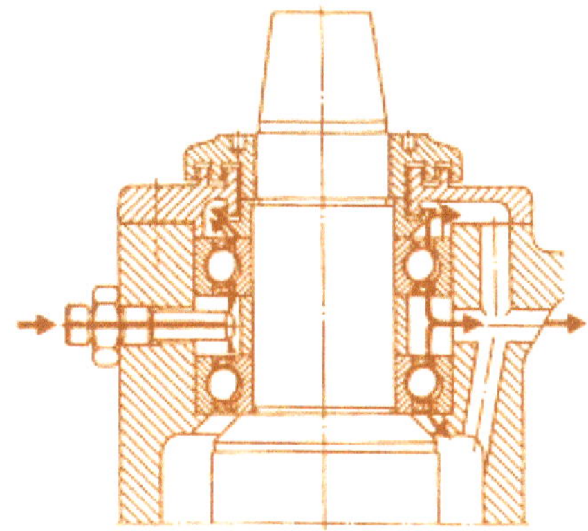

Figure 22.6 Oil-jet lubrication allows bearings to operate at lower temperature or at higher speeds than any other method of lubrication.

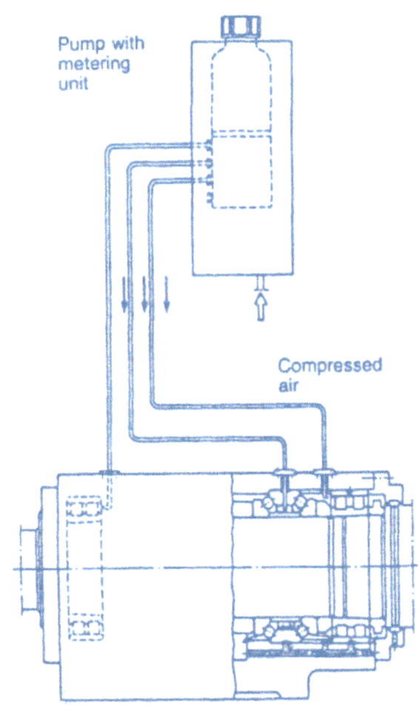

Figure 22.7 Air-oil lubrication schematic.

the lubricant film formation when using a synthetic oil may differ from that of a mineral oil having the same viscosity.

The remarks covering EP additives in the earlier segment on greases entitled "Load carrying ability," also apply to EP additives in oils.

As previously discussed, the *selection of an oil* is primarily based on the viscosity required to provide adequate lubrication for the bearing at the operating temperature. The viscosity of an oil is temperature dependent, becoming lower as the temperature rises. The viscosity/temperature relationship of

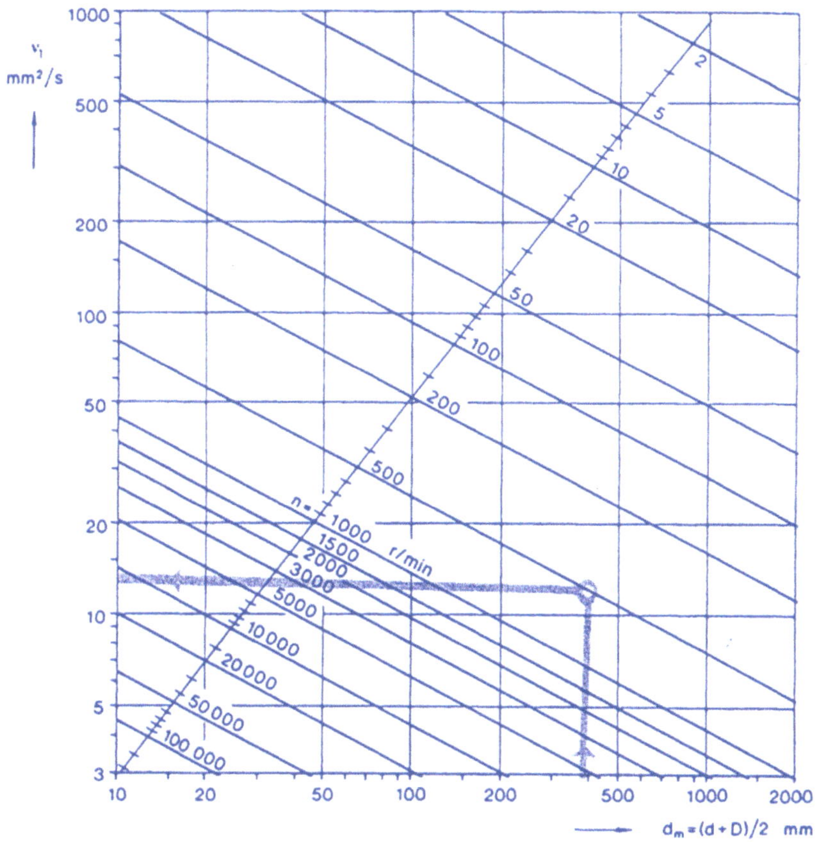

Figure 22.8 Kinematic viscosity requirement as a function of bearing mean diameter and speed.

an oil is characterized by the viscosity index, VI. For rolling bearing lubrication, oils having a high viscosity index (little change with temperature) of at least 85 are recommended.

In order for a sufficiently thick film of oil to be formed in the contact area between rolling elements and raceways, the oil must retain a minimum viscosity at the operating temperature. The kinematic viscosity υ_1 required at the operating temperature to ensure adequate lubrication can be determined from Figure 22.8 provided a mineral oil is used.

When the operating temperature is known from experience or can otherwise be determined, the corresponding viscosity at the internationally standardized reference temperature of 40°C, or at other test temperatures (e.g., 20 or 50°C) can be obtained from Figure 22.8 which is compiled for a viscosity index of 85. Certain bearing types, e.g., spherical roller bearings, taper roller bearings, and spherical roller thrust bearings, normally have a higher operating temperature than other bearing types e.g., deep groove ball bearings and cylindrical roller bearings, under comparable operating conditions.

When selecting an oil, the following aspects should be considered.

Bearing life may be extended by selecting an oil whose viscosity v at the operating temperature is somewhat higher than υ1. However, since increased viscosity raises the bearing operating temperature there is frequently a practical limit to the lubrication improvement which can be obtained by this means.

If the viscosity ratio $K = \upsilon/\upsilon_1$ is less than 1 an oil containing EP additives is recommended and if K is less than 0.4 an oil with such additives must be used. An oil with EP additives may also enhance operational reliability in cases where K is greater than 1 and medium and large-sized roller bearings are concerned. It should be remembered that only some EP additives are beneficial, however (see also under "Load carrying ability," earlier in this segment.

For exceptionally low or high speeds, for critical loading conditions or for unusual lubricating conditions please consider discussions with the applications engineering staff of major bearing manufacturers.

Example

A bearing having a bore diameter d = 340 mm and outside diameter D = 420 mm is required to operate at a speed n = 500 r/min. Since $d_m = 0.5 (d + D)$, $d_m = 380$ mm. From Figure 22.9, the minimum kinematic viscosity υ_1 required to give adequate lubrication at the at the operating temperature is 13 mm^2/s.

From Figure 22.9, assuming that the operating temperature of the bearing is 70°C, an oil having a viscosity υ at the reference temperature of 40°C of at least 39 mm^2/s will be required.

The *frequency with which it is necessary to change the oil* depends mainly on the operating conditions and the quantity of oil.

With oil bath lubrication it is generally sufficient to change the oil once a year, provided the operating temperature does not exceed 50°C and there is little risk of contamination. Higher temperatures call for more frequent oil changes, e.g., for operating temperatures around 100°C, the oil should be changed every three months. Frequent oil changes are also needed if other operating conditions are arduous.

With circulating oil lubrication, the period between two oil changes is also determined by how frequently the total oil quantity is circulated and whether or not the oil is cooled. It is generally only possible to determine a

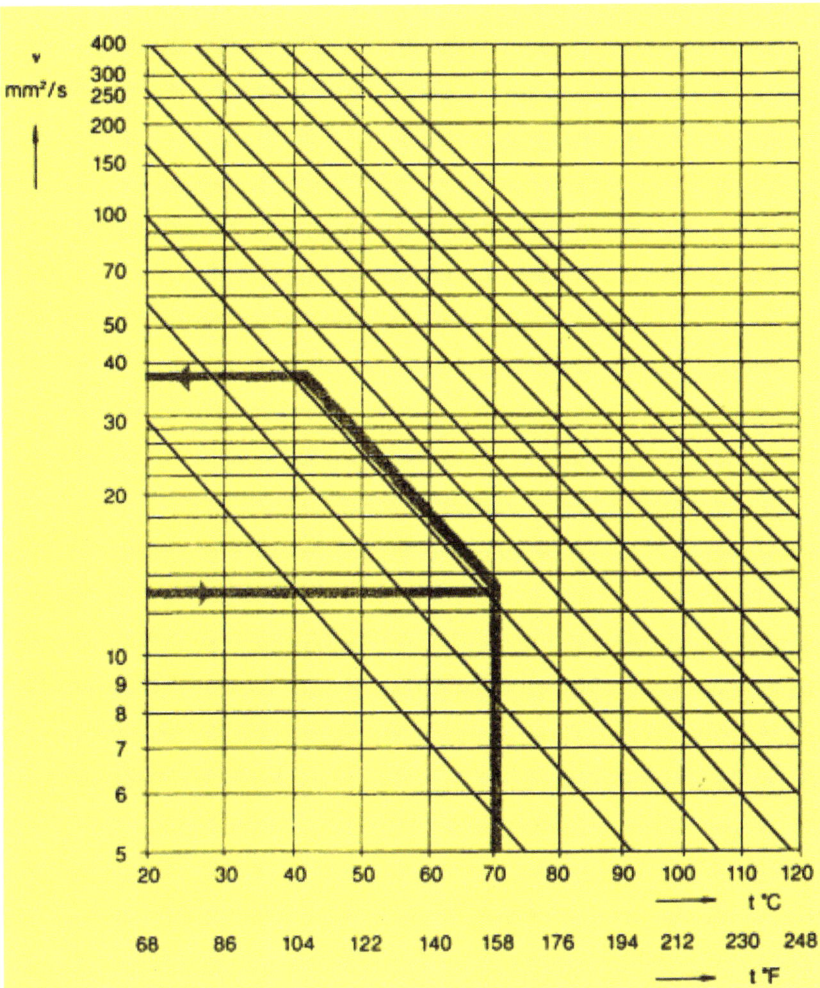

Figure 22.9 The required ISO-grade of a lubricant can be obtained from this graph.

suitable interval by test runs and by regular inspection of the condition of the oil to see that it is not contaminated and is not excessively oxidized. The same applies for oil jet lubrication.

With air-oil lubrication the oil only passes through the bearing once and is not recirculated. The same is generally true of oil mist lubrication, Figure 22.10, a superior means of conveying and applying liquid lubricants.

Spherical roller bearings present a special lubrication challenge. It is generally recommended that spherical roller thrust bearings should be oil

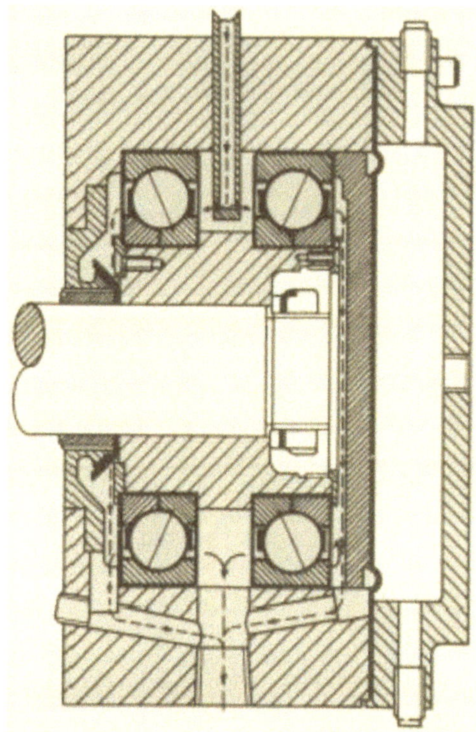

Figure 22.10 Oil mist lubrication applied to a set of split inner ring bearings. (Source: Fafnir Division of Torrington Company, Torrington, Connecticut.)

lubricated. Grease lubrication can be used in special cases, for example, under light loads and at low speeds, particularly where bearings incorporating a pressed steel cage are concerned.

When using grease as the lubricant it is necessary to ensure that the roller end/flange contacts are adequately supplied with grease. Depending on the actual application, this can best be done by completely filling the bearing and its housing with grease or by regular relubrication.

The speed ratings quoted in the bearing tables for grease lubricated bearings fitted with pressed steel cages are valid for bearing arrangements where the shaft is horizontal. For arrangements with vertical shafts, the values should be approximately halved.

Because of their internal design, spherical roller thrust bearings have a pumping action which may be exploited under certain conditions and should be taken into consideration when selecting lubrication method and seals.

More detailed information regarding the lubrication of spherical roller thrust bearings can be provided by the application engineering service groups of competent manufacturers.

In order to assure the satisfactory operation of all ball and roller bearings they must always be subjected to a given *minimum load.* This is also true of spherical roller thrust bearings, particularly if they run at high speeds where the inertia forces of the rollers and cage, and the friction in the lubricant can have a detrimental influence on the rolling conditions in the bearing and may cause damaging sliding movements to occur between the rollers and the raceways.

The requisite minimum axial load to be applied in such cases can be estimated from

$$F_{am} = 1.8\,F_r + A\left(\frac{n}{1000}\right)^2$$

(If $1.8\,F_r < 0.0005\,C_0$, then $0.0005\,C_0$ should be used in the above equation instead of $1.8\,F_r$)

Where:

F_{am} = minimum axial load, N
F_r = radial component of load for bearings subjected to combined load, N
C_0 = basic static load rating, N
A = minimum load factor, see manufacturer's bearing tables
N = speed, r/min

The weight of the components supported by the bearing, together with the external forces, often exceeds the requisite minimum load. If this is not the case, the bearing must be preloaded (e.g., by springs).

Finally, the reader may wish to use Figure 22.11, a simplified oil viscosity selection chart devised by the Fafnir Bearings Division of The Torrington Company, Torrington, Connecticut.

This chart may be used to approximate the proper oil viscosity for all bearing applications.

To use the chart, proceed as follows:

1. Determine the DN value—Multiply the bore diameter of the bearing, measured in millimeters, by the speed of the shaft, measured in revolutions per minute.

2. Select the proper temperature—The operating temperature of the bearing may run several degrees higher than the ambient temperature

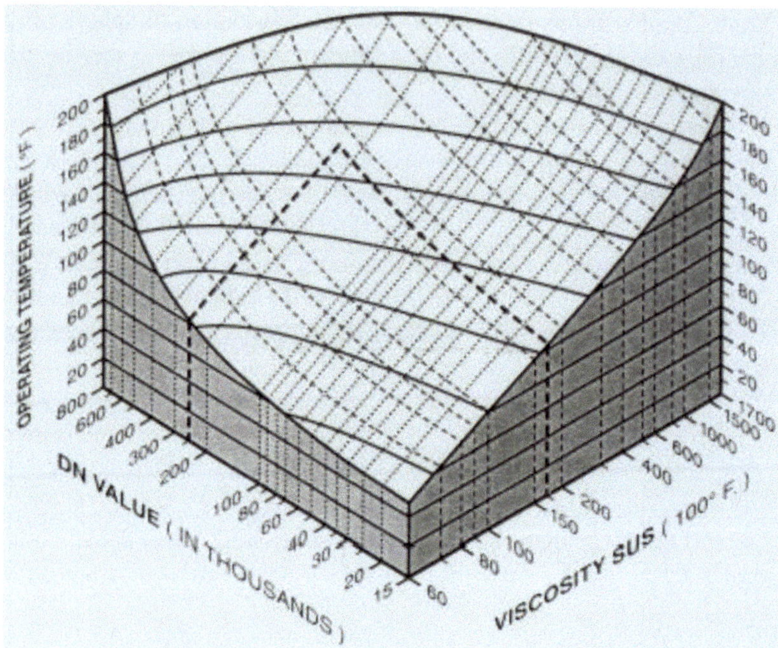

Figure 22.11 Simplified oil viscosity selection chart.

depending upon the application. The temperature scale of this chart reflects the operating temperature of the bearing.

3. Enter the DN value in the DN scale on the chart.
4. Follow or parallel the "dotted" line to the point where it intersects the selected "solid" temperature line.
5. At this point follow or parallel the nearest "dashed" line downward and to the right to the viscosity scale.
6. Read off the approximate viscosity value-expressed in Saybolt Universal Seconds at 100°F.

Typical Example

PROBLEM:
Determine the proper oil viscosity required for a 50 mm ball bearing operating at a speed of 5000 RPM at a temperature of 150°F.

SOLUTION:
Determine the DN value-the bore diameter of the ball bearing is 50 mm. Multiply this by the shaft speed in RPM; 50 mm × 5000 RPM = 250,000 DN.

Enter this value on the DN scale. Parallel the "dotted" lines to the point of intersection with the projected "solid" 150°F temperature line. At this junction, parallel the nearest "dashed" line downward and to the right to the viscosity scale. Read off the approximate viscosity of 170 SUS at 100°F.

22.3 Tilting Pad Thrust Bearings*

Tilting pad thrust bearings are designed to transfer high axial loads from rotating shafts with minimum power loss, while simplifying installation and maintenance. The shaft diameters for which the bearings are designed typically range from 20mm to over 1000mm. The maximum loads for the various bearing types range from 0.5 to 500 tons. Bearings of larger size and load capacity are considered non-standard, but can and have been made to special order.

Each bearing consists of a series of pads supported in a carrier ring; each pad is free to tilt so as to create a self-sustaining hydrodynamic film. The carrier ring may be in one piece or in halves, and there are various location arrangements.

Two options exist for lubrication. One is by fully flooding the bearing housing, the other, which is more suitable for higher speed applications, directs oil to the thrust face; this oil is then allowed to drain freely from the bearing housing.

Similarly, two geometric options exist. The first option is shown in Figures 22.12 and 22.13; it does not use equalizing or leveling links. This option is used in many gear units and other shaft systems where perpendicularity between shaft centerline and bearing faces is assured.

This design, Figure 22.14, is intended for machines where an equalized thrust bearing is specified by API requirements or where this bearing may be required for other reasons.

22.3.1 Flooded lubrication vs Directed lubrication

The conventional method of lubricating tilting pad thrust bearings is to flood the housing with oil, using an orifice on the outlet to regulate the flow and maintain pressure. A typical double thrust bearing of this type is illustrated in Figure 22.15. A housing pressure of 0.7 -1.0 bar is usual and, to minimize leakage, seal rings are required where the shaft passes through the housing.

* Sources: The Glacier Metal Company, London, England, and Mystic, Connecticut; also, Kingsbury Inc., Philadelphia, PA, and Waukesha Bearings, Waukesha, WI.

Figure 22.12 20MW naval gearbox fitted with Glacier Series II flooded lubrication standard bearings for medium speed duties. (Photo courtesy of Maag Gear Company, Zurich, Switzerland.)

Although flooded lubrication is simple, it results in high parasitic power loss due to turbulence at high speed. Where mean sliding speeds in excess of 50 m/s are expected, these losses may be largely eliminated by employing the system of directed lubrication. As well as reducing power loss by (typically) 50%, directed lubrication also reduces the bearing temperature and in most cases oil flow.

Some typical double thrust bearing arrangements using directed lubrication are shown in Figure 22.16.

It should be noted that:

- Directed and flooded bearings have the same basic sizes, and use identical thrust pads.
- Preferred oil supply pressure for directed lubrication is 1.4 bar.
- Oil velocity in the supply passages

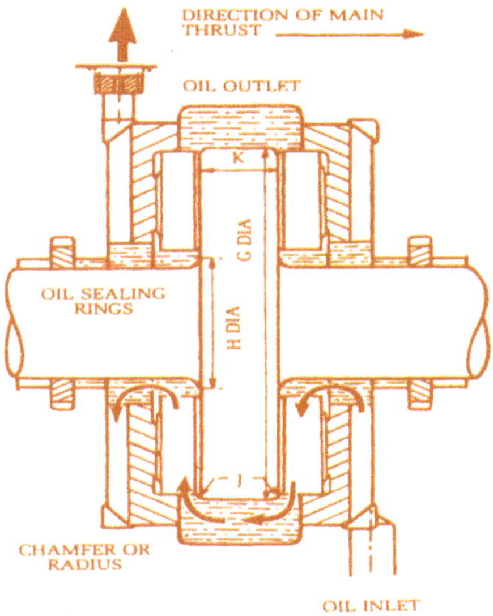

Figure 22.13 Flooded lubrication: typical double thrust arrangement.

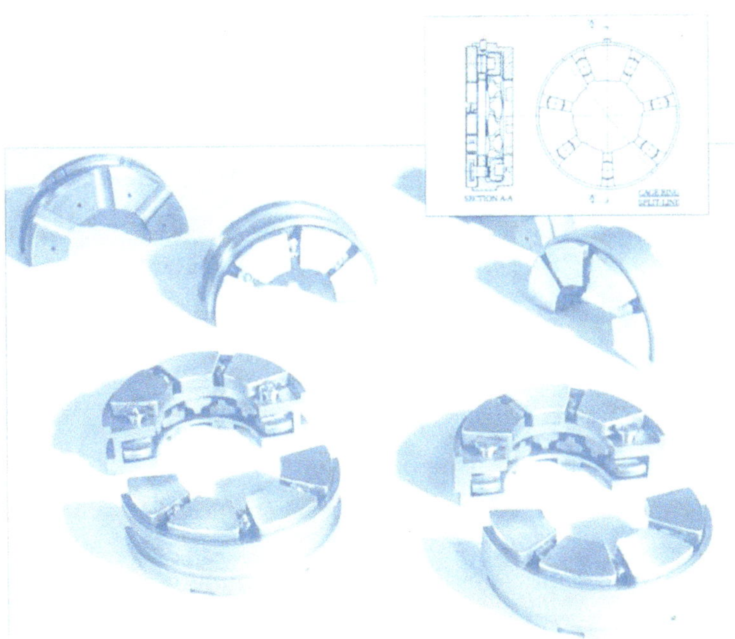

Figure 22.14 Glacier's standard 7 series bearings for both flooded and directed lubrication.

Figure 22.15 Glacier double thrust bearing size 10293 with directed lubrication installed in an ABB gas turbine. (Photo: ABB, Baden, Switzerland.)

Experienced manufacturers can offer a variety of pad materials. Some polymeric materials are capable of operating at temperatures up to 120°C higher than conventional white metal or babbitt. Also, pad pivot position can have an effect on thrust pad temperature (Figure 22.17).

All pads can be supplied with offset pivots, but center-pivoted pads are preferred for bi-directional running, foolproof assembly and minimum stocks. At moderate speeds the pivot position does not affect load capacity but where mean sliding speeds exceed 70 m/s offset pivots can reduce bearing surface temperatures and thus increase load capacity under running conditions

Thrust bearings can also be fitted with temperature sensors, proximity probes, and load cells.

In hydraulic thrust metering systems, a hydraulic piston is located behind each thrust pad, these pistons being interconnected to a high-pressure oil supply. The pressure in the system then gives a measure of the applied thrust load. Figure 22.18 shows a typical installation of this type complete with system control panel which incorporates the high-pressure oil pump and system pressure gauge calibrated to read thrust load.

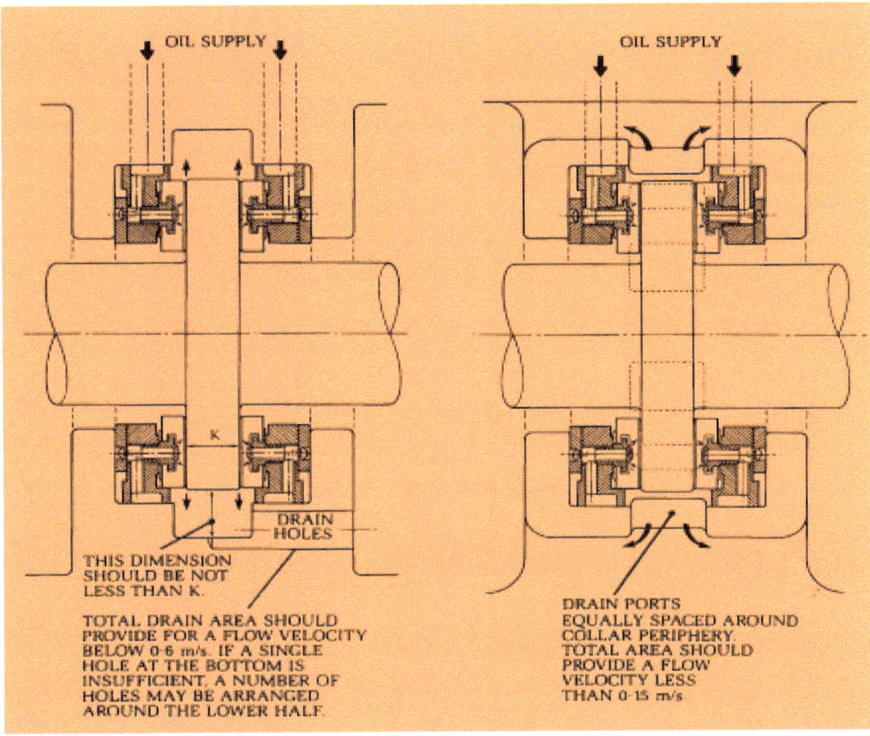

Figure 22.16 Directed lubrication: typical double thrust arrangements designed to prevent bulk oil from contacting the collar.

For systems incorporating load cells or hydraulic pistons, it will normally be necessary to increase the overall axial thickness of the thrust ring.

Finally, there are thrust bearings that incorporate hydraulic jacking provisions. These provisions ensure that an appropriate oil film exists between thrust runner and bearing pads while operating at low speeds.

At the instant of start up, the load carrying capacity of tilting pad thrust bearings is restricted to approximately 60% of the maximum permissible operating load. If the start up load on a bearing exceeds this figure and a larger bearing is not a feasible option, the manufacturer can supply thrust bearings fitted with a hydrostatic jacking system to allow the bearing to operate with heavy loads at low speeds. This system introduces oil at high pressure (typically 100–150 bar) between the bearing surfaces to form a hydrostatic oil film; Figure 22.19 shows a typical bearing of this type.

Offset pivot thrust pad.

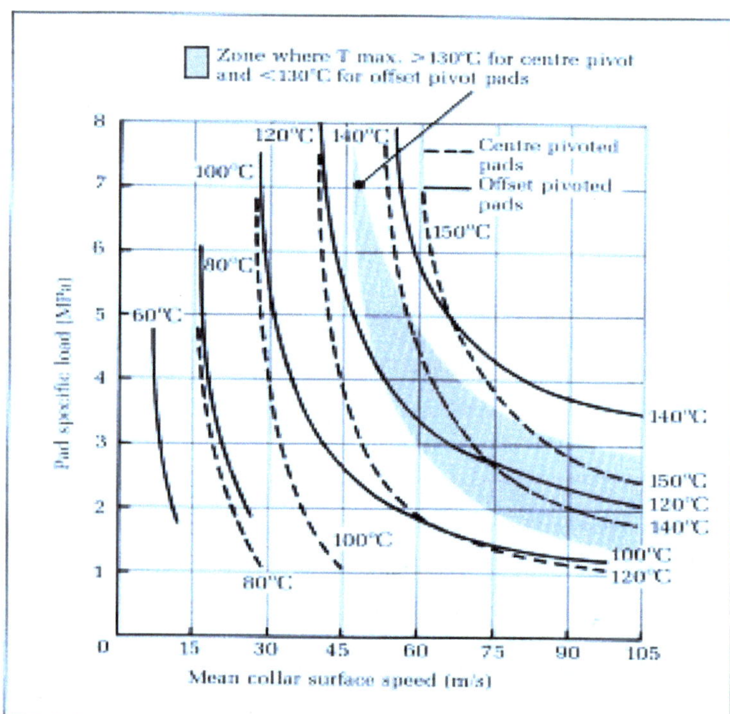

Figure 22.17 Offset pivots: effect on thrust pad temperature.

22.3.2 Bearing selection

Thrust load, shaft RPM, oil viscosity and shaft diameter through the bearing determine the bearing size to be selected.

Leading edge bearings are sized for normal load and speed when transient load and speed are within 20% of normal conditions.

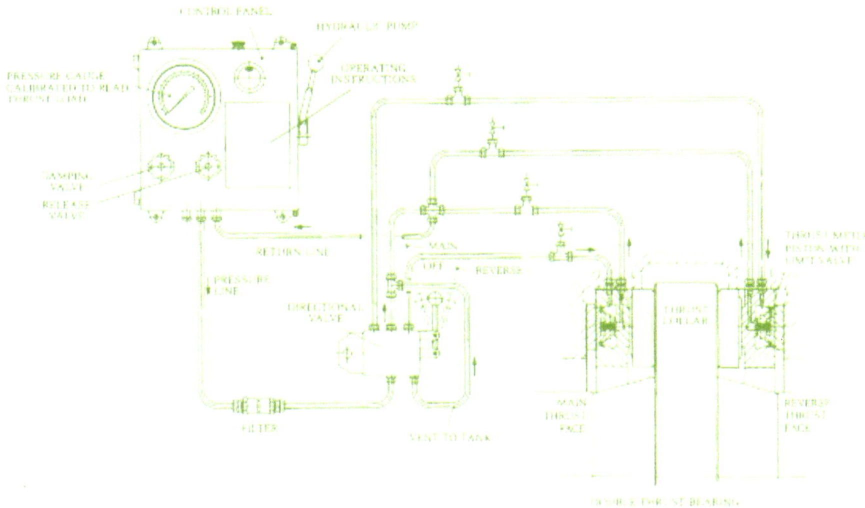

Figure 22.18 Hydraulic thrust metering: arrangement diagram.

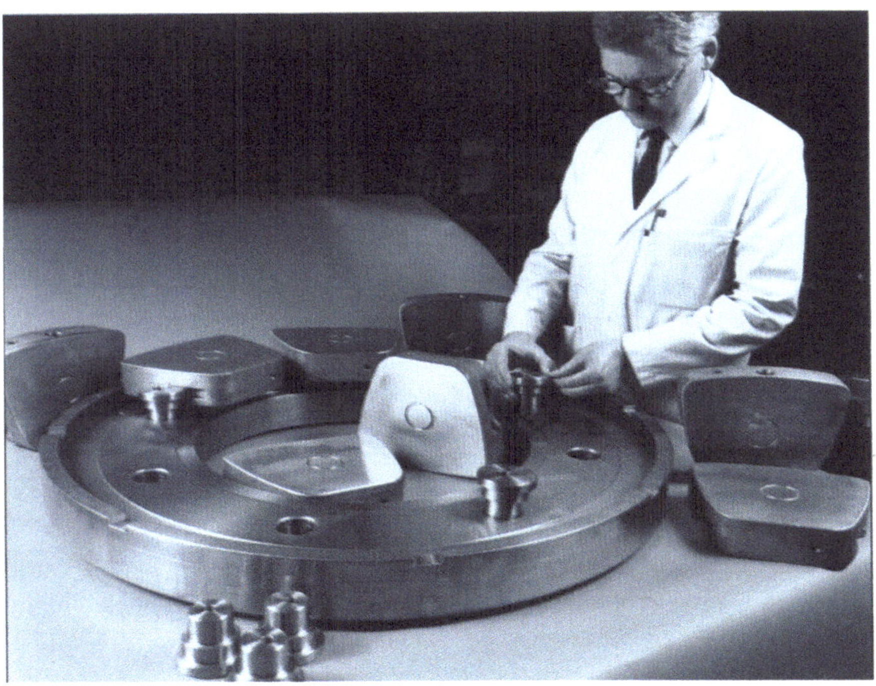

Figure 22.19 Glacier bearing featuring high pressure jacking oil for start-up and run down. Jacking oil annulus can be seen on the surface of each thrust pad.

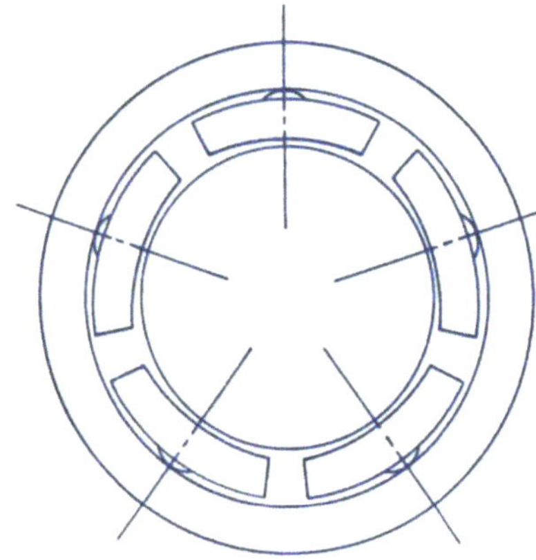

Figure 22.20 Tilting-pad journal bearing, converging geometry.

22.4 Tilting Pad Radial Bearings

Typical tilting pad journal bearings consist of three basic components: a shell, end plates and a set of pads. When the shaft is rotating, radial forces are transferred from the shaft to the journal pads through a film of oil that is self-generated between the shaft and the pads. The radial force then passes from the pads, through the bearing shell to the foundation or machine support.

To develop a hydrodynamic film in a bearing, three factors are required:

a. Viscous fluid

b. Relative motion

c. Converging geometry

The first factor, viscous fluid, is available since the bearings under consideration are fluid-lubricated (primarily with oil). Relative motion is provided by the rotation of the shaft relative to the surface of the tilting pads. Converging geometry is provided by the slight difference in the diameter of the shaft and the bore of the bearing pads (Figure 22.20). Clearances are exaggerated in Figure 22.21 for illustrative purposes.

The principle of pressure build-up in the oil film from the three factors outlined above is shown in Figure 22.21. Oil adheres to the moving (and stationary) surfaces, and thus there is flow into the converging volume. Since oil is basically incompressible, pressure builds within the converging oil film. This pressure provides a means for the oil film to transfer the load from the shaft to the pad.

The thickness of the film is of prime importance in the design and operation of a hydrodynamic oil film bearing. For the bearing and associated

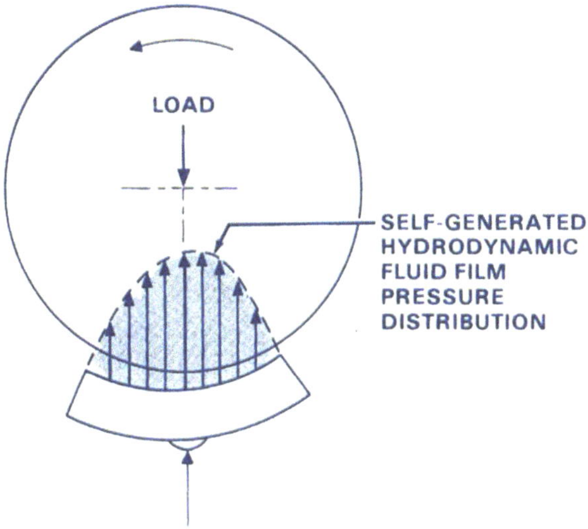

Figure 22.21 Geometry and pressure.

machinery to operate satisfactorily, it is important that the oil film fully separates the shaft and the journal bearing pad surfaces. However, during start-up and shut-down there are momentary periods when the combination of relative speed and load does not generate a full film, and at least some metal-to-metal contact results. Operating conditions such as these dictate the use of combinations of materials, such as babbitt faced journal pads operating against a steel shaft, that allow these contacts to occur without surface damage.

Modern bearings typically use tin base babbitt as the standard bearing material. Though other types of bearing material are available, each with its own advantages and limitations, tin base babbitt meets desired bearing properties such as compatibility, corrosion resistance, conformability and embeddability to such a high degree that it is widely used and accepted (Table 22.2).

Bearing instability is often a factor in selecting journal bearings. Instability refers to the problem of half-frequency whirl of the shaft within the bearing. The most serious form of this condition may occur when the operating speed is near and above twice the first critical speed of the shaft.

On simple journal bearings the displacement of the shaft within the bearing clearance in response to a radial force is not, in general, in line with the direction of that force. This lateral component of the movement of the shaft within the bearing clearance can lead to instability. This problem must be considered on lightly loaded, high speed bearing applications.

Tilting pad journal bearings are widely used because of their stability characteristics. If the load is either directly in line with a pad pivot or directly

Table 22.2 Properties of bearing alloys.

Bearing Material	Hardness Room Temp. Brinell	Min. Shaft Hardness Brinell	Compatibility*	Conformability and Embedability*	Corrosion Resistance*
Tin-Base Babbitt	20-30	150 or less	1	l	1
Lead-Base Babbitt	15-20	150 or less	1	l	3
Copper-Lead	20-30	300	2	2	5
Tin Bronze	60-80	300-400	3	5	2
Lead Bronze	40-30	300	3	4	4
Aluminum Alloy	45-50	300	5	3	l

*Arbitrary scale with 1 being the best material and 5 the worst.

between two pivots, the displacement of a shaft operating hydrodynamically within that bearing will be directly in line with the direction of load on that shaft. Thus there is no component of motion at right angles to the direction of force. Such a bearing is inherently resistant to half-frequency bearing instability.

22.4.1 Instrumentation

Temperature measurement

Changes in load, shaft speed, oil flow, oil inlet temperature, or bearing surface finish can affect bearing surface temperatures. At excessively high temperatures, the pad white metal is subject to wiping, which causes bearing failure. While computer predictions of operating temperature are typically based on extensive empirical data, the algorithms used do include assumptions about the nature of the oil film shape, amount of hot oil carryover, and average viscosity. Consequently, for critical applications, onen often uses pads with built-in temperature to allow monitoring of actual metal temperatures under all operating conditions. Either thermocouples or resistance temperature detectors (RTDs) can be installed in contact with the white metal or in the pad body near the pad body/white metal interface.

Thrust measurement

For bearings subject to critically high loads, continual thrust measurement can provide a vital indication of machine and bearing condition. On many

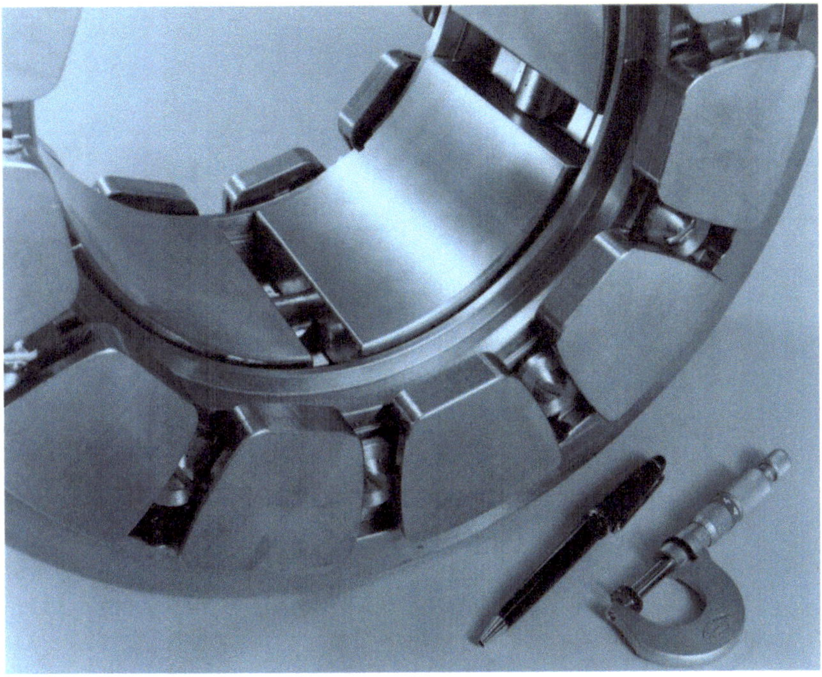

Figure 22.22 Glacier combination thrust/radial tilt pad bearing.

bearing configurations, it is possible to install a strain gauge load cell in one or more places in the bearing.

22.5 Combination Thrust and Radial Bearings

Several manufacturers produce combination bearings similar to the Glacier model depicted in Figure 22.22.

22.6 Plain Bearings

Plain bearings are machine elements transmitting forces between machine elements that move relative to each other.

A distinction is made between:

- hydrodynamic sliding bearings
  where pressure is built up in a converging lubricating gap
- hydrostatic sliding bearings
  where pressure is built up outside the lubricating gap.

- dry sliding bearings
  made of non-metallic or metallic sliding materials
- sintered bearings
  made of porous sliding materials

The service conditions may vary depending on the load and stress factors. Sliding bearings can only function properly when adequately lubricated.

22.6.1 Lubrication of hydrodynamic sliding bearings

Hydrodynamic sliding bearings are used to transfer power from a shaft to a housing via oil-lubricated bearing shells. Apart from an appropriate design, the main prerequisite for the reliable operation of a hydro-dynamic sliding bearing is that its lubrication must be tailored to suit the operating conditions. Except during starting and stopping and during slow-turn operations, a hydrodynamic sliding bearing is subject to the laws of fluid friction described in our introductory chapter.

The selection of a proper lubricant and an adequate viscosity depends on special or additional requirements, such as:

- good adhesion
- good corrosion protection
- self-lubrication with oil
- special bearing shell material
- high speed mixed friction conditions
- high and/or low temperatures
- compatibility with coatings
- compatibility with plastics/elastomers
- extended oil change intervals
- long service life
- small oil quantities
- high loads
- special steel shafts
- compatibility with the environment
- compliance with food regulations

Figure 22.23 Lowering a large Glacier medium wall bearing into a marine pro pulsion gearbox. (Photo: G EC Alsthom Gears Ltd.)

In addition, there is the possibility to optimize hydrodynamic sliding bearings by means of tribo-system coatings or tribo-system materials.

A typical hydrodynamic equipment bearing is shown in Figure 22.23. Here, a medium wall thickness, babbitted liner is fitted to a gear unit. At high speeds and light loads stability becomes a problem. Special bore profiles such as lemon bore, or lobes can give better shaft control and avoid oil film whirl.

Where machines approach or run through critical speeds bearing oil films are often the major source of damping. The right choice of oil viscosity and bearing bore profile (Table 22.3) can significantly reduce vibration amplitude. Also, bearing clearances (Figure 22.24) play an important role.

22.6.2 Sliding bearings in the mixed friction regime

One of the most difficult tasks in terms of tribo-engineering is to lubricate sliding bearings operating in the mixed friction regime. A lubricating

Table 22.3 Bearing bore options.

Bearing type	Resistance to Half Speed Whirl	Load Capacity	Stiffness	Damping	Acceptance of Rotating Load	Relative Cost	Advantages	Disadvantages	Applications	Usage
1 Plain Bore Two Axial Groove	●	●●●●●	●●	●●	●●	○	Simple to install and line up machine	Prone to half speed whirl with light load and may require very low length/ diameter ratio	All rotating machines, gearboxes and screw compressors	Wide
2 Plain Bore Circumferential Single Groove	●	●●●	●●	●●	●●●●●	○	Best for rotating load. Simple to install and line up machine	Requires high oil supply pressure. Prone to half speed whirl with light load	Blowers. Quill shaft steady bearings. some gearbox bearings	Moderate
3 Lemon Bore	●●	●●●●	●●●	●●●	●●	○○	Simple to install and line up machine	Low stiffness in horizonal direction	Large turbines and electrical machines. Pumps, gearboxes	Wide
4 Lemon Bore with bedded arc	●●	●●●●	●●●	●●●	●●	○○	Simple to install and line up machine	Low stiffness in horizontal direction	Large turbines and electrical machines	Moderate
5 Offset Halves	●●●●	●●●●	●●●●●	●●●●●	●●	○○	High lubricant pumping action gives low operating temperatures	Not suitable for reverse rotation	Turbines, pumps, electric machines, compressors, gearboxes	Increasing
6 Tilting Pad	●●●●●	●●●	●●●●	●●●●	●●	○○○	Inherently most stable of all bearings. Adapts to changing load conditions	Pivots liable to fretting with rotating or dynamic load	All rotating machines, horizontal or vertical	Increasing
7 Four Lobe	●●	●●	●●●	●●●	●●	○○	Used mainly in small sizes as a thick wall flanged bearing with taper land thrust faces	Low load capacity due to four grooves, and large lobe clearances	Small turbines Compressors Blowers Pumps	Wide

Performance Rating ●●●●● Very High ● Low Relative Cost ○ Low ○○○ High

wedge cannot form due to the low speed and the oscillating or intermittent movements.

This is where the user may have to seek guidance from manufacturers with experience encompassing lubricating greases, waxes and wax emulsions used in the mixed and partial friction regimes. A lubricating grease is recommended for temperatures above 60°C and for friction points that are relubricated via a lubrication system. A lubricating wax is preferred for small bearings to ensure a non-tacky lubricant film suitable for lifetime lubrication. Tribo-system materials and dry lubricants are an alternative to lubricating greases and waxes.

One such tribo-system material, Klüberdur, is suitable to fill the lubricating holes in metallic bushings, thus making them dry-running bearings. Pre-start lubrication with a lubricating grease or wax could be essential to adequate running-in.

Dry-running bearings (Figure 22.25) can also be manufactured from semi-finished tubes for bushings. A fluid tribo-system material is poured in the tube and the tube is rotated to ensure that the lubricant distributes evenly. The bushings are then cut in length and subject to the finishing process.

Metallic bushings can be coated with a "Klüberplast" sheet thus making them dry-running bearings. Metallic bearings with damaged surfaces can be

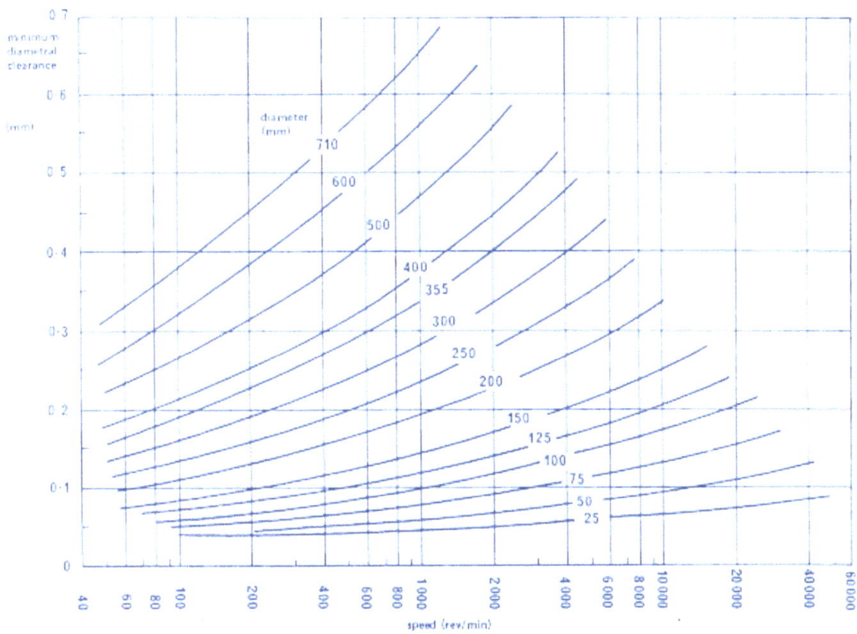

Figure 22.24 Hydrodynamic bearing clearances: recommended minima against speed by shaft diameter.

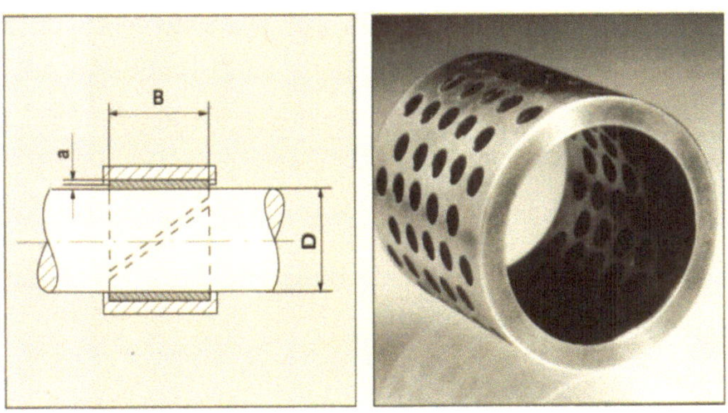

Figure 22.25 Dry-running bearings: sheet bearing (left), having lubricating holes filled (right) with "Klüberdur" tribo-system material.

repaired easily with such a coating. Except for a lateral rib of approx. 0.3 mm to guide the sheet, the bearing material is cut off in a layer over the entire width. Dry lubricants for tribo-systems ca be used to impart a running layer to smooth metal or plastic bushings.

Table 22.4 Porosity of sintered metal sliding bearings.

Material category	Porosity %	Preferred application
Sint A	25 ± 2.5	Sliding bearings
Sint B	20 ± 2.5	Sliding bearings, molded parts with sliding properties
Sint C	15 ± 2.5	Sliding bearings, molded parts

22.6.3 Lubrication of sintered metal sliding bearings

Sintered bearings are made of powder composites subject to pressure and heat. Depending on the material composition (sintered iron, steel, bronze), sintered metal sliding bearings have a different porosity. This is illustrated in Table 22.4.

Sintered metal sliding bearings have open pores which are filled with a lubricant in an immersion process. They are not operational without a lubricant and are therefore generally lubricated for life. The better the lubricant fulfills its task, the longer the bearing's service life.

The high requirements, in terms of temperature stability, corrosion and wear protection as well as oxidation resistance, are met even under difficult operating conditions, such as:

- low and/or high temperatures
- low and/or high speeds
- low noise
- low starting and running torque
- uniform operation
- very long service life
- high or low specific surface pressure

Lubricated sintered metal sliding bearings have proven effective in practical applications. Their special properties, such as low-noise behavior, high load-carrying capacity, low friction moments, constant friction values during speed changes, etc. are characteristic of impregnating fluids.

If the service life of a sintered metal sliding bearing should be increased considerably, additional lubrication with an appropriately selected specialty lubricant provides substantial advantages as compared to felt or depot grease lubrication.

23

Machine Element(s) Lubrication

23.1 Lubrication of Fastener Screws

Although mechanical fastening screws are the most frequently used detachable fastening ellement, their lubrication is often neglected. Damage due to insufficient or inadequate lubrication can easily lead to component failure, resulting in expensive maintenance or production losses.

A screw threaded connection is a power locking connection. Therefore, the criterion for the quality of a screw connection is the required preloading force which determines the extent to which the joined components are pressed together. Screwed connections that are too highly preloaded tend to fail during assembly due to elongation or breaking (Figure 23.1). If the preload is too low, the connections will fail during operation due to fatigue fracture or unintentional releasing.

The torque-controlled tightening methods currently used most frequently generate the preloading force via the tightening torque.

Apart from the assembly method, the friction behavior has a substantial impact on the clamping effect. For example, up to **90%** of the applied torque is consumed in the form of friction related to the screw head and thread. Only 10% is available to build up the required preload.

Adequate lubrication (Figure 23.2) in the form of a specially formulated screw thread lubricant will reduce head and thread-related friction; increase the available screw preload force, and optimize the functional reliability of the connection. In addition, it will ensure that the connection will not seize over time can be released without any damage.

Apart from lubricating pastes, competent lube manufacturers also offer dry lubricants for tribo-systems and tribo-system coatings for the lubrication of screws.

Appropriate screw thread lubricant products are designed to meet the following requirements:

- minimization of torsional stress

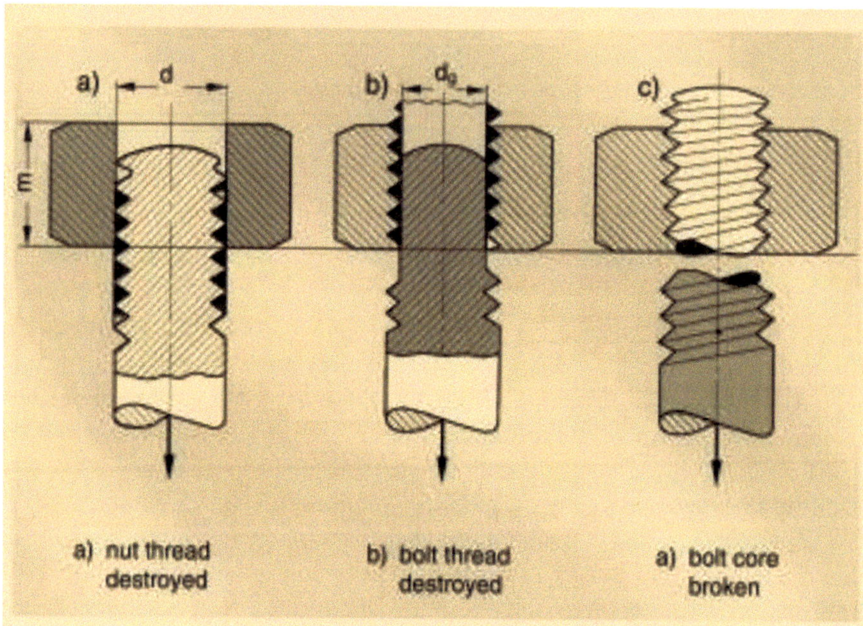

Figure 23.1 Screw connections that are too tightly preloaded tend to fail.

- constant tightening and breakaway torques
- extended corrosion protection
- resistance to aggressive media
- protection of screw at high temperatures (prevention of thread welding)
- easy and clean application

Screw lubricants are often referred to as anti-seize products and are available in many formulations based on the industrial requirement/environment in which they are being used. Table 23.1 depicts a number of available lubricating compounds and their uses.

23.2 Lubrication of Valves and Fittings

Valves and fittings are integral elements of pipe systems designed to fulfill a

- control (open/close) and
- adjustment (mixing, etc.) function.

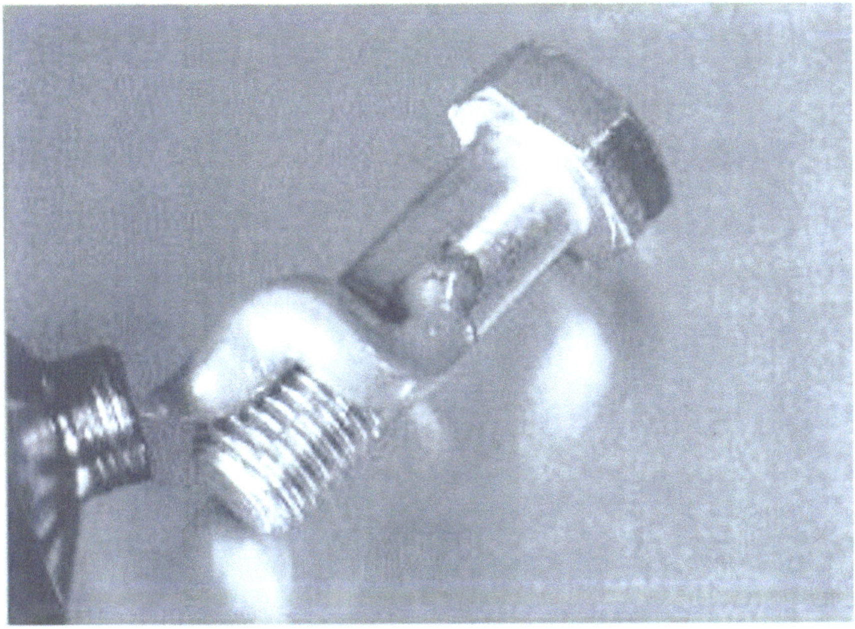

Figure 23.2 Screw anti-seize lubricant shown in tube form.

They are used in pipes transporting solids, fluids and gases.

An adequate lubricant on the individual valve components (e.g., seal, stuffing box, spindle, ceramic disks, plug, as illustrated in Figure 23.2, provides smooth operation and sealing, and minimizes wear.

To ensure long-term operational reliability of valves, lubricants should meet the following requirements:

- Resistance to ambient media
- Securing effect
- Neutrality towards other materials (metals, plastics, elastomers)
- Compliance with food regulations.

These regulations, or their appropriate US or other national codes, should be the equivalent of:

- BAM* test for an application in oxygen installations

* BAM German Federal Institute for Materials Research and Testing

Table 23.1 Typical Anti-seize thread lubricant compounds*.

Type	Description
Aircraft	A zinc and petrolatum-based compound designed for threading steel fasteners into aluminum or zinc castings. Designed to meet military and civil aviation standards that include MIL-T-5544, SAE AMS-2518, NATO S-720. MIL-T-23361 and A-A-59313 requirements
Food/Drug	Designed to meet USDA standards for incidental food contact
Nuclear	A nickel grade anti-seize product designed to minimize sulphur, halogens, and mercury. Meets MIL-A-907D and MIL-PRF-907E requirements
Marine	Designed to prevent rust and galling when in contact with fresh or salt water
Copper based	Standard industrial fastener lubricant, copper and graphite based
Nickel	Compounded from nickel, aluminum and graphite for use in high heat conditions where different fastening metals are employed e.g., brass/steel
Non-Metallic	A metal free lubricant used when copper must be avoided. Excellent for titanium, SS and many alloy steels
Moly (MOS2)	Used for high pressure (up to 500,00 psi) applications
Stainless Steel	A lithium soap-based lubricant with stainless steel and aluminum compounds

*Source: Saftlok Corporation. Lombard, IL, USA – SAF-T-EZE Anti-Seize literature

- DVGW† approvals in accordance with the pertinent drinking water regulations for an application in sanitary and drinking water valves
- DVGW approval in accordance with DIN 3536 for an application in gas installations

† DVGW German Association of Plumbers—Regulations pertaining to synthetic materials in drinking water installations

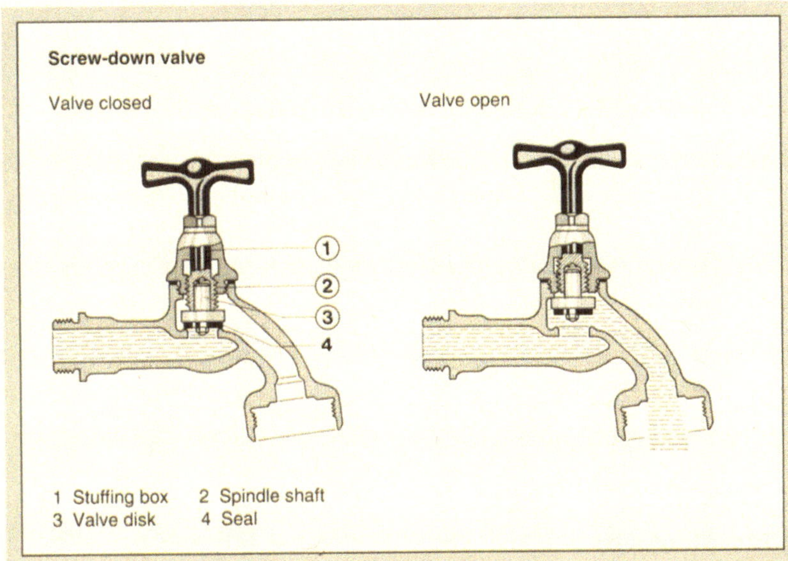

Valves (① ② ③ = lubricated)

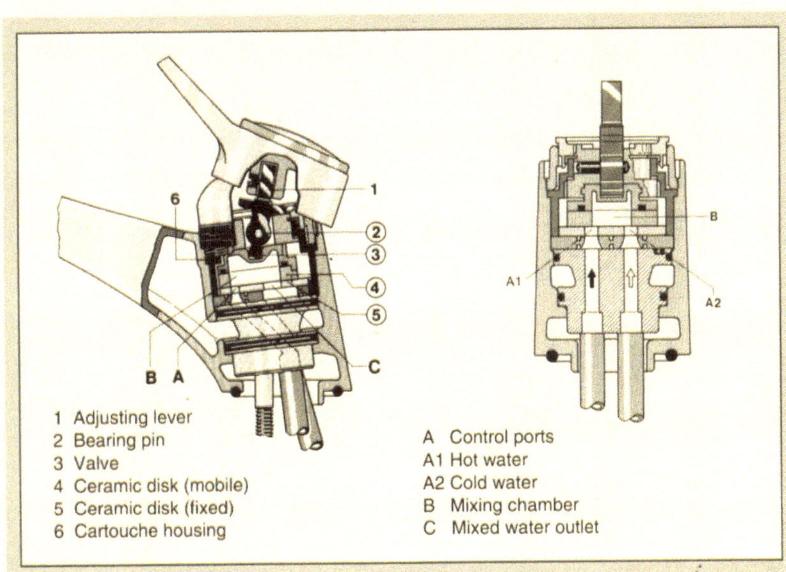

Single-lever mixer tap with ceramic disks (② - ⑤ = lubricated)

Figure 23.2 Valve components may require lubrication.

- WRC¶ approval for use in potable water supplies

23.3 Lubrication of Electrical Switches and Contacts

Switches are components consisting of one or more electrical contacts. They are actuated mechanically, thermally, electromagnetically, hydraulically or pneumatically.

Their function is to separate or close electric circuits even when subjected to high loads (Figure 23.3).

Lubricants applied in switches ensure the following advantages:

- increased life cycle
- protection against wear
- reduced switching pressure
- reduced switching noise
- reliable contacts
- protection against corrosion
- prevention of fretting
- reduction of friction forces

Competent lubricant manufacturers can offer special formulations that meet all of the following requirements and even surpass them in many respects:

- high affinity towards metals
- compatibility with plastics
- thermal stability
- purity
- constant high quality
- excellent aging resistance
- easy application

Detachable and/or movable contacts such as switches, sliding and plug-in contacts are interesting from a tribological point of view. They are normally

¶ WRC Water Research Council

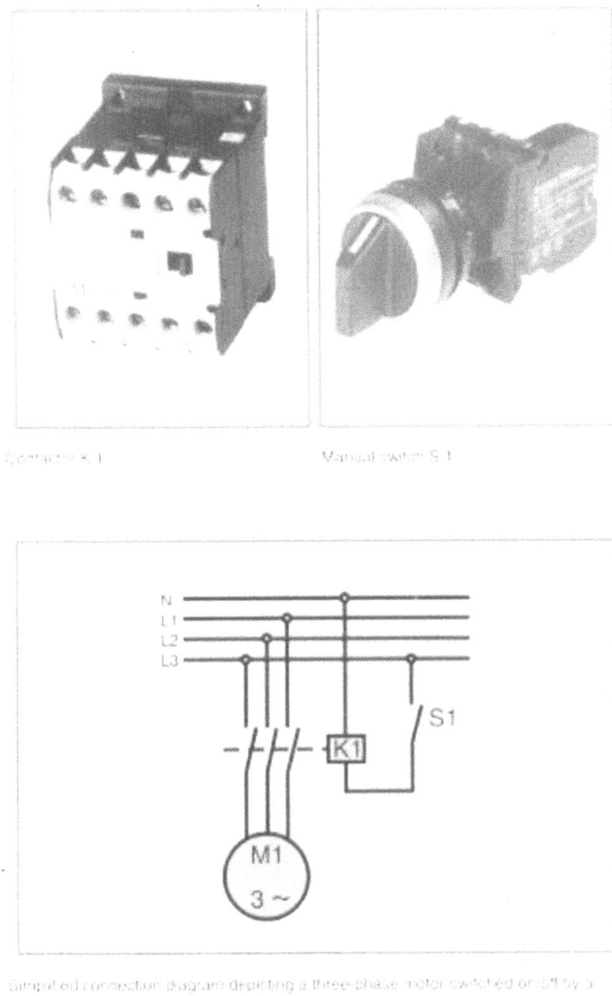

Figure 23.3 Some electrical switches and contactors must be lubricated.

made of metal alloys plus a coating depending on the application (Figure 23.4 and 23.5). Their geometries vary according to the intended use.

All contacts must be able to conduct electric power. In addition, switches must interrupt, close and insulate an electric circuit.

The resulting loads and requirements can only be met with special lubricants:

- reduction of plug-in forces

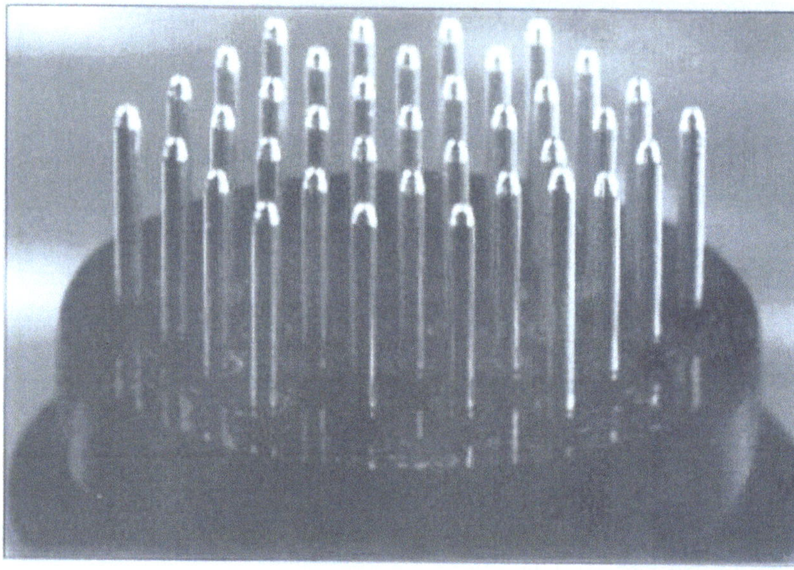

Figure 23.4 Plug contacts.

- avoidance of tribo-corrosion
- protection against oxidation
- protection against wear
- high number of actuations

Moreover, these products must also meet other important requirements such as

- excellent aging stability
- high affinity towards metals
- thermal stability
- compatibility with plastics
- purity

23.4 Lubrication of Industrial Springs

All units made of an elastic material are resilient. This capacity can be utilized by imparting a special shape.

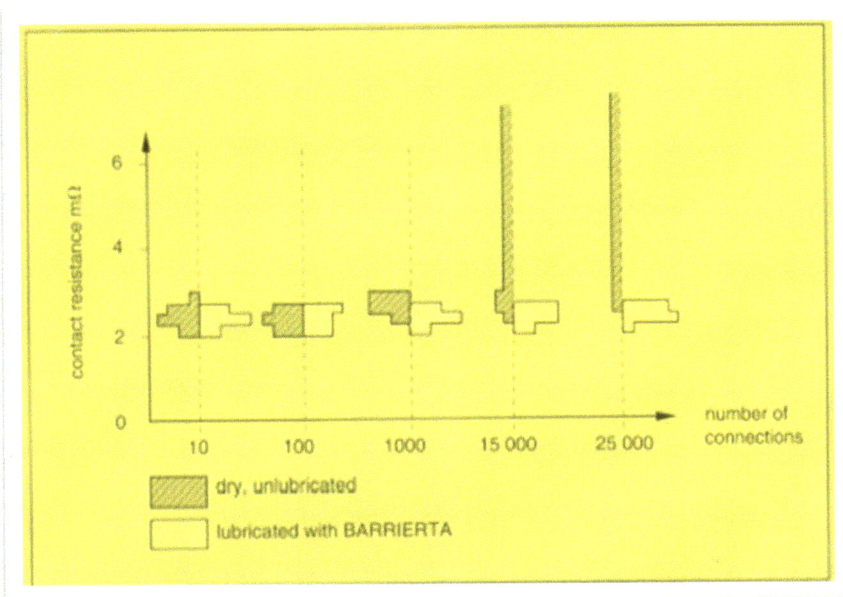

Figure 23.5 Advantages of a lubricated, gold-plated plug contact as a function of the number of connections.

Springs are used to store energy, restore components to their former position, absorb shocks, distribute, limit or measure power, maintain power-locking connections, and function as a vibration suppression element.

Coil springs and, especially, annular plate springs or belleville washers (Figures 23.6), have to meet certain requirements in terms of material and shape. Their performance can be improved by applying special lubricants and/or tribo-system coatings.

Depending on the operating conditions, an adequate lubricant will optimize the work consumed by friction.

The lubricant has to meet the following requirements:

- reduction of wear
- good adhesion in case of vibration and shock
- low friction even at extremely low or high temperatures
- excellent protection against corrosion
- good behavior if used in conjunction with plastics and elastomers
- efficient protection against tribocorrosion

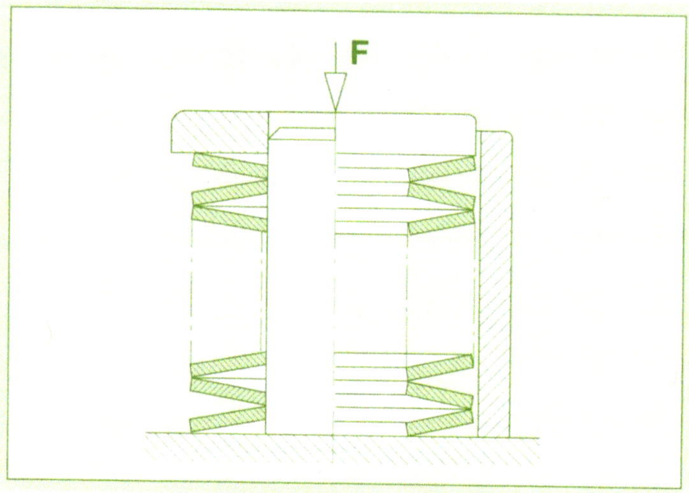

Plate spring assembly internally, externally guided

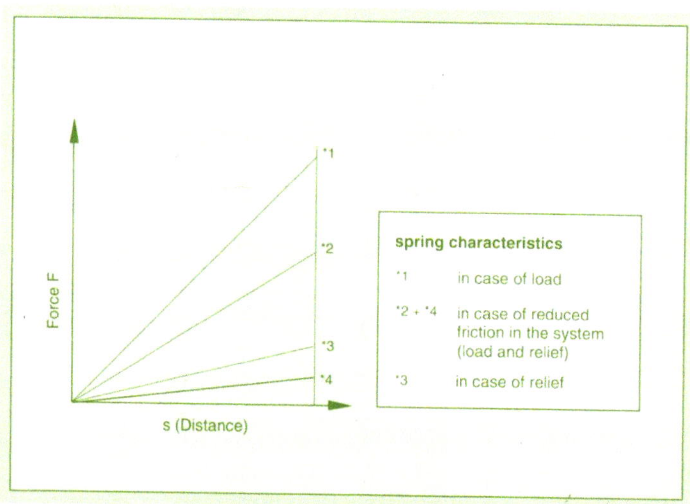

Annular spring

Figure 23.6 The performance of plate and annular springs can be improved with appropriate lubrication.

- favorable behavior towards non-ferrous metals
- compliance with food and water regulations
- uniform transmission of power

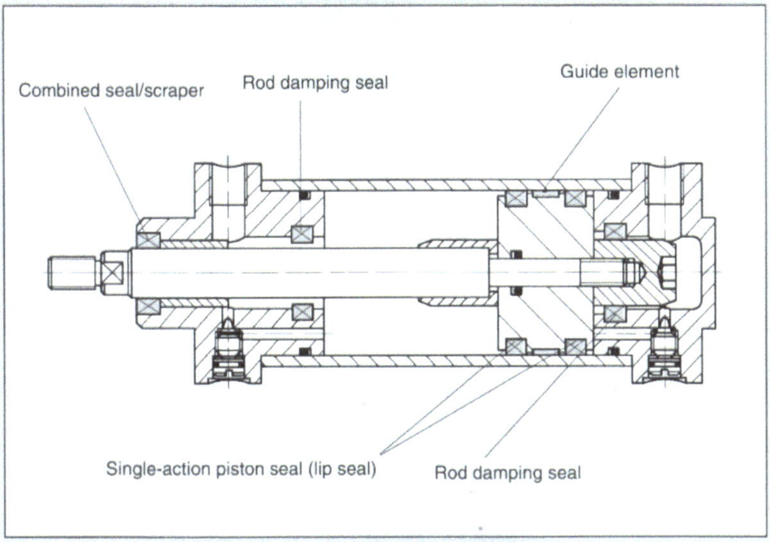

Figure 23.7 Pneumatic cylinder.

- protection against aggressive media
- prevention of stick-slip

By utilizing select lubricants it is possible to optimize individual components and even entire systems.

23.5 Lubrication of Pneumatic Components

Pneumatic cylinders (Figure 23.7), and valves (Figure 23.8) are components used in pneumatic systems. Pneumatic cylinders convert pneumatic into mechanical energy which is subsequently used to perform linear movements to move, lift or return workpieces or tools

By controlling starts, stops, directions, pressure and throughput, pneumatic valves ensure that the pressurized air carrying the energy follows the "right paths."

To ensure functional reliability it is indispensable to apply a prestart lubricant to all components performing a relative movement, for example the piston rod, pressure tube, valve elements and seals. If required, air compressor oils can be applied during operation.

Premium lubricants for pneumatic components must ensure

- an optimum sealing effect and improved efficiency

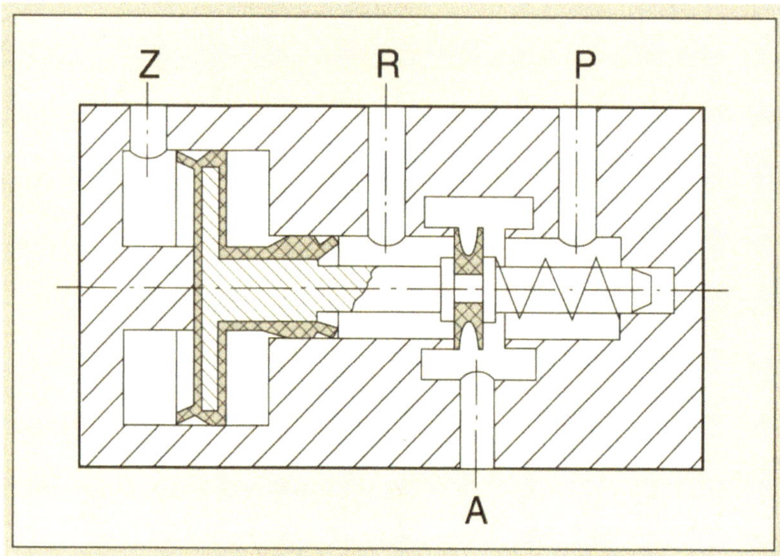

Figure 23.8 Pneumatic valves.

- increased performance
- operation without stick slip (e.g., in case of feed movements with low pistons speeds and long strokes)
- low breakaway moments (also after extended standstill periods)
- excellent adhesion and wetting properties on materials such as steel, refined steel, aluminum, brass, ceramic materials, plastics and elastomers

23.6 Lubrication of Shaft-hub Connections

Shaft/hub connections are positive or friction/power-locking connections designed to transfer torques Positive-locking connection make it possible to move the hub and the shaft in an axial direction.

A lubricant used in shaft/hub connections has to meet various requirements. The main task, however, is to prevent fretting and tribo-corrosion which would have a negative effect on the surface of the friction components. Tribo-corrosion often occurs in positive and power-locking machine elements.

Tribo-corrosion is a generic term describing the physical and chemical influences on materials. Small relative movements (micro-sliding) in the contact zone mechanically excite the surface layers, resulting in a strong reaction

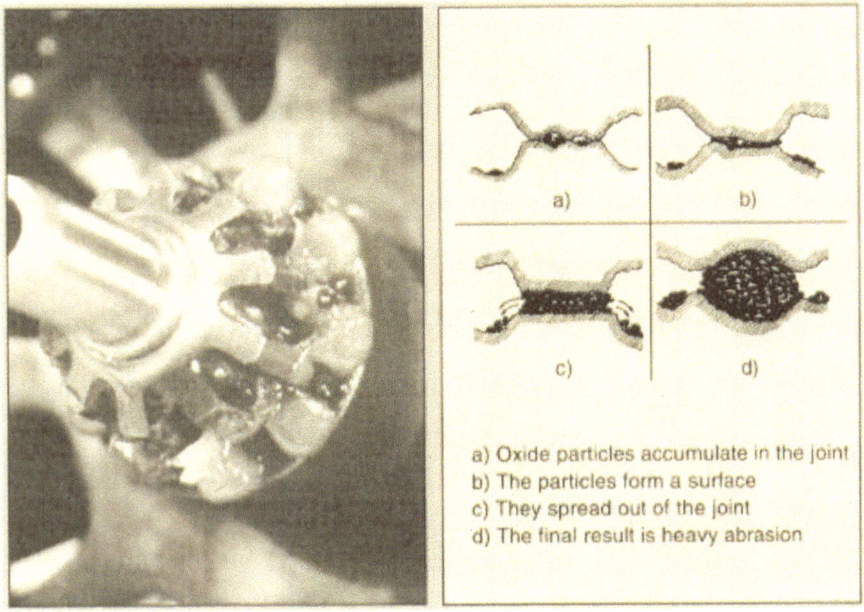

Figure 23.9 Lubrication of a multiple spline shaft and formation of fretting corrosion.

of the component material and the atmospheric oxygen. Oxidation products (wear particles) accumulate in the joints which, unless removed, will lead to malfunctions and have an impact on the axial sliding movements (Figure 23.9).

Tribo-corrosion is caused by the following load factors:

- vibrations
- micro-sliding
- oscillations
- condensation water
- atmospheric oxygen
- torque changes

Specialty lubricants, designed specifically for shaft-hub connections control these factors to such an extent that functional reliability is ensured. They also provide advantages during assembly and disassembly by facilitating:

- pressing-in,
- sliding-on,

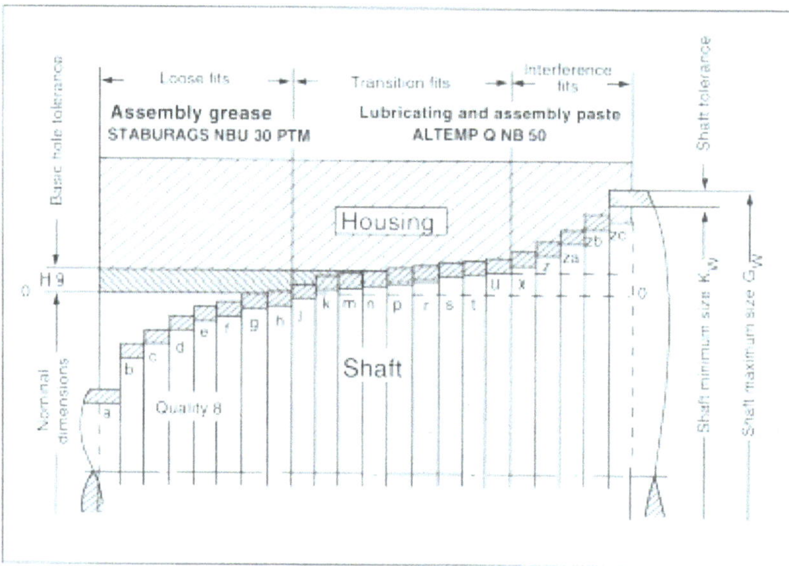

Figure 23.10 Loose fits require different lubricants than interference fits. (Source: Klüber Lubrication North America, Londonderry, New Hampshire.)

- pressing-out, and
- sliding off of the components.

For the selection of an adequate lubricant, it is important to take into consideration the type of shaft/ hub connection and the bearing fit (interference, transition or loose fit). As can be seen on Figure 23.10, the friction-locking conditions of interference fits require a different lubricant than the transition or loose fit in case of axial sliding movements.

24

Industrial Gear Lubrication

Gears are arguably one of man's oldest mechanical devices, synonymous with mechanical machinery. Gears create the impression of positive action; the coordinated-interlocked-precise application of effort to secure a desired result. Primary early use of gears was for navigation, timekeeping, grinding, etc. The automobile transmission is probably the most common use of gearing for the everyday citizen.

Gears are machine elements that transmit motion by means of successively engaging teeth (See Figure 24.1). Of two gears that run together, the one with the larger number of teeth is called the gear. A pinion is a gear with the smaller number of teeth. A rack is a gear with teeth spaced along a straight line and suitable for straight-line motion. Many kinds of gear teeth are in general use. For each application, the selection will vary depending on the factors involved. One basic rule is that to transmit the same power, more torque is required as speed is reduced. The torque is directly proportional to speed; therefore, the input and output torque for power transmissions are directly proportional to the ratio.

Gear designers must do more than just provide a mechanism that will develop the required speeds. They must endure the mechanism does not break or wear out prematurely due to the power being transmitted. Needless to say, gears must be lubricated, and the oil must be kept clean.

24.1 Lubricant Selection for Closed Gears*

Closed gears are used in many different arrangements, Figures 24.2 and 24.3. Factors affecting lubricant selection for the various arrangements are shown below (Table 24.1). and should be reviewed and evaluated to determine the required properties.

* James R. Partridge, Lufkin Industries, Lufkin, Texas; also Exxon Company, USA, Houston, Texas.

Figure 24.1 Single-helical, high-speed gear unit being assembled. (Source: Maag Gear Company, Zurich, Switzerland)

Also, Tables 24.2 and 24.3, taken from AGMA 250.04, show recommended viscosity ranges based on gear center distance. These recommendations should be used with caution since they are very loosely written.

Viscosity is probably the single most important factor in lubricant selection and relates to load, speed, and temperature.

Table 24.4 is a general guide based on the required viscosity in relation to the Operating "K" factor of the gears and pitch line speed. However, required viscosities can also be calculated from two empirical expressions:

$$V_g = 420\ (K/V)^{0.43}$$

Where: V_g = Viscosity, Centistokes
K = Operating '"K" Focor

$$k = \frac{W}{dF}\pi\left(\frac{m_G - 1}{m_G}\right)$$

W = Tangential Load
d = Pitch Diameter of Pinion
F = Effective Face Width

Figure 24.2 Double-helical, low-speed gear unit being checked at Lufkin Gear Company, Lufkin, Texas.

mG = Ratio
V = Pitch Line Velocity, ft./min.

and,

$$SSU = \frac{Vg + \left(V_g^2 + 158.4\right)^{.5}}{0.44}$$

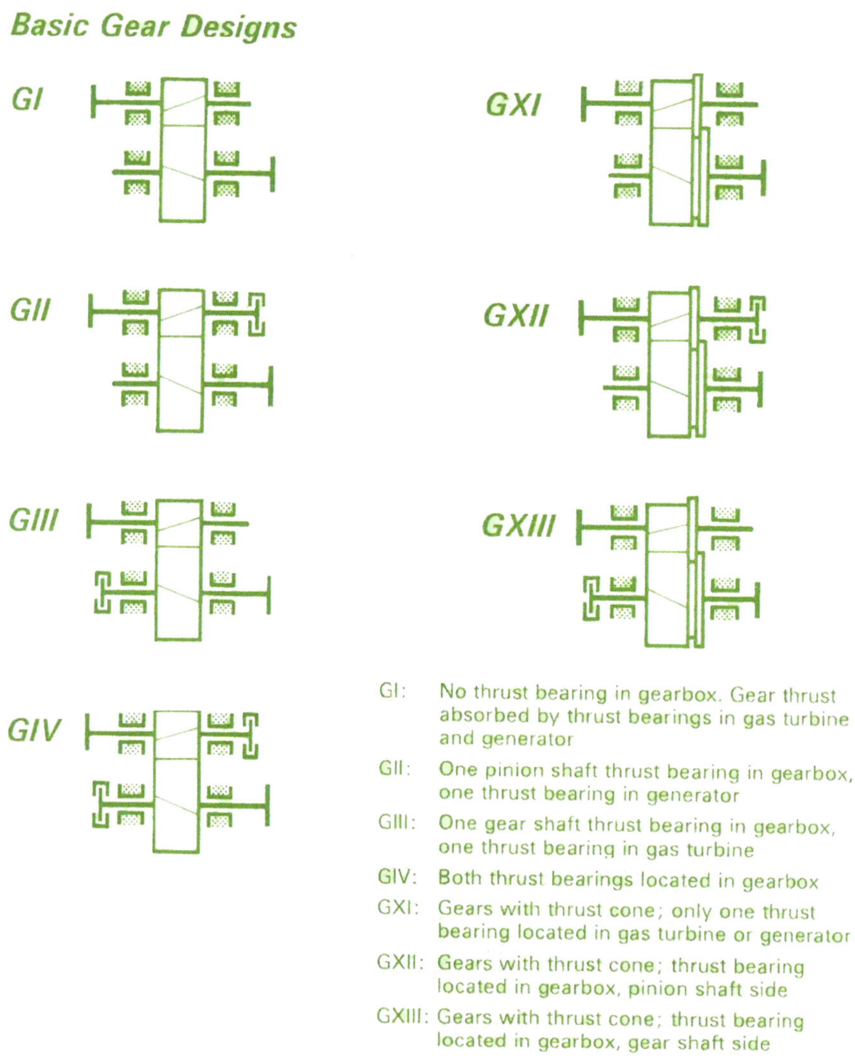

Figure 24.3 Basic gear designs. (Source: Maag Gear Company, Zurich, Switzerland)

formula was first published by shell, and all values were converted to SSU.

It should be noted that by this formula the high-speed gears (above 5000 FPM) require a heavier oil than the 150 SSU @100°F usually used, but compromises are made for bearings and sometimes seals.

As a general rule that for high speed gearing, the minimum viscosity at supply temperature should be 100 SSU.

Table 24.1 Lubricant selection factors.

Factor	Lubricant Properties
Load	Viscosity P Additives
Speed	Viscosity EP Additives
Temperature (Operating and Ambient)	Viscosity Viscosity Index Fluidity Oxidation Stability
Contamination	Demulsibility Corrosion Protection Oxidation Stability
Life	Oxidation Stability Additive Depletion
Compatibility	Synthetic (Paint and Seals) EP Additives

24.1.1 Film Thickness

Several authorities have stated that film thickness is a function of operating speed only. Based on this theory, the following formula can be used as a guide which was derived from experimental results by Crook and Arcard.

$$h_{min} = 0.33d\ V_g\ (n_p)\left(\frac{m_G}{m_G + 1}\right)^{.5}$$

h = Film thickness (micro-inches)
d = P.D. of pinion
V_g = Viscosity, Centistokes at gear blank temperature
n_p = Pinion speed, RPM
m_G = Ratio

Table 24.2 Viscosity ranges for AGMA lubricants.

Rust and Oxidation Inhibited Gear Oils	Viscosity Range ASTM System	Extreme Pressure Gear Lubricants
AGMA Lubricant No.	SSU @ 100°F.	AGMA Lubricant No.
1		193 to 235
2		284 to 347 2 EP
3		417 to 5 10 t EP
4		626 to 765 4 EP
5		918 to 1122 5 EP
6	13t5 to 1632	6 EP
7 comp.	1919 to 2346	7 EP
8 comp.	2837 to 3467	8 EP
8A comp.	4171 to 5098	

Using this formula, a 5" pinion running with a 20" gear has a film thickness of 385 and 640 microinches at 3,600 and 10,000, respectively, using 100 SSU oil at the mesh.

Due to the variables involved, film thickness calculation procedures are useful only for design comparisons and should not be used to decide that a particular gear set will not work.

24.1.2 Lubricant types

Mineral oils

Although synthesized hydrocarbons (diesters and PAOs) are rapidly gaining wide-spread acceptance, mineral oils are still the most commonly used type of gear lubricant. Containing rust and oxidation inhibitors, these oils are less expensive, readily available, and when properly cared for can have very long life. Mineral oils are ideal when gear units operate at high speed and/or low load intensity.

Extreme pressure additives

Extreme pressure (EP) additives of the lead-napthenate or sulphur-phosphorus type are recommended for high load gear drives applications. As a general rule, this type of oil performs best in low speed, highly load drives, with a medium operating temperature.

Table 24.3 Equivalent Viscosities of other systems (reference only).

AGMA lubricant no.	Equivalent ASTM-ASLE grade no.	Equivalent ISO grade no.	Metric equivalent viscosity ranges Cst at 37.8°c (100°f)
1	S215	46	41.4 to 50.6
2, 2 EP	S315	68	61.2 to 74.8
3, 3 EP	S465	100	90 to 110
4, 4 EP	S700	150	135 to 156
5, 5 EP	S1000	220	198 to 242
6, 6 EP	S1500	320	288 to 352
7 comp., 7 EP	S2150	460	414 to 506
8 comp., 8 EP	S3150	680	612 to 748
8A comp.	S4650	1000	900 to 1100

It should be remembered that EP oils are more expensive, and must be replaced more often than straight mineral oils. Some of these oils can have a very short life above 160°F (71°C) temperatures.

A good gear EP oil will have a Timken OK load above 60 pounds and pass a minimum of 11 stages of an FZG test.

Synthetic lubricants

Not to be confused with the highly desirable synthesized hydrocarbons (typically diesters and PAOs), true "synthetic" lubricants are not recommended for general gear applications due to cost, availability, and lack of knowledge of their properties. In extreme applications of high or low temperature and fire protection, they are used. The user must be careful when selecting these lubricants since some of them remove paint and attack rubber seals.

Synthesized hydrocarbons (SHC) have many desirable features such as compatibility with mineral oils and excellent high and low temperature properties. These oils can be an excellent selection when EP lubricants are required for use in high temperature applications.

Compounded oils

Compounded oils are available with many different additives. The most commonly available is a molydenum disulfide (MOS2) compound that has been successfully used in gear applications. It is difficult for a gear manufacturer to recommend these oils since some of these additives have a tendency to separate from the base stock.

Table 24.4 Viscosity, SSU @ 100°F.

	V = Pitch Line Velocity, FPM							
K	**50**	**100**	**200**	**500**	**1000**	**5000**	**10000**	**15000**
50	1900	1400	1050	700	500	250	200	175
100	2600	1900	1400	950	700	350	250	225
150	3100	2300	1700	1150	850	425	325	250
200	3500	2600	1900	1300	950	475	350	300
250	3800	2800	2100	1400	1050	525	400	325
300	4100	3100	2300	1500	1150	575	425	350
400	4700	3500	2600	1700	1300	650	475	400
600	5500	4100	3100	2100	1500	750	575	475

Viscosity improvers

Viscosity improvers in gear drives are to be used with caution. These polymer additives do great textbook things for the viscosity index and extend the operating temperature range of an oil. What must be remembered is that polymers are non-Newtonian fluids, and the viscosity reduces with shearing. A gear drive is a very heavy shear application; and as a result, the viscosity reduces rapidly if too much polymer is used.

Lubricants in gear units have basically two functions: 1) to separate the tooth and bearing surfaces, and 2) cooling. On low-speed gear units, the primary function is lubrication; on high-speed units, the primary function is cooling. This does not mean that both are not important but relates to the relative quantity of oil.

On low-speed units, the amount of oil is determined by what is required to keep the surfaces wetted. On high-speed units, quantity is generally determined by heat loss (or inefficiency). As a general rule, one GPM must be circulated for each 100 HP transmitted, which results in a temperature rise of approximately 25°F. Higher HP units use a 40°F to 50°F temperature rise and require .5 to .6 GPH per 100 HP transmitted. This is based on a 98% efficiency.

24.1.3 Lubrication of high-speed units

Oil designed for highs-speed gear drives has a dual purpose: 1) lubrication of gear teeth and bearings, and 2) cooling. Generally, between 10% - 30% of the oil is used for lubrication, with the remaining 70% - 90% used for cooling purposes.

A turbine type oil with rust and oxidation (R&O) inhibitors is preferred. This oil must be kept clean (filtered to between 40 – 25 microns), cooled, and must have the correct viscosity. Synthetic oils should not be used without the manufacturer's approval.

For some reason, the high-speed gear makes all the compromises when oil viscosity for a combined lube oil system is determined. Usually, a viscosity preferred for compressor seals or bearings is selected and gear life is probably reduced. Although bearings in a gear unit can use the lightest oils available, gear teeth ideally prefer heavier oil so as to increase the film thickness between the meshing teeth.

When selecting a high-speed gear unit, the possibility of using an AGMA No. 2 Oil (315 SSU @ 100°F) should be considered. In most cases, the sleeve bearings in the system can use this oil and, if not, a compromise 200 SSU at 100°F oil should be considered.

When 150 SSU at 100°F oil is necessary, inlet temperatures should be limited to 110°F to 120°F (43°C – 49°C) to maintain an acceptable viscosity. Oil should be supplied in the temperature and pressure range specified by the manufacturer.

Up to a pitch line speed of approximately 15,000 feet per minute, the oil should be sprayed into the out-mesh. This allows maximum cooling time for the gear blanks and applies the oil at the highest temperature area of the gears. Also, a negative pressure is formed when the teeth come out of mesh pulling the oil into the tooth spaces.

Above approximately 15,000 feet per minute, 90% of the oil should be sprayed into the out-mesh, and 10% into the in-mesh side. This is a safety precaution to assure the amount of oil required for lubrication is available at the mesh.

In addition to the above, in the speed ranges from 25,000 to 40,000 feet per minute, oil should be sprayed on the sides and gap area (on double helical) of the gears to minimize thermal distortion.

23.1.4 Types of lubrication regimes found in gear teeth

Boundary lubrication

Boundary lubrication most often occurs at slow to moderate speeds, on heavily loaded gears, or on gears subject to high shock loads. The oil film is not thick enough to prevent some metal-to-metal contact. This condition usually shows some early wear and pitting due to surface irregularities on the tooth surfaces.

When boundary lubrication is encountered, extreme pressure oils should be used to minimize wear and possible scuffing.

Hydrodynamic lubrication

Hydrodynamic lubrication occurs when two sliding surfaces develop an oil film thick enough to prevent metal-to-metal contact. This lubrication regime is prevelant on higher speed gearing with very little shock loading.

Elastohydrodynamic (EHL) lubrication

Elastohydrodynamic theory of lubrication is now accepted as very common in gear teeth. The formation of EHL films depends on the hydrodynamic properties of the fluid and deformation of the contact zone. This flattening of the contact area under load forms a pocket that traps oil so that the oil does not have time to escape resulting in an increase in oil viscosity. This increase makes it possible to use light viscosity oils in high-speed drives operating at speeds greater than 12,000 FPM.

23.1.5 Methods of supplying lubricant

Splash lubrication

Splash lubrication is the most common and fool-proof method of gear lubrication. In this system type, the gear teeth dips into the reservoir oil and in turn, distribute oil to the pinion and to the bearings. Distribution to the bearings is usually obtained by throw off to an oil gallery or is taken off by oil wipers (or scrapers) which deliver the oil to an oil trough.

Care must be taken that the operating speed is high enough to lift and throw off the oil.

In the throw-off system, the minimum speed, n_p is:

$$n_p = (70{,}440/d)^{.5}$$

Where:

d = Pitch diameter

n_p = RPM

Oil wiper systems can operate at much lower speeds which are generally determined by test. The splash system can be used up to 4000 FPM pitch line velocity. Higher speeds can be splash lubricated with special care.

Forced feed lubrication

Forced feed lubrication is used on almost all high-speed drives and on low-speed drives when splash lubrication cannot be used due to gear arrangement.

Table 24.5 Typical design attributes of a gearbox sump lubrication system.

• Large reservoir	• Safety alarms and shutdowns (Temp. and Pressure)
• Filters (Duplex or Single)	• Temperature Regulators
• Shaft driven pump	• Isolation valves
• Auxiliary Pumps (Motor and Steam)	• Heaters (Steam or Electric)
• Heat exchangers (Single or Duplex)	• Purifier (Removes water and oxidation products)
• Accumulators	• Flow Indicators
• Pressure control devices	

A simple forced feed system consists of a pump with suction line and supply lines to deliver the oil. However, lubrication supply systems for high-speed drives include many of the components listed in Table 24.5.

Many of these systems are well designed and constructed for optimum performance.

Scoring or scuffing (adhesive wear) is caused when the oil film does not prevent contact between mating surfaces. Areas touch each other due to load which results in welding of the two surfaces. As sliding continues, these surfaces break apart. These particles adhere to the surfaces, and rapid adhesive wear occurs. (See Chapter 1, Figure 1.11 and 1.12)

The flash temperature theory of this type failure indicates that the welding is caused by the high temperature generated locally in the contact area. Calculations can be made to determine the scoring risk for higher speed drives but data are not available for all oil types.

Pitting or surface fatigue comes from the formation of small sub-surface cracks that are developed by fatigue failure of the tooth surface under repeated load. As fatigue failure progresses, the surface begins to break up, and pits form. Pitting usually starts close to the pitch line in the dedendum area. (See Chapter 1, Figure 1.13)

If a gear is operating above the basic strength of the gear material, no lubricant can prevent pitting. Some pitting is corrective in nature and up to a point, is not detrimental to the gearing. Pitting in case hardened gearing usually leads to failure. Extreme pressure oils and higher viscosity can help reduce pitting.

Abrasive wear is usually caused by a very rough surface finish on the gear teeth or foreign particles in the oil. The foreign particles adhere temporarily to one surface and in turn, scratch a groove in the second surface.

Figure 24.4 Open Girth gear drive of a drying cylinder.

(See Chapter 1, Figure 1.9 and 1.10) Generally, there is very little problem with abrasive wear if the lubricant is clean.

24.2 Lubrication of Large Open Gears

Large open gears (Figure 24.4) are toothed gear systems, i.e., the gear and the pinion are not situated in a joint housing. The drive cover frequently is not oil-tight. Large open gears are mainly used in the base material industry, for example in ore and raw material processing installations, fertilizer, waste incineration and composting plants, coal-fired plants, rotary kilns, tube mills, drying, cooling and conditioning cylinders.

As the pinion and the gearwheel are supported separately, and owing to the low peripheral speed, extremely high flank load and surface roughness, such gears are mainly operated in the mixed friction regime. To ensure operational reliability it is therefore necessary to apply special adhesive lubricants having specific physical and chemical properties to form a protective layer on the tooth flanks and avoid direct contact between the metal surfaces. Competent vendors often recommend the application of running-in lubricants to rapidly reduce surface roughness when putting the gear into operation and achieve a good load distribution over the flanks and the faces.

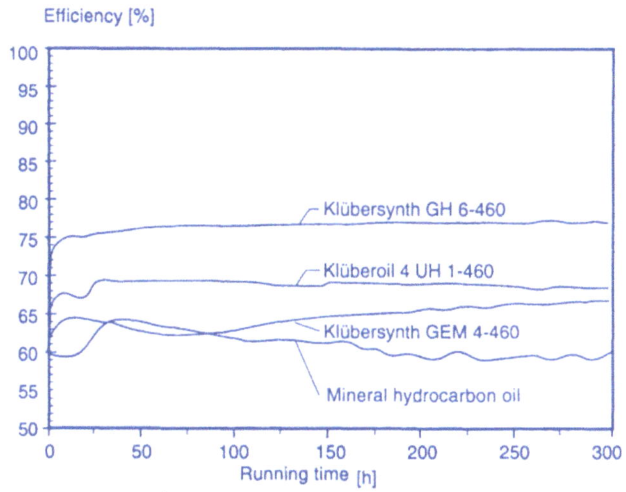

Figure 24.5 Efficiency of mineral hydrocarbon oil compared to synthetic oils. (Source: Kluber Lubricants, Germany)

24.3 Lubrication of Worm Gears

Worm gears are recognized by their cross-axis geometrical design. They have a constant transmission ratio and are used as speed and torque converters between driving engines and machines. The high sliding percentage of worm gear engagement ensures low-noise and low-vibration operation.

As compared to other gears (e.g., bevel gears), the efficiency of worm gears is relatively poor. However, they make high transmission ratios possible in a single step.

Synthetic lubricants are used to reduce the friction and power loss in worm gears up to 30%. The operational wear of the worm wheels, which usually consist of copper bronze, can be reduced substantially with synthetic lubricants containing suitable additives.

The use of synthetic lubricants in worm gears results in an improvement of gear efficiency and service life and makes them suitable for many applications.

Properly formulated synthetic gear lubricants do not only improve gear efficiency; their anti-wear additives can optimize the gear 's wear behavior. Figures 24.5 and 24.6 show the efficiency and wear curves of synthetic lubricant compared to typical mineral hydrocarbon oil. The pertinent worm gear had a center distance of 63 mm, a worm speed of 350 rpm and a drive torque of 300 Nm.

The synthetic lubricants' excellent resistance to aging allows extended lubricant change intervals. These can be three to five times longer than

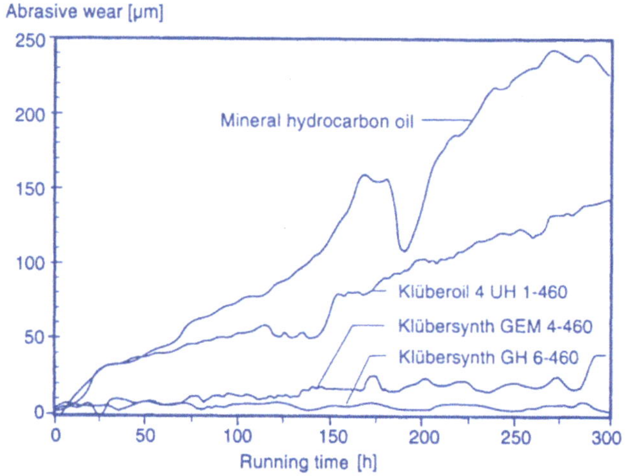

Figure 24.6 Wear curve of mineral hydrocarbon oil compared to synthetic oils. (Source: Kluber Lubricants, Germany)

intervals recommended with mineral hydrocarbon oils, which can translate to lifetime lubrication in many cases.

24.4 Lubrication of Small Gears

Small gears comprise spur, bevel and worm gears of an open, semi-closed or closed design (often not oil-tight).

Small and miniature gears are used in adjusting and control drives in the automotive industry, office machines, household appliances and machines for do-it-yourselfers, Figures 24.7 and 24.8. Their main task is to transfer movements, sometimes also power.

Due to the construction of these gears and the materials that are used—steel/steel, steel/bronze, steel/ plastic and plastic/plastic components—the lubricants have to meet various requirements, including:

- lifetime lubrication
- noise damping
- low starting torque
- low and high temperature operation
- resistance to ambient media
- compatibility with the materials used

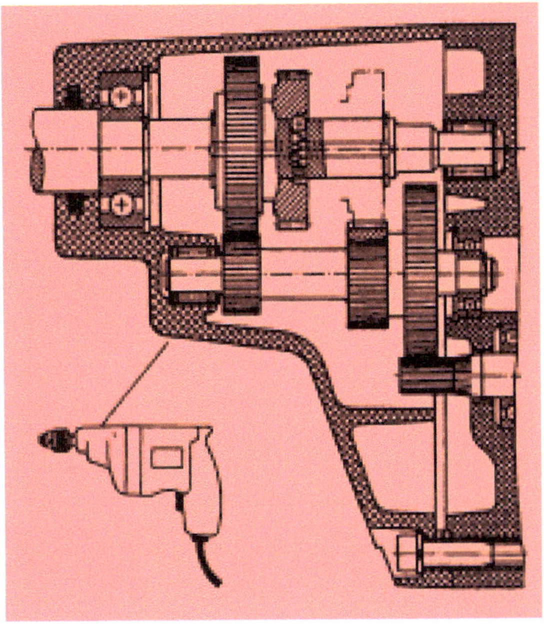

Figure 24.7 Electric hand drill, double-stage gear.

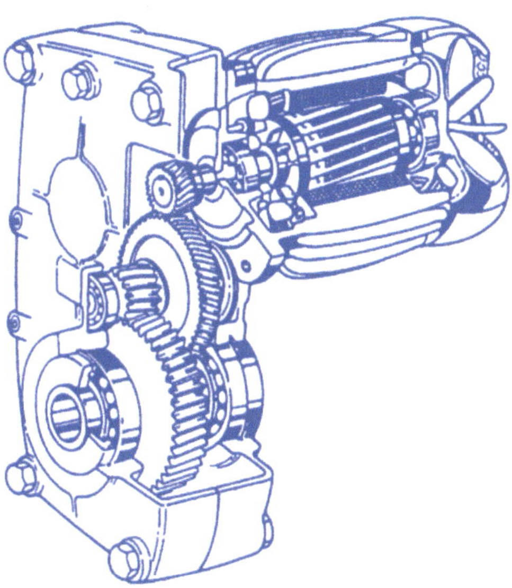

Figure 24.8 Gear motor, double-stage spur gear.

As small gears often are not oil-tight they are mainly lubricated with greases applied by dip-feed lubrication or one-time lubrication of the tooth flanks.

Dip-feed lubrication is preferred for gears in continuous operation or gears used for power transmission. One-time lubrication is suitable for gears used for the transmission of movements or gears only operating for short intervals or intermittently.

Lubricating greases of NLGI grade 000 to 0 are used for dip-feed lubrication, and of grade 0-2 for life-time lubrication. To avoid the lubricant being thrown off the gear, pastier greases are preferred in case of increased peripheral speeds. In case of dip-feed lubrication at higher peripheral speeds, however, the grease should be softer to avoid channeling.

Greases with a synthetic base oil are particularly suitable for applications where high resistance to high and low temperatures and aging are required and where the gear friction has to be low.

24.5 Testing the Performance of Gear Oils

There are three major characteristics a superior EP gear oil should exhibit: extreme pressure capability, cleanliness and demulsibility. In other words, the oil must perform under pressure for extensive time periods, must keep machine systems (and reservoirs) clean, and must separate rapidly from water.

24.5.1 Performance under pressure

Superior EP lubricants have demonstrated long service life, holding up under pressure and over time. They don't have to be replaced as often. That saves money and helps gear units last longer.

The reserve EP Capability Test measures a lubricant's load-carrying ability over time. EP—extreme pressure—oils are specially formulated to lubricate under heavy-load conditions. The longer a lubricant can maintain its load-carrying ability, the less often it must be replaced...which in turn reduces operating costs.

Before starting this test, the specific amount of a key load-carrying ingredient—phosphorus—is measured. Then the lubricant is run through the modified U.S. Steel S-200 Oxidation Test called for in U.S. Steel specification 224. (See the following section on "Superior Oxidation Stability.") Once the test is finished the amount of phosphorus is remeasured, to check for any depletion. The more that's left, the more reserve EP capability the lubricant has.

Keeping machine systems clean

If a gear oil leaves dirt and sludge in machine systems, it can eventually affect an entire plant's operation. Deposits build in the system, reducing the lubricant's effectiveness, leading to equipment failure... break-down...or even plant shutdown. Quality EP lubricants are designed to minimize these deposits.

24.5.2 Superior oxidation stability

Oils in service are exposed to oxidation—a form of deterioration that occurs when oxygen interacts with the oil. Oxidation raises the viscosity of the oil. It also leaves acidic materials which cause the soft sludge deposits or hard, varnish-like coatings. And that can lead to equipment failure. Oxidation stability is particularly important in gear oils that circulate for extended periods at high temperatures.

The primary tests for oxidation are the American Society for Testing and Materials (ASTM) D 2893 and the U.S. Steel S-200 Oxidation Test called for in U.S. Steel specification 224. Both are intended to simulate service conditions on an accelerated basis.

In both ASTM D 2893 and the modified U.S. Steel S-200 Oxidation Test, oil samples are subjected to elevated temperatures in the presence of dry air for 312 hours (13 days). First, oil samples are tested for viscosity at 100°C (212°F) and precipitation number. Then they're poured into test tubes specially fitted with air delivery tubes and flowmeters to ensure an accurate, constant flow of dry air. (The air itself is passed through a drying tower packed with anhydrous calcium sulfate to ensure that it is moisture-free.) The test tubes are immersed in a heating bath and air is bubbled through the oil. The samples are kept at a constant temperature of 95°C(203°F) (in the ASTM D 2893 test) or 121°C(250°F) (in the USS S-200 test) for 312 hours. Test samples are then removed from the bath, mixed thoroughly and tested again for viscosity and precipitation number.

24.5.3 Corrosion resistance

Many types of industrial equipment have parts made of copper or bronze, any oil that comes in contact with these parts must be non-corrosive. With EP oils, chemically active additives are needed to prevent steel-on-steel scoring and seizure. They're indispensable to applications involving steel parts—hypoid gear drives, for example.

The Copper Strip Corrosion Test is used to evaluate the corrosive tendencies of oils to copper and to check them for active sulfur-type EP additives. This test—standardized as ASTM D 130—should not be confused with

Figure 24.9 High viscosity index is mandatory for demanding marine applications.

other tests for the rust-inhibiting properties of petroleum oils, like ASTM D 665. ASTM D 130 evaluates the copper-corrosive tendencies of the oil itself, while other tests evaluate the ability of the oil to prevent ferrous corrosion.

During the standard ASTM D 130 test, special three-inch copper strips are cleaned and polished. For materials of low volatility, the strip is immediately immersed in a test tube of oil and covered with a vented stopper. The tube is held in a water bath for three hours at a temperature of 100°C (212°F). At the end of the exposure period, the strip is removed and the oil wiped off. It's compared to a specially prepared set of standardized reference strips that illustrate from Class 1 (slight tarnish) to Class 4 (heavy tarnish). Results are reported as an ASTM D 130 rating.

Another test that measures a lubricant's corrosiveness tendencies to copper-containing compounds is the Radicon Worm Gear Test. In this test, the lubricant is applied to Radicon worm and wheel gears and the clearance between them is measured. The gears are then run for 250 hours at 90°C(194°F), and the end clearance is measured.

Excellent demulsibility

Gear oils are frequently exposed to water, from damaged coolers, lines, atmospheric moisture or occasional steam sources. Water speeds the rusting of ferrous machine parts and accelerates the oil's oxidation. Gear oils must have good demulsibility characteristics for quick, effective water removal. The test for measuring demulsibility characteristics for EP oils has been standardized as ASTM D 2711.

High viscosity index required for effective marine use

A gear oil used in a marine environment must perform in a wide range of temperatures and conditions. Without this capability, at high temperatures the oil's viscosity may drop to a point where the lubricating film is broken...and that means metal-to-metal contact and severe wear. At the other extreme, the oil may become too viscous for proper circulation, or may set up such high viscous forces that proper operation of machinery is difficult.

24.6 Gear Coupling Lubrication

When users elect to use grid or gear couplings instead of non-lubricated coupling types, they should be aware of their vulnerability. The gear coupling, Figures 24.10 and 24.11, is one of the most critical components in a turbomachine and requires special lubrication consideration.

There are two basic methods of gear coupling lubrication: batch and continuous flow. In the batch method the coupling is either filled with grease or oil; the continuous flow type (Figure 24.11) uses only oil, generally light oil from the circulating oil system.

The grease-filled coupling requires special quality grease. The importance of selecting the best quality grease cannot be overemphasized. A good coupling grease must prevent wear of the mating teeth in a sliding load environment and resist separation at high speeds. It is not uncommon for centrifugal forces on the grease in the coupling to exceed 8,000Gs.

Testing of many greases in high-speed laboratory centrifuges proved a decided difference existed between good quality grease and inferior quality grease for coupling service. Testing also showed separation of oil and soap to be a function of G level and time. In other words, oil separation can occur at a lower centrifugal force if given enough time. The characteristics of grease that allow the grease to resist separation are high viscosity oil, low soap content, and soap thickener and base oil as near the same density as possible.

High speed coupling greases utilize a calcium sulfonate thickener system. This thickener system has several unique performance advantages over conventional lithium-polymer thickener systems used in many competitive coupling greases. Some of the performance benefits of the calcium sulfonate thickener are:

- Excellent corrosion prevention, even in the presence of salt mist
- Inherent EP and anti-wear protection

Grid Type

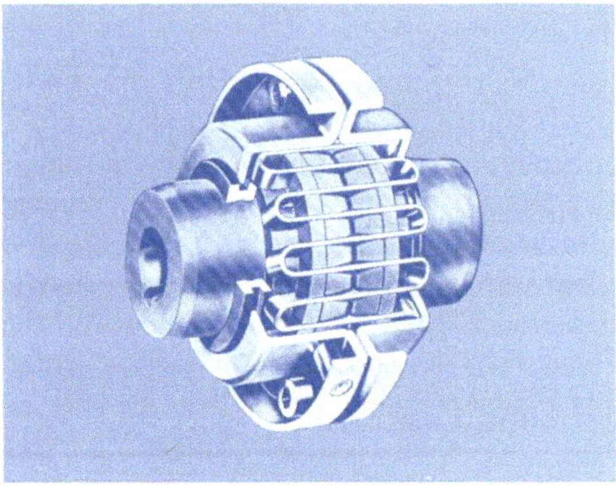

Gear Type

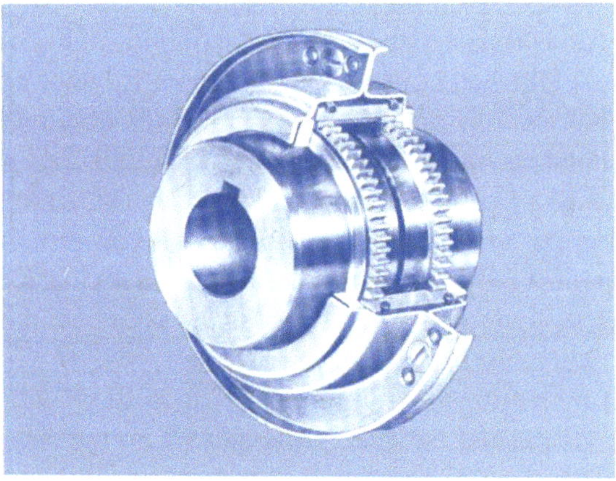

Figure 24.10 Typical couplings requiring grease lubrication.

- High dropping points
- Superior oxidation resistance
- Excellent shear stability

24.7 Lubrication of Small Geared Blowers

Large blowers often use lubricant application methods that are virtually identical to those found in rotating machinery such as compressors. That may not be the case with small positive displacement blowers that often incorporate

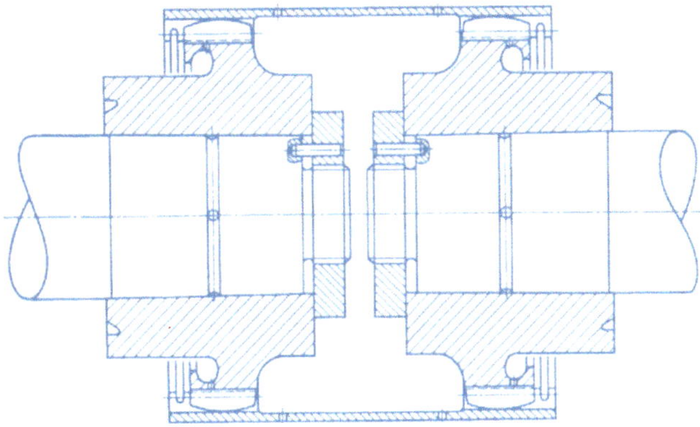

Figure 24.11 Section through a tooth-type coupling. (Source: ASEA Brown-Boveri, Baden, Switzerland)

Figure 24.12 Roots type positive displacement blower at a petrochemical plant. (Source: Lubrication Systems Company, Houston, Texas)

gears and rolling element bearings. Gears and rolling element bearings may each have their own distinct lubrication requirements and oil mist[1] may be an advantage for these machines.

Positive displacement blowers (Fig. 24.12) are often used to move air ranging in pressure from slightly negative to about 15 psig (~ 1 bar) positive. These rotary lobe machines are often used for polymer powder or pellet transfer and they are manufactured in different sizes. A number of models with 10, 12, and 12-inch shaft center to shaft centers are among the most widely used versions. Drives for these blowers include direct-connected motors as well as a number of gearbox and belt arrangements. Many positive displacement blowers are oil splash lubricated, while some of the larger ones are forced-feed lubricated.[2]

References

[1] Bloch, Heinz P. and Abdus Shamim; "Oil-Mist Lubrication Handbook"—Practical Applications," (1998), Fairmont Press, Lilburn, GA, 30047 (ISBN 0-88173-256-7)

[2] Bloch, Heinz P.; "Improving Machinery Reliability," (1998), Third Edition, Gulf Publishing Company, Houston, TX, 77520 (ISBN 0-88415-661-3)

25

Electric Motor Lubrication

Motors are electrical devices designed to convert electrical energy into mechanical energy. Mechanical energy is generated through the application of an electric current through a magnetic field that in turn causes a central shaft to spin and perform work. This shaft is located and held in place by two rolling element (anti-friction) bearings at either end of the motor frame and housing (very large horsepower motors may use journal bearings). These bearings are the only lubricated points on a motor and are virtually always grease lubricated.

With rare exception, most fractional and small horsepower motor bearings are pre-greased at the factory and sealed for life, and make no provision for external bearing lubrication. If the motor is balanced correctly, aligned, and not overloaded, the bearings will outlast the useful life of the motor.

Larger motors, are designed for heavier and often variable loads that require larger shafts and bearings. Depending on the motor manufacturer and design, external lubrication fittings are usually found on motors rated from 5hp and above. Heavier loaded motors place more load on the nearing points and require replenishment on a more frequent basis.

To ensure peak operation, a motor bearing cavity (available space between balls, raceways, cages and seals) should not be filled more than 50% capacity. Note: OEM motor bearings are delivered with a 30%-50% fill depending on the motor size and application. Figure 25.1 illustrates the temperature rise effect of a full packed bearing versus a partial packed bearing.

Because bearings are often hidden behind end plates, they are lubricated blind and as a result are susceptible to being overfilled. Once the grease has filled the cavity it has no where to go except into the motor windings where it can create an overheat condition for both bearings and motor assembly (Figure 25.2). This condition most often results in a premature failure and asset downtime.

To alleviate this condition, smaller motors can be fitted with relief style grease fittings that allow a built-in valve to open when the internal grease cavity pressure starts to rise (Figure (25.3).

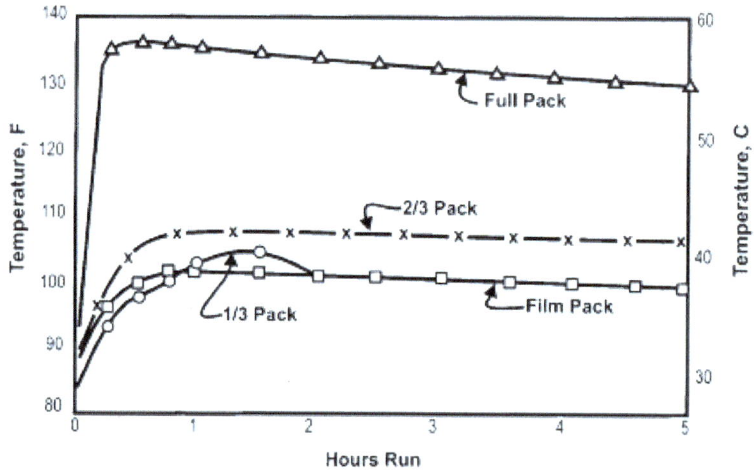

Figure 25.1 Temperature effect on fill level of grease packed bearings. Source: Exxon Mobil

Figure 25.2 Grease in motor winding.

Larger motors are designed with a drain plug positioned at the bottom of the motor end case,180° from the grease fitting. The plug is removed prior to greasing taking place. When greasing occurs, the open drain valve allows old grease to be purged from the bearing along with any excess grease that has flowed through the bearing. When greasing is completed, the bearing is allowed to purge for an additional 15-30 minutes before the grease plug is reinstalled.

Over lubricated bearings produce heat through the internal shearing action (fluid friction) of the additional grease lubricant. Over heating can also

Figure 25.3 Relief style grease fitting.

be caused through the application of an incompatible grease into the bearing, or through the introduction of solid contaminants from a dirty grease gun nozzle.

25.1 Implementing a Motor Lubrication Strategy

Depending on which motor manufacturer you talk to, the consensus is that between 70%–90% of motor failures are directly bearing related, of which, greater than 90% fail prematurely! Of course, bearings will fail if they are improperly installed or if they are subjected to misalignment stresses, but the overwhelming percentage of bearing failures are lubrication related. Recognition of this fact, along with the implementation of a simple motor lubrication strategy can result in huge savings in bearing replacement, downtime, and energy usage.

If a lubrication strategy does not exist, an excellent approach can be built using the simple 4R lubrication edict that states use of the "Right lubricant, in the Right amount, in the Right place, at the Right time". When dealing with motors, we are well advised to utilize a fifth R that represents the "Right person". This translates to a fully trained lubrication technician who understands the special lubrication needs of electrical motors.

25.1.1 Right lubricant

Lubricant selection is critical for electric motors as most multi-purpose bearing greases are not suitable for motors. If a lubricant consolidation program is in place, ensure that all motors are using a specific grade of lubricant suited for electrical motor bearing lubrication.

Most motor bearings are grease lubricated; for best results, the grease specification should meet the following requirements:

- Base oil viscosity of 90-140 cSt at 40° C
- NLGI grade 2 / 3 Note: Use a grade 3 grease when bearing dN > 250,000 (dN = mm bore x rpm)
- Melting point > 400° F (204° C)
- Low bleed
- EP additive free
- Mild anti-wear additive
- Resistant to high temperature oxidation
- Temperature stable thickener (Polyurea, and synthetic lithium complex greases are very popular for motor bearing greases)

If the motor bearing is oil lubricated using a drip oiler or mist lubrication, the oil specification choice should be based on the following requirements:

- Base oil viscosity of 90-140 cSt at 40° C
- Mild anti-wear additives (designed for mixed-film lubrication regimes)
- EP additive free (can cause corrosion in copper windings)

25.1.2 Right amount

Refer to Chapter 18 – Calculating Bearing Requirements to determine the starting amount.

A hands-on method of determining the amount of grease required uses Figure 24.1 to determine the relationship between bearing fill and bearing temperature. Using an infrared thermometer or camera to determine the motor bearing temperature under load, an approximate number of grease shots can be administered to the bearing and the temperature rise observed for the level of increase over the next 15-30 minutes. Once the number of shots is known this can be recorded and repeated for every lube cycle. Through regular temperature checks the lubrication frequency can also be determined. A word of warning, if this method is used, the motor load, speed and operational frequency must be constant. In addition, the lubrication schedule must be consistent.

Similarly, an ultrasonic grease gun (see Chapter 19 – Manual Lubrication Delivery Systems for Oil and Grease) can be used to determine the sonic

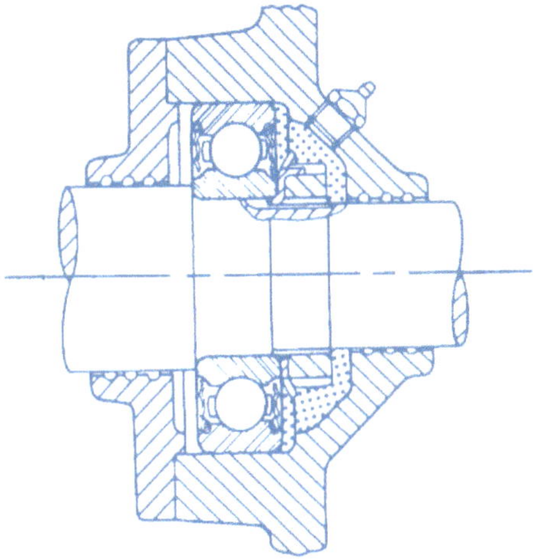

Figure 25.4 Shielded, grease-lubricated bearing (no drain).

signature that signals when the bearing requires lubrication, and the sonic signature when the bearing has reached its desired fill level. Again, operation and lubrication schedule must be consistent.

25.1.3 Right place

Most motors only require lubrication in two places, these of course being the motor shaft bearings located at either end of the motor housing. These bearings are usually ball bearings, but can be roller or plain bearings depending on the motor size and purpose. Bearings come in many different design configurations that must be understood if they are to be lubricated effectively.

How grease-lubricated bearings function in electric motors

A shielded, grease-lubricated ball bearing (Figure 25.4) can be compared to a centrifugal pump having the ball-and-cage assembly as its impeller and having the annulus between the stationary shield and the rotating inner race as the eye of the pump. Shielded bearings are **not** sealed bearings. With a *shielded* bearing, lubricant may readily enter the bearing, but dirt is restricted by the close-fitting shields. Bearings of the *sealed* design will not permit entry of new grease, whereas with shielded bearings grease will be drawn in by capillary action as the bearing cage assembly rotates. The grease will then be discharged by centrifugal force into the ball track of the outer race. If there is no shield on the back side of this bearing, the excess grease can escape (grease loss) into the inner bearing cap of the motor bearing housing.

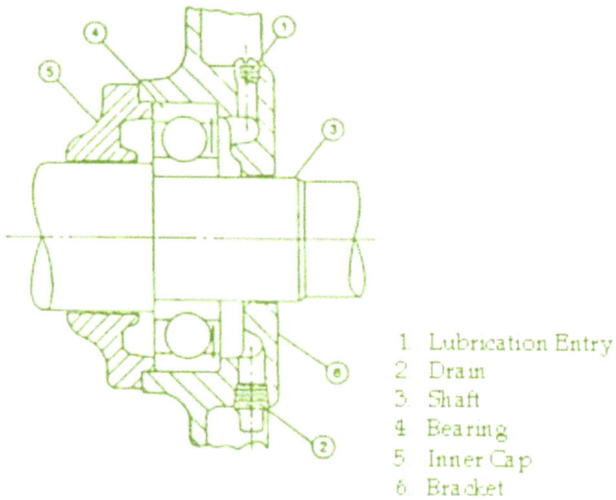

Figure 25.5 Single-shield motor bearing, with shield facing the grease cavity.

Single-shield bearings

Figure 25.5 depicts a regular single-shield bearing with the shield facing the grease supply. This is arguably the most popular bearing arrangement as it permits an extremely simple lubrication and relubrication technique, when installed. This technique makes it unnecessary to know the volume of grease already in the bearing cartridge. The shield serves as a baffle against agitation. The shield-to-inner-race annulus serves as a metering device to control grease flow. These features prevent premature ball bearing failures caused by contaminated grease and heat buildup due to excess grease. Further, warehouse inventories of ball bearings can be reduced to one type of bearing for the great bulk of existing grease-lubricated ball bearing requirements. For other services, where an open bearing is a "must," as in some flush-through arrangements, the shield can be removed in the field.

Double-shielded bearings

Some motor manufacturers subscribe to a different approach favoring the double-shielded bearing style, usually arranged as shown in Figure 25.6. The housing serves as a lubricant reservoir and are filled with grease. By regulating the flow of grease into the bearing, the shields act to prevent excessive amounts from being forced into the bearing. A grease retainer labyrinth is designed to prevent grease from reaching the motor windings on the inner side of the bearing.

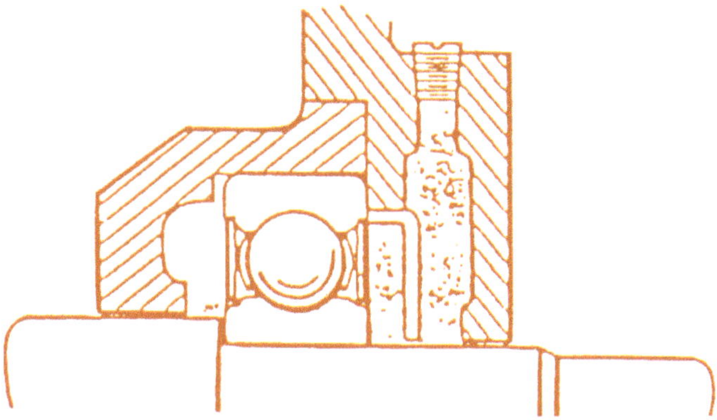

Figure 25.6 Double-shielded bearing with grease metering plate facing grease reservoir.

On motors furnished with this bearing configuration and mounting arrangement, it is not necessary to pack the housing next to the bearing full of grease for proper bearing lubrication. However, packing with grease can help to prevent dirt and moisture from entering. Oil from this grease reservoir can and does, over a long period, enter the bearing to revitalize the grease within the shields. Grease in the housing outside the stationary shields is not agitated or churned by the rotation of the bearing and consequently, is less subject to oxidation. Furthermore, if foreign matter is present, the fact that the grease in the chamber is not being churned reduces the probability of the debris contacting the rolling elements of the bearing.

On many motors furnished with grease-lubricated double-shielded bearings, the bearing housings are not usually provided with a drain plug. When grease is added and the housing becomes filled, some grease will be forced into the bearing, and any surplus grease will be squeezed out along the close clearance between the shaft and the outer cap because the resistance of this path is less than the resistance presented by the bearing shields, metering plate, and the labyrinth seal.

Open bearings

High-load and/or high-speed bearings are often supplied without shields to allow cooler operating temperature and longer life. One such bearing is illustrated in Figure 25.7. If grease inlet and outlet ports are located on the same side, this bearing is commonly referred to as "conventionally grease lubricated." If grease inlet and outlet ports are located at opposite sides, it is referred to as "cross-flow, or "cross-lubrication." Figure 25.8 depicts a cross-flow lubricated bearing.

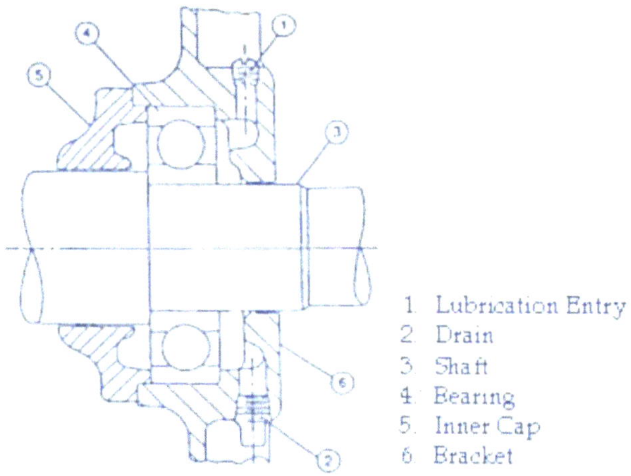

Figure 25.7 High load and/or high-speed bearings are often supplied without shield, as shown.

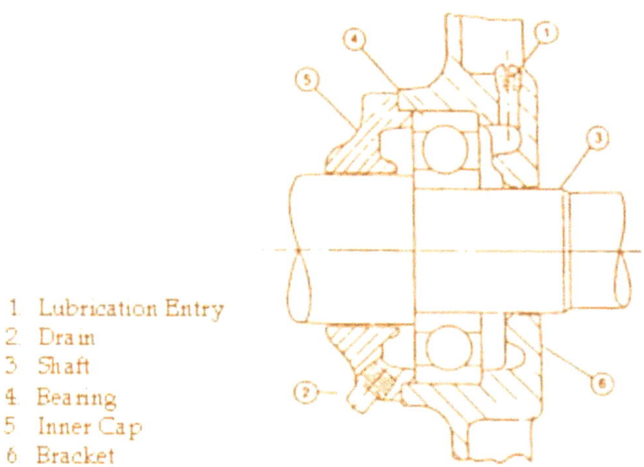

Figure 25.8 Open bearing with cross-flow grease lubrication.

Life-time lubricated, "Sealed" bearings

Lifetime lubricated, or "Lubed-for-life" bearings incorporate close-fitting seals in place of, or in addition to shields. These bearings are customarily found on low-horsepower or fractional motors, or on appliance motors that operate intermittently. Although it is often claimed that sealed ball bearings in electric motors will survive as long as bearings operating temperatures remained below 150°C (302°F) and speed factors DN (mm bearing bore times revolutions per minute) do not exceed 300,000, other studies showed that close-fitting seals can cause high frictional heat and that loose fitting seals cannot effectively exclude atmospheric air and moisture which will cause grease deterioration. These facts preclude the use of lubed-for-life bearings in installations which expect "life" to last more than three years in the typical plant environment. Moreover, this thought to be the reason why bearing manufacturers advise against the use of sealed bearings larger than size 306 at speeds exceeding 3600 RPM. This would generally exclude sealed bearings from 3600 RPM motors of 10 plus horsepower.

A 1989 guideline issued by a major bearing manufacturer gives a DN value of 108,000 as the economic, although not technically required, limit for "life-time-lubrication."

25.1.4 Procedures for re-greasing electric motor bearings

Electric motor bearings should be re-greased with a grease which is compatible with the original charge. It should be noted that the polyurea greases often used by the motor manufacturers may be incompatible with lithium-base greases.

Single-shielded bearings

To take advantage of single-shielded arrangements in electric motors, competent users have developed three simple recommendations which differ, somewhat, from motor manufacturers' idealized guidelines.

1. Install a single-shield ball bearing with the shield *facing* the grease supply in motors having the grease fill-and-drain ports on that same side of the bearing. Add a finger full of grease to the ball track on the back side of the bearing during assembly.

2. After assembly, the balance of the *initial* lubrication of this single-shielded bearing should be performed with the motor idle. Remove the drain plug and pipe. With a grease gun or high-volume grease pump, fill the grease reservoir until fresh grease emerges from the drain. The fill and drain plugs should then be reinstalled and the motor is ready for service.

   It is essential that this initial lubrication not be attempted while the motor is running. This can create a pumping action resulting in a continuing flow of grease through the shield annulus until the overflow space in the inner cartridge cap is full. Grease can then flow down the shaft and into the winding of the motor where it is not wanted. This will take place before the grease can emerge at the drain.

3. Relubrication may be performed while the motor is either running or idle. (It should be limited in quantity to a volume approximately one-fourth the bearing bore volume.) Fresh grease can take a wedge-like path straight through the old grease, around the shaft, and into the ball track. Thus, the overflow of grease into the inner reservoir space is quite small even after several relubrications. Potentially damaging grease is thus kept from the stator winding. Further, since the ball and cage assembly of this arrangement does not have to force its way through a solid fill of grease, bearing heating is kept to a minimum. Note: in tests using this method it was observed that a maximum temperature rise of only 20°F occurred 20 minutes after the grease reservoir was filled. Two hours later it returned to 5°F. In contrast, tests using a double-shield bearing arrangement caused a temperature rise of over 100°F (at 90°F ambient temperature the resulting temperature was 190°F) and maintained this 100°F rise for over a week.

Double-shielded bearings

A. Ball Bearings

   1. Pack (completely fill) the cavity adjacent to the bearing. Use the necessary precautions to prevent contaminating this grease before the motor is assembled.

   2. After assembly, lubricate stationary motor until a full ring of grease appears around the shaft at the relief opening in the bracket.

B. Cylindrical Roller Bearings

   1. Hand pack bearing before assembly

   2. Proceed as outlined in (1) and (2) for double-shielded ball bearings.

If under-lubricated after installation, the double-shielded bearing is thought to last longer than an open (non-shielded) bearing given the same treatment, because of grease retained within the shields (plus grease remaining in the housing from its initial filling).

If over-greased after installation, the double-shielded bearing can be expected to operate satisfactorily without overheating as long as the excess grease is allowed to escape through the clearance between the shield and inner race, and the grease in the housing adjacent to the bearing is not churned, agitated and caused to overheat.

It is not necessary to disassemble motors at the end of fixed periods to grease bearings. Bearing shields do not require replacement.

Double-shielded ball bearings should not be flushed for cleaning. If water and dirt are known to be present inside the shields of a bearing because of a flood or other circumstances, the bearing should be removed from service. All leading ball-bearing manufacturers provide a reconditioning service at a nominal cost when bearings are returned to their factories. As an aside, reconditioned ball bearings are generally **less prone** to fail than their brand-new counterpart bearings. This is because grinding marks and other asperities are now burnished to the point where smoother running and less heat generation are likely.

Open bearings

Motors with open, conventionally greased bearings are generally lubricated with slightly different procedures for drive-end and opposite end bearings.

Lubrication procedures for drive-end bearings:

1. Relubrication with the shaft stationary is recommended. If possible, the motor should be warm.
2. Remove fill plug and replace with grease fitting.
3. Remove large drain plug when furnished with motor.
4. Using a low pressure, hand operated grease gun, pump in the recommended amount of grease, or use 1/4 of bore volume.
5. If purging of system is desired, continue pumping until new grease appears either around the shaft or at the drain opening. Stop after new grease appears.
6. On large motors provisions have usually been made to remove the outer cap for inspection and cleaning. Remove both rows of cap bolts. Remove, inspect and clean cap. Replace cap, being careful to prevent dirt from getting into bearing cavity.
7. After lubrication allow motor to run for fifteen minutes before replacing the drain plug.
8. If the motor has a special grease relief fitting (Figure 25.3), pump in the recommended volume of grease or until a one-inch-long string of grease appears in any one of the relief holes. Reinstall the drain plug(s).

9. Wipe away any excess grease which has appeared at the grease relief port using a lint free cloth.

Lubrication procedure for bearing opposite drive end:

1. If bearing hub is accessible, as in drip-proof motors, follow the same procedure as for the drive-end bearing.
2. For fan-cooled motors note the amount of grease used to lubricate shaft end bearing and use the same amount for commutator-end bearing.

Motor bearings arranged with crossflow housings as shown in Figure 25.8, with grease inlet and outlet ports on opposite sides, are called cross-flow lubricated. Regreasing is accomplished with the motor running. The following procedure should be observed:

1. Start motor and allow to operate until normal motor temperature is obtained.
2. Inboard bearing (coupling end)
   a. Remove grease inlet plug or fitting.
   b. Remove drain plug. Some motor designs are equipped with excess grease cups located directly below the bearing. Remove the cups and clean out the old grease.
   c. Remove hardened grease from the inlet and outlet ports with a clean probe.
   d. Inspect the grease removed from the inlet port. If rust or other abrasives are observed, *do not grease the bearing*. Tag motor for overhaul.
   e. Bearing housing with outlet ports:
      (1) Insert probe in the outlet port to a depth equivalent to the bottom balls of the bearing.
      (2) Replace grease fitting and add grease slowly with a hand gun. *Count strokes of gun* as grease is added.
      (3) Stop pumping when the probe in the outlet port begin to move. This indicates that the grease cavity is full.
   f. Bearing housings with excess grease cups:
      (1) Reinstall/Install grease fitting and add grease slowly with a hand gun. *Count strokes of gun* as grease is added.

(2) Stop pumping when grease appears in the excess grease cup. This indicates that the grease cavity is full.

(3) Outboard bearing (fan end)

   a. Follow inboard bearing procedure provided the outlet grease ports or excess grease cups are accessible,

   b. If grease outlet port or excess grease cup is not accessible, add 2/3 of the amount of grease required for the inboard bearing.

(4) Leave grease drain ports open—**do not reinstall** the drain plugs. Excess grease will be expelled through the port. Consider using a short section of open pipe in lieu of the plug.

(5) If bearings are equipped with excess grease cups, replace the cups. Excess grease will expel into the cups.

Application limits for greases used in electric motor bearings

Bearings and bearing lubricants are subject to four prime operating influences: speed, load, temperature, and environmental factors. The optimal operating speeds for ball and roller type bearings—as related to lubrication—are functions of what is termed the DN factor. To establish the DN factor for a particular bearing, the bore of the bearing (in millimeters) is multiplied by the revolutions per minute, i.e.:

$$75 \text{ mm} \times 1000 \text{ rpm} = 75{,}000 \text{ DN value}$$

Speed limits for conventional greases have been established to range from 100,000 to 150,000 DN for most spherical roller type bearings and 200,000 to 300,000 DN values for most conventional ball bearings. Higher DN limits can sometimes be achieved for both ball and roller type bearings, but require close consultation with the bearing manufacturer. When operating at DN values higher than those indicated above, use either special greases incorporating good channeling characteristics or circulating oil.

25.1.5 Right time

Relubrication frequency (for grease)

Every motor manufacturer has its own *general* lubrication guide. Bearing manufacturers also provide us with a "starting point" relubrication" application graph such as the one shown in Chapter Eighteen's Figure 18.4.

RPM			HP		Ambient Temp (°F)		Regreasing Interval (Months)
1200	1800	3600	<100	>100	<140	>140	
x	x		x		x		36 - 54
x	x		x			x	24 - 36
x	x			x	x		24 - 36
x	x			x		x	12 - 18
		x	x		x		24 - 36
		x	x			x	12 - 18
		x		x	x		12 - 18
		x		x		x	6 - 9
Reduce interval in half for:							
	Every 10°C (18°F) above 70°C (160°F) bearing temp						
	Vertical motors						
	Multipurpose grease						
	Roller bearings						
	Belt drives						
Reduce interval appropriately where there is a risk of contamination							

Figure 25.9 EPRI NP-7502 motor greasing guide. Source EPRI.

The exact relubrication frequency depends on many factors as described throughout this chapter

Figure 22.9 depicts a recommended greasing frequency table developed by the Electrical Power Research Institute. The lower section of the table indicates typical operating conditions that will reduce the regreasing interval.

Table 25.1 Shows a typical motor lubrication guideline/recommendation chart provided by electrical motor companies, in this case Baldor Electric Company

25.1.6 Right person

Any person entrusted to lubricate a motor bearing must be qualified. Although as seemingly simple task, practical lubrication is an engineered task that must be taken seriously if it is to be performed correctly. Best practice dictates that lubrication tasks must be developed and performed by lubrication certified individuals/teams.

Section six is an entire section devoted to current available professional lubrication training, designation levels and certifications.

Table 25.1 Baldor Electric Company guidelines for motor relubrication.

	Rated Speed - RPM			
NEMA/(IEC) Frame Size	**3600**	**1800**	**1200**	**900**
Up to 210 incl. (132)	5500 Hrs.	12000 Hrs.	18000 Hrs.	22000 Hrs.
Over 210 to 280 incl. (180)	3600 Hrs.	9500 Hrs.	15000 Hrs.	18000 Hrs.
Over 280 to 360 incl. (225)	*2200 Hrs.	7400 Hrs.	12000 Hrs.	15000 Hrs.
Over 260 to 5000 incl. (300)	*2200 Hrs.	3500 Hrs.	7400 Hrs.	10500 Hrs.

*Bearings in 360 through 5000 frame, 2 pole motors are either 6313 or 6314 bearings and lubrication interval is shown in the table. If roller bearings are used, and the bearings must be lubricated more frequently, divide the listed lubrication interval by 2.

Service Conditions			
Severity of Service	**Ambient Temperature Maximum**	**Atmospheric Contamination**	**Type of Bearing**
Standard Severe Extreme Low Temperature	40°C 50°C >50°C* or Class H Insulation <-30°C**	Clean, Little Corrosion Moderate dirt, Corrosion Severe dirt, Abrasive dust, Corrosion	Deep Groove Ball Bearing Ball Thrust, Roller All Bearings

* Special high temperature grease is recommended

** Special low temperature grease is recommended

Lubrication Interval Multiplier	
Severity of Service	**Multiplier**
Standard	1.0
Severe	0.5
Extreme	0.1

(*Continued*)

Table 25.1 *Continued.*

Bearings Sizes and Types						
Frame Size NEMA (IEC)	(These are Bearing Description the "Large" bearings (Shaft End) in each frame size)					
	Bearing	OD D mm	Width B mm	Weight of grease to add oz (grams)	Volume of grease to be added in^3 teaspoon	
Up to 210 incl. (132)	6307	80	21	0.31 (8.4)	0.6	2.0
Over 210 to 280 incl. (180)	6311	120	29	0.61 (17.4)	1.2	3.9
Over 280 to 360 incl. (225)	6313	140	33	0.81 (23.1)	1.5	5.2
Over 360 to 5000 incl. (300)	NU322	240	50	2.12 (60.0)	4.1	13.4
Weight in grams = .005 DB						

25.2 Oil Mist for Electric Motors

Plant-wide oil mist systems were explained in Section three, chapter ten of this text. Oil mist consists of a mixture of 200,000 volume parts of clean and dry plant or instrument air and one part of lubricating oil. Since the 1970s, dry sump ("pure") oil mist has provided ideal lubrication for thousands of rolling element bearings in electric motors. In the intervening decades, this lubricant application method has gained further acceptance at many reliability-focused process plants in the United States and overseas.

The influences of bearing size, speed, and load have been recognized in an empirical oil mist applicability formula, limiting the parameter "DNL" (D = bearing bore, mm; N = inner ring rpm; and L = load, lbs) to values below 10E9, or 1,000,000,000. An 80 mm electric motor bearing, operating at 3,600 rpm and a load of 600 lbs, would thus have a DNL of 172,000,000—less than 18% of the allowable threshold value.

Major grass-roots olefins plants commenced using oil mist on motors as small as one hp (0.75 kW) in 1975. Although the largest electric motors using pure oil mist in refineries and petrochemical plants exceed 2,000 hp in size, the more typically prevailing practice among reliability-focused users is to

Figure 25.10 Oil mist routed through electric motor bearings.

apply oil mist on horizontal motors, 15 hp and larger, and on vertical motors of 3 hp and larger. In all cases, these electric motors are fitted with rolling element bearings.

API-610, the most widely used pump standard in the petrochemical and refining industries, asks for oil mist to be routed *through* the bearings (which was done in Figure 25.10) instead of *past* the bearings, Figure 25.11. Although intended for pumps, this recommendation will work equally well for electric motor rolling element bearings. The resulting *diagonal through-flow* route guarantees adequate lubrication, whereas oil mist entering and exiting on the same side might allow some of the mist to leave without first wetting the rolling elements. Through-flow is thus one of the keys to a successful installation.

Mist flow quantified

The required volume of oil mist is often expressed in "bearing-inches," or "BIs." A bearing-inch is the volume of oil mist needed to satisfy the demands of a row of rolling elements in a one-inch (~25 mm) bore diameter bearing. One BI assumes a rate of mist containing 0.01 fl. oz., or 0.3 ml, of oil per hour. Certain other factors may have to be considered to determine the needed oil mist flow and these are known to oil mist providers and bearing

Figure 25.11 Oil mist applied to the same side of a bearing is not providing optimal lubrication; much of the mist is simply flowing from entry to drain.

manufacturers. The various factors are also extensively documented in several references; they are readily summarized as:

a. *Type of bearing*. The different internal geometries of different types of contact (point contact at ball bearings and linear contacts at roller bearings), amount of sliding contacts (between rolling elements and raceways, cages, flanges or guide rings), angle of contact between rolling elements and raceways, and prevailing load on rolling elements. The most common bearing types in electrical motors are deep groove ball bearings, cylindrical roller bearings and angular contact ball bearings.

b. *Number of rows of rolling elements*. Multiple row bearing or paired bearing arrangements require a simple multiplier to quantify the volume of mist flow.

c. *Size of the bearings*, related to the shaft diameter—inherent in the expression "bearing-inches."

d. *The rotating speed*. The influence of the rotating speed should not be considered as a linear function. It can be linear for a certain intermediate speed range, but at lower and higher speeds the oil requirements in the contact regions may differ from straight linearity.

e. *Bearing load conditions* (preload, minimum or even less than minimum load, heavy axial loads, etc.)

f. *Cage design*. Different cage designs may affect mist flow in different ways. It has been reasoned that stamped (pressed) metal cages, polyamide cages, or machined metal cages might produce different degrees of turbulence. While different rates of turbulence may cause different amounts of oil to "plate out" on the various bearing components, the concern vanishes when oil mist is applied in through-flow mode.

Through-flow oil mist will accommodate all of the arrangements listed above.

Sealing and drainage issues

Although oil mist will not attack or degrade the *winding* insulation found on electric motors made since the mid-1960s, mist entry and related sealing issues must be understood and merit being included in this overview.

Regardless of motor type, i.e., TEFC, X-Proof or WPII, *cable* terminations should not be made with conventional electrician's tape. The adhesive in this tape will last but a few days and then become tacky to the point of unraveling. Instead of inferior products, competent motor manufacturers use a modified silicone system that is highly resistant to oil mist.

Similarly, and while it must always be pointed out that oil mist is neither a flammable nor explosive mixture, it would be unhealthy to allow a visible plume of mist to escape from the junction box cover. The wire passage from the motor interior to the junction box should, therefore, be sealed with Two-Part Epoxy potting compound to exclude oil mist from the junction box.

Finally, it is always good practice to verify that *all* electric motors have a small (3 mm) weep hole and that XP-motor drains are given closer attention. The latter are furnished with either an explosion-proof rated vent or a suitably routed weep hole passage at the bottom of the motor casing or lower edge of the motor end cover. Intended to drain accumulated moisture condensation, the vent or weep hole passage will allow coalesced or atomized oil mist to escape. Note, however, that explosion-proof motors are still "explosion-proof" with this passage. Reasoning on the issue should convince us that a motor with its interior slightly pressurized by non-explosive oil mist cannot ingest explosive vapors from a surrounding atmosphere.

TEFC vs WPII construction

On TEFC (totally enclosed, fan-cooled) motors, there are documented events of liquid oil filling the motor housing to the point of near contact with the spinning rotor. Conventional wisdom to the contrary, there neither were nor will there be detrimental effects with the oils used in normal industry. The motor could have run indefinitely! TEFC motors are suitable for oil mist lubrication by simply routing the oil mist *through* the bearing. No special internal sealing provisions are needed with pure oil mist filling a TEFC motor as long as the pressurized mist keeps dirty atmospheric air from entering.

On weather-protected (WPII) motors, merely adding oil mist has often been done and has generally worked surprisingly well. In this instance, however, it was found important to lead the oil mist vent tubing away from regions influenced by the motor fan. Still, weather-protected (WPII) electric motors do receive additional attention from reliability-focused users and knowledgeable motor manufacturers.

Air is constantly being forced through the windings and an oil film deposited on the windings could invite dirt accumulation. To reduce the risk of dirt accumulation, suitable means of sealing should be provided between the motor bearings and the motor interior. Since V-rings and other elastomeric contact seals are subject to wear, low-friction face seals are considered technically superior.

As is so often the case, the user has to make choices. Some low friction axial seals (face seals) may require machining of the cap, but long motor life and the avoidance of maintenance costs will make up for the added expense. Nevertheless, double V-rings using nitrile or Viton elastomeric material should not be ruled out since they are considerably less expensive than face seals.

Sealing to avoid stray mist stressing the environment

Even when still allowed under prevailing regulatory environmental regulations (e.g. OSHA or EPA), air quality and greenhouse concerns make it desirable to minimize stray oil mist emissions. It is helpful to recall that state-of-art oil mist systems are fully closed, i.e. are configured so as *not* to permit any mist to escape.

Combining effective seals and a closed oil mist lubrication system has, for many decades, represented a well-proven solution. The combination not only eliminates virtually all stray mist and oil leakage, but makes possible the recovery, subsequent purification, and re-use of perhaps 97% of the oil. These recovery rates make the use of more expensive, superior quality synthetic lubricants economically attractive.

Closed systems and oil mist-lubricated electric motors give reliability-focused users several important advantages:

- Compliance with actual and future environmental regulations

- Convincing proof that oil mist lubrication benefits electric motors and the maintenance budget
- The technical and economic justification to apply high-performance synthetic oils

PAO and diester-based "synthetic" lubricants embody most of the properties needed for extended bearing life and greatest operating efficiency. These oils excel in the areas of bearing temperature and friction energy reduction. It is not difficult to show relatively rapid returns on investment for these lubricants, providing, of course, the system is closed, and the lubricant re-used after filtration.

with other synthetic base oils and without requiring reductions in viscosity. This segment of our text recaps and incorporates the findings of this very important ASME paper. Also, it condenses more recent findings regarding the equivalent effectiveness of PAO-based synthetic lubes formulated with advanced additives.

Although it was well known that synthetic lubes reduce friction, little quantitative work had been done before 1980. It was then that Morrison, Zielinski and James rigorously documented the beneficial effects of diester fluids on the frictional power losses of industrial equipment. Because synthetic fluids are chemically different from mineral oils, one might expect effects that go beyond those attributable to viscosity relationships alone. Indeed, lubricant properties and application methods also affect lubrication effectiveness and the frictional torque to be overcome. The potential cost savings through power loss reduction appear to be quite substantial. It has been estimated that industrial machines consume 31% percent of the total energy in the United States. It has also been estimated that as much as 5% of the mechanical losses of these machines could be avoided through a combination of improved equipment design and lubricant optimization.

Bibliography

Miannay, Charles; "Grease Life Estimation In Rolling Bearings," Engineering Sciences Data (U.K.) Number 78032, November 1978.

Anonymous; "Oil Mist Arrests Bearing Failure In Aruba," *Oil and Gas Journal*, September 16, 1974.

Autenrieth, J.R.; "Motor Lubrication Experience at Phillips Petroleum, Sweeney, Texas." Documentation prepared for earlier NPRA meetings.

Aviste, M.; "Lubrication And Preventive Maintenance," *Lubrication Engineering*, Volume 37,2, February, 1981, pp. 72–81.

Bannister, K.E.; "Electrical Motors – A Perfect Canvas for Precision Maintenance" *The RAM Review*, October 2020

Bannister, K.E.; "Keeping Motors and Gearboxes in Tip-Top Shape" Efficient Plant Magazine, September 2010

Bannister, K.E; "Understanding Motor and Gearbox Lubrication" *Efficient Plant Magazine*, January 2017

BC Hydro Corporation; "Electric Motors – Energy Efficiency Guide" Paper 2007

Bloch, H.P.; "Dry Sump Oil Mist Lubrication for Electric Motors" *Hydrocarbon Processing Magazine*, March 1977.

Bloch, H.P.; "Large Scale Application of Pure Oil Mist Lubrication in Petrochemical Plants," ASME Paper No. 80-C12/ Lub-25, August 1980.

Bloch, H.P.; "Oil Mist Lubrication for Pumps and Motors" *World Pumps Magazine*, April 2015

Bloch, H.P.; "Optimized Lubrication of Antifriction Bearings for Centrifugal Pumps," ASLE Paper No. 78-AM-1D-2, April 1978.

Booser, E.R.; "When To Grease Bearings," *Machine Design*, August 21, 1975, pp. 70–73

Brozek, R.J., and Bonner, J.J.; "The Advantages of Ball Bearings and Their Application On Large-Horsepower High-Speed Horizontal Induction Motors," IEEE Transactions, Vol. IGA-7, No. 2, March/April 1971.

Clapp, A.M.; "Plant Lubrication," Proceedings of the Seventh Texas A&M University Turbomachinery Symposium, December 1978.

Eschmann, Hasbargen & Weigand, "Ball and Roller Bearings—Theory, Design and Application," John Wiley & Sons, New York, 1985 (ISBN 0-471-26283-8)

Exxon Mobil; "Guide to Electric Motor Lubrication" paper

Hafner, E.R.; "Proper Lubrication, The Key To Better Bearing Life," *Mechanical Engineering*, November 1977, pp. 46–49.

Kugelfischer Georg Schaefer and Company, (FAG); "The Lubrication of Rolling Bearings," Publication No. 81 103EA, Schweinfurt, 1977.

Miannay, C.R.; "Improve Bearing Life With Oil Mist Lubrication," *Hydrocarbon Processing*, May 1974, pp. 113–115

Miller, N.H., and Pattison, D.A.; "How To Select The Right Lubricant," *Chemical Engineering*, March 11, 1968, pp. 193–198.

Petromatic, Technical Data Bulletin (Jemalee Industries, Inc.) Grand Prairie, Texas 75051

Pinkus, O., Decker, O., and Wilcock. D.F. "How to save 5% of our energy," *Mechanical Engineering*, Sept

Reliance Electric, Cleveland, OH.; Instruction Manual B-3620-14

Siemens Corporation, E & C Newsletter, April 1982.

SKF Industries, Bulletin 144-110, "A Guide to Better Bearing Lubrication," July 1981.

Smeaton, R.W.; "Motor Application and Maintenance Hand-book," McGraw-Hill Book Company, New York, 1981.

Smith, R.L., and Wilson, D.S.; "Reliability of Grease-Packed Ball Bearings for Fractional Horsepower Motors," *Lubrication Engineering*, Volume 36, July, 1980, pp. 411–416.

Towne, C.A.; "Practical Experience with Oil Mist Lubrication," ASME Paper 82-AM-4C1, April 1982.

Morrison, F.R., Zielinsky, J., James, R., "Effects of synthetic fluids on ball bearing performance," ASME Publication, February, 1980.

26

Pump Lubrication

In the realm of lubrication, there are two definitive types of pumps: Type A is a lubrication pump specifically designed to deliver lubricant from a reservoir to a bearing point or location, sometimes under extreme pressure. The second type of pump can deliver any type of process fluid from an initial location/reservoir to a specified delivery point.

Both pump types can be manually or mechanically piston actuated; pneumatically air actuated; or by motorized gears.

26.1 Type A - Lubrication Pumps

Pumps designed for lubrication are by design, self-lubricated. As such, they are inherently maintenance free when it comes to internal bearing lubrication requirements.

26.1.1 Centralized oil delivery

Lubricant pumps designed for oil delivery are generally smaller than those designed for grease, and can employ gravity, piston or a gear drive to deliver the oil charge on a cyclic or continuous delivery. Compact in nature, oil pumps are often self-contained units within, or attached to their own reservoir. Figure 26.1 depicts a typical and extremely popular oil metering, electrical actuated piston pump style unit used on many small to medium machines.

26.1.2 Hydraulic oil delivery

When pumping hydraulic fluids, electrically driven gear pumps with larger displacements are utilized to develop the greater pressure requirements needed to actuate the hydraulic devices. Pumps are often bolted to the outside of large oil reservoir tanks as depicted Figure 26.2.

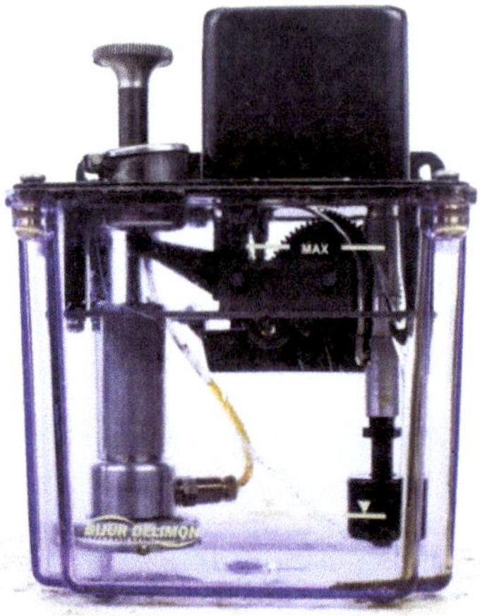

Figure 26.1 Centralized oil lubrication pump. Courtesy Bijur Delimon International.

26.1.3 Centralized grease delivery

Similar to hydraulic oil systems, grease pumps are required to develop high pressures in order to successfully move grease through a delivery system. Grease almost always employs a piston style pump. These are most evident on a manual, hand activated grease gun. For multi-point grease systems, a pneumatic driven pump to point grease unit can be employed (see Chapter 20, Figure 20.41). A typical grease pump unit with attached reservoir is shown in Figure 26.3, these pump units are typically utilized to power muti-point progressive divider systems or single/dual line parallel systems described in Chapter 20.

26.2 Type B – Process Fluid Pumps

Process fluid pumps are designed to move large quantities of fluid from one location to another, unless designed as a "loop" circuit similar to that used in a large process cooling circuit. Process pumps are generally much larger than lubricant pumps and have internal and external bearings that must be lubricated.

26.2.1 Static oil bath bearing lubrication

If pump bearings are large enough, housings can be designed to allow the static application of oil from a built-in sump housing. With a static sump, the

Figure 26.2 Large hydraulic oil pumping system (Courtesy ENGTECH Industries Inc.).

oil level is at or near the center of whichever rolling element passes through the 6 o'clock (bottom) position. (Figure 26.4) This is generically termed "oil bath" pump lubrication.

However, oil bath bearing lubrication is not for every bearing. When bearing elements are allowed to plough through an oil bath at "high" velocity, excessive heat generation is a major concern due to degraded lubricant oxidation stability caused by the elevated bearing temperatures. Therefore, static sumps—oil baths—are best suited for low-to-medium velocity bearing speeds.

Figure 26.3 Typical grease lubrication pumping unit (Courtesy ENGTECH Industries Inc).

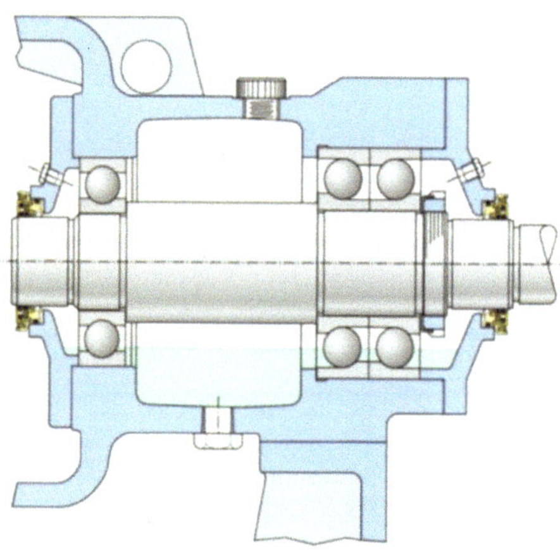

Figure 26.4 Typical sump (oil bath) housing with 6-o'clock oil level.

A widely used approximation suggests a "DN-value" of 6,000 (shaft diameter (inches) x rpm) as a threshold when bearing elements should no longer move through the oil bath and where, instead, lube oil is introduced into the bearings by other means. Traditionally, these other means have included oil rings (Figures 26.5 and 26.6), flinger discs (Figure 26.7), "jet oil spray" and oil mist (Figure 26.8).

26.2.2 Oil ring lubrication

Oil rings are found in many machines. To work correctly, they must be installed on a truly horizontal shaft system (laser alignment is preferred) and

Figure 26.5 An unrestrained oil ring can touch portions of the inside of the bearing housings and suffer abrasive damage.

not be allowed to make contact with housing-internal surfaces. Figure 26.5 shows the damage an unrestrained oil ring can inflict) Horizontal misalignment can also allow an oil ring to become wedged under the long limiter screw depicted in Figure 26.6. Immersion depth and lube oil viscosity must be kept within acceptable design range and the user must check that bore eccentricity stays within the 0.002 or 0.003 inches.

There are other issues associated with oil rings; they can get trapped and overheated behind the thrust bearing if no oil slot is provided (typical oil return slot is shown below the radial bearing in Figure 26.6). Also, beware of adding just any bearing protector seal; doing so may result in somewhat higher pressure to the right of the thrust bearing compared to the pressure in the large, often well-vented, space near the center of the bearing housing.

26.2.3 Flinger disc lubrication

Flinger discs must be carefully engineered for the intended duty and must be securely fastened to shafts. The discs allow moderate deviation from precise horizontality of shafts systems; they make contact with the oil level or are partially immersed in the bearing housing oil sump as shown in Figure 26.7.

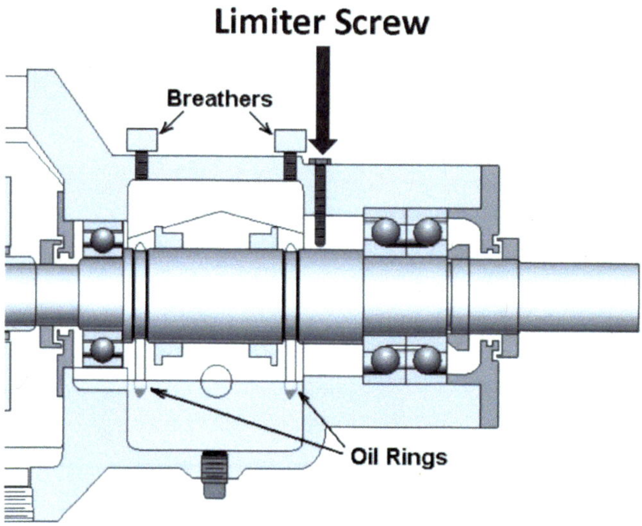

Figure 26.6 Oil rings shown in correct position.

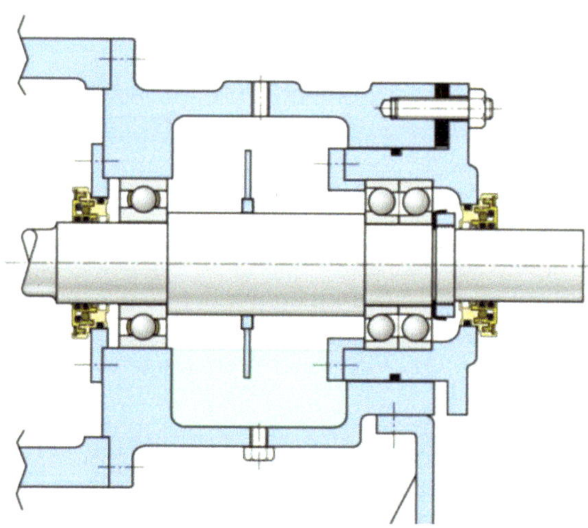

Figure 26.7 Flinger discs avoid issues with oil rings; they can be accommodated in bearing housings fitted with cartridges designed to allow access and insertion.

26.2.4 Oil spray/mist lubrication

When oil mist is employed, the mist flows through the bearing and while shaft rotation creates turbulence, atomized oil globules combine and form larger oil droplets. The coalesced oil then coats and cools the bearing. Typical

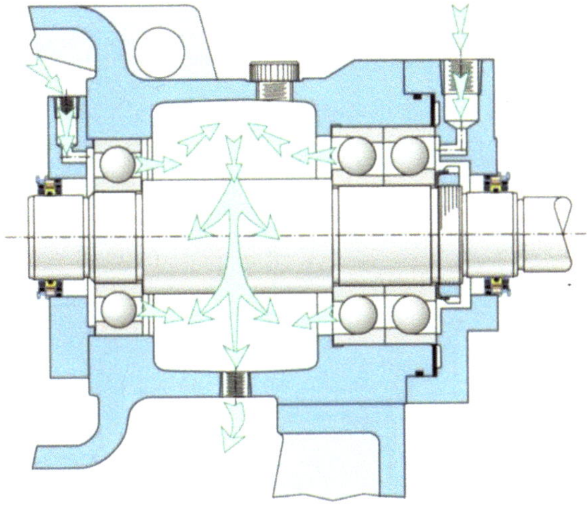

Figure 26.8 Typical oil spray/oil mist) directed into the bearing cage at two entry points provides an optimum thickness oil film for lubrication and heat removal at any bearing orientation. Note the face type bearing seals used to prevent oil loss to atmosphere through the bearings.

oil path is shown in Figure 26.8. Because the bearing housing is at slightly higher than atmospheric pressure, inward migration of atmospheric contaminants is avoided.

26.2.5 Constant level lubricators

Traditional lowest first-cost application of oil involves use of one of many available constant level lubricator devices. A typical side mounted unit is portrayed in Figure 26.8. Side-mounted constant level lubricators or "CLL-oilers" are unidirectional and should be mounted on the up-arrow rotation side of the bearing housing. The unit shown has been piped to provide an internal pressure balance between the bearing housing and the lubricator body.

CLL oilers must properly mounted to keep them airtight and ensure the oil level is always constant as a small lowering of the oil level can easily deprive a bearing of lubrication. If the pressure in a closed bearing housing increases due to a slight temperature increase, pressure will cause the oil level to go down and oil will suddenly no longer flow into the bearing. Black oil will form and the bearing will start to fail. Pressure-balanced lubricators as shown in Figure 26.8 are preferred over unbalanced types.

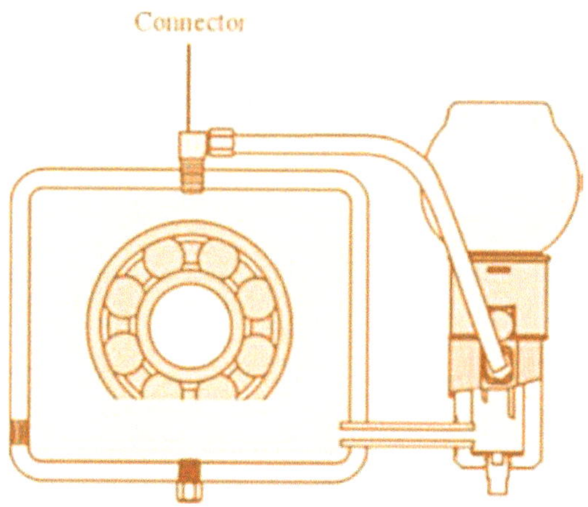

Figure 26.9 Constant level lubricator with pressure balance between bearing housing and lubricator body (Courtesy TRICO Mfg. Co).

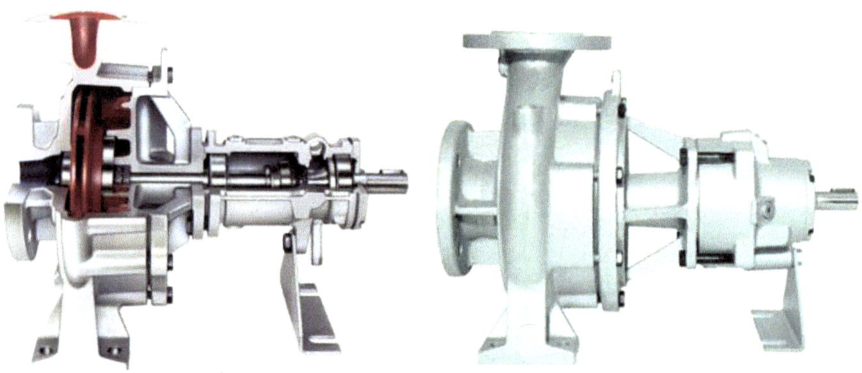

Figure 26.10 Typical direct motor driven centrifugal pump body.

26.2.6 Grease lubrication

There are many motorized pump assemblies that are grease lubricated. The most typical being a centrifugal style pump shown in Figure 26.10. The cross section shows three bearings that are typically lubricated via grease nipples on the outer casing.

26.2.7 Preferred pump lubrication method

As we have seen, there are many different ways to lubricate a pump, which way is best? Utilizing a data set of over 24,000 pumps and empirical observation

Table 26.1 Preferred lubrication method ranking table.

Oil Lubrication Method	Ranking
Oil spray, filtered/pressurized. Also, oil mist/coalesced oil coats bearing [Highest]	10
Circulation, filtered/pressurized	9
Flooded sump, w/debris	6
Sump, with slinger disc [Average Rank]	5
Sump and guided oil rings, well within concentricity specification	4
Sump and non-guided oil rings well within conc. specifications	3
Sump and non-guided oil rings out-of-concentricity [Lowest Rank]	2
Grease Lubrication Method	
Life-time PFPE-PTFE [Highest Rank]	7
Through-flow, low pressure	6
Random-flow, low pressure	4
Life-time EM polyuria	3
Random-flow, random press	2
•Random-flow, over-pressured [Lowest]	1

data a table ranking the most to least preferred lubrication method has been put together as depicted in Table 26.1 below.

References

[1] Eschmann, Hasbargen, Weigand, *Ball and Roller Bearings*, (1985) John Wiley & Sons, Hoboken, NJ

[2] Bloch, H.P., and Budris, A.R., *Pump User's Handbook: Life Extension*, 4th Ed., (2014) Fairmont Press, Lilburn, GA, ISBN 0-88173-720-8

[3] Bloch, H.P., *Pump Wisdom: Problem Solving for Operators and Specialists*, (2011) Wiley & Sons, Hoboken, NJ, ISBN 978-1-118-04123-9

[4] Bloch, Heinz P. and Abdus Shamim; *Oil Mist Lubrication: Practical Applications*, (1998) Fairmont Publishing Company, Lilburn, GA, (ISBN 0-88173-256-7)

[5] Bradshaw et al., Proceedings of the 30th TAMU International Pump User's Symposium, (2014)

[6] Wilcock, Donald F., and Booser, E.R., *Bearing Design and Application*, (1957), McGraw-Hill Company, New York, NY

[7] SKF America, General Catalog, Kulpsville, PA (2000)

[8] Aronen, Robert (see publications by Boulden Company, Conshohocken, PA and Ellange, Luxembourg)
[9] Bloch, H.P.; *Improving Machinery Reliability*, Gulf Publishing Company, Houston, TX, 1993

27

Lubrication of Wire Ropes

A wire rope is best described as a connective element employed in both static and dynamic applications servicing industry and infrastructure needs. Popular dynamic applications include mining cage hoists, elevator hoists, draglines and cranes in which the wire rope moves under tension and load to lift, hoist, and transfer motion and power. Static applications can often be found in main support systems for suspension bridges, or as a tensioned cable used to support tall free-standing structures such as a micro wave cell tower.

Depending on the application, a wire rope can be classified as a conveyor rope (cranes, winches, elevators), anchor rope (guy wires), or a load-bearing sling rope. Wire ropes (sometimes referred to as cables in gauges less than 3/8in / 10mm diameter) are primarily made of stranded wires that are combined to form ropes as shown in Figure 27.1, and Figure 27.2.

Originally developed in early 19th century Germany for the mining industry as a superior, more reliable alternative to existing metal chains and hemp rope (whose regular failure always proved catastrophic), wire rope is still manufactured in a similar manner from multiple strands of metal wire laid (wound) in a helical pattern around a center core. The center core can be made from hemp rope, plastic, fiber, or from steel (specific to aircraft cable).

The multi-strand fabrication method of wire rope provides tremendous tensile strength (100,000psi to 350,000psi depending on the grade of wire steel) for lifting and hoisting while delivering flexibility of movement needed for traction and movement recovery over curved pulley or drum surfaces. In addition, stranded wire provides superior resistance to crushing and abrasion that can result from the extreme working conditions in which the wire rope is often employed. Generally subject to tensile load, if led over return units, wire ropes are also subject to pressure, torsion and bending loads.

Wire ropes are gauged (sized) based on the number of strands surrounding the core, and the number of wires used per strand. For example, and 8x19 designated wire rope will consist of 8 strands laid around the core with each strand made up from 19 individual wires. As 152 individual wires moves across each other as the wire rope moves over the drum or sheave pulley, they

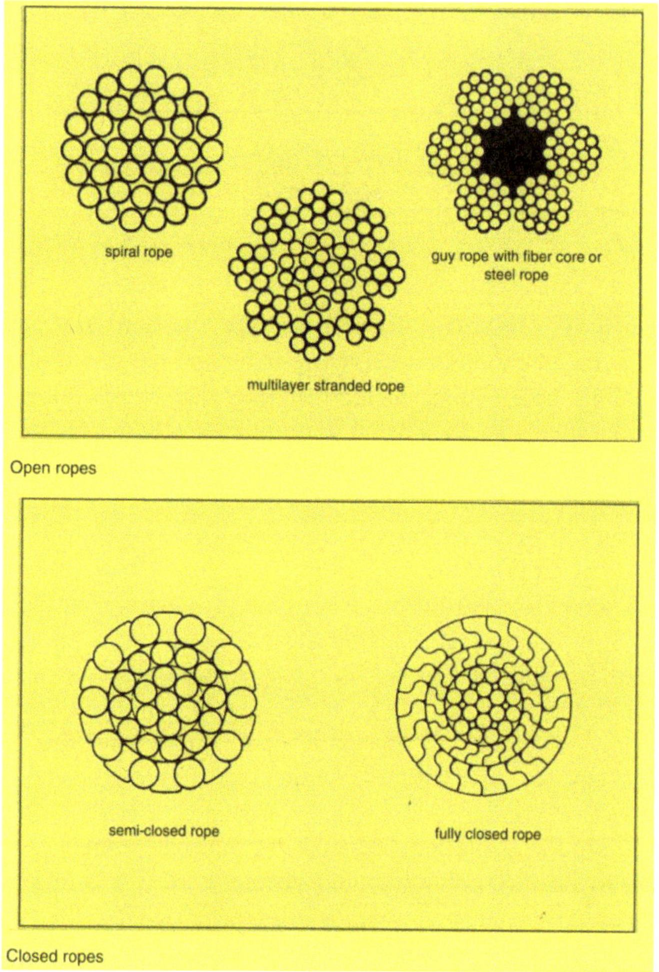

Figure 27.1 Typical open and closed style wire ropes configurations.

create both internal friction within the rope and external friction as the rope rubs against itself as it winds and unwinds on/off the drum.

If adequate lubrication is not provided on the inside, and the outside of the rope, premature wear and failure will result, with sometimes great consequence.

27.1 Wire Rope Failure

In addition to load and movement, wire ropes are often subjected to weather elements and operating conditions that can introduce contamination and

Figure 27.2 Wire rope in poor condition. Source: Photo by David Clode on Unsplash.

place heavy demands on the wire rope lubricant. Lack of a lubrication strategy can exasperate the situation and cause the wire rope to fail prematurely.

Dynamic loaded wire ropes typically fail in four ways; Fatigue, Wear, Corrosion and Core shrinkage. Fatigue is a result of many repetitive wire rope work cycles subjecting the rope to constant bending, torsional twisting and tension, which eventually lead to broken wires within the strands. These same cyclic stresses also lead to high contact pressures between the wires that in turn sets up a friction and wear cycle when the rope is inadequately lubricated. In addition, poorly lubricated ropes will allow solid contaminants between the wires as the open and close around the pulley. This contamination then sets up as three body abrasion causing accelerated wear inside the wire rope.

Because wire ropes are made from steel and can be subjected to outside elements such as moisture and acidic chemicals, unless a galvanized steel rope is employed, successful corrosion abatement will rely solely on a quality lubrication program.

Core shrinkage starts to occur when the initial lubricant charge dries out resulting in a reduced diameter and loss of support for the strands around. This is turn can cause the strands to overlap one another and lead to nicked and cut wires.

In all four-failure scenarios, effective lubrication can retard or eliminate premature failure of the wire rope.

27.2 Wire Rope Lubrication

New wire ropes come pre-lubricated from the factory, which starts to deplete once the rope is placed in service; the rate of depletion will depend on the load and working conditions.

When setting up a wire rope lubrication program, the first rule of thumb is to ensure the field lubricant is compatible with the original wire rope factory lubricant charge.

From the failure types discussed previously, a good wire rope lubricant must be able to coat the outside of the wire rope and get inside the rope in between the wires to provide a lubricant film between all the moving wires. In addition, the lubricant must provide adequate corrosion protection.

Because most wire ropes fail from the inside it is important to always use a penetrating style lubricant first. Penetrating oils are petroleum based and are designed to allow the lubricant to "creep" into the core to ensure strands and core are fully lubricated with a heavy lubricating film. This greatly reduces the friction as the individual wires move over one another. Externally, we must ensure the wire rope is protected against corrosion and rubbing wear by coating and sealing the outer surface of the wire rope using a grease or heavy oil, usually applied with a pressure stye lubricator especially designed for wire rope application

Wire rope lubricants (internal and external lubrication) ideally need to meet the following requirements:

- protection against wear
- protection against corrosion
- ensure required friction moments (friction pulley conveyor ropes)
- compatibility with rope materials
- deliver long service life
- weather resistance
- drip free
- good pumpability in lubrication systems
- availability all over the world

Not all wire rope lubricants are created equal. The correct choice of lubricant must be based on the application, load, rope construction, and working environment, all of which will require expert assistance from your local

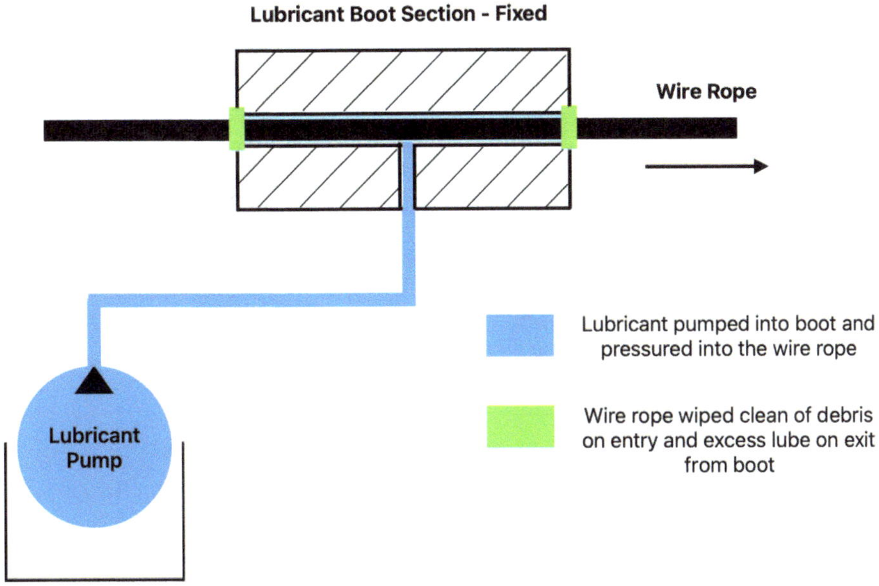

Figure 27.3 Typical wire rope clean/lubricate pressure boot system.

lubricant supplier who is best able to recommend the correct lubricant for your application.

27.3 Wire Rope Lubricant Application

If the new lubricant charge is to perform correctly a full wire rope cleaning must take place prior to re-lubrication. Ropes tend to pick up dirt in service and old lubricants can harden on the rope exterior requiring a minimum wire brush clean if only light surface debris is present, to a full steam clean for heavily soiled wire ropes. Once clean, apply the correctly chosen lubricant to the wire rope

Wire rope lubricant can be applied manually or automatically in the field. Most manual applications are performed with a brush, spray, or even a dip tank process depending on the length and size of wire rope. If the lubricant is to be manually applied, always strive to apply the lubricant at a directional change point such as a pulley sheave or drum when the rope strands naturally open up to accommodate to provide flexibility.

In the case of automatic application, a device known as a pressure boot is clamped on the rope in a tensioned straight section and the lubricant is gently pressured into the wire rope as it passes through the boot – see Figure 27.3.

Bibliography

Bannister, Kenneth E., "Don't Let Your Wire Ropes Hang You Out to Dry". *Efficient Plant magazine*, July 2018

28

Lubrication of Chains

The industrial world relies heavily on sprockets and chains to connect drive-shaft to driven shaft, and to transport goods and materials from place A to B in endless loops.

Chains provide industry a relatively inexpensive, efficient and reliable method to power transmission and conveyance. Most modern-day chains are made from steel or plastic and can be divided into four specific categories:

- *Drive chains* – a pedal bicycle or motorcycle chain, commonly referred to as roller chain,
- *Control chains* – an automotive silent "hyvo" (inverted tooth) style timing or transmission chain,
- *Transport chains* – an overhead power and free car assembly line conveyor chain.
- *Lifting chains* – a sluice gate lift, or chain hoist loop chain,

The majority of chains continue to be manufactured from formed metal sections, known as links designed to locate into rotating sprocket teeth. These links are drilled at both ends to take a bushed or non-bushed pin used to join it to the next link to allow the multi-link chain to articulate around a drive, idle or driven sprocket with close to 270 degrees of motion. Figure 28.1 shows a typical duplex roller chain drive system.

Larger conveyor chains utilize rolling element bearings where the locate/swivel pin acts as the shaft. Figure 28.2 illustrates a typical cross section of a roller type chain link pin connection.

Figure 28.1 Duplex roller chain and sprocket drive system. Courtesy: ENGTECH Industries In.

28.1 Chain Failure

Because industrial chains perform complicated movements in a variety of working conditions and environments, they create a unique set of lubrication demands due to:

- Oscillation movement of the friction components that create a permanent state of demanding sliding and rolling friction,
- Line contact of pins, bushings and rollers that create high surface pressure,
- Intermeshing of chain links and sprocket teeth that result in high shock loads.

Most chains fail due to ineffective lubrication. Arguably, dry running of the chain due to pin and bushing wear causes the highest failure rate. This condition can prevail even when lubrication is attempted but the delivery fails to

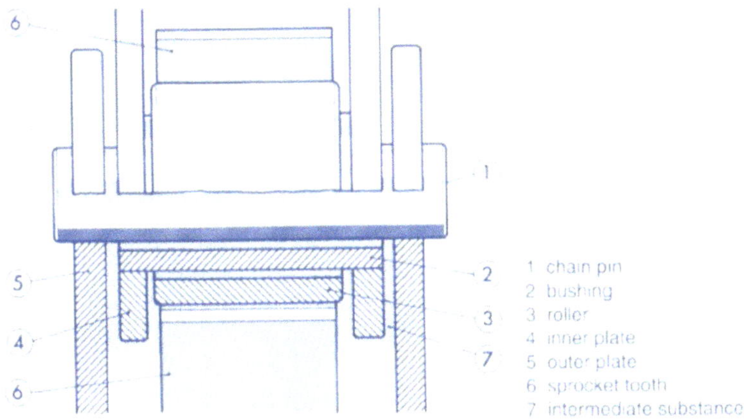

Figure 28.2 Cross section of a typical roller chain pin assembly.

place the lubricant in the bearing surface area, or through use of a non-chain friendly lubricant.

Wear causes a chain to elongate, which in turn alters the pin-to-pin link pitch that makes it difficult to mate up smoothly with the corresponding sprocket teeth. This action can result in erratic running due to the chain dragging and snatching, sometimes "skipping" a tooth that will change the driver-driven timing and cause the machine to fail dramatically! When a chain demonstrates such behavior, the tendency is to believe the chain has "stretched", but this is not the case. Chain stretch is cause by fatigue brought on by overloading the chain in tension, which in turn causes the chain to literally be pulled apart!

Despite their faults, chains compare favorably against their competition, as shown in comparison Table 28.1

28.2 Chain Lubricants

To meet the demanding service requirements for chains, combat premature wear, and stop the chain from rusting, specialized lubricants are needed to meet the many requirements that can include some or all of the following:

- Corrosion protection – ferrite metals rust quickly in the presence of moisture on unprotected surfaces. This manifests itself as a reddish-brown

Table 28.1 Drive system comparison table. Source: Tsabski, The Complete Guide to Chain, page 7.

Type		Roller Chain	Tooth Belt	V Belt	Spur Gear
Sychronization		◎	◎	×	◎
Transmission Efficiency		◎	◎	△	◎
Anti-Shock		△	○	◎	×
Noise/Vibration		△	○	◎	×
Surrounding Condition		Avoid Water, Dust	Avoid Heat, Oil, Water, Dust	Avoid Heat, Oil, Water, Dust	Avoid Water, Dust
Space Saving	High Speed Low Load	×	◎	○	○
	Low Speed High Load	◎ Compact	△ Heavy Pulley	× Wider Pulley	○ Less Durability Due to Less Engagement
Lubrication		× Required	◎ No Lube	◎ No Lube	× Required
Layout Flexibilty		◎	○	△	×
Excess Load onto Bearing		◎	△	×	◎

◎ Excellent ○ Good △ Fair × Poor

staining on the surface and throughout the oil when no anti corrosion additive is present,

- Wetting or creeping property – allows the oil to act as a penetrating oil to displace water and carry the lube to the intended and most vulnerable part of the chain that suffers the most wear,
- Adhesiveness – tactifier agents are often added to ensure the lubricant is not easily slung off the chain due to centrifugal action of the rotating sprockets and chain moving at speed,
- Temperature stability – depending on the application, a chain used in a refrigeration plant to take product through a fast freeze process, or take product through a high temperature bake oven, the oil must be able to function in both service scenarios. This requires the correct base oil type decision and oils with a high viscosity index (VI) rating,
- Low coking tendency – coking can leave residue that can fall on to conveyed product

Other selection criteria may include:

- compliance with food regulations
- environmental aspects (rapidly biodegradable)
- noise damping

28.3 Chain Lubricant Application

Chain lubricant can be applied in many different and accepted ways depending on the work application.

Manually pouring, brushing, or spraying the oil directly onto the chain whenever the chain appears to be dry can be risky business if chain failure results in unacceptable consequences. If manual lubrication is an acceptable practice, always ensure the chain is lubricated on the inside face as it moves into the sprocket so as to ensure the sprocket receives adequate lubrication at the mating faces.

Gravity drip oilers are an inexpensive way to apply lubricant to chains either in drip form directly onto the chain or onto a brush touching the chain surfaces. These types of systems, although inexpensive, require time sent on setup and regulation so as not to "flood" or over lubricate the chain.

For chains working in contained spaces, such as in an engine or in an enclosed guard system, an oil bath lubrication design in which the lower portion of the chain runs directly through a flooded sump portion of the chain enclosure is a preferred application practice. The chain should only be submerged to its running pitch line for a short section of the chain to avoid overheating and churning of the oil.

An alternative in a closed area is to employ automated oil mist or continual oil spray onto the chain in droplet form. This will not only lubricate and continually clean the chain, but also allow it to run cooler and more energy efficient.

Large slow moving conveyor chains with forged links and chain pitch up to seven inches are predominantly lubricated by fully automated dynamic chain lubricators that volley or shoot a lubricant charge directly into the pin link gaps and into the open race rolling element bearings that convey the chain through the plant. Many of these lubricators are switched on when the drive amperage reaches a determined level. Link counters then count how the number of links it takes to go one or two chain length revolutions before

Figure 28.3 Opco OP-4A Automatic conveyor chain pin oiler. Source: Opco Lubrication Systems Product Catalog, 2020.

switching the lubricator off until the cycle starts again. See Figure 28.3 for a typical automatic conveyor chain pin oiler.

Bibliography

Bannister Kenneth E, "Lubed Chains Live Longer Lives", *Efficient Plant Magazine*, July, 2018

US Tsubaki Inc., "The Complete Guide to Chain", 1997

Opco Lubrication Systems Product Catalog, Freemont, MI. USA

SECTION 5

Managing Lubricants

29

Lubricant Purchasing

Lubricants are often thought of as a machine's lifeblood and if they are to protect your machine investment, and perform as designed; their needs and requirements must be understood and met with diligence. Getting the best out of your lubricants will require a cradle-to-cradle (C2C) lifecycle management strategy.

29.1 Implementing a Cradle-to-Cradle Lubricant Management Program

In a cradle-to-cradle (C2C) lubricant management program, a lubricant commences life as base oil refined from mineral crude or synthesized stocks, in which is then blended a variety of additives to make up a proprietary lubricant product ready for delivery to market.

From the refiner / manufacturer the finished lubricant is transferred into bulk containers to continue its journey to the market supplier who offloads the bulk oil into their own storage tanks. The oil is then purchased and delivered to the end-user in pre-filled pails, drums and totes. Alternatively, depending on the client's usage, it can be delivered bulk via tanker truck for transfer into the end-user's bulk storage totes/tanks. At the end user's site, the lubricant is then stored and transferred into smaller containers from which the lubrication technician fills the machine reservoirs used.

Once the lubricant has reached the end of its service life, ideally it is collected in dedicated used oil containers used only for that oil make and specification and transported back to a short term storage area. Here, it is readied for shipment back to the oil supplier to be cleaned and recycled into marketable oil stock, and once again packaged and sold once end-users. This completes the C2C cycle as shown in Figure 29.1.

Providing we purchase the correct lubricants from the onset, each time a lubricant is stored or transferred it is at risk for solids and water contamination, both highly detrimental to bearings. As an end-user, we gain control of lubricant cleanliness upon receipt from the supplier through an effective in-house

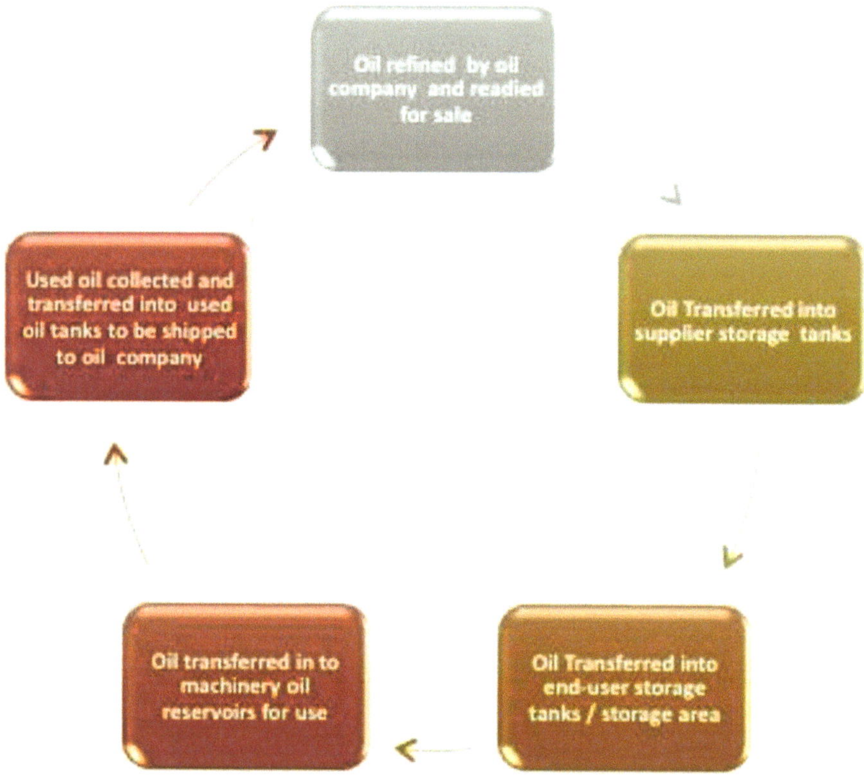

Figure 29.1 Lubricant C2C cycle.

lubricant purchase, storage and handling program. Prior to receipt, we can exercise a degree of cleanliness control through an audited cleanliness control lubricant purchase program set up in conjunction with the lubricant supplier(s).

Designing and building an end-user world class lubricant management storage and handling facility requires the end-user to view and treat lubricants with reverence using simple but effective management processes from the time lubricants are purchased, received, transferred, stored, dispensed, drained and collected for recycling. Adopting the following approach the end user will have the tools needed to put in place the best program for their facility and budget.

29.1.1 Introducing a lubricant purchase policy/program

Maintenance departments may perform their own purchasing/expediting of MRO materials, or rely on a formal relationship with the corporate purchasing department to perform that task on their behalf. In either case, it behooves the maintenance department to partner with purchasing to put in place a clear and concise purchasing process to ensure the timely purchase and receipt of the exact lubricant(s) required is successfully achieved every single time.

If any type of management software is used to track purchases within the corporation there will likely be a formal purchase request/purchase order system in place. As purchasing is a partner provider to the maintenance department, if purchases are to be received in a timely manner maintenance must meet with purchasing to review, or establish and understand the workflow procedure for the request, purchase order, and receiving process.

It is the purchasing department's job to attain requested items at the best price possible in the shortest time. If the request is non-specific, vague or unclear, for example: *please purchase 2 totes of ISO 22 Air tool oil*, you **will** likely receive the least expensive product they can find that may not entirely suit your needs. The following automotive assembly plant case study illustrates the importance of specifying your purchasing needs exactly, especially when purchasing lubricants!

29.1.2 Automotive assembly plant purchasing case study

This case study involves the maintenance and purchasing departments of an automotive assembly plant located in North America. At the time of the lubrication failure occurrence, the plant was operating a 7 day / 24 hour / three-shift operation to accommodate a running four-week backlog of vehicle orders. With an average unit selling price of $35,000 ($75,000 in 2023 $) and a line speed averaging 60 units per hour, downtime losses were unrecoverable at the rate of over $2 million sales losses per hour. ($4.5 million per hour in 2023 $)

Despite the successful order book, the corporation was in the middle of a corporate purchasing efficiency program that required the purchasing staff to individually review every purchase request and seek out less expensive alternatives, wherever possible, in exchange for a percentage personal bonus on all accumulated savings. A motivated purchasing department would now slow down every purchase as they adopted the role of a cut-price negotiator, purchasing on price rather than value. This only served to reduce the level of service to the maintenance department and to upset regular suppliers. If the original supplier didn't lower their price, the purchasing agent would then venture to seek out an alternative source based on the order specification. **If the order specification wasn't exacting** they could seek out a "similar" product, and so began the start of a very expensive lubrication failure....

Brewing a perfect storm

The maintenance purchase requisition simply stated "two totes of ISO 22 air tool oil". Because the lubrication program had recently been through a lubricant consolidation program they had consolidated their different lubricant types/SKU's by half to single digits that produced significant six-figure savings. Lubricating oils were now purchased in bulk totes from a reputable manufacturer through a reputable local supplier.

The new lubrication program had ensured that lubricants were stored separately in a lubrication crib and were adequately spill protected with a concrete berm and drain system. In addition to the lubricating oils and greases, a number of release chemical agents were also stored alongside lubricants in the lubricant crib.

To accommodate fast lubricant transfer, the maintenance department requested multiple dedicated "pogo" style pneumatic drum pumps to transfer lubricants from totes to smaller containers. At the time of the failure, the lubrication crib had only one operating transfer pump in use for all lubricants **AND** chemical fluids. Upon later questioning, purchasing would admit they had made a conscious decision to only purchase one pump to save approximately $8000.

With no purchasing confines in place, only the ISO 22 lubricant viscosity specification to go on, and a motivated purchasing department with little understanding or experience in lubrication, lubricants or lubrication equipment, the perfect storm was in place.

The consequence of saving money in the wrong places

On the day of the lubrication failure it was a productive morning for the lubrication staff within the maintenance department who had just received a new shipment of ISO 22 viscosity air tool oil that appeared to have new SKU numbers and markings on the tote. Thinking nothing of it they went about their business of dispensing a number of release chemicals and filling their corresponding reservoirs. Next, they dispensed ISO 220 gearbox oil and filled the corresponding gearbox reservoirs around the plant while the second lubrication crew now used **the same** pogo transfer pump to dispense the newly received ISO 22 air tool oil and went about their business diligently filling up air tool FRL units (Filter, Regulator, Lubricator) around the plant.

An assembly plant utilizes a lot of air tools and clean lubricated air is essential for the air tools to work properly. Within hours, one by one, assembly stations started to shut down as the air assembly tools started to fail. Maintenance was called out as the assembly line started to shut down and was shocked to find all air tools, lines and fittings were coated in a gummy plaque that couldn't be flushed or cleaned without mechanical assistance. The assembly line came to a complete halt and the second shift was sent home when the enormity of the situation was evaluated.

The third shift was told to stay home as engineers and maintenance staff frantically worked together to put in place a series of rented air compressors, new tools and flexible hose to get the assembly stations back up and working and get the line started, which they did in time for the third shift.

After suffering the loss of approximately one and a half unrecoverable production shifts additional cost was absorbed in the dismantling and cleaning of the original air tool oil system and interim rental costs.

Findings

A RCFA (Root Cause Failure Analysis) investigation found that the combination of the new air tool oil additive chemistry, purchased from a recycled oil blender was different to the original air tool oil and when combined with the gearbox oil and release chemical it acted as a catalyst that caused the resultant fluid to become highly viscous and gummy, which in turn caused the air tool bearings to overheat, degrade and rapidly fail.

Purchasing had managed to save the corporation close to a $50,000 thousand dollars in lubricant and lube pump costs but had inadvertently created an immediate downtime loss of over 700 vehicles at a retail value amounting to $25 million ($54 million in 2023 $) plus the cost of repair; air line and tool replacement costs; lost lubricant; amendment costs to the purchase program; plus the bonus paid to the purchase agents.

Lessons learned

1. Many lubricants are not compatible with one another, or other chemical fluids – Always perform a compatibility test prior to changing out lubricants for a different manufacturers product.
2. With engineered products and processes in place, always insist on "like for like" replacement/replenishment.
3. When developing lubricant purchase specifications, do not rely on viscosity alone, spell out the manufacturer and full lubricant specification on the purchase request and accept no alternatives without a lubricant trial.
4. When a lubrication management program is in place, accept no new lubricants in the plant without them first going through a lubricant trial exercise to see how it/they perform(s). This will require you to develop a

lubricant trial process designed to find out where, and why the lubricant is needed, what it will replace and how it will benefit the department.

5. Always use dedicated transfer equipment for each and every lubricant and chemical
6. Never store chemicals and lubricants in the same immediate area or allow their effluent to mix as they could create harmful gases

29.1.3 Introduce a service level agreement (SLA)

In the end, your lubricant purchases are only as good as your purchase specification and your business partnership service level agreement (SLA). An SLA is merely a signed agreement between two partners stating what services each will provide within the partnership, and what process each will follow to best facilitate each others jobs.

A typical purchasing SLA would minimally include all relevant workflow processes; a purchase request form with its minimum information requirements highlighted; and vendor requirements statements.

In addition, to further facilitate the SLA process, the maintenance department must first determine the exact lubricant(s) it requires for its equipment by performing a lubricant consolidation process. This process determines the exact lubricant specification listing for the entire plant. Maintenance/Lubrication department must then implement a lubricant trial process for any new chemical or lubricant introduction to the plant.

These processes, along with the consolidated lubricant specification, and reasons shown in the above case study "lessons learned" can be used to formulate an effective SLA.

30

Lubricant Consolidation

Foundational to any Lubrication Management program are four cornerstones commonly known as the four "R's" of lubrication; these dictate that the **Ri**ght Lubricant be placed in the **R**ight place, in the **R**ight amount, at the **R**ight time. As can be seen in the wartime poster shown in Figure 30.1, at least three of these cornerstones have been in place for the past century and are recognized as fundamental to the success of any lubrication management program.

A basic tenet of lubrication-management is to ensure the right lubricants are employed for specific equipment and, more important, specific operating conditions—"right" being an engineered choice based on performance and economy. Lubricant choice should never be based on price alone, but rather be the result of an engineered process that takes into account the bearing surface types, the ambient conditions, the working conditions, performance requirements, life cycle expectations, and economy.

Because every manufactured lubricant is a unique blend of base oil and additive packages, choosing the right lubricant is a specialized task. A task best left to the lubricant engineering specialists employed by the lubricant manufacturing companies to provide a lubrication consolidation service or program in which the least number of lubricant products are determined to fulfill the end user's needs and requirements.

30.1 Why Consolidate?

According to the Oxford dictionary, to consolidate means, "to make strong or combine". In the case of a lubricant consolidation program, the chosen lubricant manufacturer will perform an end-user plant wide audit to determine the least amount of lubrication products required to effectively provide the highest degree of lubrication protection for all moving elements within the plant or facility.

In a facility with no lubrication management program in place it is usual to find literally dozens of different lubricants in use—with many no longer in active use but still stocked and purchased from multiple lubricant vendors.

Figure 30.1 Wartime 3-R lubrication guide.

This invariably results in many duplicate purchases of similar classification – but not always compatible, lubricant products.

The consolidation audit team is tasked to group and classify all lubricants found on site as well as the current lubricants in use. These are then examined for their usage suitability in the current plant-working environment. With redundancy and duplication taken out of the equation, the remaining list is used as the starting point for cross matching to the new vendor's suite of products.

Further consolidation is often possible through understanding of the composition and design attributes of any proposed lubricants, which allows the lubrication engineer to better group and match products to the client's specific needs and requirements. This process usually drives the number of required products into the single digits resulting in considerable benefits for the end-user that include:

- **Reduced On site Inventories:** with less product SKUs, valuable inventory space is freed up and less operating capital is tied up in inventory
- **Reduced Purchasing Demand / Cost:** consolidation results most often in a single source supplier operating on a single blanket purchase order. This in turn, results in reduced purchase costs, expediting costs and administration cost that speeds up the purchasing process and virtually eliminates stock-out delays
- **Improved Handling and Storage:** less products require less dedicated transfer equipment and result in less real estate to manage, thereby reducing the chances of product cross contamination,
- **Improved Health and Safety:** less products result in less SDS /WHMIS requirements and simplify/standardize safety procedures,
- **Improved Maintenance Control:** with less lubricants to manage, preventive and predictive programs, as well as job tasks cab be simplified,
- **Improved Troubleshooting and Problem solving:** dedicated supplier relationships are more likely to produce invested working relationships in which both parties work together to solve any lubrication related problem

The cost of providing this type of engineered consolidation service is most often waived in return for an exclusive purchase contract for a mutually agreed time period (usually one to two years) so the question should not be why consolidate, but rather, why not consolidate?

30.2 Choosing a Consolidation Partner

With equipment availability, reliability, and uptime at stake, choosing a reputable lubrication vendor partner is paramount. When doing so, always involve the purchasing department to be part of the process as they can provide valuable insight and assistance to the process. Choosing a National or Multinational lubricant manufacturer is most likely to assure a major commitment to product support and research and development of the widest variety of lubricant products. However, this does not preclude that small specialized lubricant manufacturers cannot provide an equal or better commitment to product development and service. Always interview a number of lubricant suppliers to see how they have handled previous consolidation programs for other and similar clients. Reputation is made one client at a time, regardless of the size of the company; lubricants are major purchases, therefore **ALWAYS** follow through with reference checks. When making your choice take the following into consideration:

- **Value Added Services**: many services can be provided for free and must be considered when making a partner choice. Services can include:
  - **Engineering –** this would include the consolidation audit process, 24 hour general technical assistance, product manuals and technical datasheets/bulletins, personalized technical problem solving assistance,
  - **Training –** In house training programs on general lubrication practices, web based training programs, You tube®/Vimeo® training videos, etc.
  - **Product Identification labeling** – implementing lubrication management programs always require sumps, applicators and storage reservoirs to be labeled/relabeled with the lube product identification. Are these provided? If so, are they provided free of charge as part of the program?
- **Service Turnaround:** with reduced inventories on hand, turnaround on product delivery is critical and should be measured in hours. Plants that run 24/7 require 24/7 assistance,
- **Pricing:** with longer-term contracts you may be asked to negotiate a price escalation clause. Pricing should be balanced with your service needs when making your final decision,
- **Warm and Fuzzy Factor:** when entering into partnerships the warm and fuzzy factor cannot be underestimated. Test the manufacturers claims before you buy – are they nice to you after hours? Does the representative get back to you to answer silly questions in a timely manner?

Table 30.1 Current lubricant listing chart.

Lubricant Manufacturer	Lubricant Type	Location	Container	Amount on Hand
AB Lubricants	32 Hydraulic Oil	Main Lube Crib	55 gal Drum	2.5
Exxon	Unirex 2 grease	Cabinet 2	Cartridge	15
Shell	Omala 220 Gear oil	Main Lube Crib	20 Liter Pail	4

30.3 Preparing for the Consolidation Process

With a partner successfully chosen, the next step is to facilitate the consolidation process by preparing a list of current lubricants; all lubricants used and stored inside and outside the plant are to be listed on the chart as shown below in Table 30.1.

Identifying the "how" and "where" lubricants are used further speeds up the consolidation process as depicted in the Table 30.2 example shown below. Each equipment piece is identified on the form and each of its lubrication application requirements are listed complete by current lubricant type, specification, and manufacturer. The delivery application method is also identified on the form. Note; if oil analysis is performed, it too should be noted on the form.

Ideally the plant lubricant technician/specialist should be the person charged with the responsibility of making the list and reviewing it with the consolidation auditor. This process will make the data gatherer much more familiar with the equipment and its lubrication needs.

Once information is gathered, forms are handed over to the consolidation lubrication engineer who performs an audit requirements analysis to match the correct lubricant for each application. A report is then delivered to the end user, in which all applications are now matched to the new lubricant choice. If any previously used lubricants are identified as incompatible with the new lubricant, a change out procedure must be provided. A lubricant safety data sheet (SDS) for every lubricant recommended must be included in the report.

30.4 Implementing Change

Throughout the consolidation process all unwanted and unused open and closed lubricant containers can be collected for recycling as only

Table 30.2 Consolidation data gathering form example.

Equipment #: ABC123			Description: 1000 Ton Straight Side Stamping Press		
Lubrication Application		Lubricant Type	Delivery Method	Lubricant Specification	Lubricant Manufacturer
Bearings	Motor	Grease	Manual	Lithium EP2	Exxon
	Mains	Oil	Automatic - Total Loss	ISO 220	Mobil
Slides, Ways		Oil	Automatic - Total Loss	ISO 220	Shell
Compressed Air	Line Lubricator	Oil		ISO 32	ABC - Complube
Hydraulic System	Cushions	Oil	Reservoir	ISO 68	Petro-Canada

the consolidated lubricants are to be allowed on site once the program is activated.

With the report complete and new lubricants on order, the maintenance department must prepare for their delivery and perform a series of updates to manage the new lubricants, these will include:

- Arranging for the collection and return of old lubricants if containers are unopened and undamaged. Note: All lubricants have a shelf life that rarely surpasses 12 months—be prepared to either pay a restocking fee if accepted for return,
- Arranging for the disposal of old lubricants unacceptable for return. Employees could be offered the opportunity to take products (grease tubes) home for personal use with left over lubricant to be picked up by a reputable lubricant disposal company – be prepared to pay for this disposal service,
- Update the Asset Management PM system to reflect the new lubricant choices on all lubrication PM work orders,
- Update all Asset Bill of Materials in the Asset Management System
- Update the Inventory portion of the Asset management system to reflect all new lubricant products. Purge out old lubricant information
- Rearrange Lubricant storage space to accommodate any new container or dispensing equipment. Clean and free up space for other use
- Update the inventory storage labeling

- Update new lubricant labeling on equipment reservoirs and dispensing equipment.
- Update SDS manuals
- Update Equipment manuals
- Update Purchasing records

30.5 Program Monitoring

Not everyone gets it right the first time, a lubricant consolidation process is an exercise in compromise to a degree, and there may be occasions in which lubricant performance is less than desirable. Work with the lubricant vendor to detail a performance-monitoring program suitable for your work environment, and meet on a regular basis in the first three to six months to ensure all lubricants are performing at their intended level. Any exceptions can be caught early and a remedy put in place before any equipment damage is experienced.

Performing a lubricant consolidation program is a valuable exercise that results in optimized performance at the lowest cost and should be reassessed every three to five years. A consolidation exercise can considerably reduce costs by reducing the amount of different products carried in stock, which saves on operating capital, storage and handling, and administration costs. A typical company can carry over 20 lubricant types in stock that can be reduced by up to 60% depending on the type of operation, which should make the program mandatory for every plant!

31

Designing and Preparing a Lubricant Storage Area/Facility

With a new consolidated lubricant list in place, work must begin on preparing / building proper storage facilities, designed to protect lubricants from the elements, provide easy stock control, product filtration and dispensing. Utilizing the consolidation report as a starting point, perform your LOER (Lubrication Operation Effectiveness Review) for the following:

- Number of different lubricant products to be carried, stored and dispensed,
- Anticipated usage amounts for each lubricant—used to recommend the most economical purchase and storage container size. This data is also used to determine the real estate requirement for **three to six months** supply of each lubricant at the most (to ensure lubricant is always fresh), and the type of filtering, dispensing and transfer equipment needed to ensure only contamination free lubricant ends up in the bearing surface area,
- In-plant geography of where each lubricant is to be transported to for use. This facilitates to determine economical lube routes and the logistical requirements (number of lubrication technicians required, lube truck, fork lift, lubricant delivery equipment, etc. for getting the **right lubricant(s)** to the **right machine(s),**
- Machine mapping for lubrication points and access control (**right place** in the **right amount**)
- Competence training and certification of dedicated lubrication staff (**right person**)
- Processes and procedures for lubricant handling and work management
- Work Orders and schedule for lubrication work (**right time**)
- Safety Data Sheets and staff training requirements.

This review now enables the end user to commence development of a framework for their lubrication management system commencing with the specification and design of the program lubricant storage and handling facility. Table 31.1 depicts typical attributes of a lubricant storage and handling facility that must be factored into the design. A best practice design and a reasonable alternative design recommendation are shown.

31.1 Attributes of a Lubricant Storage Facility

Ideally, lubricants should be stored and dispensed in a controlled manner in a temperature controlled closed room or facility. This can be inside the plant or close by to the plant operations. If this is not possible due to expense or real estate confinement, at the very least the lubrication area must be a controlled entry space requiring it to be fenced off. Figure 31.1 shows a popular alternative that utilizes a shipping container converted into a clean storage and dispensing area, ideal for inside or outside placement for smaller facilities.

31.1.1 Location/size

As with any real estate, a good location is paramount; logistically important for receiving lubricants and dispatching them throughout the plant, larger facilities may also require controlled satellite locations. As equally important is protecting the virgin stock lubricants from the outside elements of extreme hot and cold temperatures. Large temperature swings promotes moisture and condensation in a container's headspace area; wind and rain can create a solids and water contamination environment if the lubricant containers are not stored or handled with extreme care prior to lubricant transfer. Figure 31.2. Shows full drums neatly stored on pallets in an access controlled area with no roof protection fully exposing the drums to direct rain exposure. Note: drums are stored upright (not tilted) allowing water to sit and penetrate through the bung contaminating the virgin stock oil before use. Water penetration could have been avoided by installing a simple plastic drum cap that expels water to the side. Drums stored outside are also exposed to large temperature swings through the day and night that will cause condensation inside the headroom space in the drum.

All locations must be rated for, and large enough to allow fork lift truck use to easily move drums and totes in and out of the lube room. Indoor locations are ideally temperature controlled, large enough to house 3-6 months of inventory, and contain a plant exterior wall for a dedicated shipping and receiving dock/door whenever possible.

Table 31.1 Design attributes of a lubricant storage and handling facility – aka "Lube Room".

Attribute	Best Practice Design	Alternate Design
Location	Indoors in a contained temperature-controlled environment	Outdoors in a contained environment protected from the elements
Size	Big enough to house 3-6 months supply of lubricants based on expected number and turnover rate of all lubricant products. Must be rated for fork lift truck use and big enough to store waste lubricants	Big enough to house 2-3 months supply of lubricants based on expected number and turnover rate of all lubricant products Must be rated for fork lift truck use and big enough to store waste lubricants
Ventilation	Motorized cross flow ventilation system c/w fresh air and exhaust ventilation /fan units	Prevailing wind vents and exhaust fan units
Fixtures	Storage containers and racking custom designed to suit the consolidation inventory flow, dispensing requirements and space limitations	Shelving designed to take stacked drum and pails containers
Transfer equipment / filtration	Dedicated filter transfer equipment for each lubricant type	Dedicated filter transfer equipment for each lubricant type
Spill Control	In floor central collection tank backed up with dry spill collection	Bermed or dyked areas with dry spill collection
Safety	Permanent plumbed eye wash station and portable eye wash station. Posted SDS Binders at entrance and online	Portable eye wash station Posted SDS Binders at entrance and on line
Stock control	FIFO* Stock control	FIFO stock control
Used and Waste oil control	Dedicated tanks or drop areas for each used oil product in the plant. Central collection tank for waste oil collection	Dedicated oil lay down area to store containers clearly labeled Used oil and Waste oil ready for pickup
Identification	Dedicate specific areas for each lubricant and used / waste oils and clearly label	Dedicate specific areas for each lubricant and used /waste oils and clearly label
Processes and procedures	Develop processes and procedure for sustainable operation of the facility and train all relevant staff	Develop processes and procedure for sustainable operation of the facility and train all relevant staff

*First In-First Out

31.1.2 Ventilation

Because lubricants discharge vapors that might be harmful if allowed to accumulate, a good cross flow ventilation consisting of fresh air units and exhaust fans units is very important. Outdoor facilities may require their own cross flow ventilation if due to its location the building cannot take advantage

Figure 31.1 Shipping container lube storage and dispensing unit (Courtesy Fluid Defense Systems).

of prevailing wind vents open to air (vents should be filtered using a furnace style filter) complimented with exhaust fan units.

31.1.3 Fixtures

Based on the lubricant turn over rate and storage container economics, storage and handling facilities may contain 400gal (1600 ltr) poly totes for bulk oil distribution, custom color coded steel tank bulk oil storage and dispensing units systems, drum racks designed to take palletized drums in the upright position or drum dispensing racks set on spill control platforms with the drum positioned on its side in the rack complete with a dispensing/metering valve system. Pails can also be used with lesser-used lubricants and stacked on pallets similar to the drums. Lockable fireproof cabinets should be utilized for storage of new and open grease gun and wax containers. These are also ideal for the storage of grease guns charged with grease. Separate fireproof bins should be provided for cleanable soiled rags and for smaller empty lubricant containers.

Figure 31.2 Outdoor Storage exposes lubricants to the elements (Courtesy ENGTECH Industries Inc.).

31.1.4 Transfer/Filtration equipment

To move the lubricant to the machine maintenance must transfer the lubricant from one container to another in the most non-contaminable way possible. For bulk containers, use of dedicated transfer/filter cart style dispensing units will ensure lubricant is moved from the bulk and pre packaged supplier containers to the machine reservoir and/or dedicated closed pour containers similar to that shown in Figure 2 for transfer to the lube system reservoirs.

31.1.5 Spill control

Best practice is to slope storage room floors to a low point drain where spilled product can be collected into an easy accessible common tank for recycling. Local spills can be managed with dry spill absorbent products.

31.1.6 Safety

A permanent plumbed in eye wash station is essential when handling petroleum-based products. Up to date SDC sheets for all products and SOP's are to be posted at the entrance of the lube room.

31.1.7 Stock control

As most lubricants are only rated for between 6–12 months, stock must be rotated on a regular basis following a FIFO (First In First Out) approach to stock control. Pass due date lubricants should be returned to the supplier or recycled and the stock purchasing / usage history be evaluated and adjusted accordingly.

31.1.8 Identification

Having gone to the trouble of consolidating your lubricants clearly identify a dedicated area in the facility for each and every lubricant clearly label marking in large letters (2–3 inches high) the lubricant ID. Develop a plan drawing of the facility identifying locations of each lubricant and waste tanks and post at the facility entrance.

31.1.9 Processes and procedures

Best practices are not only rooted in the design but in a sustainable operation of the facility. Be sure to develop, map put, and train all staff on all processes and procedures related to use of the storage and handling facility 5

31.2 Outdoor Storage

Outdoor storage should be avoided if possible. Weathering can obliterate the labels on containers, leading to possible mistakes in selecting lubricants for specific applications. Furthermore, widely varying outdoor temperatures, with consequent expansion and contraction of seams, may lead to leakage and wastage. The likelihood of contamination is also increased. Water can leak into even tightly closed drums by being sucked in past the bung as the drum and its contents expand and contract. See Figure 31.2.

Extreme cold or hot weather can also change the nature of some compounded oils and emulsions, making them useless. When containers must be stored outside, the following precautions are advised:

- Ensure bungs are in place tightly
- Lay drums on their sides (Figure 31.3) in a rack positioning system. Position the drums so that the bungs are at 9 and 3 o'clock, to ensure that they are covered by the drum contents, thus minimizing moisture migration and drying out of the seals.
- If drums must be placed upright without weather protection, tilt them slightly to prevent water from collecting around the bungs, or use drum covers, or spread a tarpaulin over the drums.
- Before removing the bungs, dry the drumheads and wipe them clean of any contaminant that might get into the lubricant later. The importance of keeping grit and sand out of oil used in expensive bearings must be kept in mind.

Outdoor storage should be avoided if possible. Weathering can obliterate the labels on containers, leading to possible mistakes in selecting lubricants for specific applications. Furthermore, widely varying outdoor temperatures, with consequent expansion and contraction of seams, may lead to leakage and wastage. The likelihood of contamination is also increased. Water can leak into even tightly closed drums by being sucked in past the bung as the drum and its contents expand and contract. See Figure 31.2.

Extremely cold or hot weather can also change the nature of some compounded oils and emulsions, making them useless. When containers must be stored outside, the following precautions are advised:

- Keep bungs tight
- Lay drums on their sides (Figure 31.3). Position the drums so that the bungs are at 9 and 3 o'clock, to ensure that they are covered by the drum contents, thus minimizing moisture migration and drying out of the seals.
- If drums must be placed upright without weather protection, tilt them slightly to prevent water from collecting around the bungs, or use drum covers, or spread a tarpaulin over the drums.

Figure 31.3 Twelve drum pallet rack showing bungs at 6 and 12 o'clock. The preferred bung location is 3 and 9 o'clock. (Source: MECO, Omaha, NE).

- Before removing the bungs, dry the drumheads and wipe them clean of any contaminant that might get into the lubricant later. The importance of keeping grit and sand out of oil used in expensive bearings must be kept in mind.

31.3 Indoor Storage

Storage temperatures should remain moderate at all times. The lube room should be located away from such possible sources of industrial contamination and ignition such as coke dust, cement dust, textile mill fly, and similar forms of grit or soot. It must be kept clean at all times, with regular cleaning schedules being maintained. This applies above all to the dispensing equipment, which be kept scrupulously kept clean to minimize contamination and poor functioning.

Contamination and confusion of brands are two main things to be avoided in the handling of partially emptied containers and dispensing equipment making orderliness essential.

Dispensing equipment should bear a label that matches the container from which it was filled. Labels on all equipment and containers should be kept legible at all times, as seen in Figures 31.4 and 31.5. Drying oils, such as linseed oil, should be stored in its own fireproof cabinet outside the lube

Figure 31.4 Orderly lubricant storage with adequate spill containment. (Courtesy ENGTECH Industries Inc.)

room. Similarly, linseed oil application rags should be kept in their own fireproof pail, as they are liable to self-ignite when left together in the open atmosphere. If they get into a lubrication system, the result, of course, is faulty lubrication and stoppage.

Always use dedicated dispensing equipment for each lubricant. Contamination of the rust-and oxidation-inhibited industrial oils with detergent engine oil substantially impairs the quality of the industrial oils. Trace amounts of the detergent and other alkaline contaminants can react with the acidic rust inhibitor and cause operational problems like foaming, filter plugging, and emulsion formation.

Diesel engine crankcase lubricants with zinc additives could lead to catastrophic engine failure of EMD diesel engines, in which the zinc additives attack the silver wrist-pin bushings. Similar concerns exist in certain centrifugal compressors where zinc additives may adversely affect the reliability of sealing components.

Galvanized containers should never be used for transporting oil. Many of the industrial oils used today contain additives that would react with the

Figure 31.5 Fully labelled and color coded dispensing system complete with a locable fireproof cabinet and fireproof garbage pail shown (courtesy Fluid Defense Systems).

zinc of the galvanizing to form metal soaps, which would then clog small oil passages, wicks, etc. Moreover, contamination of zinc-free diesel-engine oils could be disastrous, as mentioned in the preceding paragraph.

31.3.1 Lubricant storage policy

Every plant requires a policy or protocol regarding the storage of lubricants so they are both environmentally safe and clean from contamination. The following list will act as a guide for developing such a policy:

- Store all Lubricants in a dry, weather protected environment,
- Store all fluid lubricants on, or in, either a self-contained spill containment platform, or concrete berm designated area, capable of containing a stored fluid evacuation of 60%, or greater,
- Clearly label and match all lubricant dispensing equipment to containers (ID and color match) to minimize cross contamination of products,

- Store all greases and lubricants in small containers in a fire proof cabinet located in any of the above areas,
- All storage areas to have a Eye Wash station mounted within X ft, in accordance with local authority and plant health and safety regulations,
- All storage areas to have a Spill kit located within X ft, in accordance with current local authority and plant environmental regulations,
- All storage areas to have an appropriate fire suppressant system (sprinkler system, fire hose cabinet, extinguisher, fire blanket, etc.) located within X ft, in accordance with the current local authority and plant health and safety / fire safety regulations,
- Post all Workplace Hazardous Material Information Sheets (WHMIS) and Safety Data Sheets (SDS) information at each storage location,
- Securely close all lids to open containers when not in use,
- Store all lubricants in their allocated areas, as depicted by posted layout plan and appropriate signage
- Keep all Lubricant storage area(s) clean, and free of all debris
- Apply 5S principles to sorting and arranging and managing the lube room

31.3.2 Practical dispensing equipment

A number of procedures, practices, and equipment are already well documented which serve to mitigate the ingress and effects of contaminants in process machinery lubrication systems. These include:

- Hermetic sealing of bearing housings
- Optimum bearing selection
- Optimum lubricant/additive selection
- Documented maintenance procedures
- Personnel training & development
- Condition monitoring (vibration, oil sampling & analysis, thermography, etc.)

The extent to which organizations are adopting any or all of the above approaches is as always a function of available resources, inclusive of capital, labor and time.

Figure 31.6 A variety of open style transfer containers-clearly contaminated (Courtesy ENGTECH Industries Inc.).

Nevertheless, not all solutions to the lubricant contamination issue need be expensive or time-consuming. All too often, substantial improvements can be obtained by simply paying attention to such basics as cleanliness of transfer (dispensing) containers.

Walking through a process plant, an alert observer will often see oil cans or substitute containers with open spouts, equipment with open fill-ports ready to accept rainwater and airborne dirt, and rusty vessels which we would not dare to use as feeding bowls for farm animals. (See Figure 31.6) Even at the better facilities, the observer may find that transfer containers are not always marked with the proper oil grade, or that responsibilities and accountabilities for transfer equipment are largely undefined.

Since even the best available lubricant or hermetically sealed bearing housing will not perform under these conditions, the replacement of any questionable transfer containers with rust-proof, well-designed transfer tools should be a priority issue for modern industrial plants. With payback periods often measured in *days*, the cost-effectiveness of these tools is utterly self-evident.

Figure 31.7 State-of-the-art utility oil transfer container complete with color-coded top and ID label (Courtesy Fluid Defense Systems).

There are numerous suppliers of high quality transfer containers to choose from when looking to replace antiquated juice containers, watering cans, and water jugs never intended for contamination controlled industrial fluid transfer. Figure 31.7 shows a purpose designed lubrication oil transfer container available in a variety of styles and colors. Its unique lid and spout design keeps oil in and contaminants out. A quick-action push-pull valve incorporated in the spout allows for the adjustment of oil flow to task demands. Responsible reliability professionals consider these devices essential lubrication management tools It makes much economic sense to go back to basic steps of ensuring lube oil cleanliness before contemplating any of the other, more glamorous high-tech approaches to optimized lubrication.

31.3.3 Lubricant transfer policy

Similar to the storage policy, a policy/protocol is required for transferring fluids

- All Industrial equipment Lubricants to be transferred and dispensed by a trained Lubrication Specialist

Figure 31.8 Transferring lubricant with the "new style" oil transfer container (Courtesy Fluid Defense Systems).

- All transportation vehicle lubricants to be transferred and dispensed by trained lift truck mechanics
- Transfer all lubricants using the recommended safety equipment designated on relevant SDS sheets, and in accordance with current plant health and safety regulations; to include gloves, protective eyewear, footwear, coveralls, etc.
- Transfer all tote managed hydraulic fluids using a dedicated filter cart pumping system (one for each lubricant type), designed to be close coupled to both the tote and the receiving hydraulic system reservoir - allowing no contamination to enter during the transfer process
- All other fluid lubricants to be transferred using color-coded, dedicated, Oil-safe storage/transfer devices. These devices are to be labeled with the lubricant type
- Transfer bulk grease lubricants using dedicated air operated grease/ transfer pump units

Figure 31.9 Typical lubricant transfer/filter cart (Copyright Des-Case Corporation, 2022).

- If any pumping/transfer device is to be used to transfer lubricant from a bulk container to a dedicated Oil safe type container, it must also be a dedicated pumping/transfer device, and labeled as such
- Clean all transfer equipment on a regular basis
- Use only approved pumping/transfer devices

On larger reservoir fills a lubricant transfer/filter cart (shown in Figure 31.9) can be employed to not only perform the lubricant transfer in a timely manner, but also to "polish" the lubricant clean through a filter as the reservoir is filled. Always ensure that if a transfer cart is used, it has been dedicated for use with only that lubricant you are transferring.

31.3.4 Lubricant ID control systems

To make sure the lubricant you want in your bearings is the lubricant being pumped into your bearings, a lubricant identification control system is required.

You may of noticed how obsessed the world has become with identification gathering and use of personal data. It seems we cannot move across a border, access money, make purchases, or enter a computer program or building without setting up an ID sign-on, or presenting our ID credentials in the form of a license or job credential card.

For ID collectors, premium identification is defined as ID that shows or describes what we look like, spells out our name, current address, age or date of birth, our occupation, and often our likes and dislikes depending on how much you have decided to share with the planet at large. If we were a piece of equipment this ID data would be classified as "nameplate data". For the most part, equipment nameplate data is prolific and readily available for use—should we choose to seek it out and use it. One would think, with all the personal data sharing we contribute toward that we would be in tune with other ID systems in use in our working environment. Sadly this is rarely the case.

Equipment nameplate data is valuable machine specification data printed or stamped on a plate or sticker attached to the machine, most often in a conspicuous place. Minimally, the nameplate spells out the machine title, model and serial number; in some cases the nameplate will contain much more detailed and relevant operations and maintenance data such as operation speed ranges, pressure settings, set up data, spare part numbers for perishable items such as belts, chains, air/oil filters and if we are lucky, the lubricant ID specification and fill rates. This data is generally more detailed in the machine's operations and maintenance manual where all recommended lubricant brands, types, viscosities and fill rates are specified. In exceptional cases, the machine manufacturer nameplates the lubricant and filter requirements on the lubricant reservoirs or lube pump station(s). This is essential ID data we must seek out and review to set up a best practice lubrication program, and to continue to validate that the correct lubricant is being used at fill or change time so as to eliminate any lubricant cross contamination.

We must also be aware that this ID data could fail if a company changes its lube supplier and the original lubricant is substituted with an equivalent product and the lubricant ID is not updated.

Lubricant ID control systems are not new, they have been employed for the past 70 years or more with the most popular method employing color coding unveiled in a 1950 article in the UK Scientific Lubrication Journal by a Mr. M.J. Harrison titled "Color Codes". Harrison, an engineer with C.C. Wakefield Company (later renamed the Castrol Oil Company), detailed a symbol/color control system methodology for identifying lubricant use in an industrial plant.

In his article, Harrison recognized that a control-system methodology was needed to ensure that unskilled workers, using symbols to denote frequency of application, and colors to signify the lubricant type could carry

Table 31.2 Lubricant ID control system color chart.

Lubricant ID Control System Chart		
Symbol	**Color**	**Lubricant Description**
	Red	Oil, Gear, Mineral, ISO 220, XYZ Company, Gearlube-2000
	Blue	Oil, Hydraulic, ISO 68, XYZ Company, Hydroblue
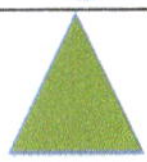	Green	Grease, Lithium, EP2, XYZ Company, Crasso 2
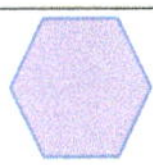	Purple	Grease, Lithium, EP1, XYZ, Crasso 1

out "factory lubrication" in a consistent manner, with scientific precision. Harrison further advocated the use of the colors on reservoirs and dedicated transfer equipment to diminish the chance of lubricant cross contamination, sound familiar?

Harrison initially promoted the three primary colors of red, blue, and yellow but in today's environment we are comfortable using both primary and the secondary color palette of green, orange and purple as evident in the numerous lubrication color identification systems now commercially available for lubricant/label identification. Figure 1 shows the color and symbolism has found new favour in today's "modernized" transfer and application of lubricants.

Even with color coding in place, the system can still fail when a lubricant is substituted and the machine specification on the machine and in the O&M manual is not updated to reflect the yellow or red lubricant change from X brand specification to Y brand specification lubricant. Furthermore, color-coding systems can also fail if the maintainer is color blind, unless the lubricant ID is accompanied by color within a control symbol and spelled out alongside the Color/Symbol on the ID nameplate located on the reservoir or at the lube control point, or on the lubricant ID control system chart depicted in Table 31.2, given to all operators, lubrication technicians, maintainers and posted on or near the lubricated equipment.

In color palette depicted in the Table 31.2 example, tells us this is a small operation using two greases and two types of oil. In more sophisticated operations using many types and viscosities the symbols can be used for viscosity and grease NLGI number differentiation, and the colors for the type

Table 31.3 Color-coded lubricant listing (partial) (courtesy ENGTECH Industries Inc).

TOX #	COLOR	Product name	Vendor	Manufacturer	Replaced Product (If relevant)	Oil ISO Grade	Grease NLGI Grade	Container	Purchase Unit of Measure	Application
										Hydraulic Oils
131966	■	MetLube H200AW	Metal Working Lubricants	MetLube	Oil Red 1308	46		Tote	Liter	Mineral base oil used for hydraulic systems
5611	□	Quintolubric	Blachford	QuintoLubric		68		Tote	Liter	Synthetic (polyester) hydraulic oil used for high-temperature hydraulic systems
										Lubricant Oil
015758	⬯	Stuart #F4611	D A. Stuart Oils of America	Stuart				Pail	Pail	Penetrating Oil
014868	⬯	LPS 2	ABC Factory Supply					Spray Bottle	Piece	General purpose light lubricating Oil
139108	⬮	Teresstic SHP 320	Metal Working Lubricants	Mobil	Mobil DTE Oil AA	320		Pail	Liter	Synthetic Oil
036126	⬯	Buehler 103330128	ABC Factory Supply	Buehler				Pail	Gallon	Machine Cutting/Polishing Oil
018957	⬮	Spartan EP 220	ABC Factory Supply	Exxon		220		Pail	Liter	Light Oil (Sulphur- phosphorous additive)
139115	⬯	Teresstic SHP 150	ABC Factory Supply	Mobil		150		Pail	Liter	Synthetic light Oil

of lubricant in use. In most cases it is relatively simple for most companies to develop their own color system that makes sense to them based on the lubricant consolidation final lubricant list. Table 31.3 shows a partial consolidation list listing each lubricants data with a color and shape.

It is important that a lubricant's color/shape ID control appear on all element that come into contact with that lubricant that can include:

- Bulk storage container (Tote, Pail, Tank, Drum, etc.)
- Small OEM containers (1 and 5 litre oil containers, grease cartridges, Single point auto lubricators
- Transfer pump, pump/filter cart
- Oil transfer container, grease gun
- Grease nipples, oil couplings
- Machine reservoirs – Oil and grease
- Used oil containers Paint lines around lay down areas for receiving areas (new and used lubricants)

31.3.5 Setting up and maintaining a lubricant ID control system

To set up and maintain a simple lubricant ID control system requires the following steps:

1. Work with your lubricant supplier to perform a lubricant consolidation program to determine the minimum number of lubricants required for use on site
2. Catalog and document all lubricant points and reservoirs by machine against the recommended lubricant consolidation list

3. Develop a lubricant ID control chart as depicted in Table 31.3 designating a symbol and color for each recommended lubricant on the consolidation list
4. Print out individual lubricant ID name plates or stickers (these are often supplied by the lubricant supplier, remove all old lubricant references from the machine and attach the new ID plates to lubricant reservoirs or relevant lubricant pump points
5. Update all PM job tasks with new lubricant descriptions,
6. Update all relevant lubricant specifications in the O&M manual
7. Update the inventory control system to close out old products and enter new lubricant product information,
8. Update all equipment bill of materials (BOM) to now reflect only new lubricants,
9. Update any previous symbol/color code references to reflect the new ID control system chart,
10. Purge all old lubricant transfer containers and grease guns and replace with new dedicated color-coded transfer equipment and transparent colored grease guns to match the new lubricant ID control,
11. Flush all old lubricant pumps using the oil supplier's recommended flushing oil and procedure,
12. Put in place a change notice requirement process to remove, replace, or add a lubricant in the plant or to the ID control list,
13. Train all persons involved in the engineering, purchase, issue, storage, or application of lubricants in the plant on the Lubricant ID control system. To prevent cross contamination of lubricants during the transfer process all lubricants in non-OEM marked storage container are required to be labeled for environmental, safety, and good management reasons. This includes machine reservoirs and all storage, transfer and handling containers.

Figure 31.10 shows some of the color ID pieces commercially available and 31.11 show how labels that utilize the color/shape/language method of display are easily attached to storage/transfer devices. Also shown are grease nipple ID caps and matching color tape for reservoirs and grease guns, etc.

Figure 31.10 Lube Label utilizes a yellow color, a shield shape and a description of the lubricant (courtesy ENGTECH Industries Inc.).

31.3.6 Receiving lubricants

In smaller facilities, most purchase lubricants are received in smaller containers and drums as shown in Figure 31.4. These are relatively easy to receive and with proper handling are put into stock with relative ease.

Larger facilities, or facilities that consume a lot of one particular lubricant due to their process, will consider purchasing their lubricants in bulk totes similar to that shown in Figure 31.12. Small totes, can be filled by transfer from a drum, but larger totes like the 400-gallon (1600 liter) style shown, are usually filled on site by a tanker truck or replaced full on an exchange program.

Prior to initial shipment, oil is tested for base oil viscosity, flash point, additive package composition and concentration known as the "treat rate", and its level of contamination or cleanliness; if the batch meets the lubricant design requirements it is given a Certificate of Analysis (COA), which is copied to the lubricant purchaser. This COA document is important as it acts as a baseline measurement for all corresponding quality checks prior to machine "point of use".

To ensure your supplier provides you with the cleanest bulk lubricant product when filling tote container follow these simple rules before implementing bulk lubricant delivery in your new storage and handling facility.

Figure 31.11 Lube Labels hang off a dispensing tank utilizing a blue circle and red hourglass with their written descriptions to identify two different lubricants metered side by side (Courtesy Fluid Defense Systems).

- Write into your contract that you will receive a lubricant Certificate Of Analysis (COA) for each lubricant delivered. Keep this document on file until the batch of lubricant has been used before archiving
- Never assume all lubricants are delivered as per their COA document specification,
- Set up a delivery acceptance agreement with the supplier to deliver lubricant based on the COA and/or a set of internal minimum cleanliness - usually a minimum ISO cleanliness level of 18/15/12 and viscosity specifications (within +/- 10% of COA specification),

Figure 31.12 400-gallon (1600 Ltr) refillable tote sitting on a dedicated transfer pump station (Courtesy Fluid Defense Systems).

- Establish an oil quality analysis test acceptable to you, the end user, and your supplier. Then develop a service level agreement (SLA) that outlines the lubricant condemn levels and remedial action requirements should the lubricant fail the quality test on delivery,
- Perform quality testing regularly taking a bulk sample after the tanker truck lines have been flushed prior to transfer, and from the center of any supplier pre-filled containers.

31.3.7 Fluid cleanliness delivery/Acceptance procedure for all tote managed lubricants

Now that you have set up a bulk delivery program, the totes and bulk lubricants must be monitored on a regular basis following a policy/procedure similar to that shown below.

- Upon purchase order issue, the first lubricant delivery is to be delivered in clean virgin sealed totes. The filled lubricant will have a minimum lubricant cleanliness rating of ISO 18/15/11 or cleaner as specified,
- Totes are to be dedicated specifically to your company only, and marked clearly with:
  - Product name
  - Viscosity
  - Approval or Tox # (Toxicity specification) details
  - Date the tote was put in service
- The tote cap is to be sealed directly after filling at the supplier and the seal number recorded on the accompanying invoice,
- The driver to inspect the cap seals on all full totes and note any damage. If a compromised seal is found, the driver will notify the end-user Lubricant Specialist and together they will inspect the tote for contamination. If the tote is *believed to be contaminated* the tote is to be removed from service and a credit issued for the unused lubricant. The tote is to be taken off site, drained, and thoroughly cleaned, inspected, refilled, sealed and returned to service. If the cap is determined to be acceptable, the driver will reseal the cap and date accordingly,
- Upon subsequent bulk filling of dedicated totes situated at the user's plant, the supplier's driver will assess each tote for physical damage

and both internal / external cleanliness prior to filling. IF the driver believes the current condition of the tote will compromise the 18/15/14 ISO lubricant cleanliness during the filling process, the tote will be removed from service. The driver will notify the user lubrication specialist and inspect the tote together and a credit will be issued for any substantial residual fluid found in the tote. The tote is to be taken off site, drained, and thoroughly cleaned, inspected, refilled, sealed and returned to service,

- On occasion, the user lubrication Specialist will perform lubricant analysis for quality control purposes, checking for fluid cleanliness level and water presence in the lubricant totes. If significant water or cleanliness levels higher than 20/18/14 are found, the Lubricant Specialist will contact the vendor to show the results and the tote will be removed from service. The tote is to be taken off site, drained, and thoroughly cleaned, inspected, refilled, sealed and returned to service,
- Once the totes are determined fit for refilling, they will receive product from a clean bulk fill nozzle. All product delivered is to be dispensed through a 5 micron absolute filter and metered through a temperature compensated pumping system,
- Once filled, the totes are to be immediately capped and sealed, unless a fluid sample is to be taken by the Lubrication Specialist at this time. The seal #'s are to be recorded on the invoice and for FIFO stock control purposes. Driver may be asked to assist the Lubrication Specialist in the arranging and maintaining of the tote FIFO inventory control,
- All polyethylene (see thru plastic) totes installed for bulk fluid containment will not remain in service longer than 6 months, and are to be replaced when emptied after 6 months with clean fresh totes,
- All metal totes are to have a protected exterior tank level sight gauge/ tube installed,
- The driver is to maintain a record of dates, volumes, deliveries, and removals of totes, product and seals. The record is to include any problems, requests, and proposed solutions incorporated, and to whom they were addressed. The driver is expected to share copies of this record with the user lubrication specialist.

31.3.8 Stock rotation procedure

In today's consumer savvy world, many people are aware of the consumable item "best before date," found on most product packaging and labeling materials. Provided as a "when to use by" guide, these dates are based on a manufacturer's projected shelf life of a product when stored as directed, per information on its packaging.

Of course, using a product after expiration of its "best before date" doesn't necessarily mean that it's not usable. It means that while the quality may have begun to decline, the product is still generally safe to use. At that point, the consumer must exercise judgment through sensory perception, *i.e.*, usually the good old "sniff" test with food, to determine if the product is still good and usable.

Industry also utilizes many consumable products with recommended shelf lives. For such products, the "best before date" is used as the basis for effective rotation, control, and management of stored or inventoried goods. Unfortunately, lubricants aren't regulated in the same way many of its chemical cousins and other consumer products are. Consequently, lubricants in our plants and facilities are rarely provided with a specified "best before date." This lack of information, however, doesn't mean that a robust strategy to store and use lubricants in an unspoiled state can't be deployed successfully (and with minimal effort on the part of the end user).

31.3.9 Why lubricants degrade

Lubricants are tailored products composed of base oil blended with a set of prescribed additives designed to meet a specific set of operational demands and conditions. When needlessly exposed to air, water, or dirt, or stored in extreme or rapidly changing temperatures and/or environments, a lubricant's additive packages will quickly deplete. This will cause the unprotected base oil to degrade as it begins to oxidize, thicken in viscosity, and lose its lubricating properties. Once the degradation process starts, the lubricant's shelf life and usefulness diminishes rapidly.

As with most products, quality components make for longer-lasting products. Lubricants manufactured with high-quality refined mineral or synthetic base stocks tend to age better than those made with inferior, lower-grade mineral base oils. In addition, lubricants containing corrosive extreme-pressure (EP) additives will degrade much faster than those containing anti-corrosion, rust and oxidation (R&O) additives.

Yes, lubricants do have a shelf life. That life is initially determined by how the product is stored and transferred throughout its staged journey from the refinery to the machine. Once it has been delivered and accepted by a qualified lubrication specialist/technician shelf life each tote, barrel, and pail must be entered into a lubricant-inventory log noting manufacturer or supplier, product and batch number, and storage-container volume size. At this point, each stored container is assigned a sequential stock-rotation number. That number, in turn, is written in large numbers on the container itself, or on a securely attached label, and recorded in the lubricant log. Once the lubricant is color-coded it is now ready for use.

Controlling contamination is best achieved by storing lubricants in an indoor temperature-and-humidity-controlled environment. If indoor storage isn't possible, and lubricants are to be stored outdoors, they must be protected from the elements, particularly rain. Large temperature changes witnessed during "shoulder seasons," *i.e.,* the beginning of spring and fall, in northern climates can cause condensation in lubricant drums. If any water contamination is suspected, the lubricant must be lab tested (ASTM D1744/D95) for water and for viscosity change (ASTM D445) prior to use. If +/- 10% of difference is found from the COA baseline, the lubricant will require centrifuging to extract the water. As previously mentioned this can be avoided by investing in outdoor drum-storage protection lids or containers.

As lubricants are transferred into machine reservoirs, product usage must be recorded and tracked in the lubricant log. To maximize shelf life, the lubricant stock must be rotated using a FIFO (First In – First Out) method. This is a relatively simple operation since all containers, at this point, will have been sequentially numbered and all usage logged. Usage logs are essential in building trend reports that determine and identify any lubricant usage anomalies requiring investigation. More important, usage logs can help determine the average turnover rate of each lubricant type and how long individual containers are stored prior to usage. This is very important information with regard to the control and optimization of lubricant shelf life. Manage lubricants as follows:

- Check lubricant tote manifest for all new product arrivals during the previous week,
- If no new product arrival, check to ensure totes are still lined up chronologically by delivery date, with the last tote to be delivered (newest delivery date) staged furthest from the door, and the "First In" (Oldest) delivery date closest to the door ready for immediate delivery in to the plant
- If new product has been shipped, or existing totes have been rearranged, check all tote delivery dates and use a fork lift truck to move and position the totes into delivery date chronological order by staging the

newest product arrival furthest from the door, and the "First In" (Oldest) delivery dated tote closest to the door, ready for immediate delivery in to the plant

- Complete FIFO check work order and return to Planner
- If any product is over six months old it should be taken out of the stock rotation

Lubricant manufacturers and suppliers/distributors often have differing opinions regarding the storage shelf life of their products (that is, if you can get them to commit to a timeframe). In the end, it is always the end-user's treatment of a lubricant that dictates the product's shelf life. The best defense is to always ensure the product is fresh and continually rotated by following a simple rule set:

- Always carefully monitor usage by container size and lubricant type.
- Purchase container size based on a three to six-month usage stock rotation. Note: Should an extended storage time be preferred (larger containers receive bigger bulk discounts), rigorously monitor them by testing the lubricant for degradation against the COA specification each month past the three-month period until a suitable extended stock rotation is determined.
- When storing and transferring lubricants in a plant environment, the following simple recommendations can help minimize contamination and maximize lubricant life in machine reservoirs:
    - Use only dedicated storage tanks, pumps, and transfer equipment, one set per lubricant. Label all equipment with the appropriate lubricant identification
    - Ensure storage tanks (including at point reservoirs) have their fill caps and breathers securely in place
    - Use desiccant-type breathers to minimize moisture contamination.
    - Implement a regular cleaning PM-work order for all tanks, reservoirs, and transfer equipment.
    - Transfer and filter oil using dedicated filter carts with quick connect couplings for pump transfer wherever possible to minimize contact with the outside air, dirt, and water contamination.

- Do not use open transfer containers that could double as a watering can.
- Never leave lubricant containers open after transfer has taken place.

Lubricants do have a shelf life. And that life is primarily in the control of the end-user. Performing basic testing, housekeeping, and taking a common-sense approach to the control and transfer of lubricant stocks will give these crucial products, once they're in use, the best chance possible to do exactly what they've been designed to do.

31.4 Guidelines for Designing a Lube Storage Facility

Ensuring your equipment and machinery receive quality lubricants from the onset requires understanding of your lubrication needs, your commitment to successfully manage storage, stock rotation, dispensing lubricants, and building the appropriate lubricant storage facility that will accommodate the corporation's lubrication needs for the next ten years.

There are three main factors that play influence the finished design of a purpose-built lubrication facility, or Lube room. The first two are simple; budget and real estate. The third is a little more introspective but arguably the most important - common sense. Common sense is needed to drive the design decision choices relevant to your immediate and future needs, balanced with the demands of your working environment.

Commencing the lubricant-storage design process requires answers to three simple questions:

1. Are you planning to expand operations in the next five years?

   If your company is planning to expand operations in the near future, you may want to consider room for growth in the original lubricant-storage design and location,

2. What are you planning to store in your lubricant-storage facility?

   Before management can commit to real estate and funding, they will need to understand the scope of the project. This requires a shopping list of **MUST HAVE** requirements, and a secondary list of **Nice to Have** requirements that can be placed into a second stage expansion if management is not willing to release the funding for all requirements immediately.

Figure 31.13 Dirty Tote that should not be accepted for filling before it is cleaned (Courtesy ENGTECH Industries Inc.).

Lubricant storage facilities can store numerous items, these will include:

- New lubricants—oils (lubricating, hydraulic, cutting), greases, pastes, and waxes
- Used lubricants—used oils for recycling that must be stored separately by oil type, and viscosity
- Bulk-storage tanks for new lubricants—may be custom-designed and color-coded tanks complete with specialized metering and dispensing attachments or they made be simple large-capacity tote bins
- Shelving—for storage of oil drums and grease pails
- Fireproof cabinets—for storing dispensing containers, sprays, and small specialty-lube products
- Spill-containment equipment
- Motorized and manual transfer-cart parking area
- Lube-truck parking area
- Forklift parking and maneuver area
- Fire-safety equipment
- Eyewash safety equipment.

The amount of storage space required for lubricants and chemicals will greatly depend on annual usage factor and product turnover rate. This can be assessed more easily once a consolidation program has taken place.

Now that the number of products and usage are known, decisions can be made based on economy of bulk purchase (this requires a larger footprint) versus the additional cost of purchasing in smaller, more manageable container sizes that will always assure fresh lubricant. Although most lubricant products have shelf lives of more than a year if stored in a clean, dry, temperature-controlled environment, always strive to rotate stock within a three-to-six month timeframe.

3. What plant real estate are you willing to commit to your lubrication-storage facility?

   Now that we have an idea of the fixtures required to manage the lubricant traffic-flow through the plant on a daily basis, we are in a position to start the lubrication facility preliminary design so that the real estate can be appropriated and a build budget prepared for funding purposes.

31.4.1 Design process

For a lube storage facility to operate successfully, it must have the following attributes:

- Direct road access for bulk oil and drum delivery, as well as used and waste oil pickup (oil pickup could be in different location if real estate is not available not part of the building design)
- Lift truck maneuverability access to promote FIFO (first-in, first-out) stock rotation and to move totes and new/used drums in and out of the facility
- Protection from outside elements such as rain, snow, wind, excess cold and heat
- Electricity, water and sewage connections
- Ability to support a concrete slab and footing walls (berms, low-point drains, high-point entrance thresholds) for environmental-spill containment.

The real-estate decision may end up with a segregated outdoor facility or an integrated indoor facility. Each has their pros and cons. For example, in Northern climates, cold winters require insulated and heated outdoor facilities. Indoors facilities are naturally heated to plant temperature, but air exchange and filtering must be piped to the outside of the plant. In both cases, good lubrication-room design points will include:

- Good lighting—intrinsically safe (fixtures designed to eliminate sparking)
- Air-quality sensing
- Spill-control design
- Temperature control
- Access control
- Ventilation air-exchange system
- Eyewash station(s) and hand wash station
- Grounded fixtures
- Fire-proofing
- Fire-suppression systems

- Fire-emergency plan logged with local fire department.

Location will have an impact on the initial facility build cost and ongoing operating costs, but in most cases, these pale in comparison to the cost of poor lubrication practices. A business case will likely need to be prepared.

With a build-specification-requirement list in hand, this information can be passed to a lubrication-storage-facility design specialist company who will be able to develop a design proposal that's assured to best meet your site's needs. You will likely only get one chance to design and build your lubricant storage facility, use it wisely.

31.5 Used and Waste Oil Management

There was a time when the terms "used oil" and "waste oil" meant the same thing and could be used interchangeably; no more. Federal, State, and Local environmental regulations have effectively redefined both terms as distinct oil states that must be dealt with in very different ways.

Because legislation differs among authorities and jurisdictions, it is the responsibility of plant owners/operators to contact appropriate authorities for clarification regarding regulations under local law defining the management, and disposal of the used and waste oils at their sites.

31.5.1 Identifying used oil

Used oil is generally defined as a product refined from crude oil or any synthetic oil that has been used in a machine and, as a result of such use, is contaminated and unsuitable for its original purpose due to the presence of impurities (water or dirt) or the loss of original properties (loss of additives).

Like virgin stock oils, used oil should be thought of as a resource that can be reprocessed *in situ* with an industrial filter cart to clean and polish the oil whilst in the machine reservoir. Or, it can be shipped to an oil recycler where it will be treated using settling, dehydration, filtration, coagulation, and centrifugation to remove contaminants and, if needed, refortified with additive package and placed back into service—all at a fraction of the cost of new oil, with no disposal management and associated fees.

Alternatively, used oil can be re-refined into lubricant or fuel oil products that can legally be sold as new oil. Re-refined products must be processed to meet the same stringent requirements and standards set for their virgin-oil counterparts. Once the re-refining is completed, the products are considered brand new oils. Less expensive to manufacture and purchase, re-refined products conserve virgin-oil stocks—10 barrels of crude are conserved for every barrel of re-refined new oil made from used oil—and minimize the negative environmental impact of oil disposal.

Typical used-oil candidates for re-refining include:

- Compressor oil
- Electrical insulating oil (except that likely to contain PCBs)
- Crankcase (engine) oil
- Gear oil
- Hydraulic oil (non-synthetic)
- Industrial process oil
- Neat (undiluted) metalworking fluids and oils
- Refrigeration oil
- Transfer oil
- Transformer oil
- Transmission oil
- Turbine oil

In some jurisdictions, used oil is allowed as a fuel oil and can be burned for heat. For little effort, and a little due diligence in managing used oil, a company may receive a small payment for the used lubricant, or at the very least, free pickup and no disposal fees. Although used oil is generally considered a commodity, in a handful of states it is viewed as a hazardous material and, as such, must be treated as hazardous waste when stored for disposal. Plants must check with their local authorities in this regard.

31.5.2 Identifying waste oil

Waste oil differs from used oil in that it reflects new oil that has become contaminated and, consequently, is deemed no longer useful for service. In the view of many jurisdictions, such oil is a hazardous waste. Used oil, cross-contaminated with chlorinated products or other chemical products, must be treated as a hazardous liquid and disposed of accordingly. Once again, it's imperative for lubrication personnel to check with their local authorities to understand the legislative definitions and requirements.

Figure 31.14 Outdoor used and waste oil storage facility example (Courtesy ENGTECH Industries Inc.).

Management tips

Collecting used and waste oil on site is a natural occurrence in any industrial plant and is allowable in all jurisdictions. There are, however, regulations regarding its labeling, storage, spillage control and disposal.

Figure 31.14 demonstrates a typical outdoor storage area for the collection of used and waste oils in a plant. Although it shows a designated area, it exposes a very ineffective and expensive oil management approach that contravenes most of today's regulations in the following ways:

31.5.3 Used or waste oil tanks must be clearly labeled and accessible

We can clearly see two grates in the photo but we have to assume if they are a pour through grate or if they are just a spill catch grate. In this case, the grates cover two large and separate below-grade tanks that are not clearly identified. In addition, grated pits are generally classified as "confined spaces" and must be clearly identified as such. Many jurisdictions today do not allow such open pits. Only one of these two restricted-access pit tanks is labeled as "Waste Oil," a fact that's partially obscured by the barrels.

Given the proximity of the two pits to each other, poor access to the rear one, and their uncontrolled exposure to outside elements, virtually all regulatory agencies would immediately classify oil pumped from both tanks as hazardous waste, requiring costly disposal procedures.

Recommendations

- Decommission the pits.
- Install two above ground steel tanks in accordance with regulations, designating one for used oil and the other for waste oil. Work with the local oil disposal company to determine its haulage capacity to size the tanks accordingly. They may even lease tanks a part of the contract.
- Clearly label each tank in accordance with local regulations.
- Move tanks into a controlled indoor space or cover the area to protect from outside elements.
- All tanks are to be bunded (placing the tank inside a leak proof berm (concrete, asphalt, or steel/plastic) controlled catch-basin area. The bund must equal or exceed the volume of the largest tank in bunded area.
- Padlock tanks shut when not in use.

Dedicated oil-transfer containers must be used to control cross contamination.

In the photo example the company has a variety of different-sized open pails containing non-descript oils and what appears to be a white chemical product. Once again, all fluids are exposed to the elements and to each another. This automatically makes all of them hazardous waste. The only way to be sure used oil does not become contaminated with hazardous waste is never to mix it with anything else and store used oil separately from all solvents, chemicals, and other incompatible products.

Recommendations

- List all oil and non-oil products used in the plant and work with your oil-disposal partner to decide which products are to be treated as recyclable used oil, waste oil, and hazardous materials (chemicals and non-oils).
- Use closed, dedicated containers for used oil, waste oils, and other products stored in the same area.

- Log any bulk transfer of oils into the tanks.
- Record all products being held in the area on a manifest and log their release with signature from the disposal company.
- Retain all records in accordance with the company's record-retention schedule.

Spill controls are mandatory

Although the photo above also shows evidence of a contained spill around the oil pallet, the contaminated spill material hasn't been removed and is now itself an uncontained, contaminated oil product.

Recommendations

In accordance with most safety legislation, every oil-storage facility will generally be required to retain, display and keep current the following information and equipment:

- Spill contingency plan and procedures
- Spill-control equipment
- Fire plan
- Emergency Response Plan (ERP) that includes an evacuation plan

If a site's oil-storage building is indoors or in a closed area, it will require ventilation as regulated by local building codes.

31.5.4 The cost of doing business

Disposing hazardous waste can be time-consuming and costly. Research local oil recyclers and hazardous-waste haulage companies to determine what they charge for their services. Some will handle both oil reclamation and disposal of hazardous waste. Such organizations should be able to work with your site to set up a value-based program that adheres to all local regulations.

Oils spill materials and Oily rags must also be managed and disposed of in the correct manner according to your local environmental legislation.

With waste oil taken care of in a responsible manner, the Cradle-to-Cradle oil life cycle is complete.

Bibliography

Harrison, M.J. "Color Codes", *Scientific Lubrication Journal*, UK, 1950

Bannister, Kenneth E., "Understand Lubricant ID Systems" *Efficient Plant Magazine*, 2017

Bannister, Kenneth E., "Manage Used and Waste Oil Wisely" *Efficient Plant Magazine*, 2017

Bannister, Kenneth E., "Part III: Take a fingerprint to Footprint Approach to Asset Management (Fluid Lifecycle)", *The RAM Review E-zine*, 2020

Bannister, Kenneth E., "Design Lube Storage with These Guidelines", *Efficient Plant Magazine*, 2018

Bannister, Kenneth E., "Do Lubricants Have a shelf Life?", *The RAM review E-zine*, 202

32

Contamination Control

Lubricants rarely get the respect they deserve until they are recognized as an integral part of a machine's design. Original equipment manufacturer (OEM) machine designers have long recognized and respected the value of choosing the correct lubricant for the job and spend considerable time in conjunction with numerous lubricant company engineers to choose a lubricant selection that will adequately perform in a given set of ambient and machine conditions, and allow the machine to function as per its design specification.

The machine designer also recognizes that a lubricant is a consumable weak link in the design and as such is subject to many stresses, such as temperature and contamination that can significantly degrade its ability to protect the machine. To combat temperature excesses, a lubricant with a suitable viscosity and viscosity index is chosen. To ensure the lubricant has a chance of an extended life cycle, a good designer will incorporate a variety of contamination control devices in the lubrication system design. Equally important in the lubricant life cycle equation is the role of the equipment owner, who must also recognize the lubricant as an integral part of the equipment design and be diligent in controlling contamination elements such as dirt and water around the machine, and care for the contamination control devices built into the machine.

32.1 The Contamination Effect

In 1986, the National Research Council of Canada (NRC) published a landmark study titled "A Strategy for Tribology in Canada". While it primarily focused on natural resource industries and agriculture its findings were remarkably similar to the Jost report, published some twenty years earlier in the UK. One of the standout statements made by the report was the recognition of failure due to ineffective lubrication practices regarding lubricant cleanliness in the following statement:

> ***"In primary industries 82% of wear induced failure was particle induced failure from use of dirty lubricants, with the greatest wear caused by particles whose size equaled the oil film thickness".***

(See Section 1, Lubrication Fundamentals, Figure 1.10, 3-body abrasion).

The study went on to illuminate the apathy surrounding lubrication failure with the following statement:

> ***"77% of respondents believed their current level of friction and wear could be reduced but no action was being taken as current levels were erroneously accepted as normal!"***

32.1.1 Three types of lube oil contamination identified

There are three primary contamination threats to lubricants, and by extension, the bearings they are designed to protect. Of these, the first one, dirt, is usually filterable; hence, it can be readily controlled. However, dirt is often catalyzed into sludge can be virtually eliminated.

The second contaminant, process-dependent dilution, can be witnessed in internal combustion engines and gas compressors where hydrocarbons and other contaminants blow past piston rings or seals and are captured within the lube and/or seal oils. Dilution results in reduced viscosity, lower flash points, and noticeable reduction of lubrication efficiency.

The last one, water, is perhaps both the most elusive and vicious of rotating machinery enemies. In lube oil, water acts not only as a viscosity modifier but also actively erodes and corrodes bearings through its own corrosive properties and the fact that it dissolves acid gases such as ones present in internal combustion engines. Moreover, water causes corrosion of pumps, and rusts cold steel surfaces where it condenses.

Furthermore, in some systems water promotes biological growth which, in itself, fouls oil passages and produces corrosive chemicals. Ironically, the maintenance department directly and indirectly causes much of the contaminant ingression into its lubricants.

Ironically, the maintenance department directly and indirectly causes much of the contaminant ingression into its lubricants

32.1.2 Contamination sources

The maintenance department as described below, can manage contamination, in all forms, once it recognizes the contamination source. The three main contamination sources are built in, ingested, and generated

32.1.3 Built-in

Not all original equipment manufacturers (OEM) are diligent in ensuring all the manufacturing process debris and assembly fluids have been cleaned prior to first time equipment start up. The same is true for any service / upgrade work performed on the equipment throughout its life cycle. It is the responsibility of the maintenance department to ensure a thorough clean up has taken place prior to lubricant fill on any equipment when any work affecting lubricated areas has been performed.

32.1.4 Ingested

Ingestion is the primary source for any second type process-dependent dilution contamination, often the result of the ambient conditions the host machine works in. In specific cases machine design may dictate that ingested contamination occurrences will take place requiring the lubricant cleanliness strategy to take this into account if the design cannot be change (more regular oil changes). This is also the case when excessive wear (wear induced combustion chamber blow-by) allows contamination to accelerate.

Ingested contamination is also the result of carelessness through forgetting to reinstall breathers, or tank opening caps.

32.1.5 Generated

The third source is a result of how the lubricant performs to protect the bearing surfaces that comes down to correct lubricant choice, preventive and predictive maintenance strategies and effectiveness of the maintenance efforts.

We can see from the type of contaminants that affect lubricant and equipment performance listed in Figure 32.1, most can be prevented from entering in the system and those that can't be prevented can be managed with the right strategy.

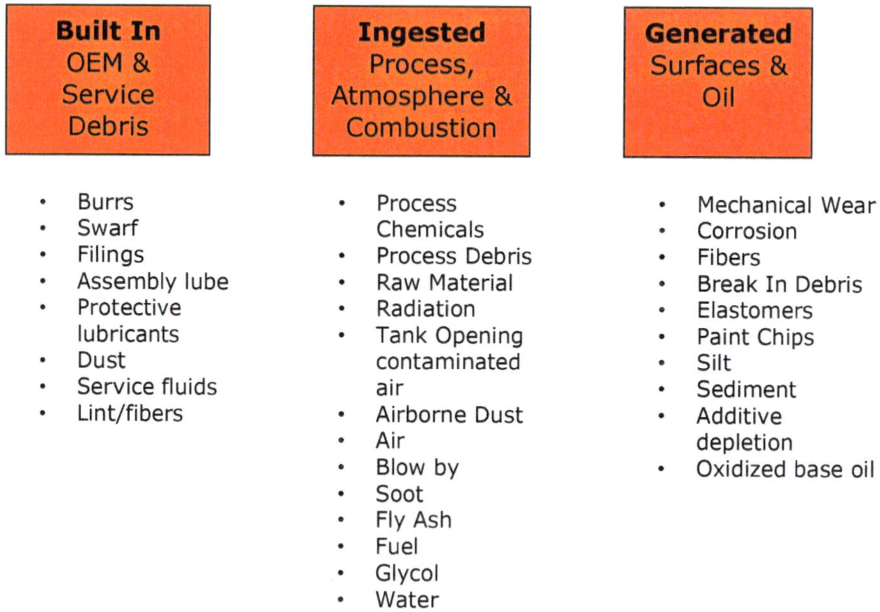

Figure 32.1 Contamination ingression causes.

32.2 Solids Contamination

Solids contaminants are usually present in the form of dust, silica, silt, sediment, lint, fibers, etc., more commonly referred to as "dirt". The larger the particulate size, the greater damage they will cause if allowed to accumulate in the lubrication system and lubricated bearing areas.

Figure 32.2 compares the solids contamination effect on roller bearing life expressed in MBTF (Mean Time Between Failure). We can see from the chart that we need to clearly focus on reducing and controlling contaminants in the less than 5 Micron range as life expectancy of the bearing effectively will triple.

The Macpherson contamination curve Figure 32.3 clearly shows the relationship between particulate size and the expected life cycle of a bearing. We can see that filtering contaminants to 4 Micron in size will triple the bearing life cycle has we only filtered to 14 Microns.

One of the primary tests performed with an oil sample is the ISO 4406:1999 test to count and measure the size of contaminant particulate into three groups of > 4 microns, > 6 microns, and >14 microns. These counts are then used to determine the lubricant cleanliness by measuring the number and size groupings of particles found in a 1ml representative sample of

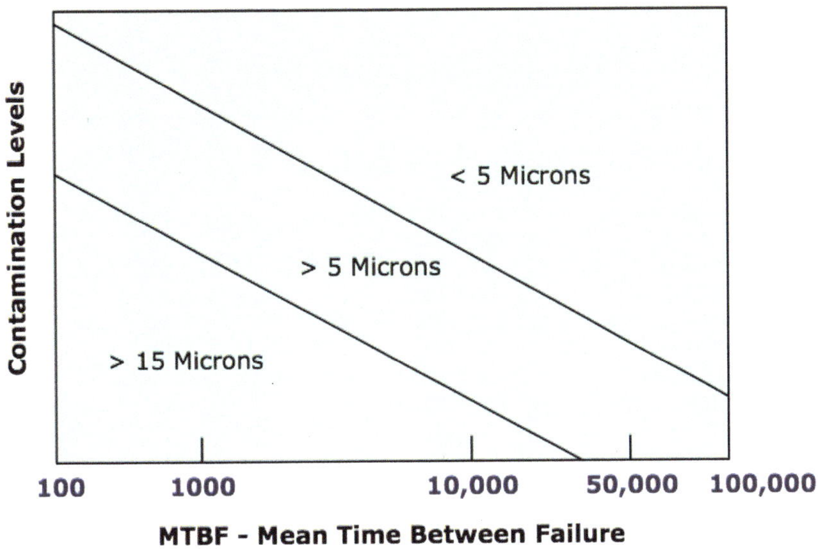

Figure 32.2 The effect of solids contamination on MTBF in hours (Courtesy ENGTECH Industries Inc.).

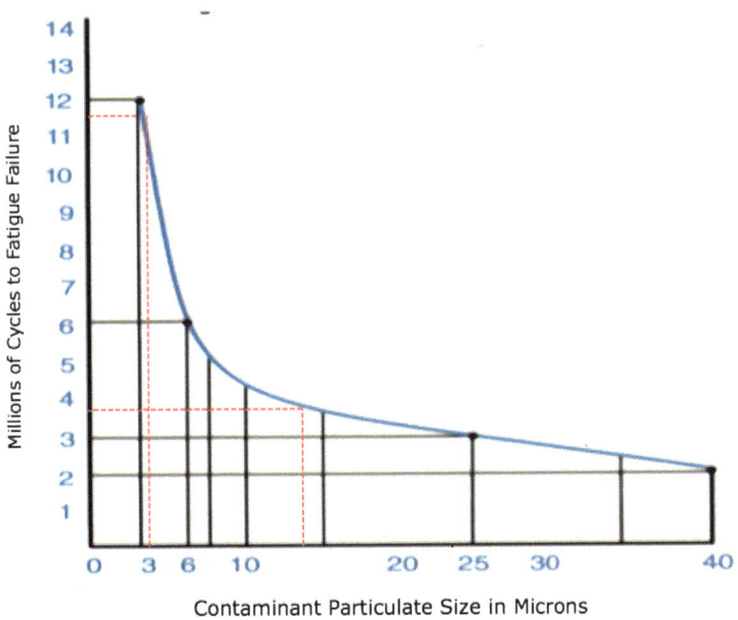

Figure 32.3 Macpherson contamination curve.

Table 32.1 ISO particle concentration range table.

Particles per mL		
>	< including	Range #
80,000	160,000	24
40,000	80,000	23
20,000	40,000	22
10,000	20,000	21
5,000	10,000	20
2,500	5,000	19
1,300	2,500	18
640	1,300	17
320	640	16
160	320	15
80	160	14
40	80	13
20	40	12
10	20	11
5	10	10
2.5	5	9
1.3	2.5	8
0.64	1.3	7
0.32	0.64	6
0.16	0.32	5
0.08	0.16	4
0.04	0.08	3
0.02	0.04	2
0.01	0.02	1

oil, which are then compared to Table 32.1 to determine the representative sample range and come up with the lubricant sample's ISO 4406 cleanliness number.

The example particulate count shown in Figure 32.4 are 2130 of the > 4 micron, 620 of the > 6 micron and 72 of the >14 micron. When applied to the range chart we end up with an 18/16/13 ISO fluid, making it reasonably clean oil close to what one would expect to find when new. Note: one micron = 1/1000th mm or 1/100,000th inch

Note: Not all lubricants are as clean when delivered and regular cleanliness checks of virgin stock oils should be part of any lubrication management program—especially when bulk oils are used. Figure 32.5, shows the

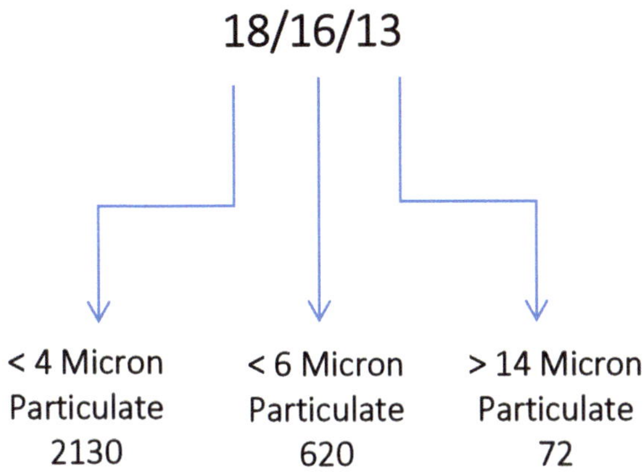

Figure 32.4 Particulate count sample.

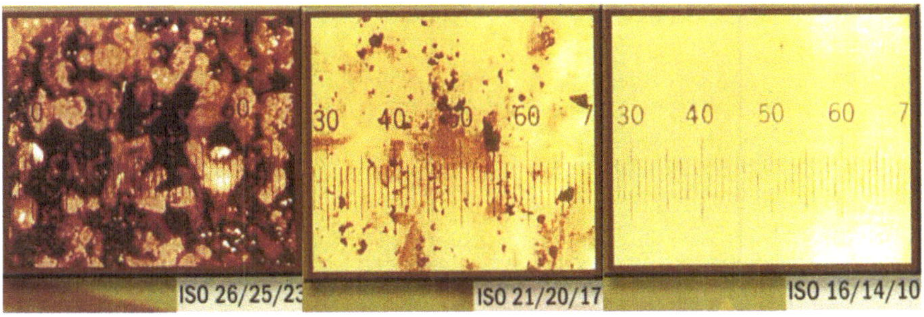

Figure 32.5 Fluids cleanliness under a microscope (Courtesy of Fluid Defense Systems).

Note: For more detailed information on particle and moisture analysis see Monitoring Oil Contamination in Chapter 35, Lubricant Condition Testing - Oil Analysis

physical difference under a microscope between a very dirty oil (26/25/23), a dirty oil (21/10/17), and a clean oil (16/14/10).

32.3 Water Contamination

In oil systems associated with process machinery, water can, and will, often exist in three distinct forms: free, emulsified, and dissolved. But, before examining the effects of water contamination, it may be useful to more accurately define these terms.

Free water is any water which exists in excess of its equilibrium concentration in solution. This is the most damaging water phase. Free water is generally separable from the oil by gravity settling.

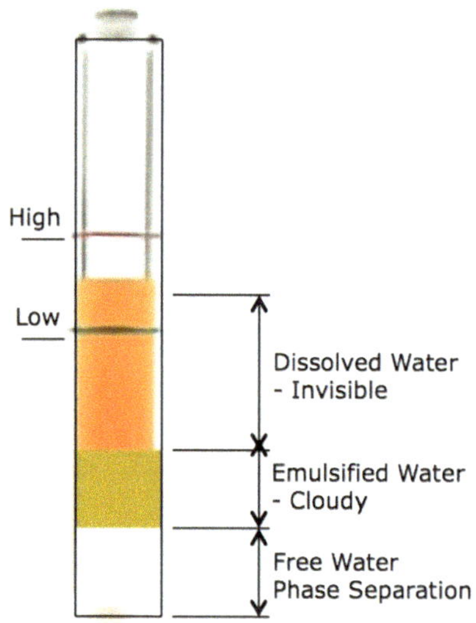

Figure 32.5 Oil reservoir sight gauge showing all three states of water presence (Courtesy ENGTECH Industries Inc.).

Emulsified water is a form of free water which exists as a colloidal suspension in the oil. Due to electro-chemical reactions and properties of the oil/water mixture in a particular system, some or all of the water that is in excess of the solubility limit forms a stable emulsion and will not separate by gravity even at elevated temperatures. In this respect, emulsified water behaves as dissolved water, but it has the damaging properties of free water and modifies the apparent viscosity of the lubricant.

Dissolved water is simply water in solution. Its concentration in oil is dependent upon temperature, humidity and the properties of the oil. Water in excess of limits imposed by these conditions is free water.

One of the best ways to recognize water ingression is with a sight gauge on a reservoir. When water is present the color of the oil changes as can be seen in Figure 32.5.

Figure 32.6 Clearly demonstrates the bearing life is reduced rapidly when water ingression rises from 0.0015% to 0.002%. Lubricating oils typically saturate around 0.04% - 400ppm; Hydraulic oil typically saturates at the 0.03% -300ppm level, whereas Transformer oil saturates at just .005% – 40ppm.

For corrosion to occur, water must be present. Free water, in particular, will settle on machinery surfaces and will displace any protective surface oil film, finally corroding the surface. Emulsified water and dissolved water

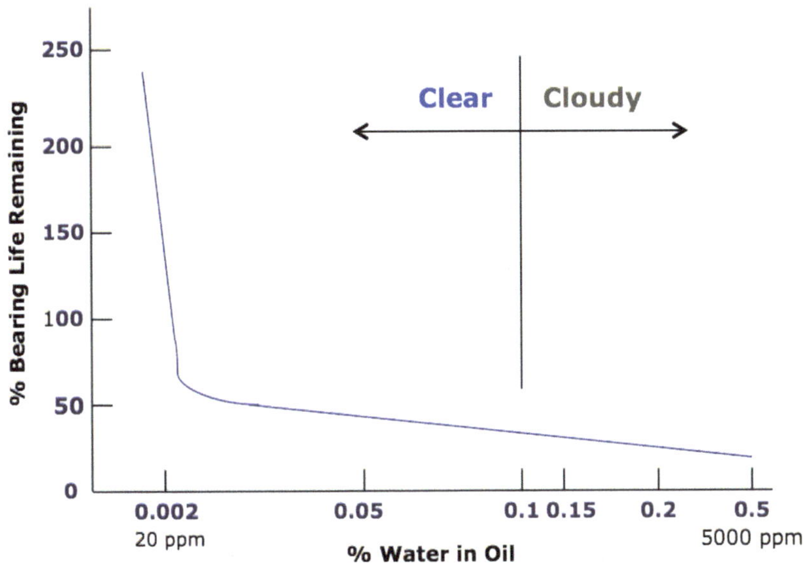

Figure 32.6 Water contamination effect on bearing life (Source: Timken Bearing Co. – 1986).

may vaporize due to frictional heat generated as the lube oil passes through bearings. Very often, though, the water vapors recondense in colder pockets of the lube oil system. Once recondensed, the free water continues to work away at rusting or corroding the system.

Larger particles generated by corrosion slough off the base metal surface and tend to grind down in the various components making up the lube system, i.e. pumps, bearings, control valves, and piping. The mixing of corrosion products with free and emulsified water in the system results in sludge formation which, in turn, can cause catastrophic machinery failures. Suffice it to relate just one of many examples of water-related damage to major machinery.

When a steam turbine at a medium-sized U.S. refinery failed catastrophically, the initial problem was attributed to coupling distress and severe unbalance vibration. When the coupling bolts sheared, the steam turbine was instantly unloaded and the resulting over-speed condition activated a solenoid dump valve. Although the oil-pressurized side of the trip piston was thus rapidly depressurized, the piston stem refused to move and the turbine rotor sped up and disintegrated. The root cause of the failure to trip was found to be water contamination of the turbine control oil. Corrosion products had lodged in the trip cylinder and, although enveloped in control oil,

Table 32.2 Water ingression effects on base oils.

Result	Symptoms
Increased Aeration	Foaming, Air entrapment, Cavitation
Increased Hydrolysis / Oxidation	Acid formation, Viscosity thickening, Varnish, , Rust, Poor filterability, Flash vaporization, EHD film strength loss
Reduced Dielectric Strength	Reduced insulating property

the compression spring pushing on the trip piston had been weakened by the presence of water.

As mentioned earlier, water is an essential ingredient for biological growth to occur in oil systems. Biological growth can result in the production of acidic ionic species and these enhance the corrosion effects of water. By producing ionic species to enhance electro-chemical attack of metal surfaces, biological activity extends the range of corrodible materials beyond that of the usual corrodible material of construction, i.e., carbon steel.

While corrosion is bad enough in a lube oil system, erosion is worse as it usually occurs at bearing surfaces. This occurs through the action of minute free water droplets explosively flashing within bearings due to the heat of friction inevitably generated in highly loaded bearings.

Additive loss from the lube oil system is another issue to contend with. Water leaches additives such as anti-rust and anti-oxidant inhibitors from the oil. This occurs through the action of partitioning. The additives partition themselves between the oil and WPPM (1%) water phase in proportions dependent upon their relative solubilities. When free water is removed from the oil by gravity, coalescing or centrifuging, the additives are lost from the oil system, depleting the oil of the protection they are designed to impart.

32.3.1 Water ingression

Water ingression can affect the base oil and the additive package in different ways as depicted in the Tables 32.2, and 32.3.

32.3.2 Ideal water levels difficult to quantify

Manufacturers, operating plants and the U.S. Navy all have their own water condemnation limits as illustrated in Table 32.4. Extensive experience gained

Table 32.3 Water ingression effects on lubricant additives.

Additive	Result
Antioxidant, Rust Inhibitor	Acid, Sludge Sediment (floc), premature depletion
Rust Inhibitor, Detergents, De-Emulsifying agent	Sludge, Sediment, Bacteria, Poor oil/water separation, Premature depletion
Borate EP	Loss of EP capability
ZDDP Anti Wear	Hydrogen sulfide, Sulfuric acid

by a multi-national petrochemical company with a good basis for comparison of competing water removal methods points the way here. This particular company established that ideal water levels are simply the lowest practically obtainable and should always be kept below the saturation limit. In other words, the oil should always have the ability to take up water rather than a propensity to release it.

32.3.3 Methods employed to remove water

Centrifuge

Centrifuges have been used for decades. They operate on the principle that substances of different specific gravities (or densities) such as oil and water can be separated by centrifugal force. Centrifuges achieve a form of accelerated gravity settling, or physical separation. At a given setting, centrifuges are suitable for a narrow range of specific gravities and viscosities. If they are not used within a defined range, they may require frequent, difficult readjustment.

They will not remove entrained gases such as hydrogen sulfide, ethane, propane, ethylene, etc., or air. Although centrifuges provide a quick means to separate high percentages of free water, they are maintenance intensive because they are high-speed machines operating at up to 30,000 RPM. More importantly, they can only remove free water to 20wppm above the saturation point in the very best case, and none of the dissolved or emulsified water. In fact, centrifuges often have a tendency to emulsify some of the water they are intended to remove.

Table 32.4 How guidelines on allowable water contamination vary. The table applies to turbomachinery lubricants.

Guidelines Issued by User or Manufacturer	Max. Allowable Water Content Quoted	Sampling Frequency	Purifier Operation
Chemical Plant, Texas	4000 wppm	Monthly	Intermittent
Consulting Firm, Europe	2000 wppm (steam turbines only)	—	–
Refinery, Europe	1000 wppm (BFW and cooling water pumps)	Monthly	Intermittent
Refinery, Kentucky	1000 wppm	—	As Required
U.S. Navy (MIL-P-20632A)	500 wppm	—	Continuous
Consulting Firm, USA	500 wppm gas turbines only	–	–
Chemical Plant, Canada	200	Visual	As Required
Utility, Michigan	100	Monthly	Continuous
Chem. Plant, Texas	100	6 Months	Intermittent
Chem. Plant, Louisiana	100	Monthly	Intermittent
Refinery, Texas	100	—	Intermittent
Consulting Firm, USA	100	Weekly	As Required
Utility, Europe	100	—	Continuous
Refinery, Texas	100	—	Continuous
Utility, Michigan	100	—	—
Turbine Mfr., Japan	100	—	—
Compressor Mfr., Pennsylvania	40	—	—
Refinery, Louisiana	—	—	Intermittent

Coalescers

Coalescers are available for lube oil service and have found extensive use for the dewatering of aircraft fuels. Unfortunately, coalescers remove only free water and tend to be maintenance-intensive. More specifically, a coalescer is a type of cartridge filter which operates on the principle of physical separation.

Figure 32.7 Portable oil centrifuge (Source: Kleanoil, 2012).

As the oil/ water mixture passes through the coalescer cartridge fibers, small dispersed water droplets are attracted to each other and combine to form larger droplets. The larger water droplets fall by gravity to the bottom of the filter housing for drain-off by manual or automatic means. Since coalescers, like centrifuges, remove only free water, they must be operated continuously to avoid long-term machinery distress. The moment they are disconnected, free water will form and begin to cause component damage. Again, because they are based on a physical separation principle, they are only efficient for a narrow range of specific gravities and viscosities.

As previously mentioned, coalescers are used in thousands of airports throughout the world to remove water from jet fuel. Water, of course, freezes at high altitudes. Refining of jet fuels is closely controlled to rigid specifications which allows successful water removal by this method. There are several disadvantages to coalescers. They are only efficient over a narrow range of specific gravities and viscosities. They do not remove dissolved water which means they must be operated continuously, and it is expensive to change elements.

Filter/Dryers

Filter/Dryers are also cartridge type units which incorporate super-absorbent materials to soak up the water as the wet oil passes through the cartridges. They remove free and emulsified water, require only a small capital expenditure, and are based on a very simple technology. However, they do not remove dissolved water, and their operation might be quite costly because the anticipated usage rate of cartridges is highly variable due to the changing water concentrations. The amount of water contamination at any given time would be difficult, if not impossible, to predict. Additionally, high cartridge use creates a solid waste disposal problem.

Vacuum oil purification

Vacuum oil purifiers have been in use since the late 1940s. A vacuum oil purifier operates on the principle of simultaneous exposure of the oil to heat and vacuum while the surface of the oil is extended over a large area. This differs from the other methods we have discussed in that it is a chemical separation rather than a physical one. Under vacuum, the boiling point of water and other contaminants is lowered so the lower boiling point constituents can be flashed off. Typical operating conditions are 170°F (77°C) and 29.6" Hg (10 Torr). Because water is removed as a vapor, there is no loss of additives from the oil system. The distilled vapors are recondensed into water to facilitate rejection from the system. Non-condensables such as air and gases are ejected through the vacuum pump.

As illustrated in Figure 32.8, the typical components of a vacuum oil purifier are an inlet pump; a filter typically rated at 5 microns; some method of oil heating such as electric heaters, steam, hot water, or heat transfer fluid; a vacuum vessel; and a vacuum source such as a mechanical piston vacuum pump or water eductor. Vacuum oil purifiers may, or may not, incorporate a condenser depending upon the application. A discharge pump is employed to return the oil to the tank or reservoir, and an oil-to-oil heat exchanger may be employed for energy conservation.

Vacuum oil purification is the only extended range method capable of removing free, emulsified, and dissolved water. Since vacuum oil purifiers can remove dissolved water, they can be operated intermittently without the danger of free water forming in the oil. Furthermore, they are the only method of oil purification which will simultaneously remove solvents, air, gases, and free acids.

In virtually all instances, a cost justification study by medium and large users of industrial oils will favor well-engineered vacuum oil purifiers over centrifuges, coalescers, and filter/dryers. Cost justification is further

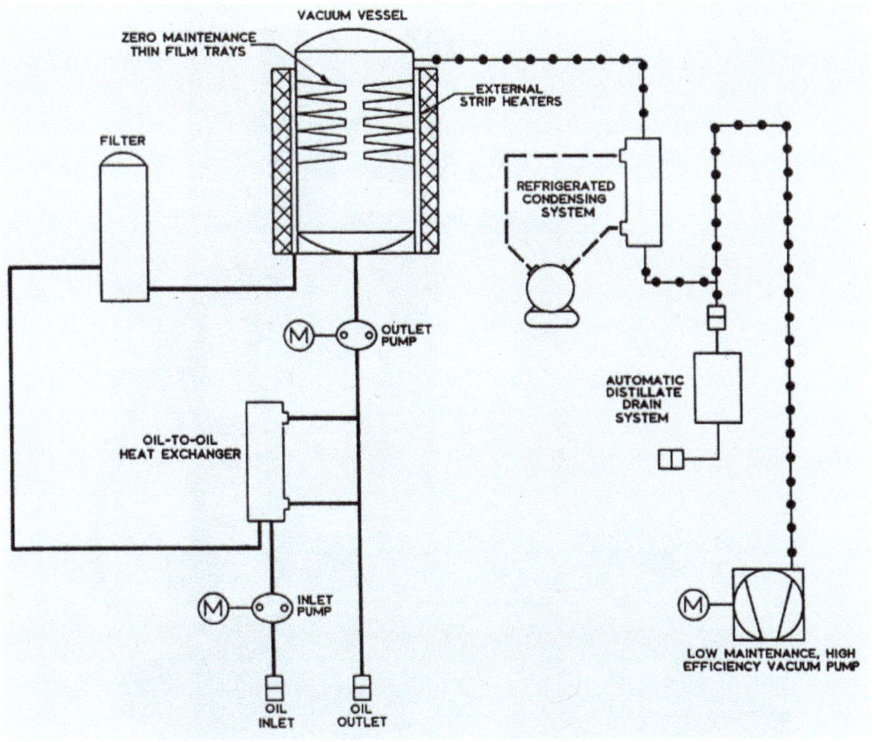

Figure 32.8 Schematic view of vacuum dehydrator. (Source: Allen Filters, Inc., Springfield, Missouri)

influenced in favor of vacuum oil purifiers by bottom-line analyses which look at the cost of maintenance labor and parts consumption.

Vacuum distillation separation

Vacuum distillation is a process derived from the distillation process used by oil refineries. It works by "expanding" the oil in a four-stage cyclical process that 1) heats the oil, 2) vaporizes the oil, 3) Cools the oil and finally condenses and reconstitutes the oil.

The staged process is performed at a controlled and reduced pressure of less than 1 atm that produces a lower boiling point (Vacuum 27" Hg = a 57C (135F) boiling point) that quickly separates the oil from the water.

This process effectively removes 100% free water and 90% dissolves water, including any solvents, refrigerants and fuels in the oil. An example of a portable Vacuum distillation unit is shown in Figure 32.9.

Figure 32.9 Portable vacuum distillation unit (Courtesy Des-Case Corporation).

Bibliography

Allen, J.L.; "On-Stream Purification of Lube Oil Lowers Plant Operating Expenses," *Turbomachinery International*, July/August 1989, pp. 34, 35, 46.

Bloch, H.P., *Improving Machinery Reliability*, Volume 1, Gulf Publishing, Houston, 1998.

Schatzberg, P. and Felsen, I.M., "Effects of water and oxygen during rolling contact lubrication," *Wear*, 12, 1986

Schatzberg, P. and Felsen, I.M., "Influence of water on fatigue failure location and surface alteration during rolling contact lubrication," *Journal of Lubrication Technology*, ASME Trans., F91, 2, 1969.

Bannister, Kenneth E., ICML Certification Series: Industrial Lubrication Fundamentals Article Series, "*Lubrication Management &Technology" magazine*, 2014

Bannister, Kenneth E., ICML/IS O MLT/MLA/LCAT Level1 Training Materials authored for *ENG*TECH *Industries Inc*, 2016–2023

National Research Council of Canada (NRC); "*A Strategy for Tribology in Canada*," 1986

33

Lubricant Condition Testing – Oil Analysis

Today's machinery and equipment can be maintained to achieve useful operating lives many times those attainable just a few years ago. For oil lubricated machinery there are many opportunities in what is commonly referred to as proactive maintenance.

By carefully monitoring and controlling the conditions of the oil, many root causes of failure can be systematically eliminated. For decades now it has been, and still is, universally accepted that oil analysis plays an important, central role in any asset management program that includes lubricated machines and systems. If a lubricant condition testing program is to succeed, the user organization must define what their goals are to be.

Some people see oil analysis as a tool to help them time oil changes. Others view it in terms of its fault detection ability. Still, others apply it to a strategy relating to contamination control and filter performance monitoring. In fact, when a program is well-designed and implemented, oil analysis can do all of these things and more. The key is defining what the goals will be and designing a program that will effectively meet them. One might refer to it as a ready-aim-fire strategy. The *ready* has to do with education on the subject of oil analysis and the development of the program goals. The *aim* uses the knowledge from the education to design a program that effectively meets the goals. The *fire* executes the plan and fine-tunes through continuous improvement.

33.1 Detecting Machine Faults and Abnormal Wear Conditions

In the past, success in fault detection using oil analysis has been primary limited to automotive reciprocating engines, power train components, and aviation turbine applications. The small sumps associated with this machinery helped to concentrate wear metals, coupled with the rapid circulation of lubricating oil kept the debris in uniform suspension, making oil analysis trending more dependable.

Table 33.1 Application of lube oil analysis. (Courtesy Noria Corporation)

	Root Cause Detection	Incipient Failure Detection	Problem Diagnosis	Failure Prognosis	Post Mortem
What Oil Analysis Is Telling You	*When something is occurring that can lead to failure-root cause conditions*	*When an early-stage fault exists that is otherwise going unnoticed - e.g., abnormal wear*	*What the nature of a problem is that has been observed.* *Where is it coming from?* *How severe is it?* *Can it be fixed?*	*That a machine is basically worn out and needs to be fixed or replaced*	*What caused the machine to fail?* *Could it have been avoided?*
What You Monitor	*Particles, moisture viscosity, temperature additives, oxidation, TAN/TBN, soot, glycol, FTIR, RBOT*	*Wear debris density temperature, particle count, moisture, elemental analysis, viscosity, analytical ferrography*	*Wear debris, elemental analysis, moisture, particle count, temperature viscosity, analytical ferrography, vibration analysis*	*Elemental analysis, analytical ferrography, vibration analysis, temperature*	*Analytical ferrography, ferrous density, elemental analysis*
Maintenance Mode	Proactive	Predictive	Predictive	Breakdown	Breakdown

As oil analysis has matured over the past decades so has the level of reported success with wear debris analysis for detecting machine anomalies in every machine type and oil, including hydraulic fluids. Table 33.1 provides a simple overview of the application of oil analysis, specifically wear debris analysis, in machine health monitoring. The various specific methods are discussed in later sections of this chapter.

33.2 Performing Condition-based Oil Changes

Each year huge amounts of oil are disposed of prematurely, all at a great cost to the world's economy and ecology. This waste has given rise to a growing number of companies to discontinue the practice of scheduled oil changes by implementing comprehensive condition-based programs in their place. This, of course is one of the principal roles of oil analysis. One might say your oil is talking, but are you listening?

By monitoring the symptoms of oil when it tires and needs to be retired, we are able to respond to the true and changing conditions of the oil (see Table 33.2). And, in some cases it might be practical to consider reconditioning the oil, including the reconstruction of depleted additives.

Table 33.2 Parameter change and its effect on remaining oil life. (Courtesy Noria Corporation)

↓ Decreasing Value ↑ Increasing Value	Percent Oil Life Remaining					
	100%	50%	25%	10%	5%	0%
Viscosity*	–	–	–	↑	↑	↑
TBN (crankcase oil)	–	↓	↓	↓	↓	↓
TAN	–	↓	↓	–	↑	↑
FTIR Oxidation	–	↑	↑	↑	↑	↑
Darkening	–	↑	↑	↑	↑	↑
Pungent Odor	–	–	↑	↑	↑	↑
Surface Tension	–	–	–	↓	↓	↓
Oxide Insolubles	–	–	↑	↑	↑	↑
RBOT Life	–	↓	↓	↓	↓	↓
Filterability	–	↓	↓	↓	↓	↓
Specific Gravity	–	–	–	↑	↑	↑
Sludge	–	–	-	↑	↑	↑

Many oil analysis test results now provide a forward-looking prediction of residual life of the oil and additives. Dirty oils can be cleaned in situ on the machine using a filter cart. Distressed oils, in cases, can be conveniently fortified or changed without disruption of production. And, those fluids that degrade prematurely can be reviewed for performance capability in relation to the machine stressing conditions.

33.3 Monitoring and Proactively Responding to Oil Contamination

While the benefits of detecting abnormal machine wear or an aging lubricant condition are important and frequently achieved, they should be regarded as low on the scale of importance compared to the more rewarding objective of contamination/failure avoidance (see Figure 33.1).

Whenever a proactive maintenance strategy is applied, three steps are necessary to ensure that its benefits are achieved. Since proactive

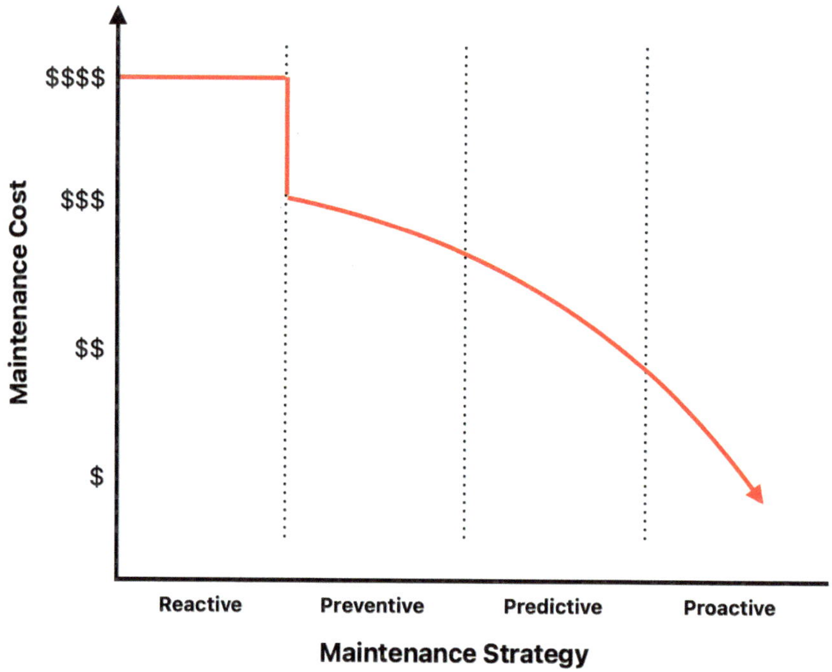

Figure 33.1 Some maintenance strategies are more costly than others.

maintenance, by definition, involves continuous monitoring and controlling of machine failure root causes, the first step is simply to set a target, or standard, associated with each root cause. In oil analysis, root causes of greatest importance relate to fluid contamination (particles, moisture, heat, coolant, etc.) and additive degradation

However, the process of defining precise and challenging targets (e.g., high cleanliness) is only the first step. Control of the fluid's conditions within these targets must then be achieved and sustained. This is the second step and often includes an audit of how fluids become contaminated so their entry points can be systematically eliminated. Often better filtration and the use of separators will be required.

The third step is the vital action element of providing the feedback loop of an oil analysis program. When exceptions occur (e.g., over target results) remedial actions can, and should, be commissioned immediately. Using a proactive maintenance strategy, contamination control becomes a disciplined activity of monitoring and controlling high fluid cleanliness, not a crude activity of trending dirt levels.

Finally, when the life extension benefits of proactive maintenance are combined with by the early warning benefits of predictive maintenance, a comprehensive condition-based maintenance program result. While proactive maintenance stresses root-cause control, predictive maintenance targets the detection of incipient failure of both the fluid's properties and machine components like bearings and gears. It is this unique, early detection of machine faults and abnormal wear that is frequently referred to as the exclusive domain of oil analysis in the maintenance field.

33.4 Oil Sampling Methods Examined

As in all aspects of life, the end result of any endeavor is only as good as the effort put in to the exercise and the quality of elements used to create the result. Such is the case in lubricant and wear particle analysis in which results accuracy is highly dependent upon the care and method used to procure and deliver a quality used oil sample into the hands of a laboratory to commence analysis.

Procuring and delivering an analysis ready, superior quality used oil sample requires discipline and consistency as depicted in Table 21.3: 7-Best Practice Principles of Oil Sampling, and is the result of choosing the best procedure, method, hardware, and sample location. The samples choices for each piece of sampled equipment will likely differ based on whether the sample is taken from a pressurized or non-pressurized system, whether the machine or gearbox is designed or set up for best practice sampling techniques, the consistency of the sampling methods and techniques, the training of the person taking the sample, and the sampling cleanliness protocol used, achieved by extracting the sample in the most appropriate place.

The best sample choices for the equipment being sampled are driven by three main objectives to 1) maximize sample data density, 2) minimize sample data disturbance, and 3) maximize sampling consistency.

33.4.1 Maximize data density through sampling point selection

Each and every oil sample carries a unique time stamped composition signature of base oil chemistry, additive package level and chemistry and wear particle type, size and count used to compare against a virgin oil sample to determine the oil's chemical condition and the machine's moving parts condition at that moment in time—the more representative the sample, the more accurate the diagnosis. Therefore, every sample must contain the maximum amount of data density (representative data) it can, best achieved by extracting the sample in the most appropriate place.

For pressurized systems, e.g., hydraulic, and recirculating lubricating oil systems, oil is pumped from a reservoir under pressure, through a series of

Table 33.3 7-Best practice principles of oil sampling. (Courtesy ENGTECH Industries Inc.)

1	Sample when machine is running under load
2	Sample Upstream of filters, and downstream of lubricated bearings
3	Before taking actual sampling, use the 10x flush rule
4	Always pre flush sampling equipment - pump/hose
5	Only use virgin clean sample bottles
6	Always sample at the same spot, using the same procedure and equipment
7	Forward sample and completed sample details to lab within 24 hours

filters in a piping distribution system to the bearing surface areas form where it is returned back to the reservoir to be once again filtered and cooled ready for recirculation. Maximum data density is always found downstream of the lubricated bearings and upstream of the return filter where it is laden with any contaminants just washed from the bearing surfaces. To assure the most representative sample, take it:

- When the machine is running at temperature and under regular working condition load,
- From a live fluid zone meaning no dead pipe legs (static areas) or line ends,
- From a sample port connected to an elbow used to create a turbulent zone and ensure a colloidal (well mixed) sample

In circulating oil systems such as the one shown in Figure 33.2, the best (primary) location is a live zone of the system upstream of filters where particles from ingression and wear debris are the most concentrated. Usually, this means sampling on fluid return or drain lines. Figures 33.3 shows typical sampling on a low pressure return lines. In the case of vented vertical drains from bearing housings there is not a solid flow of oil (air and oil share the line) making sampling more difficult. In such cases, a hardware adapter called a sample trap can be effectively installed to “trap” the oil for easy sampling.

In applications where oil drains back to sumps without being directed through a line (e.g., a diesel engine and wet-sump bearing and gear casings), the pressure line downstream of the pump (before filter) must be used. Figure 33.4 shows various options for sampling pressurized fluid lines.

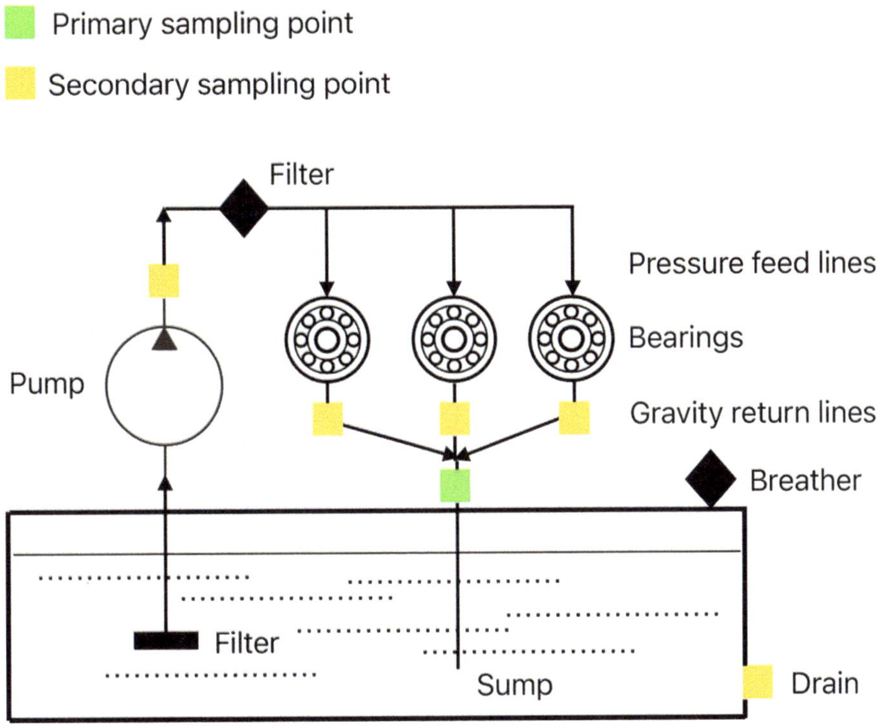

Figure 33.2 Circulating system recommended oil sampling points.

Where possible, always avoid sampling from dead zones such as static tanks and reservoirs. Splash, slinger ring, and flood-lubricated components are best sampled from the drain or casing side using a short inward-directed tube attached to a sample valve (see Figure 33.6). It may be necessary to use a vacuum pump to assist the oil flow for high viscosity lubricants.

33.4.2 Typical procedure for extracting an oil sample

Once a sampling point is properly selected and validated, a sample must be extracted without disturbing the integrity of the data. When a sample is pulled from turbulent zones such as at an elbow, particles, moisture, and other contaminants enter the bottle at representative concentrations. In contrast, it is well-known that sampling from ports positioned at right angles to the path of the fluid flow in high velocity, low viscosity fluids cause particle fly-by. In such cases, the higher density particles follow a forward trajectory and fail to enter the sampling pathway.

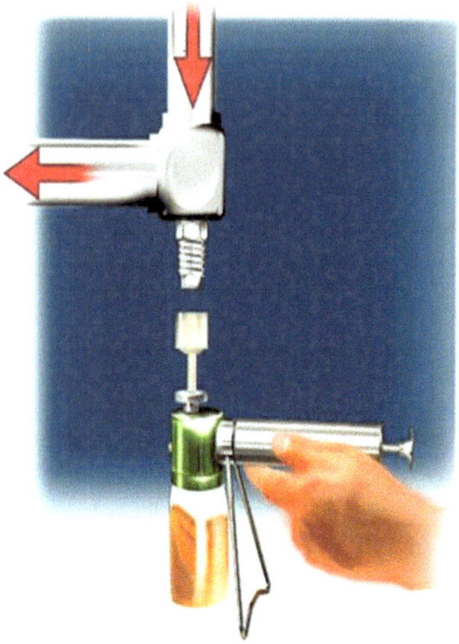

Figure 33.3 Low pressure sampling. (Courtesy Noria Corporation)

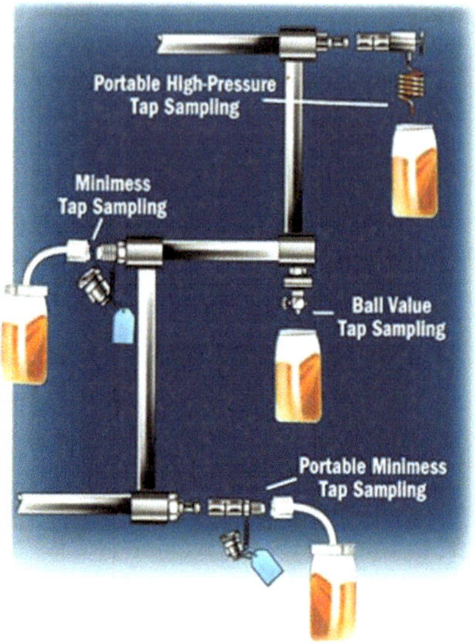

Figure 33.4 High pressure sampling options. (Courtesy Noria Corporation)

Machines should always be sampled in their typical work environment, ideally while they are running with the lubricant at normal operating temperature. Likewise, during (or just prior to) sampling, machines should be run at normal loads, speeds, and work cycles. This helps to ensure that the wear debris that is typically generated in the usual work environment and operating conditions is present in the fluid sample for analysis.

Sampling valves should be flushed thoroughly prior to sampling. If other portable sampling hardware is employed, these devices need to be flushed as well. Once the flushing is complete the sample bottle can be filled. However, never fill a sample bottle more than three-fourths full. The headspace in the bottle (ullage) permits adequate agitation by the lab.

With many non-circulating systems, static sampling may be the only option. Often this can be done effectively from drain ports if a sufficient volume of fluid is flushed through. Alternatively, drop-tube vacuum samplers could be used (Figure 33.6). Care should be taken to always sample a fixed distance into the sump. Using a rod with a marked standoff from the bottom of the tank is a reliable way to do this. Flushing of the suction tube is also important. Never reuse suction tubes to avoid cross contamination and mixing of fluids.

Static sampling using a vacuum sampler can be improved by installing a quick-connect sampling valve to which the vacuum tube is attached.

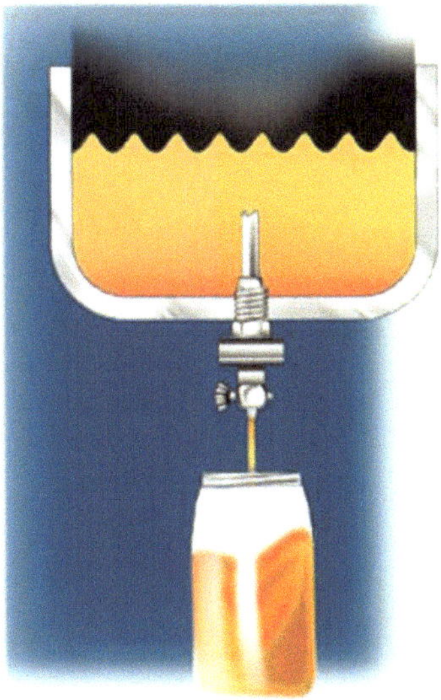

Figure 33.5 Drain sampling with tap, can be connected to a vacuum sampler. (Courtesy Noria Corporation)

Often this will require drilling and tapping, preferably in the wall of the sump or casing. It is best if the valve can be located near return lines and where turbulence is highest. Generally, it is desirable to install a short length of stainless-steel tubing inward from the valve.

Another ideal sample method employs a combination Pilot tube/level gauge device affixed at the correct reservoir sample level. As most reservoirs do not come with such devices, it will require an aftermarket purchase and installation.

33.4.3 Minimize sample data disturbance— don't contaminate the contaminant!

One of the main objectives of oil analysis is the routine monitoring of oil contamination. Therefore, in order to do this effectively, considerable care must be taken to avoid "contaminating the contaminant." Once atmospheric contamination is allowed to contact the oil sample, it cannot be distinguished from the original contamination.

It is important that the oil sample data is not disturbed or become contaminated as a result of the sampling and sample handling process itself. For example, reservoir sludge, dirty sample/drop tubes, or dirty sample bottles, etc. can all distort the data readings if not minimized.

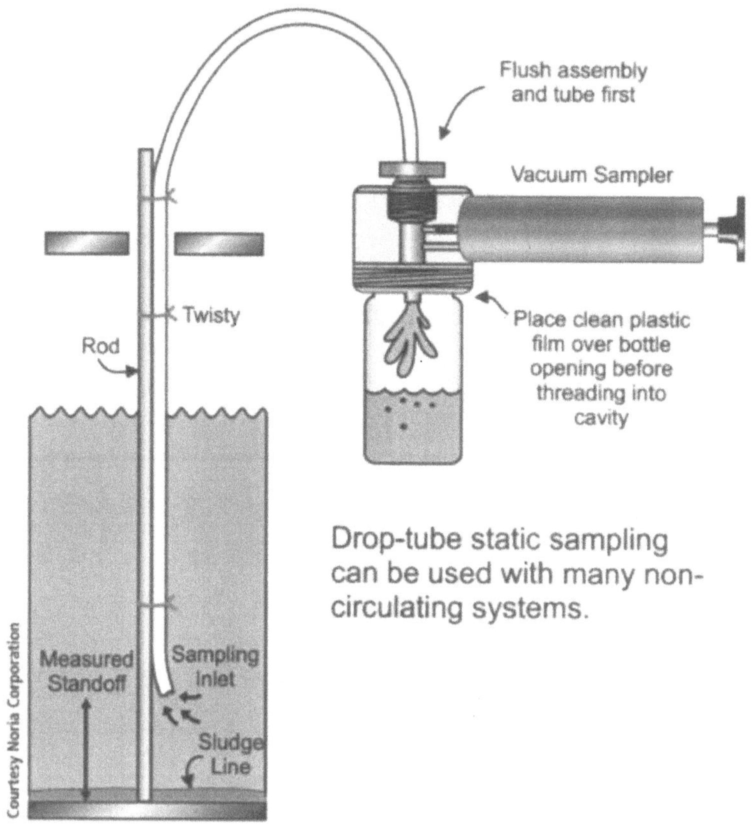

Figure 33.6 Drop tube static sampling arrangement using a vacuum sampler.

Simple, effective tactics for managing data disturbance, sometimes known as interference include:

- Clean hands, clean sampling port/area, clean sampling equipment,
- Use only virgin sample bottles designed for oil analysis sampling, glass is preferred but the most expensive (NO washed-out jam jars pill bottles!)
- Fill the sample bottle to only 60% to 70% allowing the lab headspace to agitate and successfully re-suspend the solids for testing purposes
- Perform the 10x flush rule for every sample where the sample valve and tube (when used) are flushed with approximately 10 x the sample volume space required with the system oil being sampled into a non-sample container before the real sample is taken

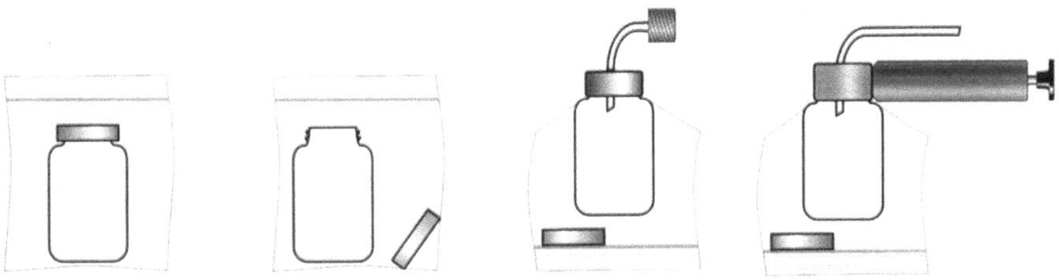

Figure 33.7 Zip-Lock® bags prevent contamination of samples.

Avoid sampling methods that involve removing the bottle cap, especially where significant atmospheric contamination is present. One effective method that ensures that particles will not enter the bottle during sampling is a procedure called "clean oil sampling." It involves the use of common zip-lock sandwich bags and sampling hardware such as vacuum pumps and probe devices. Below is an outline description of this procedure:

Step One: Obtaining a good oil sample begins with a bottle of the correct size and cleanliness. It is understandable that the bottle must be at a known level of cleanliness and that this level should be sufficiently high so as not to interfere with expected particle counts. Some people refer to this as signal-to-noise ratio, i.e., the target cleanliness level of the oil (signal) should be several times the expected particle contamination of the bottle (noise). For more information on bottle cleanliness refer to ISO 3722.

Step Two: Before going out into the plant with the sample bottle place the capped bottles into very thin zip-lock sandwich bags, one per bag (Figure 33.7). Zip each of the bags such that air is sealed into the bag along with the bottles. This should be done in a clean-air indoor environment in order to avoid the risk of particles entering the bags along with the bottles. After all of the bottles have been bagged, put these small bags (with the bottles) into a large zip-lock bag for transporting them to the plant or field. Sampling hardware such as vacuum pumps and probe devices should be placed in the large bag as well.

Step Three: After the sampling port or valve has been properly flushed (including the sampling pump or probe if used) remove one of the bags holding a single sample bottle. Without opening the bag, twist the bottle cap off and let the cap fall to the side within the bag. Then move the mouth of the bottle so that it is away from the zip-lock seal. Do not unzip the bag.

Step Four: Thread the bottle into the cavity of the sampling device (vacuum pump or probe). The plastic tube will puncture the bag during this process, however, try to avoid other tears or damage to the bag (turn the bottle, not the probe or pump, while tightening). If a probe device is used, it is advisable to break a small hole in the bag below the vent hole with a pocket knife. This permits air to escape during sampling.

Step Five: The sample is then obtained in the usual fashion until the correct quantity of oil has entered the bottle. Next, by gripping the bottle, unscrew it from the cavity of the pump or probe device. With the bottle free and still in the bag, fish the cap from the bottom of the bag onto the mouth of the bottle and tighten.

Step Six: With the bottle capped it is safe to unzip the bag and remove the bottle. Confirm that the bottle is capped tightly. The bottle label should be attached and the bottle placed in the appropriate container for transport to the lab. Do not reuse the zip-lock bags.

Three levels of bottle cleanliness are identified by bottle suppliers: clean (fewer than 100 particles >10μm/ml), Super-clean (<10μm/ml), and Ultra-clean (<1μm/ml). Selecting the correct bottle cleanliness to match the type of sampling is important to oil analysis results.

33.4.4 Maximize sample data consistency— oil sample frequency

To ensure high quality sample results that can be trusted, the sampling protocol must assure sample consistency through:

- Develop an engineered oil sampling program in which every sampling port and method is documented and regular sampling frequencies are set up in a work order system. Commencing a program, bearings are usually start sampled on a 500hr frequency, industrial hydraulic systems on a 700hr frequency, with light duty gearboxes on a 1000hr frequency and heavy duty gearboxes on a 300hr frequency,
- Use the oil sampling program to develop and train oil sampling standard operating procedures,
- Always sample from the same location,
- Regularly sample virgin oil stocks when new lubricant stock arrives on site,
- Use the same laboratory for sampling and ask for a dedicated lab technician(s) to perform your sampling

- Always fill in the sample data form, including sampling date and time stamp
- Send off sample to the lab within 24 hours of sampling. Over 24 hours, retake the sample.

33.5 Oil Sampling Frequency

The objective of oil analysis and condition monitoring in general, is to find bad news. The objective of proactive maintenance is not to have any bad news to find. The machine and oil will generally give off silent alarms when problems first occur. In time, as the severity increases, these alarms are no longer silent and even the most rudimentary condition monitoring methods can reveal the problem. Of course, at this point, a great deal of damage may have already occurred and, it is likely too late to arrest the problem on the run; the machine may have to be taken apart and repaired.

One of the extraordinary benefits of oil analysis is its incredible sensitivity to these silent alarms and the detection of incipient failures and faults. It operates on the very basic principle that you cannot hear an alarm unless you are listening for an alarm—restated, you can't catch a fish unless your hooks in the water. Too often we hear about oil samples being taken every six months or annually; yet, on the same machinery we see vibration readings taken every month. Often is the case when results are received and lay around for days, unread. Once you have an alarm state, the process has worked and is all for nought unless immediate action is taken when results are received.

Scheduled sampling intervals are common in oil analysis. The frequency may be keyed to drain intervals or operating hours. Table 33.4 conservatively recommends intervals based on operating hours for different machine classes. Proper selection of sampling frequencies considers machine and application-specific criteria such as those below:

33.5.1 Penalty of failure

Safety, downtime costs, repair costs, and general business interruption costs must be considered.

33.5.2 Fluid environment severity

Operation and fluid environment conditions influence both frequency and rate of failure progression. Influencing factor include pressures, loads, temperature, speed, contaminant ingression, and system duty.

Table 33.4 Conservatively recommended oil sampling intervals for different equipment categories.

Application	Hours
Diesel engines - off-highway	150
Transmissions, differentials, final drives	300
Hydraulics - mobile equipment	200
Gas turbines - industrial	500
Steam turbines	500
Air/gas compressors	500
Chillers	500
Gear boxes - high speed, heavy duty	300
Gear boxes - low speed, heavy duty	1000
Bearings - journal and rolling element	500
Aviation reciprocating engines	25-50
Aviation gas turbines	100
Aviation gear boxes	100-200
Aviation hydraulics	100-200

33.5.3 Machine age

In general, the chances of failure are greatest for machines going through break-in and after major repairs and overhauls. Likewise, the risk increases as a machine approaches the end of its expected life.

33.5.4 Oil age

Infant oils and old oils are at highest risk. Infant oils are those that have just been changed and are less than 10% into expected life. Old oils are showing trends that suggest additive depletion, the onset of oxidation, or high levels of contamination.

33.6 Selection of Oil Analysis Test

Once proper oil sampling has been mastered it is time to analyze the oil. Because each test that is conducted by an oil lab adds cost to the program, it is important that an optimum selection of tests be defined. There are generally two types of tests; routine and exception. A routine test is a scheduled test that is repeated with each scheduled sample such as tests for viscosity, moisture, and particle count.

An exception test is triggered by a previously non-complying condition or test result. It is conducted to either confirm a conclusion (diagnosis/prognosis) or seek further information that might identify the cause or source of the problem. Exception tests might, for instance, include specialized tests for confirming oil oxidation or abnormal machine wear. Table 21.5 shows how routine tests can be combined with exception tests to provide comprehensive test bundles by machine application.

To be thoroughly effective, a well-designed oil analysis program must encompass three categories of routine tests: (1) fluid properties, (2) fluid contamination, and (3) fluid wear debris.

33.6.1 Fluid properties analysis

This essential type of oil analysis helps ensure the fundamental quality of the lubricating fluid. The standard to which a used oil's properties should be routinely compared are the new oil's properties; a listing of each of the new oil properties should be a standard feature of used oil analysis reports. Examples of common tests include viscosity, total acid number, total base number, infrared for oxidation, emission spectroscopy for additive elements, flash point, specific gravity, and rotating bomb oxidation test (RBOT).

33.6.2 Contamination analysis

Despite the use of filters and separators, contaminants are the most common destroyers of machine surfaces that ultimately lead to failure and downtime. For most machines, solids contamination is the number one cause of wear-related failures. Likewise, particles, moisture, and other contaminants are the principal root cause of additive and base stock failure of lubricants. It is important to perform basic tests such as particle counting, moisture analysis, glycol testing, and fuel dilution as directed by a well-designed proactive maintenance program.

Table 33.5 Selecting oil analysis tests by application. (Courtesy Noria Corporation)

Test or Procedure	Paper Machine Oils	Motor & Pump Bearings	Diesel &Gas Engine	Hydraulics	Air & Gas Compressors	Chillers and Refrigeration	Transmissions, Final Drives, Differentials	Industrial Gear Oils	Steam Turbine Oils	Gas Turbine Oils	EHC Fluids
1. Particle Count	O,L	O,L	L	O,L	O,L	O,L	L	O,L	O,L	L	O,L
2. Viscosity a. 40°C	O,L	O,L		O,L	O,L	O,L	L	O,L	L	L	O,L
b. 100°C	-	-	L	-	-	-	-	-	-	-	-
3. TAN	L	E(5a)	-	L	L	L	L	L	L	L	O,L
4. TBN	-	-	L	-	-	-	-	-	-	-	-
5. FTIR a. Ox/Nit/Sul	L	L	L	L	L	L	L	L	L	L	
b. Hindered Phen	-	L	-	L	L	-	-	L	L	-	-
c. ZDDP	-	L	-	L	L	-	L	L	-	-	-
d. Fuel Dil/Soot	-	-	L	-	-	-	-	-	-	-	-
6. Flash Point	-	-	E(2b,5d)	-	L*	-	-	-	-	E(2b,5d)	-
7. Glycol-ASTM Test	-	-	E(14b)	-	-	-	-	-	-	-	-
8. Ferrous Density	E(1)	E(1)	L	O,L	O,L	O,L	L	O,L	E(1)	E(1)	O,L
9. Analytical Ferrography	E(8,14a)	E(8,14a)	E(8,14a)	E(8,14a)	E(8,14a)	E(8,14a)	E(8,14a)	E(8,14a)	E(8,14a)	E(8,14a)	E(8,14a)
10. RBOT	-	-	-	-	L	-	-	-	L	L	-
11. Crackle	O,L	O,L	L	O,L	O,L	L	O,L	O,L	O,L	-	O,L
12. Water by KF	E(11)	E(11)	E(11)	E(11)	E(11)	E(11)	E(11)	E(11)	E(11)	-	E(11)
13. Water Separability	L	-	-	-	L**	-	-	-	L	-	-
14. Elemental Analysis a. Wear Metals	L,E(1)	L,E(1)	L	L,E(1)	L,E(1)	L,E(1)	L	L,E(1)	L,E(1)	L	L,E(1)
b. K, Na, B, Si	L	L	L	L	L	L	L	L	L	L	L
c. Additives	L	L	L	L	L	35-L	L	L	L	L	L

*Gas Compressors Only **Air Compressors Only

Table 35-5 Key

O = On-site routine test (small on-site lab or portable instrument)

L = Fully equipped oil analysis laboratory

E = Exception test keyed to a positive result from the test in parenthesis

33.6.3 Fluid wear debris analysis

Unlike fluid properties and contamination analysis, wear debris analysis relates specifically to the health of the machine. Owing to the tendency of machine surfaces to shed increasing numbers of progressively larger particles as wear advances, the size, shape, and concentration of these particles tell a revealing story of the internal or state condition of the machine.

Cost-effective oil analysis can generally be done when on-site oil analysis tools are available. For many machines, the particle counter serves as the best first line of defense. Only when particle counts exceed preset limits is exception testing performed. The best exception test is ferrous density analysis, such as a ferrous particle counter. When ferrous levels are high, a failure condition exists, triggering yet further testing and analysis. In addition to on-site particle counting, on-site moisture analyzers and viscometers also assess important root cause contributors.

33.7 Monitoring Changing Oil Properties

Today there are a growing number of organizations transforming their lube programs from scheduled to condition-based oil changes. In fact, many companies claim that they easily pay for the cost of oil analysis from savings achieved through reduced lubricant consumption. Such progressive goals as these, place a greater burden of precision on the selection of oil analysis tests and alarm limits to reveal non-complying lubricants.

It is not uncommon for plants to interpret oil analysis results independent of the lab. Essentially, the lab is relied on to provide accurate and timely data, leaving both interpretation and response to the plant; i.e., personnel close to the equipment with knowledge of application and operating conditions. The use of modern oil analysis software can greatly assist in this such programs.

In order to reduce oil consumption, two plans must be implemented. The first plan is proactive in nature and relates to the operating conditions the oil lives in. By improving the oil's operating conditions its expected life can increase many times. For example, based on the Arrhenius Law (10 degrees C rule) a reduction of operating temperature of just 10 degrees C (18 degrees F) can double an oil's oxidation stability and double the oil change interval in many instances. The inverse equation is also true in that for every 10 degree C raise in bearing or oil temperature will reduce its life by half accordingly.

The second plan to reducing oil consumption is predictive in nature and relates to the timing of oil changes. Basically, through oil analysis, key physical properties can be trended to help forecast the need of a future oil change. Restated, by listening to the oil, it will tell us when it needs to be changed. And, if the need of an oil change occurs prematurely, then an assessment of the oil's operating conditions (cleanliness, dryness, coolness, etc.) and oil formulation should be made. The nature of the degradation will provide basic clues defining the solution.

There are numerous modes of degradation of lubricating oil. These change many of the properties of the fluid. In order to recognize these changes, it is important that the correct properties be monitored, realizing that over-kill is wasteful. What follows is a discussion of common oil degradation modes and the properties that can best reveal them. In all cases, it is important to get a base signature of the properties of the new oil so as to benchmark the trended change. These reference properties should remain as a permanent record on the oil analysis report and include additive elements, neutralization numbers, -infrared units (unless spectral subtraction is used), RBOT minutes, viscosity, flash temperature, VI, and color.

33.7.1 Viscosity stability

Viscosity is often referred to as the structural strength of a liquid. It is critical to oil film control plus is a key indicator to a host of ailing conditions relating to the oil and machine. As such, it is often considered a critical "catch-all" property in oil analysis. Essentially, when viscosity remains in a controlled narrow band one can assume that a great many things that could be going wrong are, in fact, not going wrong. Conversely, when viscosity falls outside of the band an exception test is usually needed to identify the nature and cause of an abnormality. Monitoring viscosity thus represents a first line of defense against many problems.

Because viscosity is so important it is often monitored on-site by the reliability team. It is used as an acceptance test for new oil deliveries and to verify the correct lubricant is in use. When viscosity changes with in-service lubricants, the cause is either oil degradation or oil contamination. Oil degradation relates to changes to the base oil and additive chemistry (molecular changes). Contamination of an oil can either thicken or thin the oil depending on the viscosity and emulsifying characteristics of the contaminant (see Table 33.6).

In oil labs, viscosity is typically measured using kinematic viscometers. ISO viscosity grades shown on lubricant spec sheets are based on kinematic viscosity in centistokes (cSt) at 40 degrees C. Another way to represent

Table 33.6 The origin of viscosity changes.

	Decreased Viscosity	Increased Viscosity
Changes to base oil (molecular changes)	• Thermal cracking of oil molecules • Shear thinning of VI improvers	• Polymerization • Oxidation • Evaporative losses • Formation of carbon & oxide insolubles
Additions to base oil (contamination)	• Fuel • Refrigerant • Solvents • Wrong oil (low viscosity)	• Water (emulsions) • Aeration • Soot • Antifreeze (glycol) • Wrong oil (high viscosity)

kinematic viscosity is Saybolt Universal Seconds (SUS). This is determined with use of a U-tube kinematic viscometer. In this device, the oil is allowed to drain by gravity through a capillary at constant temperature. The drain time (efflux time) is measured and translated into centistokes. Viscosity varies nearly proportionally to drain time. Because gravity is involved, kinematic viscosity characterizes both the oil's resistance to flow (absolute viscosity) and specific gravity.

On-site oil analysis labs frequently use absolute viscometers to obtain a precise indication of base oil condition. Unlike kinematic viscometers, absolute viscosity measures only an oil's resistance to shear or flow (not specific gravity). Figure 33.8 shows an absolute viscometer designed for plant-level use. It employs a capillary in its tip, through which the oil flows under constant pressure and temperature. An inline plunger moves outward with the flow. The speed of this plunger, measured electronically, varies nearly proportionally to absolute viscosity.

Viscosity is typically trended at 40 degrees C although for high temperature applications, such as crank-case lubricants, a 100-degree C trend is sometimes preferred. Both temperatures are needed to determine the oil's Viscosity Index (VI). However, the VI rarely is trended for routine condition monitoring. Monitoring viscosity at 40-degrees C, for most industrial applications, will provide the most reliable early indication of base oil degradation and oxidation.

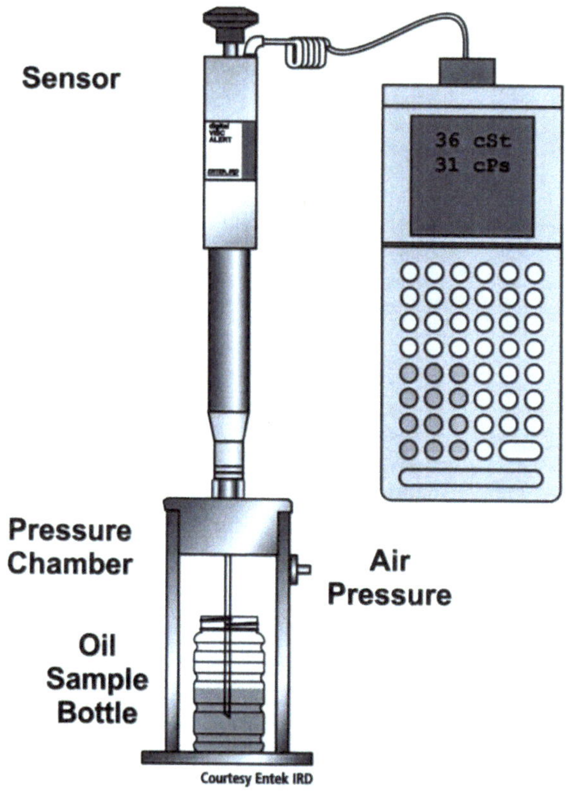

Figure 33.8 Absolute viscometer for plant use.

33.7.2 Oxidation stability

When oil oxidizes the base oil thickens and discharges sludge and acidic materials; all detrimental to good lubrication. Oxidation is uncommon in applications when conditions are such that oils are relatively cool, dry, and clean. And, this is especially true for low viscosity oils such as hydraulic fluids and turbine oils that have higher oxidation stability. However, when operating conditions are severe, oil oxidation can be a recurring problem. Where a proactive solution cannot be applied (controlling oxidation root causes or the use of resistant synthetics) it is best to monitor the progress of oxidation. Monitoring the depletion of oxidation inhibitors provides an early, forecastable trend, but this may not be practical in some applications. The technologies used to monitor the depletion of oxidation inhibitors include:

1. Infrared spectroscopy (FTIR) can pick up trend changes in phenolic and ZDDP inhibitors. However, only a few of the laboratories report

additive depletion with FTIR because of unreliable reference oils and occasional inferences from contaminants

2. Total acid number (TAN) is sensitive to both mass-transfer and decomposition depletion of ZDDP inhibitors. Interpretation of the trend takes practice and a good new-oil reference

3. Elemental spectroscopy can show reliable mass-transfer depletion trends in ZDDP inhibited oils

4. Rotating Bomb Oxidation Tests (RBOT) provide a highly forecastable trend on additive depletion. Because of the time needed to run this test it is expensive and usually saved for exception testing or special circumstances. (See Table 33.7)

5. Voltametry is a new technology that has shown particular promise in trending the depletion (mass transfer and decomposition) of phenolic and ZDDP inhibitors.

If trending the depletion of oxidation inhibitors is not practical then oxidation itself must be monitored. The problem with this approach relates to the fact that oxidation can progress rapidly in stressful conditions once the antioxidant has depleted. Simply stated, with oxidation, "the worse things get the faster they get worse." If the goal is a condition-based oil change, this translates to the need to monitor with sufficient frequency to catch the problem in the incipient stages, not after the oil throws sludge and destructive lubrication has occurred.

The most common and reliable methods to detect and trend oil oxidation are the following:

1. If a reliable new oil reference is available to the laboratory, infrared analysis (FTIR) is dependable for mineral oils and many synthetics including organic and phosphate esters. The acids, aldehydes, esters, and ketones formed during oxidation are detected by FTIR in mineral oils and PAO synthetics.

2. Total acid number (TAN) will quantify the growing acid constituents in oxidizing oils.

3. Because oxidation results in polymerization of the base oil and the discharge of oxide insolubles, the viscosity will increase.

4. Color-bodies form in oxidized oils resulting in a marked darkening of the oil's color.

5. Oxidized oils give off a sour or pungent odors, similar to the smell of a rotten egg.

Table 33.7 Additive depletion tests.

Additive	Monitoring Method	Effectiveness
ZDDP (antioxidant, anti-wear, and corrosion inhibitor)	FTIR - (@ ▯950 wavenumbers)	Fair (early detection)
	FTIR - oxidation (1750 wn)	Good (later detection)
	RBOT	Excellent (early detection)
	TAN (downward trend)	Fair (early detection)
	TAN (upward trend)	Good (later detection)
	Voltametry	Excellent (early detection)
	Elemental Spectroscopy	Excellent (early detection)
Rust Inhibitors	Elemental Spectroscopy	Fair (early detection)
Foam inhibitors	Elemental spectroscopy	Fair (interferences from dirt)
Sulfur Phosphorus - EP	Elemental spectroscopy	Excellent
Molybdenum Disulfide - EP	Elemental spectroscopy	Good
Borate - EP	Elemental spectroscopy	Excellent
VI Improver	Viscosity at 40°C and 100°C	Excellent
Dispersants	Blotter spot test	Good
Detergents	TBN	Excellent
	Elemental Spectroscopy	Excellent
Hindered Phenol (antioxidant)	FTIR	Fair
		Courtesy Noria Corporation

33.7.3 Thermal stability and varnish tendency

The thermal failure of an oil can be localized or uniform. Localized thermal failure occurs when the bulk oil temperature remains generally suitable for the selected lubricant but oil is exposed to hot surfaces such as the discharge valves of reciprocating compressors, or hot surfaces in internal combustion engines and turbomachinery. Another common cause of localized thermal failure is associated with entrained air that is permitted to compress, such as occurs to air bubbles passing through a high-pressure hydraulic pump. The air bubble implosion causes heat to concentrate generating microscopic specks of carbon. These carbon insolubles later condense on machine surfaces, forming what is commonly called varnish.

The varnish tendency of an oil is often difficult to detect due to the fact that the majority of the physical properties of the oil are unaffected. For instance, oxidation may occur without change in viscosity, TAN, or FTIR. However, sophisticated labs having experience with hydraulic fluids will employ specialized tests such as ultracentrifuge, FTIR for nitration, and submicron membrane tests. Other, less reliable, indicators include oil color and paper chromatography (blotter spot test).

The uniform thermal failure of an oil results from excessively high operating temperatures due to any of a number of reasons. However, the most common reasons include overloading, inadequate oil supply, failure of a heat exchanger, and the use of a high watt-density tank heater. When any of these conditions occur, the oil fails by evaporation (thickening), carbonization (coking, carbon stones, etc.), or cracking (thinning) in extreme cases. Regardless of origin, the uniform thermal failure of the oil is serious and threatens the reliable operation of the lubricated machine.

An oil's thermal stability it often measured using the Cincinnati Milacron test (ASTM D 2070-91). However, because this test takes a week to complete, it is generally impractical for routine used-oil analysis. Other ways to evaluate thermal failure include viscosity analysis, ultracentrifuge, total insolubles, and oil color. Less reliable indicators include oil odor (either a burnt, rancid odor or no odor at all) and paper chromatography.

33.7.4 Additive stability

Additive monitoring is one of the most challenging and evasive areas of used-oil analysis. The reasons for this are many and complex. As a starting point, it is worthwhile to review how additives deplete during normal use and aging.

It is generally accepted that there are two forms of additive depletion, both are common and can occur simultaneously. The first form of depletion is known as decomposition. Here the additive mass stays in the oil but its molecular structure changes resulting in an assortment of transformation products (other molecules). In some instances, the transformation products may possess properties similar to the original additive, but in most cases performance is degraded or is completely lost. This sacrificial form of depletion is common to what happens over time to oxidation inhibitors, as described previously under "oxidation stability."

The second form of additive depletion is called mass transfer. This type of depletion is often the easiest to detect because the entire mass of the additive transfers out of the bulk oil. And, as such, any measurable property of the additive leaves as well. For instance, if the additive is constructed with phosphorous, a down-ward trend of phosphorous in the used oil is a reliable indication of its mass transfer depletion. Conversely, an unchanging level of phosphorous in used oil in no way confirms that decomposition depletion has not occurred. With decomposition the elements of the additive remain suspended in the oil.

Mass transfer of additives occur in normal operation, usually as a result of the additive doing the job it was designed to do. For instance, when a rust inhibitor attaches itself to internal machine surfaces it depletes by mass transfer. It is common for additives to cling to various polar contaminants in the oil, such as dirt and water. The removal of these contaminants by filters, separators, and settling action causes a removal of the additive as well. And, over time, aging additives can form floc and precipitate out of the oil due to decomposition and long cold-temperature storage. The insolubles formed will migrate out, often ending up on the bottom of the sump or reservoir.

Table 33.7 describes common methods used to monitor additive depletion. It is worth restating that the use of elemental spectroscopy to trend additive depletion is only effective where mass transfer is involved. It is not uncommon, therefore, for oil labs to condemn an oil with only a 25 percent reduction in the concentration of telltale additive elements, e.g., zinc and phosphorous in the case of ZDDP.

33.8 Monitoring Oil Contamination

Contamination can be defined as any unwanted substance or energy that enters or contacts the oil. Contaminants can come in a great many forms, some are highly destructive to the oil, its additives, and machine surfaces. It is often overlooked as a source of failure because its impact is usually slow and imperceptible yet, given time, the damage is analogous to eating the machine up from the inside out. While it is not practical to attempt to totally eradicate contamination from in-service lubricants, control of contaminant levels within acceptable limits is accomplishable and vitally important.

Particles, moisture, soot, heat, air, glycol, fuel, detergents, and process fluids are all contaminants commonly found in industrial lubricants and hydraulic fluids. However, it is particle contamination and moisture that are widely recognized as most destructive to both oil and machine.

33.8.1 Particle contamination

There is no single property of lubricating oil that challenges the reliability of machinery more than suspended particles. It would not be an exaggeration to refer to them as a microscopic wrecking crew. Small particles can ride in oil almost indefinitely and because they are not as friable (easily crumbled) as their larger brothers, the destruction can be continuous. Many studies have shown convincing evidence of the greater damage associated with small particles compared to larger. Still, most maintenance professionals have misconceptions about the size of particles and the associated harm caused.

These misconceptions relate to the definition people apply to what is clean oil and what is dirty oil. And, it is this definition that becomes the first of the three steps of proactive maintenance; the need to set appropriate target cleanliness levels for lubricating oils and hydraulic fluids. The process is not unlike a black box circuit. If we want a change to the output (longer and more reliable machine life) then there must be a change to the input (a lifestyle change, i.e., improve cleanliness). For instance, it's not the monitoring of cholesterol that saves us from heart decease, instead it's the things we do to lower the cholesterol. Therefore, the best target cleanliness level is one that is a marked improvement from historic levels.

This leads us to the second step in proactive maintenance, the lifestyle change. By effectively excluding the entry of contaminants and promptly removing contaminants when they do enter, the new cleanliness targets are frequently easily achieved. Concerns that filtration costs will increase may not materialize due to the greater overall control, especially from the standpoint of particle ingression.

The third step of proactive maintenance is the monitoring step, i.e., particle counting. If this is done frequently enough, not only is proactive maintenance achieved but also a large assortment of common problems can be routinely detected. As such, particle counting is another important "catch all" type test, like viscosity analysis. Because of the obvious value, the particle counter is probably the most widely used on-site oil analysis instrument. It is not uncommon to find organizations testing the cleanliness of their oils as frequently as weekly.

The activity of routine particle counting has a surprising impact on step number two. When the cleanliness levels of lubricants are checked and verified on a frequent basis a phenomenon known as the "invisible filter" occurs, which is analogous to the saying, "what gets measured gets done." Because a great deal of dirt and contamination that enters oils comes from the careless practices of operators and craftsmen, the combined effect of monitoring

In this example, you can see how the particles measured at the given micron levels are assigned the specific code based on where that value falls in the table. For this example, the ISO code would be 20/17/13.

Table 1

	PARTICLES/ML	ISO CODE
>4 microns	9,721	20
>6 microns	1,254	17
>10 microns	326	
>14 microns	73	13
>21 microns	12	
>38 microns	5	
>70 microns	0	
>100 microns	0	

MORE THAN (p/ml)	UP TO AND INCLUDING (p/ml)	ISO CODE
80,000	160,000	24
40,000	80,000	23
20,000	40,000	22
10,000	20,000	21
5,000	10,000	20
2,500	5,000	19
1,300	2,500	18
640	1,300	17
320	640	16
160	320	15
80	160	14
40	80	13
20	40	12
10	20	11
5	10	10
2.5	5	9
1.3	2.5	8

Figure 33.9 ISO contaminant code (ISO 4406) Courtesy Noria Corporation.

with a modicum of training can go a long way towards achieving cleanliness goals.

The ISO Solid Contaminant Code (ISO 4406) is probably the most widely used method for representing particle counts in oils. The current standard now employs a three-range number $R_4/R_6/R_{14}$ system (see Figure 33.9). The first range number corresponds to particles larger than 4 microns, the second range number for particles greater that 6 microns, and the third range number for particles larger than 14 microns. From the red particle count chart, we see that as the range numbers increment up one digit the represented particle count, roughly doubles.

While there are numerous different methods used to arrive at target cleanliness levels for oils in different applications, most combine the importance of machine reliability with the general contaminant sensitivity of the machine to set the target. This approach is shown in Table 33.8. The Reliability Penalty Factor and the Contaminant Severity Factor are arrived at by a special scoring system that is included with the Target Cleanliness Grid.

There are many different types of automatic particle counters used by oil analysis laboratories. There are also a number of different portable

Table 33.8 Contaminant severity factor (CSF).

Reliability Penalty Factor (RPF) / Cost, Safety, and Business interruption Penalty From Failure	1	2	3	4	5	6	7	8	9	10
10	19/16/13	18/15/12	17/14/12	16/13/11	15/12/10	14/11/9	13/10/8	12/10/8	11/9/7	10/9/7
9	19/16/13	19/16/13	18/15/12	17/14/11	16/13/10	15/12/10	14/11/9	13/10/8	12/9/7	11/9/7
8	20/17/14	20/17/13	19/16/13	18/15/12	16/13/11	15/12/10	14/11/9	13/10/8	12/9/8	12/9/7
7	20/17/14	20/17/14	19/16/13	18/15/12	17/14/11	16/13/10	15/11/9	14/11/9	13/10/8	12/10/8
6	21/18/15	21/18/14	19/16/13	18/15/12	17/14/11	16/13/10	15/12/9	14/11/9	13/10/8	12/10/8
5	21/18/15	21/18/15	20/17/14	19/16/13	18/15/12	17/14/11	16/14/11	15/13/11	14/11/10	13/11/9
4	22/19/16	22/19/16	20/17/14	19/17/14	18/15/13	17/14/11	16/14/11	15/13/10	14/12/9	13/11/9
3	22/19/16	22/19/16	21/18/15	20/17/14	19/16/13	18/15/12	17/14/11	16/14/11	15/13/10	14/12/9
2	23/20/17	23/20/17	22/19/16	21/18/15	20/17/14	19/16/13	18/15/12	17/14/11	16/14/11	15/13/10
1	24/20/17	23/20/17	22/19/16	21/18/15	20/17/14	19/16/13	19/16/12	18/15/11	17/14/11	16/14/11

Contaminant Severity Factor (CSF)

Sensitivity of Machine to Contaminant Failure

particle counters on the market. The performance of these instruments can vary considerably depending on the design and operating principle. Particle counters employing laser or white light are widely used because of their ability to count particles across a wide range of sizes (see Figure 33.10). Pore blockage-type particles counters have a more-narrow size range sensitivity; however, they are also popular because of their ability to discriminate between hard particles of other impurities in the oil such as water, sludge, and air bubbles (see Figure 33.11).

Figure 33.12 shows how particle count trends vary depending on the machine application and the presence of a built-in filter. Because particle counters monitor particles in the general size range controlled by filters, equilibrium is usually achieved, i.e., particles entering the oil from ingression minus particles exiting from filtration will leave behind a steady state concentration. When filters are properly specified and ingression is under control this steady state concentration will be well within the cleanliness target. Systems with no continuous filtration, e.g., a splash-fed gearbox, an equilibrium is not really established (there is no continuous particle removal). This causes the particle concentration to be continuously rising. Still, contamination control can be achieved by periodic use of portable filtration systems, such as a filter cart.

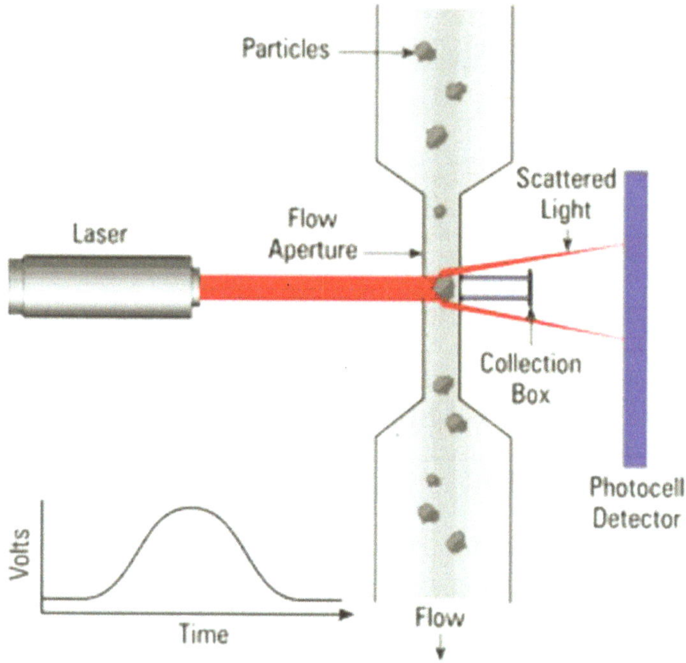

Figure 33.10 Light scattering particle counter. (Courtesy Noria Corporation)

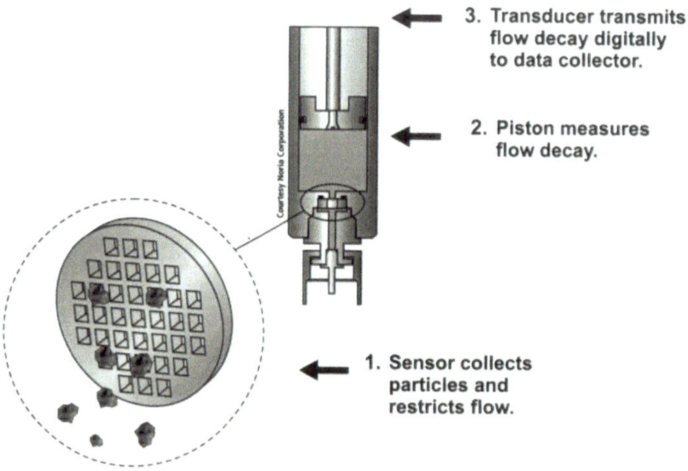

Figure 33.11 2 Pore blockage-type particle counter.

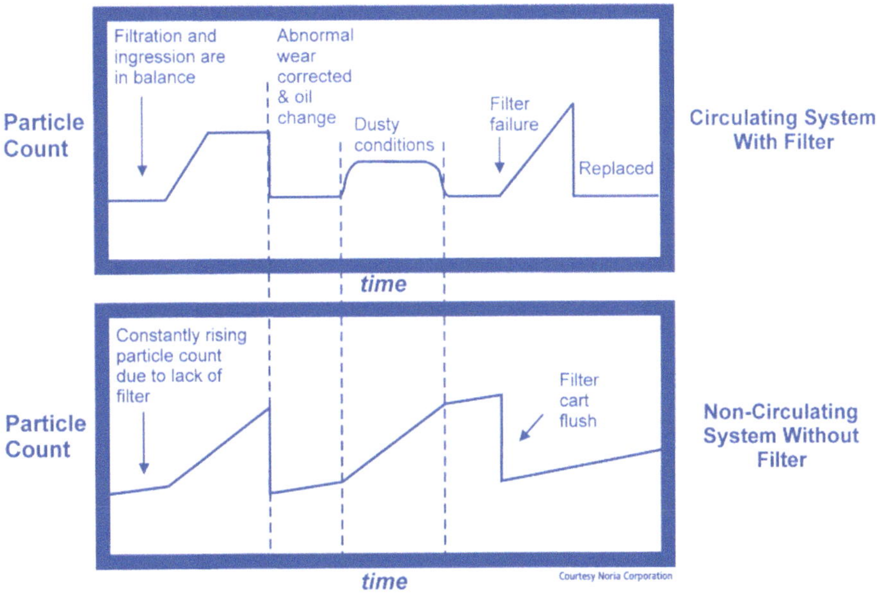

Figure 33.12 Particle count trend graphs.

33.8.2 Moisture contamination

Moisture is generally referred to as a chemical contaminant when suspended in lubricating oils. Its destructive effects in bearings, gearing, and hydraulic components can reach or exceed that of particle contamination, depending on conditions. Like particles, control must be exercised to minimize water accumulation and resulting destruction to the oil and machine.

Once in the oil, water is in constant search of a stable existence. Unlike oil, the water molecule is polar, which greatly limits its ability to dissolve. Water may cling to hydrophilic metal surfaces or form a thin film around polar solids co-existing in the oil. If a dry air boundary exists, water molecules may simply choose to migrate out of the oil to the far more absorbent air interface. If water molecules are unable to find polar compounds on which to attach, the oil is said to be saturated. Any additional water will create a supersaturated condition causing the far more harmful free and emulsified water. The temperature of the oil as is shown in Table 21.9 also influences the saturation point.

With few exceptions, the chemical and physical stability of lubricants is threatened by small amounts of undissolved suspended water. In combination with oxygen, heat, and metal catalysts, water promotes oxidation and hydrolysis. An overall degradation of the base oil and its additives results.

Table 33.9 Make-up of water concentration in oil at different temperatures. (Courtesy Noria Corporation)

			Water Concentration					
Temperature			100	ppm	500	ppm	1000	ppm
°F		°C	D	F&E	D	F&E	D	F&E
32	R&O	0	28	72	28	472	28	972
	Gear		100	0	100	400	100	900
68	R&O	20	72	28	72	428	72	928
	Gear		100	0	200	300	200	800
104	R&O	40	100	0	170	330	170	830
	Gear		100	0	500	0	500	500
140	R&O	60	100	0	350	150	350	650
	Gear		100	0	500	0	1000	0
158	R&O	70	100	0	500	0	520	480
	Gear		100	0	500	0	1000	0

D = Dissolved Water Amount (ppm)

R&O = ISO 32 Rust and Oxidation Inhibited oil (e.g., turbine oil)

Gear = Gear oil

F&E = Free and Emulsified Water (ppm)

The harmful effects of water on the life of rolling element bearings and other contact zones when boundary lubrication prevails are well documented. According to SKF, "free water in lubricating oil decreases the life of rolling element bearings by ten to more than a hundred times… "And, it is well known that water promotes corrosive attack on sensitive machine surfaces discharging harmful abrasives into the oil.

The omnipresence of water in the environment makes it difficult to completely exclude it from entering and combining with the oil. However, its presence can be greatly minimized and controlled through good maintenance practices. And, just like particle contamination, a proactive maintenance program needs to be established to control water. This should start with the setting of a target dryness level for each different oil application. By investigating the sources of water ingression, a plan can be implemented to exclude the water. Occasional removal by water absorbent filters, vacuum dehydrators or air stripping units (see Chapter 34) may also be necessary.

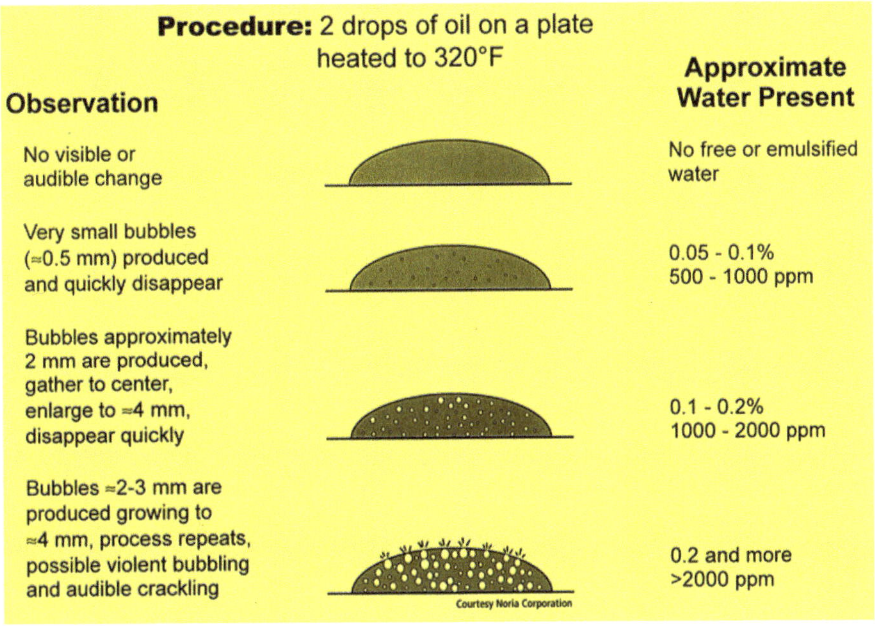

Figure 33.13 Crackle test for water contamination.

A simple and reliable test for water is the crackle test (a.k.a. the sputter test). In the laboratory two drops of oil are placed on the surface of a hot plate heated to approximately 320degree F. The presence of free or emulsified water in the oil will result in the formation of vapor bubbles and even scintillation if the water concentration is high enough. Although generally used only as a go/no-go procedure, experienced lab technicians have learned to recognize the visual differences associated with progressive concentrations of water contamination, see Figure 33.13.

Other widely used methods to detect water include the following:

1. Dean & Stark apparatus is occasionally used by laboratories and involves a procedure of co-distilling the water out of the oil and establishing the water content volumetrically (ASTM D 4006).

2. Karl Fischer titration is commonly used by laboratories as an exception test should initial presence of water be detected by crackle or infrared analysis. Two Karl Fischer procedures exist, volumetric titration (ASTM D 1744) and coulometric titration (ASTM D 4928).

3. Infrared spectroscopy can reliably measure water concentrations down to about 0.1 percent. This lower limit may not be adequate for many oil analysis programs.

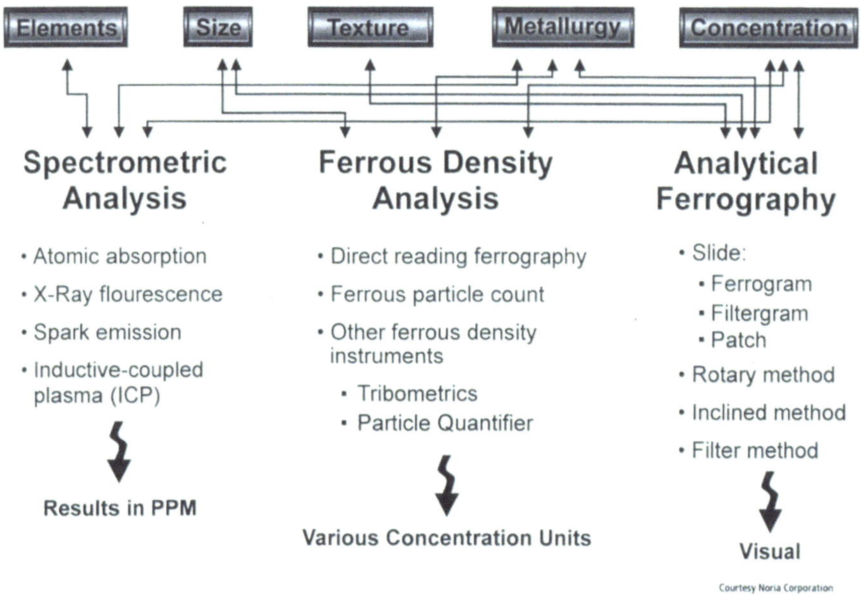

Figure 33.14 Three common categories of wear particle detection.

33.9 Wear Particle Detection and Analysis

Where the first two categories of oil analysis (fluid properties and contamination) deal primarily with the causes of machine failure (proactive maintenance), this category emphasizes the detection and analysis of current machine anomalies and faults, i.e., the symptoms of failure. The oil serves as the messenger of information on the health of the machine. Basically, when a machine is experiencing some level of failure the affected surfaces will shed particles, releasing them into the oil. The presence of abnormal levels of wear particles serves as problem detection where their size, shape, color, orientation, elements, etc. defines the cause, source, and severity of the condition.

33.9.1 Elemental spectroscopy

Figure 33.14 illustrates the three common categories of wear particle detection and analysis. The oldest and most widely used of these methods is elemental analysis, done today primarily with optical emission spectrometers. The procedure involves applying high heat to the oil. Particles in the oil will totally or partially vaporize in the presence of heat, producing incandescent emission of light. The light is diffracted such that spectral intensities at different wavelengths can be measured. Specific wave-lengths are associated with certain elements and the spectral intensities define the concentration of the elements.

The typical output from elemental spectroscopy is concentration units (parts per million) across 10 to 25 common elements such as iron, copper, lead, aluminum, etc. By comparing the major, minor, and trace metals to the metallurgical chart of the machine a fingerprint of the probable sources of the wear can be established. Many of the laboratories do wear metal interpretation with the help of sophisticated software programs and extensive metallurgical databases. Note: A sample of the "*potential sources of metals in oil*" can be found in the Appendix section.

Most oil analysis laboratories offer elemental spectroscopy as standard with all samples analyzed. Both spectrometers and technology vary somewhat, which translates to variations in detection range and sensitivity. The precision of these instruments is also influenced by the size of the wear particles suspended in the oils. During analysis, small particles vaporize more completely while large particles (> 10 microns) are difficult to measure. This particle-size bias leads to occasional errors, some serious (false negatives).

A popular way to reduce the particle-size error is to use electrode filter spectroscopy. This capability is available with spark-emission spectrometers at many of the large commercial laboratories. By pushing the particles into the interstices of the disc electrode a more complete vaporization of larger particles is achieved (possible sensitivity to 20 microns). A special fixture is required to process the sample through the electrode prior to analysis. Because the oil is washed through the electrode during preparation, a separate test is performed on the oil alone to measure dissolved metals and additive elements.

33.9.2 Ferrous density analysis

The most serious wear particles of all are generated from iron and steel surfaces. In fact, in most oil-lubricated pairs, at least one of the two surfaces is a ferrous surface. And, it is usually the ferrous surface that is the most important from the standpoint of machine reliability.

This means that the oil analyst requires a dependable reading of the ferrous particle concentration at all sizes, an important issue considering the particle-size bias associated with elemental spectrometers. Therefore, in order to ensure that abnormal wear of iron and steel surfaces doesn't go undetected, ferrous density analyzers are widely employed, both in commercial and on-site laboratories. These instruments provide a first line of defense by reliably detecting free-metal ferrous debris. Example instruments include:

1. Direct Reading Ferrograph: reports results in Wear Particle Concentration units
2. Particle Quantifier: reports an index scale
3. Wear Particle Analyzer: output in micrograms/ml
4. Ferrous Particle Counter: assigns a percent ferrous to particle count ratio

33.9.3 Analytical ferrography

Elemental spectroscopy and ferrous density analysis are just two of many different ways to detect problems in machinery. Thermography and vibration monitoring are also effective at detecting specific faults and modes of failure. Once there is an initial indication of a fault by any of these methods, the process must continue to:

1. Isolate it to a single component
2. Identify the cause
3. Assess how severe or threatening the condition is, and, finally
4. Determine the appropriate corrective action

When problems are detected and analyzed early, they can often be arrested without downtime or expensive repair. In fact, root causes of the most common problems are usually correctable on the run. The key is the timing of the detection. An important part of timing is a regimen of frequent sampling.

Successful analysis of a current wear-related problem requires many pieces of information and a skilled diagnostician. To this end, the practice of analytical ferrography has received recently prominence. Unlike other common instrumentation technologies, analytical ferrography is qualitative and requires visual examination and identification of wear particles. Numerous properties and features of the wear debris are inventoried and categorized. These include size, shape, texture, edge detail, color, light effects, heat treatment effects, apparent density, magnetism, concentration, and surface oxides.

This information is combined with other information obtained by particle counting, ferrous density analysis and elemental spectroscopy in defining a response to items 1-4 above. Figure 33.15 presents a general overview of

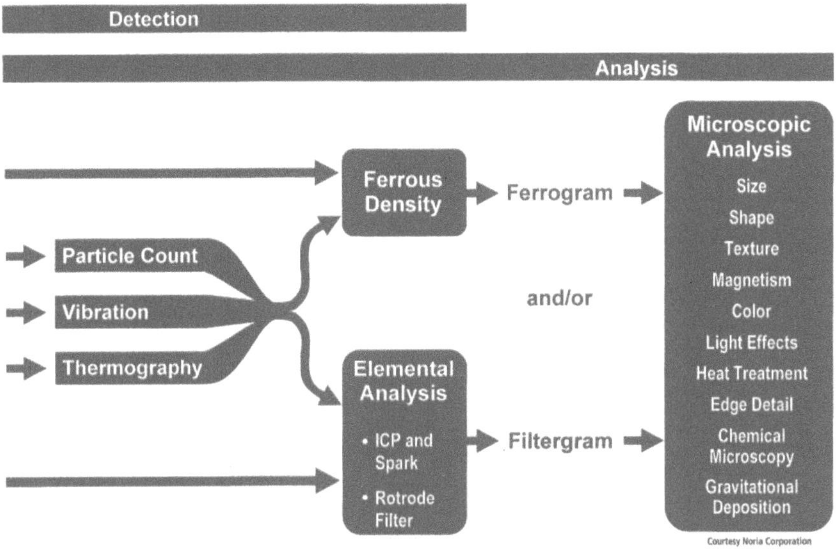

Figure 33.15 Combined detection and analysis process.

the combined detection and analysis process. Here are analytical ferrography is represented by microscopic analysis. Two methods are commonly used to prepare the particles for viewing by the microscope.

If a high level of ferromagnetic debris is detected by ferrous density analysis then a ferrogram is typically prepared. The process involves slowly passing solvent-diluted oil down the surface of an inclined glass slide. The instrument involved is called a ferrogram maker. Beneath the slide is a strong magnet. Ferromagnetic particles become quickly pinned down onto the slide and oriented to the vector lines of the magnetic field.

Non-magnetic debris deposit gravimetrically in random fashion, although larger and heavier particles settle first. Approximately 50% of the non-magnetic particles wash down the slide and do not deposit. A ferrogram of wear is shown in Figure 33.16.

In those cases where low levels of ferromagnetic particles are detected but high nonferrous debris is found (by a particle counter or elemental analysis) a filtergram is preferred. Unlike the ferrogram, the filtergram does not use a magnet and therefore all particles are randomly deposited regardless of size, weight, or magnetic attraction. This is accomplished by passing an exact quantity of solvent-diluted oil through a membrane of about three-micron pore size. No particles are hidden from observation except those too small to be retained by the membrane. The single disadvantage of the filtergram is the

Figure 33.16 Examples of particles identified by analytical ferrography. Source Engtech Industries Inc.

difficulty of distinguishing ferrous from non-ferrous debris. The skillful eye of an experienced technician can usually over-come this drawback.

33.10 Interpreting Test Results

Most machines are highly complex, consisting of exotic metallurgy and intricate mechanisms. The numerous frictional and sealing surfaces usually employ varying contact dynamics and loads, all sharing a common lubricant. Failure to gain knowledge of the many internal machine details and using them as a reference base for interpreting data may lead to confusion and indecision in response to oil analysis results. A good approach is to build a three-ring binder with index tabs for each machine type.

Include in this binder photocopied pages from the service and operation manuals plus other accumulated information. For example:

- Identify types of bearings in use and their metallurgy
- Identify input and output shaft speeds and torques
- Identify type of gears in use, speeds, and loads. Determine gear metal hardness, surface treatments, alloying metals
- Locate and identify all other frictional surfaces, such as cams, pistons, bushings, swash-plates, etc. Determine metallurgy and surface treatments
- Locate and identify coolers, heat exchangers and type of fluids used
- Obtain fluid flow circuit diagrams/schematics
- Locate and determine the types of seals in use, both external and internal
- Identify possible contacts with process chemicals
- Record lubricant flow rates, lubricant bulk oil temperatures, bearing drain and inlet temperatures, and oil pressures
- Record detailed lubricant specification and compartment capacity
- Record filter performance specification and location

In many cases oil analysis data can be inconclusive when used alone. However, combined with sensory inspection information, a reliable and more exact determination is possible. Likewise, the application of companion maintenance technologies (like vibration and thermography) can help support a conclusion prior to expensive machine tear-down or repair. Table 21.11 represents a two-page summary of combined analytical and inspection/sensory indications of frequently encountered problems.

33.11 Simple Steps to Implementing an Industrial Oil Analysis Program

When doctors need to know more regarding the condition of your health, they are likely to start with a series of primary diagnostic blood tests. Similarly, oil analysis, sometimes referred to as "wear particle analysis", is a mature condition based / predictive maintenance approach used to determine the lubricated machine and it's lubricating oil current condition by testing a small sample of oil from the machine's operational lubricating system. The test laboratory then compares its working sample to a virgin stock sample through a series of laboratory tests, and secondly to previous (history) sample results from the same sample point (predictive trending) to determine the current "wellness" of both machine and oil.

Oil analysis is a highly effective and inexpensive method used to determine when to change oil based on its condition, to predict incipient bearing failure so that relevant action can be taken in a timely manner to avert failure and to diagnose bearing failure should it occur. Despite being around since the 1940's and its proven track record for effectiveness, oil analysis is still misunderstood and overlooked as a proactive / predictive maintenance strategy in many of today's industrial plants! Setting up and implementing an industrial oil analysis program is relatively easy and should be an included strategy in any industrial plant that purchases, stores, dispenses, changes, uses, or recycles lubricants as part of its manufacturing or maintenance process.

A successful oil analysis program can pay for itself in a matter of weeks as it delivers benefits in multiple ways that include:

- Oil change intervals (often extended) optimized to the machine's ambient conditions and operational use requirements,
- Probable reduction in lubricant inventory purchase costs and spent lubricant disposal costs,
- An understanding of how bearings can, or are failing in their operating environment so that bearing failure can be controlled or eliminated,
- Increased asset reliability, availability, and production throughput.

Note: program success is better assured if a work management approach is already in place that can assure completion of corrective action in a timely manner whenever the oil analysis report recommends such action.

Similar to all successful change management programs that must be rolled out across the organization an oil analysis program will benefit from a piloted phased implementation. Taking a stepped approach allows management and workforce alike to become accustomed to the new sampling and reporting processes and to quickly iron out any problems prior to rolling out the program en masse.

Step One: Appoint a program champion

All programs require a "go to" decision making person who advocates on the program behalf and is committed to making the program implementation a success; the champion should be at a supervisor or manager level.

Step Two: Choosing a suitable pilot area/machine

Oil analysis begins with sampling the oil and can include both lubricating and hydraulic fluids. Choosing a suitable program pilot will depend on the type of industry and business operation. Typical starting points to look at can include:

- Critical product, process, line, or major piece of equipment. *Criticality determined by constraint and/or lack of back up, downtime costs, product quality, etc.*
- Mechanical equipment with moving components that include lubricant reservoirs for recirculating oil transmission systems, mechanical and/or hydraulic in design.

Step Three: The lubricant audit

A lubricant audit is required to identify what lubricants are currently employed in service at the plant and will require the following checks:

- Check work order system PM job plans for lubricant specification(s),
- Check on or near the lubricant reservoir for lubricant identification labels or stickers,
- Check for matching MSDS (Material Safety Data Sheets).

Note: If a Lubrication Operation Effectiveness Review (LOER) or Lubricant consolidation program has already taken place (as described in Chapter 18: Lubricant Purchase, Handling & Storage) refer to lubricant listing charts.

If a discrepancy is found at this stage, outside assistance from a lubrication expert or lubricant supplier may be required to identify lubricants in use are the correct lubricants specified for the application.

Step Four: Choose a laboratory

Not all oil analysis laboratories are created equal, making your choice of laboratory an important step. Most oil analysis reports are divided into three major sections that report on 1), the sampling and virgin oil specification data 2), the spectral analysis testing results for wear elements identified as

Table 33.10 Oil analysis data interpretation/problem identification. (Courtesy Noria Corporation.)

Problem Area	Analytical Indications[a]	Inspection/Sensory Indications[b]
Wrong Lubricant	- Change in viscosity, VI, flash point, additive elements, FTIR[b] specta, TAN[d]/TBN[d] - Change in wear patterns	• Change in oil gage or bearing temperature • Bearing distress or noise • Hard turning of shaft
Antioxidant Depletion	• Decreasing TAN[d], RBOT oxidation life, and Zn/P content • Increasing viscosity, TAN[d], particle count • FTIR: decreasing antioxidant, increasing oxidation, sulphation, and/or nitration	• Oil darkening • Pungent odor • Hot running
Dispersancy Failure	• FTIR,[b] low TBN[d] • Increasing particle count, pentane insoluble • Defined inner spot on blotter test	• Filter inspection: sludge on media, filter in bypass • Black exhaust smoke • Deposits on rings and valves
Base Oil Deterioration	• Increasing viscosity, TAN[d], particle count, and/or ferrous particles • Decreasing TBN[d] • Change in VI and lower dielectric strength	• Poor oil/water separability • Air entrainment/foaming • Pungent odor, sludge/vamish formation • Blotter spot yellow/brown, oil darkening
Water Contamination	• Increasing viscosity, TAN[d], Ca, Ma, and/or Na • Rapid additive depletion/failure • Crackle test, VISA[e], KF[f], FTIR[b] • Reduced dielectic strength • Blotter test: sharp or star burst periphery on inner spot	• Oil clouding/opacity, water puddling/ separating, sludging, foaming • Evidence of fretting wear/corrosion • Filter: paper is wavy, high pressure drop, short life. Ferrogram shows rust. • Valve stiction, orifice silting, bearing distress/failure, noisy pump/bearings
Coolant Contamination	• Increasing viscosity, copper, particle count, wear metal, Na, B, and/or K • FTIR[b]: glycol • Crackle test, VISA[e], KF[f]	• Bearings dark charcoal color, distressed • Dispersancy failure, sludging, varnishing • Blotter test: sticky, black center • Filter plugs prematurely, oil has mayonnaise consistency, white exhaust smoke
Fuel Dilution	• Low oil viscosity, flash point • Additive and wear metal dilution (elemental analysis) • FTIR[b]/gas chromatography for fuel • Rising particle count and wear metals	• Rising oil levels and oil gage temperatures • Blotter test: halo around center spot • Blue exhaust smoke (collapsed rings), plugged air filter, defective injectors • Oil has diesel odor, overfueling conditions

Air Entrainment	- Increased viscosity, TAN(d), water, and/or FTIR(b) for oxidation - Silicon de-foamant levels too high/low - Blotter test: coke-like carbon on patch	• Oil clouding/foaming, increase in oil gage temp. • Spongy/slow hydraulics, cavitation of pump/ bearing, noisy operation
Abrasive Wear Conditions	• Increased silicon, aluminum, particle count and/or ferrous particles • Water contamination • Ferrogram has cutting wear, silica particles	• Scratch marking or/polishing of frictional surfaces • Cutting wear on blotter/patch/filter media • Filter/breather/seal failure
Corrosive Wear Conditions	• Increased TAN(d), particle count, spectrographic iron & bearing metals, water • Decrease in TBN(d) • Ferrogram shows submicron debris at ferrogram tail, rust particles, metal oxides	• Fretting, pitting and etching on contact surfaces • Transient electric currents, high engine blowby • Rust on patch or filter media
Failed Filter	• Increasing silicon/aluminum, particle count, ferrous particles, and/or elemental iron • Ferrograms show green looking particles, cutting wear, filter fibers	• Valve silt lock, noisy bearings • Unchanging or high delta P of filter • Frequent bearing failures, high levels of bottom sediment
Overheating	• Increasing ferrous particles, particle count, flash point, viscosity, or oil specific gravity • Ferrograms show friction polymers, oxides, bluing/tempering of particles, sliding wear particles, bearing particles, e.g., babbitt	• Bearing distress/failure • Hot spots and high bearing metal temp. • Evidence of coking/sludge • Bumt/rancid odor, high oil gage temp.
Misalignment, Imbalance, Overloading	• Ferrograms densely loaded with black-iron oxides, dark metallo oxides, severe cutting and sliding wear, tempered particles, large, chunky particles, or bearing metals • Increase in viscosity, TAN(d), particle count, and/or ferrous particles • Depletion of Zn and P	• Engine lugging/stalling, black exhaust • Raised oil, bearing metal, or jacket-water temperature • Dark, foul-smelling oil, bearing distress/ failure, hard turning of shaft • Abnormal vibration, noise • Blotter test: coke, metal chips • Metal chips on filter, highly loaded chip detectors
Impending Failure of Bearing, Gear, Pump, etc.	• Exponential increase in particle count and number of wear particle concentration • Increase in iron or bearing metals • Ferrogram shows rate increase in spheres, dark metallo oxides, particle bluing, spalling/ chunks, severe sliding/galling particles, cutting wear	• Shaft wobble, vibration, acoustic changes, blue exhaust smoke, hot spots, hard turning shaft and/or high bearing metal temperatures • Patch/blotter shows coking
(a) Not all of the identified indications would be expected for each problem area, (b) Fourier Transform Infrared Spectroscopy (c) Total Acid Number (d) Total Base Number (e) Vapor-Induced Scintillation Analysis (f) Karl Fischer		

lubricant additives or contaminants 3), additional physical test results for viscosity, water, glycol, fuel, soot, acidity, etc. and 4), the resulting conclusions and recommendations.

Some laboratories specialize in engine oil testing that focus more heavily on the physical tests for water, glycol, fuel and soot; some laboratories will specialize in industrial sample testing that focus more heavily on the wear particle testing and some physical testing for viscosity, water and acidity and post mortem testing for root failure causes using ferrographic testing. In addition, there are also laboratories that have lab technicians specialized in both areas of testing.

Interview and tour a number of different laboratories to understand their specialty area and how they will look after and process your samples. Key to any testing program is receiving results in a timely and consistent manner, especially when critical equipment is involved. When interviewing laboratories be sure to rate their sample "turnaround" time and how they can assure testing consistency (usually through use of dedicated technicians testing your samples). Working with a laboratory should be viewed as a long-term relationship as the laboratory builds and analyzes your complete data history and makes conclusions and recommendations based not only on your current sample versus its virgin sample counterpart, but also on an understanding of your plant ambient conditions and overall trending history of each sample.

Step Five: Setting up the pilot sampling program

A good laboratory will work with you to set up your sampling program, sell or attain sampling point hardware, extraction pumps, quality sample bottles and train your staff to take consistently "clean" oil samples.

The best oil samples contain maximum data density with minimum data disturbance, meaning the sample should best represent the oil condition and it's particulate levels as it flows through the system, or as it sits in the reservoir. For example, extracting a sample from the bottom of a reservoir in a non-pressurized gearbox lubrication system the particulate fall out will be dense due to large wear particles and/or sludge accumulation and not correctly represent the remaining 80% - 90% of reservoir lubricant that performs the actual lubrication of the gears.

In pressurized recirculating lubrication systems samples are best taken when the machine is running and at operating temperature from a "live" fluid zone where the lubricant is freely flowing. Whenever possible, the sample should be extracted from an elbow taking advantage of the data density caused by the fluid turbulence. Sample points are best located downstream of the lubricated areas to catch any wear elements before they are filtered out by inline pressure or gravity filters. The following sections in this chapter will detail the where and how of sampling techniques. You will be required at this time to sample and send off virgin oil samples of all lubricants to be checked in the pilot program. These samples are used as a benchmark for the laboratory to measure and understand what additive ingredients and lubricant state represents a normal status so that additive depletion and wear elements can be easily identified in operating samples.

Outside assistance/training from a lubrication expert or oil analysis laboratory is advisable when setting up the pilot sample points.

Step Six: Setting up your work management approach to sampling

Lubricant sampling must be performed in a consistent manner on a frequent basis lending itself well to work management through a maintenance/asset management work order system. Utilizing the written sampling procedure as a job plan, physical sampling can be scheduled effectively through the Preventive Maintenance scheduling software.

Extracting and sending a sample to the laboratory is only the first half of the oil analysis process, someone (usually the Planner, if one exists) has to receive the results electronically by email, read the recommendations and take any necessary corrective action and/or file the laboratory report electronically to history—usually as an attachment to the PM sampling work order. This will require a workflow procedure to be developed and trained to all maintenance staff involved in the program.

Step Seven: Commence sampling and your program roll-out

An oil analysis program will immediately identify major problems with contamination and wear in the first sample set. Sample trending can begin on the third set of samples with which the user can start to identify/predict any negative trend toward potential failure and schedule corrective action before failure occurs. Ideally, a pilot program should be allowed to run for approximately three months or more to show basic results before tweaking the process and rolling out to the next area within the plant.

Once your program is working and providing results, a larger corporation may wish to investigate investing in an in-house staffed laboratory that will deliver faster results turnaround.

33.12 Importance of Training

When a well-intentioned oil analysis program fails to produce the expected benefits it is often thought that a main contributing factor is an attitude of indifference among those involved. While this is occasionally true, the problem is generally much more fundamental and deep-rooted. Unless maintenance professionals have an understanding of the purpose and goals of oil analysis and are literate in the language of oil analysis, they cannot be expected to carry out its mission.

This mission is accomplished through a liberal amount of training and education. However, the effort should not simply be concentrated on a single individual but should involve all those that benefit from and contribute to machine reliability. Instead, training and education should be directed at several different functions including craftsmen, operators, engineering, and management. Below are a few subjects for which seminars and training classes are generally available:

1. Lubrication fundamentals and their application
2. Mechanical failure analysis
3. Proactive maintenance and root cause
4. Analysis and toubleshooting of hydraulic systems
5. Lubrication and maintenance of bearings and gear units
6. Oil analysis fundamentals
7. Oil analysis data interpretation
8. Filtration and contamination control
9. Wear particle analysis and machine fault detection

Once these fundamentals are in place, oil analysis can move forward enthusiastically, beginning with the development of its mission and goals. And, instead of indifference to oil analysis exceptions, rapid-fire corrections are carried out and measures are taken to preempt their recurrence. In time, unscheduled maintenance is rare and oil analysis exceptions become fewer as the machine operating environment becomes more controlled.

Finally, as the many elements of oil analysis and proactive maintenance merge together into a cohesive maintenance activity, the benefits should not be allowed to go unnoticed. Unlike many applications of new technology, proactive maintenance seeks non-events as its goal and reward. These non-events include oil that continues to be fit-for-service, machines that don't break down, and inspections that don't have to be performed. This quiet existence is the product of a highly disciplined activity but, let us remember, the activity risks being thought of by the casual observer as unneeded. Therefore, the plant's expenditures for proactive maintenance and the highly attractive benefits of proactive maintenance must be measured, monitored, and the outcome displayed for all to view.

Bibliography

Fitch, J.C., Jaggernauth, Simeon, "Moisture...the Second Most Destructive Lubricant Contaminant, and its Effects on Bearing Life," *P/ PM Magazine*, Dec. 1994

Fitch, J.C., "Clean Oil Sampling," *Practicing Oil Analysis Magazine*, July 1998

Fitch, J.C., "The Ten Most Common Reasons Why Oil Analysis Programs Fail and the Strategies That Effectively Overcome Them," *Diagnetics, 1995*

34

Safety

34.1 Implementing a Lubrication Safety Program

Part of any best-practice lubrication-management program is the proactive minimization of associated safety risks surrounding the 1:purchase, 2: lubricant use, and 3: disposal of lubricants. Three specific areas being foundational to the development of an effective lubrication safety plan.

34.1.1 Purchasing

When asked to provide an inventory list of all brands and formulations of lubricants (oil, grease, paste, solvent, kerosene, gasoline) currently carried and in open and unopened states within your plant, could you provide an accurate report within minutes complete with an up-to-date Safety Data Sheet (SDS) for each lubrication product in the plant? If you have already invested in a lubrication safety program, congratulations! You will likely be one of the few able to answer yes with confidence.

Because we are able to easily purchase lubricant products from any local hardware or automotive store, we can be easily lulled into a sense that all lubricants are all safe to handle. This is not always the case. To mitigate risk associated with unknown or suspect lubricants and fluids, we must refer to the consolidation list where we should find the lubricant toxicity number and refer to its SDS sheet to understand how we must handle and behave around the lubrication product.

Part of the lubricant consolidation program is to identify the risks associated with handling lubricants and if possible, cross reference and replace existing harmful products with lubricants that are safer to handle, use, and dispose. This process not only drastically reduces the number of lubricant SKUs tracked within the plant; it puts in place a control system that confines all lubricant purchases to an approved list. Toxicology is managed alongside up-to-date SDS documents, which spell out how each product is to be stored, handled, used, and disposed of in an environmentally friendly manner.

34.1.2 Lubricant use

When handling and using lubricants, personal injuries can occur if common sense is not part of the process. Many lubricants are delivered and stored in 55gal. drums that can weigh in excess of 400lb (180kg). Drums are not easily handled and back strain is common when trying to manually move them manually. If a forklift is used, great care must be taken to ensure the drum is positioned and/or secured correctly on a wooden pallet so it can't fall off in mid-air and cause injury from collision or explosion on impact.

Where oils are purchased in bulk, explore using totes that are filled on site. They are sturdy and can be placed in a permanent location, eliminating the need to rotate and move drum stock. For grease, consider buying smaller 5-gal. pails. These are easily transported on a two-wheel hand movers hand dolly or on a purpose-built lube truck like the one shown in Figure 34.1.

Lubrication activities require a technician to fill machine-lubricant reservoirs, swap out breathers, change filters, take oil samples, and often manually apply lubricants directly to machine bearings with grease or oil gun under pressure. If the lubrication function cannot be performed without dismantling or entering through a machine's envelope perimeter, or removing guards to gain access to bearing points, always perform lock-out/tag-out (LOTO) procedures before starting work.

While it may seem inherently safe to perform perimeter-based lubrication where all bearing fill ports, and lube points are located and accessible outside of the moving danger zone, **ALWAYS** exercise extreme caution, focusing full attention on the job at hand. Respect the machine at all times.

LOTO - Lock Out – Tag Out

If lubrication points and reservoir fill ports are within the machine movement zone, or you are required to work in a confined space area the equipment will be required to be shut down and locked out before work can proceed.

The purpose of LOTO is to ensure the safety of the lube technician or maintainer working on the machine within its moving zones. By de-energizing the electrical power and padlocking the main electrical power source, the worker is assured no-one can inadvertently turns the power back on while

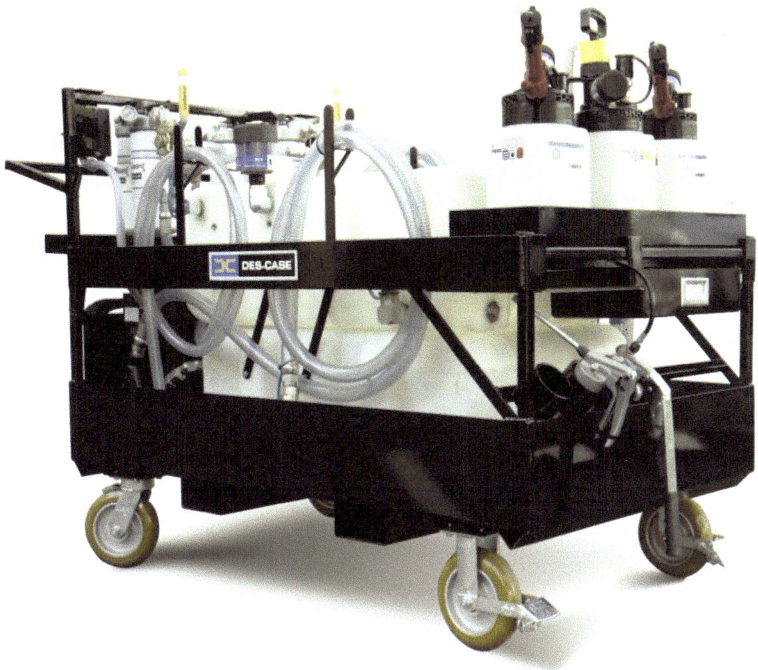

Figure 34.1 Purpose-built manual push lube truck. (Copyright Des-Case Corporation, 2022)

working on the machine. Figure 34.2 shows a typical TAG lockout station from which a worker can take a colored lock and lock out both the main panel and the machine panel with the same color safety padlock. A red and white lockout tag is also places with the lock and filled ion by the worker giving the reason for the lockout, the section of the machine being worked on, the expected downtime, and the name and contact information of the worker.

If more than worker is working at the same time on the machine, each worker will have their own color safety padlock which are attached through a clamped on multipoint lockout hasp (red clamp style device with six holes shown in lower LH corner of Figure 34.2. The machine cannot be started until all workers have removed their own safety padlocks form the hasp.

Basic safety gear

When it comes to the handling and transfer of chemicals and lubricants, the safety dress code for personnel must be specific and very clear. The need for the correct personal protective equipment (PPE) is absolute. Minimal PPE

Figure 34.2 LOTO station board. (Courtesy ENGTECH Industries Inc.)

for anyone handling, transferring, or cleaning up lubricant spills will at a minimum include the following:

- eye protection (glasses or goggles, depending on whether the product splashes easily)
- safety boots with oil-resistant soles
- long-sleeve, close-fitting coveralls that are cleaned regularly
- hardhat
- gloves to protect skin from dermatological rashes and high-pressure injection hazards.

34.2 Lube-application Safety Specifics

34.2.1 Pressure injury incidents

Few maintainers and lubrication personnel aware that the types of simple electric/compressed air grease guns available from their local hardware and big-box stores can deliver lubricants at nozzle pressures up to and greater than 5000 psi. Nor are they aware that an innocuous hand-grease-gun can develop pressures up to 15,000 psi.

A 2020 study titled, *Management of High-Pressure Injection Injury of the Hand in the Emergency Department,* by S.O. Sanford and T.J. Mills, noted that, "High-pressure injection injuries occur when a high-pressure injection device, such as a paint or grease gun, injects materials into the operator [skin]. This injury most commonly occurs in the dominant hand and index finger." As the researchers further explained, "The injection typically occurs to the fingertip when the operator is trying to wipe clear a blocked nozzle, or to the palm of the hand when the operator is attempting to steady the gun with a free hand during the testing or operation of the [grease gun] equipment."

Similar injuries can occur when performing condition and leak checks on other types of pressurized hoses and lines (especially flexible line material). Alas, it's not uncommon to see maintainers, and machine operators in TPM (Total Productive Maintenance) environments running their hands over high-pressure flexible lines to feel for leaks or soft spots that, if found, can easily inject fluid into a curious hand. Leak checks are best performed using a piece of stiff card, NEVER your arm, hand, or fingers.

Proper grease-gun training instructs the operator never to place his or her finger directly over the flexible delivery tube or nozzle tip when activating the trigger. If cleaning the nozzle tip, always take the non-dominant hand off the trigger or pump lever. Unfortunately, too many companies fail to provide adequate grease-gun training to their maintainers and machine operators.

Injection through the skin can occur at less than 1,000 psi. The higher the pressure, though, the more damage that can occur to the structure of the hand or finger. (If you have ever received an inoculation in your arm through use of a medical pressurized jet injector, you may recall the "pressure" felt fairly low, as such devices operated between 1,000 psi and 2,000 psi. Imagine, however, what pressures of 5,000 psi to 15,000 psi could feel like or do to a body part.). (Keep in mind that hydraulic injection can also take place using a simple power washer to clean off a lubrication system)

It can be easy to ignore a pressure-injection injury and, ultimately, begin second-guessing if it did or did not occur. Don't, it can prove lethal! According to the cited Sanford and Mills study, "A high pressure injection injury should be considered a surgical emergency." To see why, refer to the 2018 YouTube safety video titled, *Hydraulic Injection*, from Northwest Linemen College (see link below). It visually demonstrates how, if it's not quickly addressed medically, a seemingly small injection site, with a slight area of redness, can turn into a devastating injury within mere hours that can lead in some cases to amputation and death.

The Sanford and Mills study found the overall incidence of amputation stemming from pressure-injection hazards was 48%, with solvents causing the greatest damage followed by grease injection. In 25% of all incidences, amputation occurred.

Similarly, an earlier medical study performed at the New York Methodist Hospital by J.T. Snarski, MD and R.H.Birkhan, MD found that high-pressure grease guns/systems were responsible for 57% of all injection injuries, seconded by hydraulic-fluid injuries. Furthermore, the researchers found that the overall incidence of medical amputation resulting from such injuries was 48%, and closer to 100%, when injection pressure was greater than 7,000 psi. Even if affected limbs can be saved, long-term outcomes of the injured individuals are not always good. For example, the Sanford and Mills study found that among those who underwent successful surgical debridement, 50% ended up with a reduced range of motion; injuries adversely affected daily-living activities; grip strength was reduced by 35% in 75% of cases; and all patients continued to suffer from some level of neuropathic pain.

Knowing what to do in a pressure related incident

If you suspect a high-pressure injury has occurred, you (or the person in question) must seek professional medical treatment immediately. All studies show that the quicker these injuries are recognized and professionally treated, the higher the chance of limiting damage and saving an affected limb.

The Sanford and Mills study listed the following details that physicians require to determine the severity (and treatment) of pressure-injection injuries:

- Type and viscosity of the injected material (a copy of the fluid-type SDS is essential here).
- Time interval between injury and treatment (log the time that the injury occurred).
- Amount of material injected and velocity of injectant.
- Pressure of the appliance (or system).
- Anatomy and distensibility (swelling amount) of the site of injection.

If at all possible, use a smart phone, tablet, or other device to photograph or record a video of the injury, as soon after the incident as you can. Also photograph or record a video of the injection device.

In its safety video, Northwest Lineman College recommends the following first-response plan for pressure-injection type injuries:

- Treat Immediately (seek medical attention).
- Don't let victim drive.
- Don't leave victim alone.
- No food or drink.
- Immobilize and elevate the wound.

If you feel that you (or others) have ingested lubrication fluid, have skin or eye irritations through exposure, or believe you (or someone else) *may have* suffered a pressure-injection injury, seek medical advice immediately. Since time is of the essence in these situations, it is always better to be safe than sorry.

34.2.2 Breathing incidents

Working with lubricants and solvents, especially in a confined space area can sometimes create toxic fumes that can temporarily cause breathing problems. These of course are workplace dependent so care must be taken to understand the products we work with and the quality of the air we breathe when in a confined space.

Confined space air quality should always be tested, especially when working on a lube system in a pit under a machine, e.g., large straight side stamping press. Always test air quality using an oxygen sniffer that is lowered down into the are to detect for harmful gases and oxygen levels BEFORE entering the space. A normal atmosphere oxygen level is 21%. The Occupational Safety and Health Administration's (OSHA) Respiratory Protection Standard paragraph (d)(2)(iii) considers that "any atmosphere with an oxygen level below 19.5% to be oxygen deficient and immediately dangerous to life or health. Known potentially hazardous areas to breathing should be marked with the symbol shown in Figure 34.3.

Work can still be carried out by using a breathing respirator like the one shown in Figure 34.4. When setting up oil mist systems a breathing respirator is highly recommended, Use of a breathing respirator will require individual training and certification

34.2.3 Eye and skin incidents

There are times that lubricants and chemicals can splash onto our skin and into our eyes. If this does occur and lubricant does enter the eye it will likely start to irritate immediately and will require to be rinsed using an eye wash station.

Figure 34.3 Breathing hazard warning. (Courtesy ENGTECH Industries Inc)

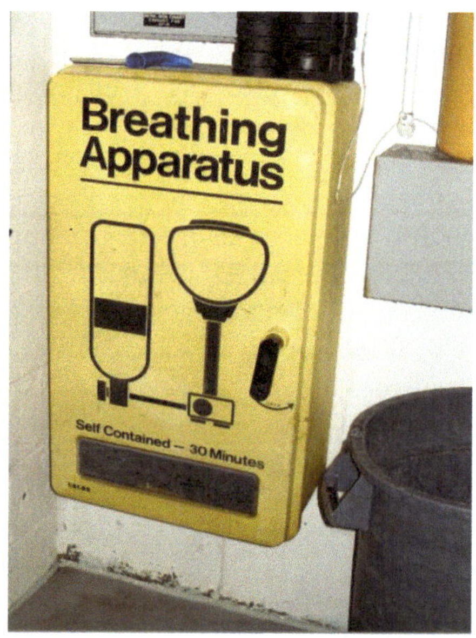

Figure 34.4 Breathing respirator. (Courtesy ENGTECH Industries Inc)

Eye wash stations must be in place in a lube room and in various areas of the plant. Familiarize oneself with their locations before they are needed in an emergency. There are two main types of eye wash station, a permanent plumbed in unit that may include a full body shower with a pull chain

Figure 34.5 Squeeze bottle eye-wash. (Courtesy ENGTECH Industries Inc.)

Figure 34.6 Eye station with shower. (Courtesy ENGTECH Industries Inc.)

activation like that shown in Figure 34.6, or a non-plumbed in squeeze bottle type like that shown in Figure 34.5. Both types need to be on a preventive maintenance schedule so that they are always clean, operable and free of obstructions and ready for use. We can see that the station in Figure 34.6 has an obstruction that could heed use of the unit in an emergency. In these cases, a yellow outline "no place zone" needs to be painted on the floor to remind people of its importance.

As each plant or workplace supports the use of a unique inventory of lubrication fluids and chemicals, determining the correct PPE requires maintenance to list all lubricants and chemicals on site. This list is then used to gather an up-to-date set of Safety Data Sheets (SDSs) for each lubricating fluid and chemical used in the plant/workplace from the lubricating fluid supplier(s). Once this has been done, these documents can be referenced to build a specific PPE-requirement table listing for each fluid, based on what's spelled out in "**Section 8: Exposure Control/Personal Protection**" of its SDS.

Whenever a safety incident occurs regarding lubricants and their safety, ALWAYS refer to the Safety Data Sheet (SDS),

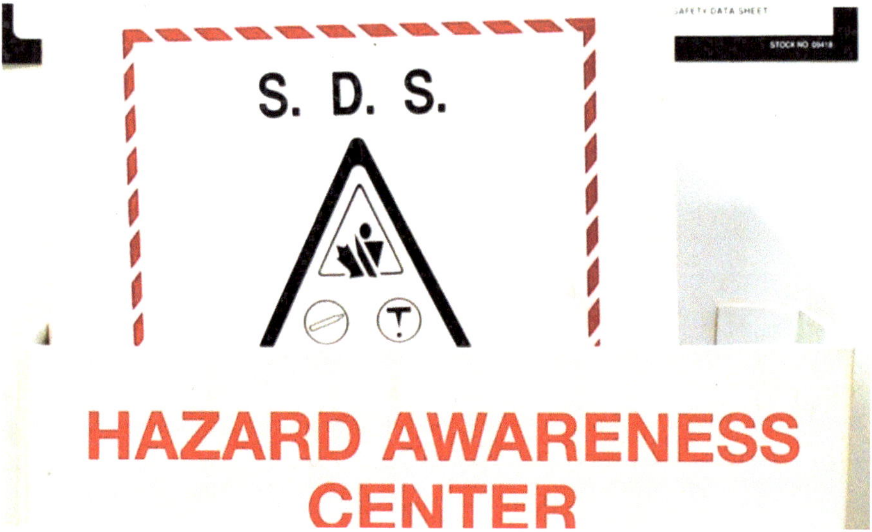

Figure 34.7 Typical SDS manual containing current SDS sheets for ALL lubricants and Chemicals used in the plant. (Courtesy ENGTECH Industries Inc.)

34.3 Safety Data Sheets (SDS)

Safety Data Sheets are mandatory documents that play a vital role in workplace-safety and lubrication-management programs. These are the documents where you go to find crucial information about lubricants and related products if you or others were to accidentally ingest or come into skin contact with them, or an uncontained spill into water or soil occurred.

Formerly referred to as a Material Safety Data Sheet (MSDS) it has undergone some changes and is now simply referred to as a Safety Data Sheet (SDS). The SDS is a mandatory, 16-section, fact-sheet-style document for each potentially harmful product that meets the international requirements of the United Nations' Globally Harmonized System (GHS) of chemical classification and labeling.

The Occupational Safety and Health Administration (OSHA, osha.gov, Washington) is the U.S. regulatory agency that legislates under the Hazard Communication Regulation to ensure all employees in a workplace coming into contact with potentially harmful substances—such as lubricants and related products—have ready access to current SDS documents, typically compiled in a print manual that is conspicuously placed at the entrance of the lube room and in various locations around the plant, (see Figure 34.7), and online.

34.3.1 Understanding SDS elements

An SDS reflects 11 individually described, mandatory elements (Sections 1 to 11), and five nonmandatory—but GHS-required—elements (Sections 12 to 16). The information is similar to that found in an MSDS, but is now presented in a more consistent format. Here, based on OSHA requirements, and using a standard air-tool oil lubricant as an example, is an overview of the 16 sections.

Section 1: Identification. This section identifies the lubricant by the common name found on its labeling and any other synonyms used to identify the product worldwide, as well as the detailed contact information for the manufacturer or importer, and an emergency phone number. This section also describes the recommended use of the product, *i.e.*, lubricant, and any restrictions on its use.

Section 2: Hazard Identification. This section identifies any hazards associated with the lubricant's use and details the appropriate warning information for the product. Details of the standard air-tool oil lubricant in our example would include:

- *Classification Category*: Category 1- Aspiration hazard
- *Signal Word:* Danger
- *Hazard Statement:* "May be fatal if swallowed and enters airways"
- *Pictogram:* Icon or infographic that represents the aspiration hazard visually (Figure 34.3)
- Precautionary Statements: If swallowed "Immediately call a poison center or doctor.
- Do NOT induce vomiting. Store locked up."
- *Other Hazard Statement:* "6% of the mixture consists of ingredient(s) of unknown acute toxicity." (or similar statement referring to the percentage of unknown toxicity that is present in the product's ingredients).

Section 3: Composition/Information on Ingredients. This section describes the ingredient substances and mixtures contained in the product that must also include any impurities and stabilizing additives. If the concentration of ingredients is a trade secret, a statement must be made here. Details of our example air-tool oil lubricant would include:

- *Chemical Name*: Mineral oil, petroleum

- *CAS Number*: Unique identifier, *i.e.*, the product's Chemical Abstracts Number (CAS#8042-92-3)
- *Statement*: "The exact percentage (concentration) of composition has been withheld as a trade secret in accordance with paragraph (i) of 1910.1200."

Section 4: First Aid Measures. This section covers the initial care and recommendations that can be provided by untrained responders to an individual exposed to the lubricant. Details of our example air-tool oil lubricant would include:

- *First Aid Measures by Exposure Type*: Inhalation - "If breathing is difficult, remove victim to fresh air and keep at rest in a position comfortable for breathing." And,"Seek medical advice/attention if you [the victim] feel unwell."
- *Most Important Symptoms and Effects*: Inhalation - "May cause respiratory tract irritation."
- *Recommendations for Immediate Medical Care and Special Treatment*: Note to Physician "Treat symptomatically. Symptoms may not appear immediately. In case of accident or if you [the victim] feel unwell, seek medical advice immediately." (Show the label or SDS when possible.)

Section 5: Fire-fighting Measures. This section covers recommendations for fighting a fire caused by a lubricant. Details of our example air-tool oil lubricant would include:

- *Extinguishing Media:* Suitable -"Treat for surrounding materials." Unsuitable -"None known"
- *Specific Hazards:* Products of combustion include, but are not limited to, oxides of carbon.
- *Protective Equipment and Precautions for Firefighters:* "Wear full firefighting turnout gear (full Bunker gear) and respiratory protection (SCBA). Keep upwind of fire."

Section 6: Accidental Release Measures. This section provides appropriate response recommendations for spills, leaks, or releases of the product, including information on containment and cleanup practices. Details of the example air-tool oil lubricant would include:

- *Personal Precautions*: "Use personal protective equipment recommended in Section 8. Isolate the hazard area and deny entry to unnecessary and unprotected personnel."

- *Environmental Precautions*: See Section 12 for ecological information.
- *Methods and Material for Containment and Cleaning Up*: "Contain and/or absorb spill with inert material (sand, vermiculite), then place in a suitable container. Do not flush to sewer or allow entering waterways. Use appropriate Personal Protective Equipment (PPE)."

Section 7: Handling and Storage. This section provides guidance and precautions for safe handling and conditions for safe storage. Details of our example air-tool oil lubricant would include:

- *Handling*: "Avoid contact with skin and eyes. Avoid breathing vapor or mist. Do not swallow. Handle in accordance with good industrial hygiene and safety practice. Wash hands after use. Do not eat, drink, or smoke when using this product."
- *General Hygiene Advice*: "Launder contaminated clothing before reuse."
- *Storage Conditions:* "Keep container closed when not in use. Store in a dry, cool, and well ventilated area. Keep out of reach of children."
- *Incompatible Materials:* "None known."

Section 8: Exposure controls/Personal Protection. This section specifies the exposure limits, controls, and protective measures required to minimize exposure. Details of our example air tool oil lubricant would include:

- *Control Parameters:* "No data available."
- *Appropriate Engineering Controls:* "Apply technical measures to comply with occupational exposure limits. Use ventilation that is adequate to keep exposure (including airborne levels of dust, fume, and vapor, among others) below recommended limits."
- *Individual Protection Measures*:
- *Respiratory Protection* "Not required under normal use conditions. Ensure adequate ventilation, especially in confined areas."
- *Skin and Body Protection:* "Wear suitable protective clothing. Wear proper, compatible, chemical resistant protective gloves."
- *Eye/Face Protection:* "Safety glasses or goggles are recommended when using product."
- *General Work/Hygienic Practices*: "Handle in accordance with good industrial hygiene and safety practice. Do not eat, smoke, or drink where material is handled, processed, or stored."

In clear language Safety Data Sheets are critical components of any workplace-safety and lubrication-management program. They spell out, in clear language how lubricants should be stored and handled in the working environment, as well as how they react with other lubricants, and instruct on the dangers and remedial action required in the event of a spill or contact.

Sections 1 through 8 of a lubricant's SDS are the most helpful for personnel who need to quickly find lubricant information. The remaining sections contain scientific and technical details on physical and chemical properties (Section 9); stability and reactivity (Section 10); toxicology (Section 11); ecology (Section 12); disposal (Section 13); transport (Section 14); and regulatory issues (Section 15).

All maintainers and lubrication technicians are required to be trained on how to read and use an SDS, and to have easy access to print copies of an SDS manual located in conspicuous places at a site, *i.e.*, in or near first-aid or eyewash-stations, lubricant storerooms, and/or maintenance shops, as well as in an online manual.

It is also important to note that workers themselves are responsible for knowing and understanding the properties and associated risks surrounding the lubricants with which they work, and to make use of appropriate safeguards.

34.4 Disposal of Lubricants

See Chapter 33 section on "Used and Waste Oil Management"

Bibliography

Bannister, Kenneth E., "Lube Safety Comes in Threes", *Efficient Plant Magazine*, 2018

Bannister, Kenneth E., "Respect Safety Data Sheets", *Efficient Plant Magazine*, 2018

Bannister, Kenneth E., "Better Safe Than Sorry with Pressurized Fluids", *The RAM Review E-zine*, 2021

Sanford, S.O., Mills, T.J. (Ed.). (2020, March 20). *Management of High-Pressure Injection Injury of the Hand in the Emergency Department.* Medscape https://emedicine.medscape com/article/826620. (Accessed March 5, 2021)

Northwest Lineman College. (2018) "*Hydraulic Injection*". Available at: www.youtube.com/watch?v=09n23cY65bc (Accessed March 5, 2021

Snarski, J.T., Birkham, R.H. (2004, Oct 15) "Non-Operative Management of a high-pressure water injection injury of the Hand" www.cambridge.org/core/services/aop-cambridge-core/content/view/2103EE7D3453DB-43FE63774DB214AE18/S1481803500013099a.pdf/nonoperative_management_of_a_highpressure_water_injection_injury_to_the_hand.pdf (Accessed March 6, 2021

SECTION 6

Implementing World Class Lubrication

35

Implementing World-Class Lubrication

We know that in the elements of lubrication management have come under review as part of quality and document management programs through the ISO 9000 quality assurance standard, alongside the ISO 14000 environmental management standard.

More recently, the 2014 introduction of the ISO 55001—Asset Management standard has shone a bigger spotlight on the practice of lubrication through its focus on assets. Unlike its predecessor PAS 55 (Publicly Available Specification), which was solely physical asset focused (machine-centric) serving as the model for ISO 55001, lubrication is still only a secondary thought.

35.1 Working to a Lubrication Standard

ISO 55001 views an asset as "an item, thing or entity that has potential or actual value to an organization." It views asset management as something that "involves the balancing of costs, opportunities and risks against the desired performance of assets, to achieve the organizational objectives."

Traditionally, lubrication has always been defined as an act of performing lubrication, or as a physical sub-system of a machine consisting of a reservoir, pump, metering valves, lubes lines and control system belonging to a parental asset. Under ISO 55001 the definition takes on new meaning when associated with life cycle management of the asset. For example: maintenance and set up of the lubricant delivery system; procurement, storage, and handling of lubricants; analysis and reporting of lubrication related failures; competence and training of persons performing and managing the lubrication program; lubrication job plans, SOPs (standard operating procedures) and workflow can all come under the ISO audit microscope if they affect the operation and life cycle of valued assets—as they should in any lubrication management program. However, lubrication has always been viewed as the poor cousin in maintenance and as such most people working with lubrication are not specifically trained, certified, or dedicated to the task. Lubrication is all too often relegated to a sub-system or secondary task, even though ineffective lubrication practices are responsible for up to 70% of all bearing failures that adds up to $Billions of unnecessary GDP loss annually!

Machines cannot exist with lubrication. Bearings, chains, gears, motors, slides will all fail in a very short time frame if they are not lubricated. Lubrication IS the lifeblood of any machine and is deserving of being treated as THE major player in keeping that machine healthy and productive for a long, sustainable life. Just as the world of asset management was recognized in 2014 with a dedicated world standard, it is rather fitting that 100 years after the world's first centralized lubrication system was introduced by Joseph Bijur, 2023 witnessed the launch of a dedicated world standard for the "*optimized lubrication of mechanical physical assets*." All thanks the International Council for Machine Lubrication who launched their ICML 55® standard to effectively manage all things lubrication.

The ICML 55® standard is designed to align with ISO 55001, although it was created and developed independent of ISO 55001 and as such is not endorsed by the International Organization for Standardization. The ICML 55.1 states in its disclosure page that it "*represents, to the most thorough degree possible, the consensus of the machine lubrication community [worldwide] pertaining to the scope and best practices associated with Lubrication Management*".

The ICML is a vendor-neutral, non-profit organization founded in 2001 to serve global industry as a world-class authority on machinery lubrication to advance the optimization of asset reliability, utilization, and costs through effective lubrication.

Since 2001, ICML has worked diligently to give voice to the world of lubrication, where it matters, on the shop floor. It has built a huge worldwide membership of certified lubrication personnel and managers through its training partnerships and certification and badge programs. All geared to ensure machinery lubrication is designed and performed by certified people who truly understand the impact lubrication has on machine health, machine uptime and life cycle, while being kind to the environment.

Effective lubrication management isn't only about physical lubrication, it also engages with and affects:

- Data management
- Document management
- Condition monitoring

- Risk management
- Quality management
- Environmental management
- Life cycle costing
- Sustainable development
- Non-destructive testing
- Machine Inspection
- Value management
- Shock and vibration
- Training and qualification of personnel
- Equipment management
- Energy management
- Financial management

35.1.1 Interpreting the standard

There are many practical reasons for adopting the ICML 55 standard. The primary purpose if as standard is to provide a framework or foundation on which a lubrication management program can grow and flourish.

Many existing work management systems and data repositories are choked full of Meaningless Unrelated Data (MUD) making it difficult to mine the data and turn it into meaningful information that allows management to make confident and effective decisions regarding the management of people and assets.

If we lose confidence in our data, it is of little value or relevance for historical or due diligence purposes, which means we must address how we collect, compile and use that data. It also means we must assess and address the validity and value of our current lubrication management approach and activities, which include the use of resources, materials, tools, instruments, data, processes, procedures, plans, methods and activities and formulate an lubricant management approach that allows us to mitigate risk and meet our department and corporate objectives.

35.1.2 ICML 55 Standard Series

Within the ICML 55 Standard umbrella there are four standards. These are:

- ***ICML 55.0*** - The purpose of ICML 55.0 is analogous to the 30 second elevator pitch. Its intent is to provide an overview of a lubrication management system and processes applicable for the effective management of physical assets related to lubrication; its principles; and terminology. It also serves to provide context for the ICML 55.1 (audit requirements) and ICML 55.2 guideline for the optimized lubrication of physical assets.
- ***ICML 55.1*** – This the standard that a lubrication organization will be audited against. This standard sets out the auditable requirements of twelve specific interrelated areas that make up the Lubrication Program Plan and form the basis for the twelve auditable elements of ISO 55outlined in Figure 35.1 below.
- ***ICML 55.2*** – is essentially a detailed blueprint for implementing a best practice lubrication program that meets and surpasses the certification requirements of ICML 55.1. This standard is designed to illustrate the value of each of the twelve auditable areas (the why?) and provides many practical "How To" examples to fulfil the requirements including examples of what an auditor would look for to prove compliance in an audit.
- ***ICML 55.3*** – ICML 55.3 is an auditor guideline developed to assist an internal assessor preparing a company for an audit to evaluate audit readiness, and to give professional third-party auditors a framework from which to approach the ICML 55.1audit.

For those requiring copies of these standards contact ICML at www.icmlonline.com

35.2 Developing the Lubrication Program

The first objective is to understand and adopt the Lubrication Management System (LMS) as shown in Figure 35.2. The LMS is a collective series of linked elements designed to meet the audit requirements set out in twelve auditable areas shown in Figure 35.1. The LMS is a living system that can be monitored and updated in response to organization plans, objectives, assets, policies as they drive the need for improvement change.

Within the LMS, the lubrication management plan id developed to convert organizational objectives and strategies into lubrication management

Figure 35.1 Twelve ICML 55 Auditable areas. (Courtesy ICML)

objectives and strategies, from which the lubrication plan, or program is designed.

A lubrication program is much more than collection of work orders, oils, greases and lube systems. It is an engineered methodology designed to put in place a structured approach to understanding the current lubrication approach, and upgrading it to a best practice approach for managing lubrication assets and methods specifically designed to meet the machine's need

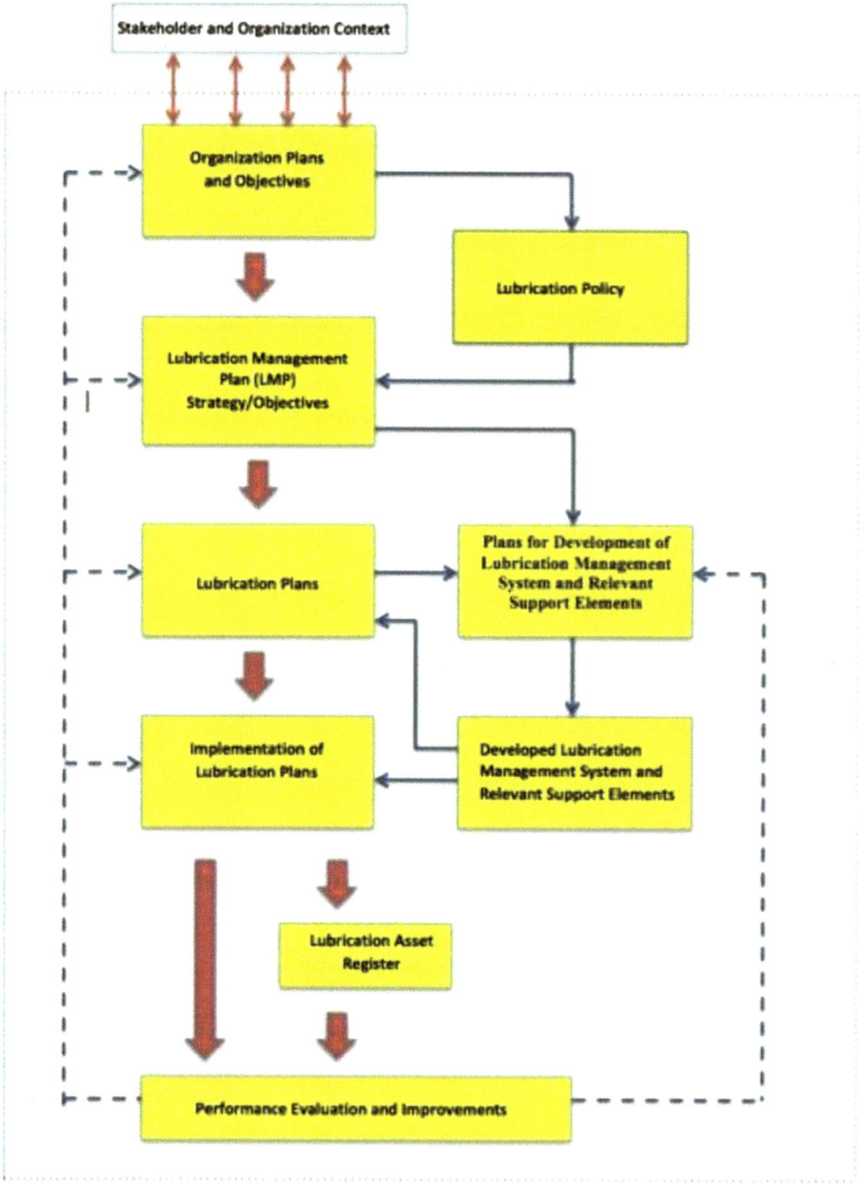

Figure 35.2 A lubrication management system (LMS). (Courtesy ENGTECH Industries Inc.)

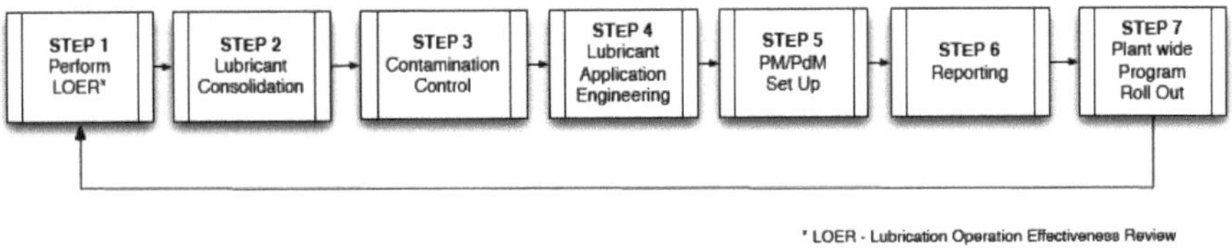

Figure 35.3 7-Step lubrication program journey. (Courtesy ENGTECH Industries Inc.)

while aligning with all stakeholder (maintenance, production and corporate) needs and objectives.

Typical objectives include:

- Accountability (corporate)
- Increased availability (production)
- Increased reliability (Maintenance)
- Throughput (Corporate and Production)
- Enhanced Regulatory Compliance (All)
- Safety (All)
- Due Diligence (All)
- Energy Reduction (Corporate, Maintenance)
- Carbon Footprint Reduction (All)
- Cost Reduction (All)

The lubrication program is developed using a seven-step process similar to that depicted in Figure 35.3.

STEP 1 – Perform a LOER

A Lubrication Operation Effectiveness Review, or LOER^{TM} is a third-party investigative audit styled process designed to investigate, review and analyze the current state of lubrication. This is accomplished by examining objective data from work management systems and perceptive feedback from interviews with lubrication, maintenance, and client/stakeholders (production, engineering, stores, purchasing, HR, management, etc.).

Typical review elements include:

- Audit of plant equipment for current lubrication application methods and wellness issues

- Review plant equipment of optimized lubricant application opportunities
- Review hydraulic systems for wellness issues
- Review listing and storage methods of current lubricants stored in the facility
- Identification of opportunities and benefits to performing a lubricant consolidation program
- Review storage and handling methods currently employed in the plant
- Revies lubricant transfer and documentation methods employed
- Review safety implications of current Lubrication practices
- Review contamination levels and control method effectiveness
- Review current auto-lube delivery systems, assess state of repair and calibration effectiveness
- Review current manual lubrication practices (grease gun application) and their effectiveness
- Assess operator TPM opportunities and visual factory methods specific to lubrication systems and their ongoing maintenance
- Review current lubrication PM tasks for effectiveness and objectivity
- Review the current planning and scheduling of lubrication related tasks
- Review current oil analysis predictive practices for effectiveness
- Review department business goals and assess plant equipment condition to make recommendations on a suitable pilot application area to launch the new lubrication management program
- Review current lubricant purchasing activities and make recommendations to implement a controlled lubricant purchasing policy
- Review and make recommendations on current lubrication approach and assess lubrication training needs
- Review current staffing configuration and make recommendations on the current use of dedicated lubrication personnel
- Review corporate energy and sustainability program and determine how the current approach succeeds or requires improvement to meet the requirements

- Review mission and vision alignment
- Review KPI reporting and measurement
- Review complaints of poor lubricant performance and make recommendations to resolve issues

NOTE: an LOER™ is a must do first step for those preparing for an ICML 55.1 certification

STEP 2 – Lubricant Consolidation
Refer to Section Five – Chapter Thirty-Two – Lubricant Consolidation

STEP 3 – Contamination Control
Refer to Section Five – Chapter Thirty-Four – Contamination Control

STEP 4 – Lubricant Application Engineering
Once lubricants have been consolidated and a contamination control policy put in place, it is now time to address the design and set up effectiveness of all lubricant delivery/application systems.

Many of the major issues surrounding lubrication hardware issues should have been identified in the LOER process; these are a natural starting point to begin a systematic address and improvement of all lubrication delivery systems in the plant. This next step begins with lubrication mapping.

Lubrication mapping is recognized as a basic fundamental building block essential to any best practice lubrication management program, although not new—LM was first mass introduced by the automotive companies in the early 1900's—the practice is not common in a modern-day plant.

Classic Lubrication Mapping (LM) is the practice of identifying, classifying, and charting out the physical location of all filter, reservoir and lubrication delivery points found on each machine (industrial), or vehicle drive train/chassis (automotive) used to facilitate an end user lubrication regime. The present-day interpretation of industrial LM has been expanded to include lubrication resources (lubricant storage, transfer, delivery, filtration, environmental spill and safety item product inventory), lubrication workflow routes, and line identification marking for all centralized machine delivery systems.

35.2.1 Benefits of lubrication mapping

Where asset reliability is concerned, it is no longer acceptable to state, "lubricate as necessary" on any machine PM work order. Building and executing an effective lubrication PM job task requires the job planner and the lubrication technician to understand each machine's lubrication requirements, alongside the logistics of efficiently managing and delivering lubricant to the required bearing point to satisfy the 4R principles of lubrication. LM is an inexpensive and highly cost-effective initiative that facilitates the commencement of a lubrication management program and delivers many benefits that:

- Provide a documented "as built" inventory and location schematic of each machine's lubrication points (Right place) and requirements (Right amount) used to build the PM job task and determine the appropriate lubricant delivery system choice if no centralized or automated system is already in place,
- Facilitates the consolidation of lubricants (Right lubricant) that results in:
  - The correct lubricant choice for the end user operating conditions,
  - Reduced lubricant inventory savings,
  - Reduced lubricant storage real estate requirements,
  - Reduced lubricant purchasing and handling costs,
  - Reduces incidence of lubricant cross contamination,
- Facilitates training and rapid assimilation of new hires to perform lubrication in a consistent and correct manner,
- Facilitates the development of efficient lubrication routing (the "ganging" of multiple lubrication PM work orders based on use of similar lubricant type to increase lubrication technician "wrench time" effectiveness).

35.2.2 Implementing a lubrication mapping initiative

If your company or department has performed a lubricant consolidation exercise in the recent past, you will likely already have much of the information required to complete a full LM initiative. If this is not the case, implementing your LM initiative is best performed in following three stages, these being: 1) Machine mapping, 2) lubricant and lubricant resource mapping, and 3) workflow and delivery route mapping.

Prior to any mapping exercise gather up as much information about the lubrication systems and points on the current machine inventory, these will include:

- Lubrication Schematic drawings or take offs (Operation and Maintenance (O&M manual),

- Machine engineering drawings (engineering department),
- Machine Bill of Material (BOM) lists (asset management work order system),
- Vibration monitoring route plans (maintenance planning department--these identify critical bearing point locations),
- Oil sampling route plan (maintenance planning department)
- Machine inventory list (maintenance planning department),
- Lubricant (oil and grease) purchase list, (stores or purchasing department)
- Lubricant filters, breathers, spill kits, and safety items used on Lubrication PM work orders (stores or purchasing department and maintenance planning department)
- SDS sheets for all lubricants

Next step is to develop a data gathering form to document and collect the data for each machine. You will need a digital camera and a machine marker to complete the data collection. Prior to inspection, ensure the machine is de-energized and locked out, and be sure to wear appropriate safety gear. Collect the following machine lubrication data:

1. Machine data such as asset number, description, and location. Photograph the machine from all four corners and above, if possible
2. List all reservoirs for gearboxes, auto-lube systems, or hydraulics. Note if there are tags identifying lubrication information. Photograph each reservoir and note if there is a min-max fill level indication, identify any installed breathers and filters; identify fill port and drain port locations; identify reservoir capacity (measure and calculate if not known); identify oil analysis points
3. Identify, photograph, and with a machine marker number all bearing points on the machine. Note the bearing type and size and if possible, the running speed. Identify if bearings are oiled or greased and what lubricant is currently in use. If on the PM work order, list the current amount of lubricant asked for
4. Identify if bearings are lubricated automatically, semi-automatically (centralized with manual pump or grease gun) or manually. Furthermore, check if lube lines are marked or tagged for identification and troubleshooting purposes.

Figure 35.4 Poorly set up auto lube system. (Courtesy ENGTECH Industries Inc.)

With physical machine mapping completed, stage two can commence. Here we use the lubricant consolidation report to identify which system uses what lubricants. In addition, we also identify, document and photograph all stocked breather and lube filters for inclusion on the machine map.

Stage three examines the current lubrication system design, setup, and calibration. An evaluation of the design effectiveness can determine if grease nipples are better served by an automated centralized system; if the current oil lubrication system requires updating to include modern breathers, reservoir level markings, oil analysis ports, improved control system, etc.

Most automated lubrication systems are incorrectly calibrated, often over lubricating the bearings - see Figure 35.4. Proper calibration of automated system can make huge differences to improve bearing/machine life cycle and reduce energy requirements that convert into greenhouse emission savings as shown in the case study one example in Figure 35.5.

35.2.3 Combating bearing failure

In the 2014 "State of the Nation Lubrication Practices" study conducted by ATP network's *Maintenance Technology* magazine, found that "60% of

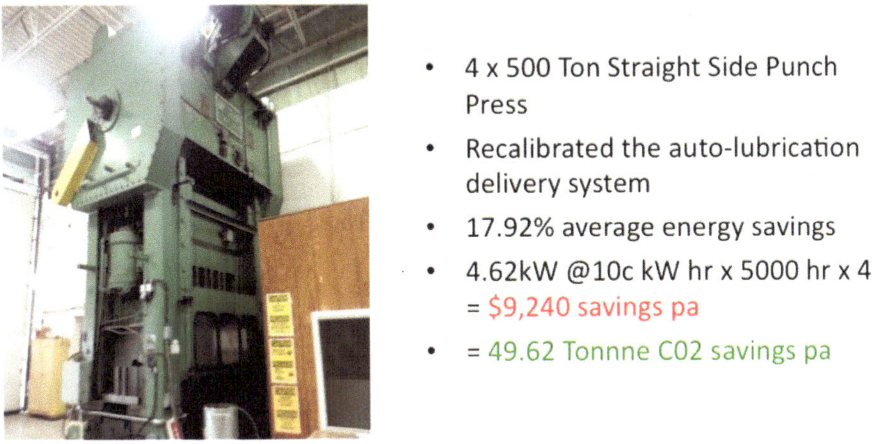

Figure 35.5 Lubrication system calibration savings example. (Courtesy ENGTECH Industries Inc.)

respondents did not track or report their lubrication related machine failures or downtime occurrences."

Quite simply, knowledge and awareness of how bearings fail in an operating environment is the first step to implementing any lubrication management program! Thirty-five years of empirical bearing failure data analysis has shown that failure can occur due to one or a combination of the following top ten conditions/causes (not in any order):

1. Lack of lubrication training,
2. Lack of lubrication application engineering,
3. Lack of cleanliness—use of dirty or contaminated lubricant transfer equipment and/or lubricants
4. Bearing is over lubricated
5. Bearing is under lubricated
6. Bearing is dirt or water contaminated by manufacturing or asset cleaning process
7. Infrequent oil/filter changes
8. Bearing lubricant is cross contaminated with an incompatible lubricant
9. Bearing is lubricated with the incorrect lubricant
10. Bearing is mounted out of square or misaligned when set up

Nine of the top 10 bearing failures are due directly or indirectly to ineffective lubrication practices. Implementing a root cause analysis of failure (RCAF) program to weed out the real reasons your bearings are failing will give immediate direction and focus for your lubrication management program.

When an engineer takes on a machine design project in which moving parts are involved, the engineer is expected to choose the appropriate bearings. Optimal bearing choice is dependent on many factors that include:

- Budget
- Application (radial, axial, and planar)
- Load
- Speed
- Space
- Clearance and fit
- Length of machine warranty
- Bearing reliability specification
- Lubrication entry design and method
- Expected operating conditions
- Expected ambient conditions

When a machine goes into service, the reliability baton is passed on to the end user's maintenance and reliability group who must work with the engineer's final design and an all too often-vague machine operations and maintenance (O & M) instruction manual, which rarely spells out good lubrication requirements/instructions.

Many maintainers will recognize the catchall work order instruction "lubricate as necessary," placing full responsibility on the end user to singularly develop a lubrication strategy for each bearing that is suitable for the conditions in which the machine is expected to perform. In the unlikely event the maintenance group literally starves the bearing of lubricant in the first year, most bearings will surpass their warranty period and the manufacturer will rarely get to learn if their design was/is truly robust and reliable.

Rotating and linear moving equipment employ lubrication as an integral part of their design; a well-designed and maintained lubrication system can ensure long bearing life, machine accuracy, and energy conservation as it serves to reduce friction through separation of moving surfaces, cools down the bearing surfaces, washes away debris, and protects the surfaces from outside contamination. Regular checks and monitoring of the lubricant and lubrication system integrity, followed up with timely action when a potential or actual failure state is found is foundational to any proactive lubrication program. Taking advantage of condition based and predictive strategies for managing lubrication potential failure and alarm states is essential to assure best practice asset reliability and life cycle longevity.

Failure can be broken down into three distinct stages. The first stage is known as potential failure and occurs when a change in condition or performance is detected. For example, if we determine the normal, or design lubricant operating pressure, is between 80psi and 100 psi, any plus or minus pressure deviation outside of that parameter is now seen as a change in condition or performance indicating an alarm level 1 potential failure condition, which in this case could interpret to a problem in the pump, line integrity, seal integrity, or change in lubricant viscosity. Subsequent readings are used to confirm a trend in condition and the rate of decay toward alarm state 2, known as the functional failure, which can be predicted and translated into a tangible window of time in which maintenance can troubleshoot and repair the problem with minor or no consequence to the machine or operational performance. This is known as the P-F curve (Potential to Functional failure). Be aware, if this is P-F warning is ignored, the equipment will continue to rapidly decay into the third stage, downtime state of catastrophic failure! (see Figure 35.6).

Whereas a catastrophic failure is easily defined as a total non-operative state, functional failure is an interpretation based on the consequence to the person viewing the failure. For example, a seal leak at any rate is an immediate functional failure for any safety or environmental engineer if the fluid is allowed to leak unabated. The leak becomes a maintenance functional failure when the seal leaks at a rate faster than it is practical to keep the reservoir filled to the low level limit. For most production engineers, functional failure occurs only when minimum production levels and quality levels are breached, which may only be reached once significant internal equipment damage has been sustained! When managing alarm states, all of these functional failures must be taken into account.

To assess potential lubrication related machine failure, the maintenance technician or reliability engineer must regularly capture, plot and trend a variety of current lubrication state indicator readings. Depending on the type of lubrication system and mode of failure, lubrication system indicators are used to monitor six major lubrication system indicators that include lube

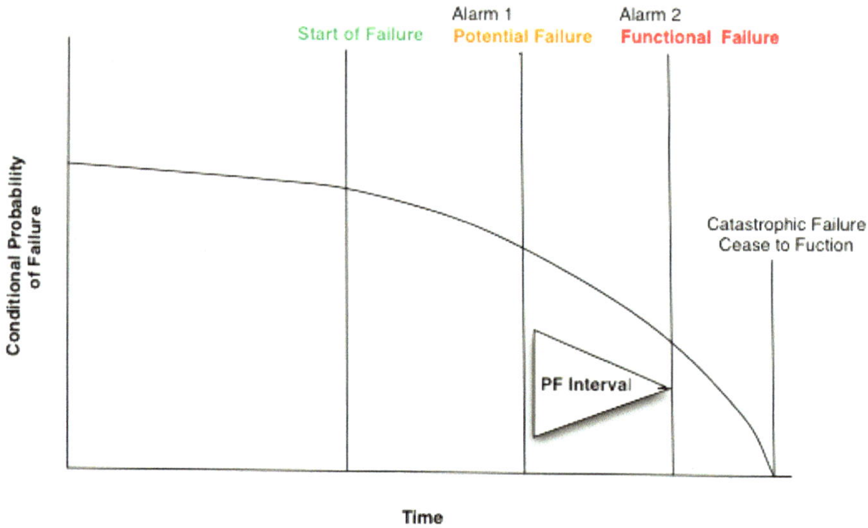

Figure 35.6 Typical P-F curve. (Courtesy ENGTECH Industries Inc.)

pressure, flow, temperature, lubricant quality (can be partially assessed by on line monitoring, or by a third-party laboratory performing sample testing using oil wear particle analysis), filter bypass condition and/or system volume. A seventh system indicator, energy draw (amperage), can also be assessed on line to indicate friction build up due to a lack of lubricant at the bearing surface(s).

The method of current state data capture and quality of trending will vary depending on whether the on line system monitoring devices are passive or active devices. The effectiveness of any device or test method is dependent on the process used to capture data, and the process used to take action when a potential or actual failure state has occurred.

35.2.4 Passive monitoring control

Passive devices, also known as "eyes on" devices rely on regular operator/maintainer interaction to perform a visual check to take readings and manually determine if a change in state has occurred. Typical passive monitoring devices include in line temperature gauges, pressure gauges, and reservoir fill level windows. The quality of data and associated potential and actual failure control is dependent on reading consistency and frequency. Most passive gauges are not interactive and require the viewer to capture the reading every time and interpret based on experience for every reading. This creates a lot

Figure 35.7 Manual marked gauge with needle showing outside safe area. (Courtesy: ENGTECH Industries Inc.)

of data capture that results in excess administrative work, and interpretive inconsistency when using operators or maintainers with differing skill levels.

Tip #1 – Mark up all passive devices with engineered Hi-Lo limits (see Figure 35.7 and 35.8), and only capture readings by exception when the needle displays outside of the Hi-Lo safe operating window. This now makes the device visually interactive, reduces the amount of administrative data capture, and can be consistently read and interpreted by all skill levels.

35.2.5 Active monitoring control

Active devices most often have some form of sensor control that once triggered by an alarm condition can output an external electronic or electrical signal. Depending on the level of control required, the alarm signal can

Figure 35.8 Commercial marked gauge with needle in safe area. (Courtesy: ENGTECH Industries Inc.)

be used to activate an audible or visual alarm such as a buzzer or warning light on an operator panel. Simple alarms however, still require an operator to understand and relay the problem to maintenance in the form of a work request. The danger of any simple system is the alarm can be interpreted as a nuisance to the production process and temporarily disconnected at source, bypassing any monitoring control if maintenance does not tend to the alarm in a timely manner.

As machine control systems have become more sophisticated through use of computer control and SCADA (supervisory control and data acquisition) systems, alarm states are now able to sound a local alarm (if still required) whilst incidence data is gathered remotely, trended, and proactively reported to the maintenance or reliability engineering office for action. In many cases the data acquisition signal can also be hooked up directly to the CMMS (computerized maintenance management system) to print out an investigative work order request in the maintenance department.

The following tips show how active monitoring control devices can be used to automate maintenance actions at the machine level.

Tip #2 – For automated conveyor lubrication, the lube system can be set up to turn on and off using an ammeter-controlled switch connected to the drive motor. Whenever amperage draw hits a high threshold point the lubricator is turned on and allowed to run for X chain revolutions, or turn off when amperage draw reaches a low threshold point certain. This ensures the chain is properly lubricated based on condition and load whilst optimizing energy use.

Tip #3 – When large hydraulic oil filtration systems use a bypass monitoring sensor alarm, instead of turning on a panel light and allowing the lube to internally bypass the filter element, the signal is used to activate a solenoid operated valve placed before the filter to externally divert the lubricant through a "twinned" secondary oil filter loop with no loss to production. The same signal is used to send a work order to change the original filter.

Tip #4 – In critical systems, an alarm state can be used to shut down the lube system and activate a bac- up lubrication system whenever a problem occurs. Once again, the signal can also be used to send a work order to change the blocked filter.

More and more newer equipment installations utilizing controlled lubrication monitoring devices are choosing to utilize active sensors that are IIOT 4.0 (Industrial Internet of Things) and remote control ready. However, the alarm state is only the first step; it is imperative that all lubrication management programs have quality work flow and business processes in place that ensure that when potential failure is encountered, maintenance attends to the assets needs prior to functional or catastrophic failure occurring.

When design conditions are set, machine availability and reliability is dependent on how well the maintenance group understands 1) the current bearing design and its lubrication requirements, 2) how the machine bearings will fail in their working environment, and 3) their own ability to design, implement and execute a relevant asset/bearing lubrication management program.

35.2.6 Regular PM/Operator maintenance

Daily lubrication system checks are essential for ensuring the lubrication system is operating as designed and that there is lubricant in the system. This is often best performed on a daily basis by the equipment operator who visually checks the entire system in a quick system walk-around and only notifies the maintenance department when an exception is found. Most lubrication work instructions to an operator are generally modelled after the PM work order used by a maintainer or lube specialist. Check functions can include:

- Check reservoir fill level —is the level between the Lo and HI mark on the reservoirs (some systems use a RAG—Red/Amber/Green indicator system to denote when to fill the reservoir— refer to chapter five – Designing a lubrication delivery system
- Check for and notify immediately of any system leaks.
- Check for apparent system damage including line crush and any over-pressure indicator signal denoting backpressure in the system caused by a damaged or blocked bearing or line.
- Check controller warning signals/lights are all off
- Check any pressure filter (used in recirculating oil and hydraulic systems) is not showing a red flag signal indicating the filter is full and in bypass mode.

Putting together a World Class Lubrication program truly is a collaborative effort, Relative to the cost of machine failure, repair and downtime, the effort is minimal and hugely cost effective.

Bibliography

International Organization for Standardization, ISO 55001: 2014, Switzerland, ISO, 2014

International Council for Machinery Lubrication, ICML 55.0: 2023, USA, Rivers Publishing, 2023

International Council for Machinery Lubrication, ICML 55.1: 2023, USA, Rivers Publishing, 2023

International Council for Machinery Lubrication, ICML 55.2: 2023, USA, Rivers Publishing, 2023

Bannister, Kenneth E., "Map Your Way To Successful Lubrication", *Efficient Plant Magazine*, 2017

Bannister, Kenneth E., "7-Step to Lubrication Excellence" Training program, ENG***TECH*** Industries Inc, 2012–2023

Bannister, Kenneth E., "Managing Alarms with these Techniques", *Efficient Plant Magazine*, 2019

36

Lubrication Work Management

Effective work management is about creating an environment that recognizes and takes advantage of proactivity. An environment that promotes communication excellence between clients, operators, or stakeholders, who often provide that "first alert" warning of machine failure, and a response that is interactive and timely. This requires excellence in our maintenance planning and scheduling work management efforts.

The maintenance work management process is triggered either by a PM work order, or by an observation of a machine problem that triggers a work request that ends up the lap of the planner who then decides how to respond to the request. If that request is acted upon with a work order, the job must then be scheduled in a timely manner.

36.1 The Work Request

A maintenance work order is developed in two distinctive stages. Stage one involves receiving, understanding and accepting the "ask" or request for work from the maintenance client or stakeholder. This is commonly called a Maintenance Work Request, or MWR. Stage two involves taking the MWR and converting it into a Maintenance Work Order, commonly referred to as an MWO, or simply a WO. In this stage, the simple request is built out into a complete action plan, or job plan in preparation for scheduling and performing the work in a timely manner.

36.1.1 Stage One – The maintenance work request (MWR)

Every day, the maintenance department must confront a minefield of Work Requests they are expected to respond to in a timely manner. The level and quality of response will depend greatly upon three main factors:

1. The manner in which the request is allowed to be delivered,
2. The amount of information contained within the request,

3. The manner in which the request is accepted

Traditionally, maintenance has accepted work requests in many different ways that include casual face-to-face conversation, telephone conversation, telephone voicemail, email, text, or informal "passed on" paper requests.

More recently, the use of computer and phone applications that include Twitter®, or Facebook® are also being utilized for work requests. In the absence of any formal request/acceptance process a "poorly detailed" one-line request with little or no supporting details can be readily accepted, by any delivery method with little or no scrutiny, and immediately queued into the Work Order system for scheduling. Many maintainers/lubrication specialists are likely familiar with the scrawled "matchbook" written request passed to the maintainer on the run, often without explanation, almost always categorized by the requestor as "URGENT!"

Often, duplicate requests by the same requestor, or multiple requests for the same problem by different requestors, on different delivery platforms, are not recognized and sifted out at the request stage and are allowed onto the work queue that can wastefully result in multiple maintainers responding to the same problem, multiple times.

Without direction or clear process, human nature dictates we take the path of least resistance. Following the "we accept and respond to everything and anything" work acceptance model allows the requestor to become almost anonymous and pass on the responsibility of the problem without effort or conviction, and forces maintenance into an automatic reactive maintenance approach. This *modus operandi* can substantially increase the maintenance workload and unnecessarily overburden the maintenance work backlog.

Fortunately, with little effort, the problems described by this poor work acceptance model can be alleviated, even eliminated, through the introduction of a simple work order acceptance business process, backed up with a well-designed Work Request form.

Building a quality work acceptance model

Getting control of the MWR system starts with controlling how the request is delivered to the maintenance department. Converting a request to a work order requires written instruction, therefore expediency dictates that all requests be formalized into a written format if they are to be accepted by the maintenance department. Although this can be performed manually on paper, many corporations will likely go direct to electronic written requests via email or through a specific internet/intranet request application/platform.

Utilizing a formalized and controlled written communication protocol simplifies training requirements and guides the requestor to better describe the problem they are seeing and reporting, provide minimum information requirements to allow the Planner to accept the request, and to build a quality job plan that allows the correct trade to be dispatched in a timely manner based on the true nature of the problem.

As individuals, we don't always comprehend a situation or problem in the same way and base its level of magnitude or importance based on our knowledge and tolerance for adversity. This creates a problem for the maintenance planner when trying to interpret the criticality of the situation to build a suitable job plan and response strategy. The following examples are real life "one-liner" MWRs for the same problem, requested by different requestors.

Example: A reservoir has been overfilled and created an 6-inch diameter pool of oil underneath the reservoir that was not been cleaned up. The requests included the following:

1. "Leaking lube reservoir"
2. "Oil spilt on machine and floor requires cleanup"
3. "Broken lube system spurting oil on shop floor"

We can see in the example that in request #2, that the word flood not an exact description of the event, but without any other knowledge at hand, it forces an immediate priority #1 response by the Planner. Similarly. request # 3 is a purely guesswork on the requestor's behalf that indicates a serious lubrication maintenance problem. Request #2 is the most accurate, but in the mind of a safety engineer is still a #1 priority!

To effectively identify a repair strategy a Planner must understand what the requestor is seeing, (this can be best accomplished with a photograph), hearing (unusual noise, an audio file taken on a cell phone can help immensely), smelling, feeling (vibrations, heat), taste, and what their intuition tells them. In the example above, a simple cell phone photograph or video would have ruled out flood immediately and would have likely resulted in sending a non-skilled person to clean up and report back if there was a leak or not, saving precious skill trade time chasing a clean-up..

When maintenance provides a formal work request process template forms for collecting minimum information requirements, the requestor is guided into giving the basic information required for maintenance to accept the Work Request with confidence, and respond in the appropriate time frame. If the request is incomplete, the maintenance department simply does not accept the Work Request and returns it to the requestor for completion. This part of the model requires fortitude on behalf of the Maintenance department, and cooperation from other departments. This simple change immediately eliminates virtually all nuisance calls as requestors are less inclined to extend the effort for a frivolous event.

Many by now are familiar with, and have been exposed to the concepts and principles of Quality Assurance / Quality Management (QAQM) systems driven by the ISO/QS 9000 quality system. The basis of any QAQM system revolves around clear communication and documentation; affected parties work together to understand the communication problem, develop procedures in which clear expectations and guidelines are set out, and complete the process by documenting the results with a step-by-step procedure and business process map. The following 5- step process details the simple change process:

Step 1: Understanding the problems

Maintenance meets with other departments to discuss the shortfalls of the present work request system and documents their concerns, whilst voicing the Maintenance department concerns.

Step 2: Building the Work Acceptance Structure

Maintenance builds a process map (Flow Diagram) that shows Maintenance will now only accept written or email requests (no more verbal work requests)

Step 3: Building the new Work Request

Maintenance develops a Minimum Information Work Request document that can be electronically delivered to a third party help desk, or directly to a maintenance planner. The following 10-Point Work Request contains the mandatory details required to make a qualified planning approach to the work:

1. Date and time of request - *important for Mean Time to Response/Repair – MTTR tracking*

2. Requestor name and # – *Maintenance needs to know who to call for further information if this is a true emergency, or request is incomplete and to know if the requestor activates a political priority response*

3. Contact name and #: *May be the requestor or a different person who needs to be contacted by the maintainer to get access to the building or machine*

4. Exact geographical location of problem (Building, Floor, Room, Column, or GPS locate for liner assets) – *this information reduces*

response time by understanding travel time, and allows us to review any relevant floor plan drawings for job plan development

5. Asset # - *allows maintenance to immediately review asset history and immediately assign the work order to the correct account*
6. Exact location of problem on asset (Lubrication system) – *again allows the work order to be assigned at the correct level for capturing work information and for reviewing history and Bill of Materials*
7. Problem symptoms – *what is happening that causes the requestor to believe there is a problem?*
8. Primary sensory information – *important for determining the true nature and criticality of the problem, these include:*
   a. *Vibration – visual, feeling, touch*
   b. *Noise – hearing*
   c. *Unusual odors – smell, taste*
   d. *Smoke - visual, smell*
   e. *Heat – feeling, touch, smell*
   f. *See – Description of symptom, Photos, video or audio file attachments*
9. Immediate impact on equipment operability/availability (Downtime, Limp Mode, No Effect) – *allows the planner to prioritize the work more effectively*
10. Immediate impact on person or building safety - *allows the planner to prioritize the work more effectively*

Step 4: Presenting the New Approach

With new tools in place, Maintenance must meet with its partners and stakeholders and present the new Work Acceptance approach. The roll out plan is discussed along with a mutually agreed Work Request drop off point and procedure.

Step 5: Implementation

The program is rolled out and monitored carefully

This simple proactive approach provides a relatively painless change management process, allowing maintenance service levels to increase substantially. The requestor now knows the minimum information required to expedite a faster response, and subsequently is a significant partner in providing a faster resolution to problems that affect asset availability and reliability.

36.1.2 Stage Two – The maintenance work order (MWO) – Work

The work order structure

A maintenance work order (MWO) is primarily made up of three distinct sections. The first section contains the general information required to raise a work order and make it an entity within the asset management system (AMS) software. The second section, is arguably the most important MWO section defined as the job plan, or work instruction section. The third section is reserved for post work data collection and comments.

The job plan/work instruction section is the major instructional component of the MWO and it is the primary responsibility of the planner to build a comprehensive work instruction set or package that includes identification of the trade group(s) and specific trade(s) who will perform the work detailed in the job plan/work instruction set. To complete the work package, the planner will also detail a list of required spare parts (if known), specialty tools, and permits to complete the job on site. Finally, to assist the scheduler to gain access to the asset for maintenance purposes, pull together a weekly and daily work schedule for each trade group, tradesperson and schedule any required contractor assistance, the planner must also assess and document the estimated time to complete the work on the work order. Once this is complete, the work order status can now be from U (Unplanned) to PC (Planned Complete, Ready to Schedule) status, in doing so the status change once again time stamps the work order as it moves from the inactive work order backlog on to the active work order (to be scheduled) backlog. These time stamps allow us to look at planner performance by monitoring the mean time to convert an unplanned WO to a planned WO so as to determine if the planning department is understaffed.

36.1.3 Breaking down the work order – Job plan section

Most poorly planned work orders suffer most in the work instruction area that is the primary communication area to the Maintainer/Lubrication Specialist. One of the tasks of a LOER is to uncover work orders that contain little or NO work instructions IN this case, the MWO becomes a vehicle for charging time (Usually for payroll and back charges) and purchases at the expense of gathering crucial reporting information.

If the planner is unable to ascertain enough information to understand the exact work requirements, and is unable to view the failure or maintenance problem virtually or first hand, an investigative WO is issued with a one line instruction that always commences with the action command "Investigate.....". Its primary purpose is to instruct the maintainer to assess the problem and act as a surrogate "eyes-on" planner to develop and document a relevant job plan and work instruction so parts can be ordered and the work scheduled according to the failure consequence.

In pre-planned resulting from an investigation work, or preventive maintenance work the planner's responsibility is to facilitate the work process and ensure, as much as possible, ensuring the tradesperson performing work on location has all the pieces required to complete the work within the estimated time frame. The following informational fields are to be found in the Job plan section of the work order that act together to build a complete informational maintenance work package.

Safety requirements

Due diligence dictates that we all adopt a safety-first approach toward work; the work order can't make the maintainer work safe, but it can remind the maintainer—every time—of the minimum expected safety equipment to be used and protocols to be followed throughout the work process, these will change depending on the work to be performed, the location of work, and the trade group performing the work.

Typical safety requirements will list the standard safety equipment. E.g. gloves, eyewear safety boots, etc., and can also include specialized equipment such as gas analyzers used when working around chemicals, gases or in confined spaces such as pits and vessels.

When working with energized equipment, the work order will serve to remind the maintainer about the need to follow any required safety standards such as lock out/tag out (LOTO) requirements. Any actions required by the maintainer to follow the standard can also be included in the actual job plan section narrative.

Parts requirements

The parts section of the work order requires the planner to list all known parts required to complete the job. In a planned complete work order the planner will ensure that all listed parts are in stock and pre-assigned to the WO number awaiting pick up by the maintainer. In advanced planning environments, parts are picked in advance by the MRO stores group, kitted together under the WO# and readied for pick up by the maintenance group in a defined staging area near the maintenance department, or pre-delivered to the job site in preparation for and prior to the maintainer's arrival in site. In the case of lubrication work, parts will include lubricant, filters, breathers, etc.

Permit requirements

In most companies, health and safety standards dictate that permits must be issued for certain types of work performed so as to ensure certain safety protocols are followed. These can include *Hot Work* permits if welding or hot metal cutting is to be performed, or a *Confined Space* permit if work is to be performed in a designated confined space area. These permits also alert the plant management staff of the type of work being performed in their area ahead of time.

Tool requirements

Tools listed in this area usually apply to occasional used common tools, expensive diagnostic tools, or special tools not expected to be carried in a maintainer's toolbox. These are shared tools and usually managed in a tool program by the inventory management group and require a WO number to pre allocate the tool for pick up or staging as with spare parts.

Reference documentation

Part of the Maintenance Work request process is gather documentation and pertinent information relevant to the failure from the requestor. This will include any photographs taken of the event at the time or subsequent to the event. In addition, any sensory information provided by the requestor at the time of the event discovery can be documented here, these can include anything heard such as unusual noise, (any audio file taken on a cell phone can be reference here), or unusual smells, feelings (vibrations, heat), taste, and what their intuition tells them.

Also included in this documentation area will be references to drawings, redlines, as-builts, especially where problems are suspected inside a machine or behind a wall. References can be actual drawings, photocopied drawing sections physically bundled as part of the work package, or drawings referenced electronically to be accessed by the maintainer's laptop, tablet or cell phone on site.

Job plan time estimate

A job plan time estimate is usually given in Hours and portion of hours. Most planners like to round off time to quarter or half hour increments for easier scheduling.

How Job Plan estimates are calculated vary from company to company. The recommended method is for the planner to only allocate time for the work to be performed on the work order. Additional time for travel (this can vary throughout the day due to traffic when work is not in the same location as the maintenance department as with multi-site facilities), picking up parts and tools from their staging areas, time to log in with the department foreman, etc. can then be added by the scheduler. When the work order is closed it is the difference between the estimated time allocated to perform work and the actual time taken to perform the work that is used to calculate a planner's work estimation performance using variance reporting. Lubrication PM lends itself to the development of Lube routes for maintaining systems that use the same lubricants.

Job plan/work instruction

This section contains the actual work instructions to be performed by the lubrication specialist.

There are two distinct methods of preparing work instruction. The first is for Preventive and Predictive work that is pre-planned and designed to repeat multiple times throughout the year using the same job plan or instruction set. The second is for unique demand type work that is a single occurrence and must be planned for every occurrence. The techniques and tools used for both types of work instruction are similar but different and are the topic of the next making work orders work installment.

Developing objective and effective work instruction is a science. The quality of work performed by a maintainer when following a planner's instruction will vary greatly depending on the objective requirement set within each line of instruction. If direction is vague, for example, "grease all bearings", the result will differ considerably based on the maintainer's knowledge of grease types of grease, and the amount of grease dispensed into the bearing. Furthermore, if the planner modifies the directive slightly to "grease all bearings as required" the end result will vary even more as the maintainer/Lubrication Specialist now has the added responsibility to decide if the bearing requires grease. This type of directive is deemed subjective and more reminder than instruction.

Lubrication is a science and we know that mixing lubricants (grease types), over greasing, under greasing, or not greasing a bearing at all when needed can have a decided effect on the life of the bearing, the machine it serves, and the consequence of its failure. This is also true for all fluid top up/changes, filter changes, belts, pulleys, chain and sprocket lubrication, adjustment and change outs; all crucial engineered consumables, that from the basis of most PM work, and whose life cycle and performance depends almost entirely on how the maintenance instruction set is interpreted and carried out. Words do matter!

The simplicity and power of the objectively written checklist

In Atul Gawande's book "The Checklist Manifesto – How to Get Things Right" he describes how the power of instructional language, delivered in a simple, correct checklist format was heralded as the "the biggest clinical invention in thirty years".

Gawande was tasked by the World Health Organization (WHO) to find a solution that would reduce the amount of post-surgical complications and deaths due to surgeries performed on the wrong person or body part, and to post surgical foreign bodies mistakenly left inside patients. In answer to this phenomenon, Gawande implemented a beautifully simple set of hospital checklists now credited for helping save tens of thousands of lives, pain and suffering worldwide over the past few decades. If you have been to an emergency room lately, how many times did they ask you your name and what you were in for while using a checklist in front of them?

Gawande didn't reinvent the wheel when he came up with his objective checklist system, he modeled it after the Boeing flight checklist system designed in the 1930's after the fatal crash of its futuristic Model X299 airplane.

It was October 30th, 1935 at the Wright Patterson AFB in Dayton Ohio where Boeing was about to show off its prototype model X299 to the USAAC (United States Army Air Corps) as part of an air corps flight competition between Boeing, Martin and Douglas aircraft companies. On the line was a commencement contract for an immediate 65 aircraft and the Boeing X299 was considered a "shoe-in" as it was bigger, faster, had twice as many engines (4), twice the range of the competition, and could carry five times the army's requested payload.

Boeing was relying on their veteran test pilot, Major Ployer T. Hill to fly the aircraft in a simple circle and flyby maneuver. The aircraft took off smoothly that morning rising sharply to 300 feet when suddenly it stalled, lost control, and crashed killing Hill and one other crewman on board. Because of this, the Boeing X299 did not complete the competition requirements and was disqualified.

Ironically, the initial cause was thought to be a maintenance issue (some accounts still hold this to be true). A subsequent crash investigation revealed the actual cause to be a simple pilot error in which the vastly experienced Hill had simply forgotten to release a new locking mechanism on the rudder and elevator controls required to keep the powerful engine draft in check while the aircraft was on the ground. What happened next changed the course of the aviation industry and much more.

Normally, such a crash would result in more intensive pilot training, but Boeing decided to take a new and innovative approach to the problem in which it simplified the training process while simultaneously overcoming any pilot ineptitude and ignorance through the development of a simple aviation pilot checklist. Due to the complicated nature of the aircraft, the checklist ensured that any future pilot would go through the same exacting preflight, taxiing, takeoff and landing processes every time the aircraft flew—a checklist procedure so successful that is modeled almost a century later on every aircraft flying today, aircraft infinitely more complex than those of the 1930's.

Despite Boeing's and the military's disappointment that day the army was impressed enough to take another look at its bidding process rules and found a loophole that enabled them to purchase 13 of the aircraft for evaluation purposes. This resulted in the aircraft seeing military service a short while later under its new name, the Boeing B-17 "Flying Fortress". Arguably the greatest bomber in aviation history with almost 13,000 B-17 bombers built seeing much of its service over Germany and the Far East leaving a checklist legacy that continues to result in safer flights today.

In both the WHO and USCC examples, checklists proved they help provide protection against elemental errors that can affect the health and lifecycle of the checked asset, item, thing, or person. In essence, checklists turn already experts into experts who perform consistently and flawlessly every time!

The PM maintenance checklist

There are two distinct methods of preparing work instruction. The first is for Preventive. Predictive, and Legislated work that is pre-planned and designed to repeat multiple times over a calendar period utilizing the same job plan and instruction set.

When dealing with highly trained individuals such as pilots, surgeons, and maintenance professionals the idea of working with and adhering to a simple checklist is seen as a difficult sell and according to Gawande, many view the use of such a simple tool as embarrassing. Overcoming this bravado and misguided thinking requires a checklist be written in a concise manner that does not challenge or insult the professional requirement of the work. According to Daniel Boorman of Seattle, Washington, developer all Boeing aircraft checklists for all of Boeing's aircraft for the past twenty plus years, the secret of a good checklist is in how it is written. Using simple and exact language that is familiar to the profession, a well-designed checklist will fit

Item	Instruction
1	*Inspect air compressor for proper operation. Change filter, check belts and lubricate as necessary*

Figure 36.1 Typical air compressor PM instruction set. (Courtesy ENGTECH Industries Inc.)

the flow of work, encourage the user to read each point out loud, and allow the user to detect potential failure prior to occurrence. Boorman goes on to advocate use of two styles of checklists, the **DO-CONFIRM** checklist, and the **READ-DO** style checklist choosing the appropriate style for the task.

The **DO-CONFIRM** style checklist allows the professional to perform work from memory and experience and then causes them to stop and confirm each work task was carried out. In a **READ-DO** style checklist, the user carries out their task as they check them off, Boorman likens this type of checklist to a recipe.

In the maintenance field, we are probably most familiar with the operator checklist written in a **DO-CONFIRM** style, used as a due diligence check and commonly performed in a TPM (Total Productive Maintenance), or 5S environment. As the skilled workforce diminishes, maintenance departments are increasingly tasked to do more with less. Coping successfully with this transition means applying innovative strategies and tools to ensure work is performed consistently and faster. One such strategy already proving the result is worth the effort is the adoption of the checklist for maintainer and operators, for both planning and work execution purposes, modeled on the successful aviation and medical industry formats.

Figure 36.1 shows a typical Generic PM instruction to perform a basic inspection of an industrial air compressor written in a subjective manner most maintainers will recognize.

Figure 36.2 we see the same requirements objectively written in a checklist format that is a combination of the **DO-CONFIRM** style (see instructions shown in bold type with only a confirm check option) and a **CHECK-DO** style similar to **READ-DO** (See instructions in non-bold type that have both a confirm and a No-Go option). This work order style delivers more detailed and explicit in what must be checked and follows a natural workflow or sequenced events. When there are engineered decisions to be made and adhered to, i.e., lubricant types, air pressure ranges, etc. these are specific. Note; the instruction does not tell the maintainer how to remove or install the guard, lubricate or make adjustments, only what to look for to take action.

Item or Sequence	Instruction	Confirm?	No-Go?
1	**Lock out unit as per State Safety Regulation XXXX**		
2	**Remove v-belt guard**		
3	**Vacuum all dust and debris from unit**		
	CHECK:		
4	Is oil reservoir fluid level showing between the Hi-Lo fluid level indicators? If below Low mark add ABC ISO 32 oil to Hi-level mark. If too High, drain off oil to just below Hi-level mark		
5	Are V belt(s) free of cuts, abrasions, fraying, and glazing condition(s)?		
6	Are V-belt(s) correctly tensioned within the allowable slack zone? Use Belt tension tool to ensure belt tension is within XX in and XX in		
7	Do V-belt sheave pulleys show less than 0.032" light exposure using a V-belt profile gauge and flashlight?		
8	Are all mounting bolt markers in line? If No-Go, retighten and torque loose bolt to XX fti/bs		
9	**Reinstall Guard**		
10	**Replace Air filter (identified in parts area of work order)**		
11	**Clean auto-drain assembly and drain moisture from air receiver**		
12	**Remove Lock out**		
13	**Start up Unit and run for 10 minutes**		
	CHECK while running:		
14	Safety valves closed with no Hissing sound?		
15	Unit appears leak free?		
16	Is electrical system ammeter needle positioned in the marked operation zone depicted on the gauge glass?		
17	Air supply pressure is showing greater than 60 psi?		
18	**Shutdown unit and cycle on/off 2 times**		
19	Does unit manually cycle on/off without problems?		
20	**Complete all sections of work order including relevant comment**		

Note: Once each sequence item is completed, place a "tick" mark in the corresponding Confirm check box area. When performing the PM CHECK procedures, if the answer to the check question is YES, confirm and move on to the next question. If the answer is NO, mark your initials in the corresponding "No-Go" checkbox area, take a digital photograph of any problem areas, note findings on the work order comments section, complete PM if possible and return this work order to your supervisor immediately

Figure 36.2 Revised air compressor PM instruction set in a checklist format. (Courtesy ENGTECH Industries Inc.)

Whenever a check is required in the Check-Do section of the instruction set, a No-Go box is provided to advise the Supervisor, Planner, and Reliability Engineer additional work has already taken place or needs to take place to remedy the problem found. The note text on the bottom of the checklist advises the maintainer of the checklist procedure to follow.

Both work instructions (36.1 and 36.2) expect the same result, the difference in how they are written is evident and it is very apparent as to which instruction set will achieve consistent results every time regardless of which maintainer performs the work. Once a PM work instruction is completed it can be considered for use as a template and modified for similar asset types and renamed.

The Non-PM maintenance checklist

The second type of work order using a checklist format is the non-PM repair, corrective, or demand style work order designed for work that has been planned in advance of the work to be performed. This work is more likely to be unique in format and machine specific. Unlike the PM style work instruction, a non-PM Checklist will be predominantly a **DO-CONFIRM** style checklist.

Unlike the PM template, the majority of non-PM work is unique requiring a different work instruction for every job. However, Savvy planners get to know what work tends to be similar and/or repetitive and will classify job plans into categories by work type, machine or asset type, location, etc. and build a job plan library that they can search review for similar work and use a previous or similar instruction set as a starting template to build a new work instruction thereby saving considerable time. Many asset management program software packages already have this job library or job bank feature.

36.1.4 Work order completion

W. Edward Deming once stated "In God we trust, all others must bring data!"

Data is a four-letter word for some and the value of collecting and recording data and evidence regarding an asset failure event is often poorly understood. Old school thinking dictates, "get the asset back up and running, and get on to the next job ASAP!" The reality is that Asset Management Systems (AMS) and Computerized Asset Management Systems (CMMS) both employ the word "management" and rely almost completely on the capture of post-work data to capitalize on its designed ability to convert and translate data into meaningful evidence-based informational trends and reports used daily by planners, schedulers, foremen, managers, clients and stakeholders to confidently perform agile evidence-based decision-making.

Once the maintainer leaves the job site, the opportunity to reliably collect and record the scene is lost forever. Think of the criminal justice system in which CCTV or cell phone photographic evidence always trumps the hearsay memory of the eyewitness after the fact. The longer the time between the event and recall, the more hearsay evidence is mistrusted and based on character rather than conclusive data based on photographic evidence or a written account performed at the actual time of the event.

From the perspective of asset life cycle reliability, availability, and maintainability (RAM), collection and documentation of findings, work details and data at the job site is arguably **THE** most critical point in the maintenance work process without which, the AMS or CMMS is relegated to a mere work order print and repository engine with little or no management value.

Findings/Comments/Field notes

This is where the maintainer gets to execute and incorporate best practice action through the documentation and capture of qualitative notes and improvement ideas that spell out a narrative of findings, evidential materials, conclusions and recommendations; all or any of which can be incubated at any time in the future.

How the maintainer's notes and comments are laid out will depend greatly on the design layout and flexibility of the AMS/CMMS. Sometimes the software will allow text to be formatted into a list that can be cross-referenced to an electronic companion document, photo, SCADA reading/report, alarm report, scanned field notes, etc. Older software may be rudimentary and offer only a text field with a fixed number of allowable characters. Such constraints can be managed with a data collection protocol/SOP in which the accompanying data is placed in a separate controlled maintenance data file referenced in the work order close out comment area.

Photograph attachments

With cell phones, tablets, and inexpensive data storage memory, photographs are now so easy to take, document, link and store. Photographs taken from multiple different angles allow a planner to remotely view the work site and plan work remotely.

Photographs taken before and during the failure investigation or repair (combined with a box of actual failed parts) allow a reliability engineer to perform a remote failure investigation; compare similar failures to determine patterns, and to work with the planner to determine if the current Preventive strategy is working.

Photographs taken after the repair is complete prove the work has been performed and that the job site was left clean.

When taking photographs of the apparent damage or workplace scene, photographing from different angles help document the scene from both the requestor's point of view and from a maintenance planning perspective. For example, when a requestor calls in with the somewhat ambiguous request "Area is Flooded" will require a set of photographs of the area containing the liquid in question, representing what the requestor sees as the problem as well as photographs from the scene based on the maintenance viewpoint, documenting both the damage at the leak's source e.g. a damaged lube line, as well as any residual damage caused by the leak that may need to be repaired. All photos should include a reference scale of measure, e.g., a six-inch rule.

If the client /stakeholder is unable to be physically informed at the time of work completion a photograph of the completed work scene should be sent immediately to their contact address to inform them the work has taken place and their asset is once again in service.

Parts used

Even if parts are issued with the planned work order, all unused parts need to be adjusted for on the work order and returned. In addition, any additional non-stock or spot buy parts purchased for the work and charged to the work order must be captured on the work order and entered in to the work history.

Work completion time

Once the repair finished, the job site is cleaned up, and the work order details entered, the work order is complete and must be time stamped.

This time stamp is used to calculate the ACTUAL time to complete work, used to compare against the ESTIMATED time for time variance reporting, as well as to calculate the Mean Time to Repair (MTTR) indicator.

For time reporting, maintenance must have a clear set of rules to determine how actual Work time is recorded and how non-work time is managed, recorded and charged back to maintenance or the client. Time spent to carry out the completion of a work order can include:

- Actual Work time
- Travel time to the job, the next job, or for parts
- Waiting time for parts/tools
- Waiting time for machine, client or access to asset

The actual planned estimated work time is the only constant for planning purposes, all other times need to be accounted for scheduling and can vary on every job. How the time is managed, collected and charged will depend on the type of business, geography, how ell maintenance is set up to provide parts and tools, and the service level agreement expectations between maintenance and the client/stakeholder.

Failure symptoms

Some AMS/CMMS systems provide drop down or pick lists that allow the user to choose the most appropriate choice. The "symptoms" list is usually a good candidate for this type of software feature. Keeping in mind there may be one or more choices that are relevant that need to be accommodated for this can be a useful list for helping determine the root failure cause. For example: heat (bluing/scorch marks), leak, jamming, excessive vibration, looseness, vandalism, etc.

Failure cause

It is often very difficult to assess the true cause of failure without further investigation of parts, production data, historical maintenance data, etc. and is generally not the not the purview of the maintainer, but rather that of the reliability engineer to perform a root Cause analysis of Failure (RCAF)

However, the maintainer is eminently qualified to identify if the failure was maintenance related/caused, or if it was a non-maintenance related/caused failure at this stage. Employing this simple, yet powerful choice to filter all work history into two reports can instantly determine and differentiate work that the maintenance department controls (maintenance related/caused) versus the work it merely manages (non-maintenance related/caused work) and has little or no control over.

Failure effect

Whenever maintenance work is performed the client/stakeholder will judge the maintenance department's performance based on numerous indicators: how they are informed; if the work is completed satisfactorily; how long it took; and most importantly, how it affected their ability to carry on working. The service maintenance delivers are primarily asset availability and reliability that minimum loss of service to the client /stakeholder.

By documenting one of four service loss scenarios to determine the failure effect at the time of event provides a simple data filter that allows maintenance to provide proactive reporting on its level of service to clients and stakeholders backed up with tangible evidence of the event and if the event was maintenance caused or not. These simple four report or data filters (No loss of use / Partial loss of use (slowdown or partial downtime) / Full loss of use) form the basis of a powerful, yet simple client/stakeholder service level agreement in tangible terms easily defined by the maintainer at the job site.

Bibliography

Bannister, Kenneth E., "Maintenance Work Orders: The Request", *The RAM Review E-zine*, 2020

Bannister, Kenneth E., "Maintenance Work Orders: The Work Order", *The RAM Review E-zine*, 2020

Bannister, Kenneth E., "Maintenance Work Orders: The Request", *The RAM Review E-zine*, 2020

Bannister, Kenneth E., "Maintenance Work Orders: Work Order Completion", *The RAM Review E-zine*, 2020

Gawande, Atul, "The Checklist Manifesto – How to Get Things Right", *Metropolitan Books*, 2009

37

Key Performance Indicators

Implementing a quality work management system that drives out usable program documentation, management information and reporting metrics is a critical requirement in both ICML 55.1: section 5:12 Program Management and Metrics, and ISO 55001: Section 9.1.1.1 (Performance Measures).

When mechanical/hydraulic devices and systems fail, the majority of their failure modes can be directly, or indirectly, attributed to ineffective lubrication practices.

Direct-lubrication failure modes are most often attributed to applying to the bearing surfaces too little, too much, or the incorrect lubricant. Indirect-lubrication failure can be attributed to lubricant cross contamination; the mixing of two or more incompatible lubricants, and poor solid and fluid contamination control. When the latter two are allowed to infiltrate lubricant and bearing surface areas, they invariably result in rapid surface wear, seal deterioration, and major detrimental change to a lubricant state and its efficacy.

Remaining failures are primarily due to design issues (overloading), assembly, or set-up issues (warranty-related failures). Assessing the validity and success of any lubrication program, therefore, requires the capture and analysis of performance data surrounding the maintenance and availability of lubricated systems and assets.

ICML 55.1 demands that "*The organization shall develop metrics [KPI's] that will monitor the effectiveness of all twelve lubrication program elements described in Section 5.0*". ISO 55001:9.1.1.4 demands that "*The asset-management system should use data from monitoring and measurement to identify patterns and obtain information regarding its performance. The data should be used to evaluate whether the organization's policy and objectives are being achieved, as well as identifying corrective actions and areas for improvement.*"

37.1 Key Performance Indicators (KPI) for Lubrication

Key Performance Indicators (KPIs) allow us to set an end-goal target, or objective, based on multiple business needs, or objectives, and track whether we are successful in meeting or surpassing them. KPIs also allow us to analyze and understand, in a laser-like fashion, where we need to focus improvement efforts.

Managing with performance measurement means that reliable data must be collected on a regular basis. The obvious maintenance conduit for tracking performance is the work management system in which assets, type of work performed, and failure types are tracked simultaneously and reported. Lubrication work management can be as a work type within a general maintenance-management system such as a computerized maintenance-management software system (CMMS), enterprise asset management (EAM) system, or a dedicated Lubrication Work Management Software system.

KPIs rely on system reports pertaining to labor, work type, occurrence events, and factors such as cycle, time, cost, inventory, and trend alarms, to populate the KPI data requirements

In addition to program performance, KPIs also test how well a software system is implemented and used by the quality and accuracy of data extracted. Once the management system is able to produce the data required by the KPI calculation(s), reporting routines can be set up in the system to provide updates on a regular basis

Where no dedicated lubrication-management-software system is used to singularly track lubrication-related work, identifying lubrication-related work is simply a matter of capturing and classifying all lubrication work orders as a specific lubrication work-order type. For example, many maintenance systems typically sort their work orders prior to release into lubrication (work specific to activities such as lubrication-system maintenance and servicing, filling of reservoirs, and filter changes), mechanical, electrical, facility, and general/service maintenance-related work.

Failure occurrences are tracked based on the same classifications and, in addition, are tracked upon completion as to whether they are maintenance-caused or related issues over which maintenance/lubrication technicians exercise full control, or non- maintenance caused or related issues (issues that maintenance and the lubrication technician cannot prevent but must manage as part of their mandate. Examples include graffiti, operator damage, and denied access to an asset.) In addition, consequence of failure can also be simply tracked by categorizing the result of any breakdown or corrective work performed as no loss of service, minor service loss (as long as one shift), and major service loss (more than one shift).

To understand and report on the effectiveness of a lubrication program, the cost of all production downtime loss and the reactive cost of all corrective, repair, and breakdown work related to bearing and bearing-surface components (seals, pistons, valves, way failure/loss must be compared to the cost and effort of the proactive work related to the filling, cleaning, and maintenance of lubrication-delivery systems, hydraulic-fluid systems and management of lubricants. Almost all maintenance related-failures are preventable.

KPIs allow a department or corporation to turn simple data collected on a work order into meaningful management information. This is often referred to as "*data-driven decision making*". KPIs are primarily used to assess risk, performance, and cost. They are further categorized into strategic/long-term, tactical/mid-term, and operational/immediate.

Calculating KPIs

The following represent the most commonly used lubrication KPI calculations. KPIs can be expressed in terms of a single equipment piece or rolled up hierarchically to represent the manufacturing line, area, product, or plant. They can also be classified by lubricant type or lubrication delivery system type, oil or grease system. |

KPI #1: Asset(s) (machine/line) % availability (based on lubrication-related failures)

$$\frac{\begin{array}{c}\text{Total \# lubrication hours from date A to B } - \\ (\text{Total minor} + \text{major service loss hours from same date A to B})\end{array}}{\text{Total \# lubrication hours from Date A to B}} \times 100$$

KPI #2: Lubrication-related maintenance cost as a percentage of total maintenance cost

$$\frac{\begin{array}{c}\text{\$Cost of all lubrication} - \text{related maintenance work} \\ (\text{parts and labor}) \text{ from date A to B}\end{array}}{\text{Total cost of all maintenance work } (\text{parts and labor}) \text{ from date A to B}} \times 100$$

KPI #3: Lubrication-related maintenance cost as a percentage of total operating cost

$$\frac{\begin{array}{c}\text{\$ Cost of all lubrication} - \text{related maintenance work} \\ (\text{parts and labor}) \text{ from date A to B}\end{array}}{\text{Total department or corporate operating cost from date A to B}} \times 100$$

KPI #4: Mean time between lubrication failure (MTBLF)

$$\frac{\textit{Total operational time (shift/day/hours/minutes) from date A to B}}{\textit{Total number of lubrication-related failures from date A to B}}$$

KPI#5: Lubrication PM (LPM) work-order compliance (can also be used with other lubrication work-order types)

$$\frac{\textit{Total \# issued LPM WOs from date A to B} - \textit{Total number incomplete LPM WOs} > \textit{X days Scheduled}}{\textit{Total \# issued LPM WOs from date A to B}}$$

KPI#6: Lubrication Work-order aging report

Total number of lubrication work orders open for XX days (30, 60 ,90, etc. since issue date

KPI#7: Lubrication work-spend analysis report

Total $ cost of all lubrication-related work from date A to B

KPI#8: Overall lubrication (program) effectiveness (Production based)

OLE is a derivative of OEE (overall equipment effectiveness). The only change in the calculation is in respect to availability based on lubrication-related failure as opposed to all maintenance failures.

Lubrication-related availability x rate of throughput x rate of quality

OLE can also be expressed in terms of a percentage of OEE to demonstrate the overall effectiveness of the lubrication program on the manufacturing process

The old adage "*What gets measured, gets done*" still holds true today. Measuring your lubrication-program performance allows everyone involved in your operations to intimately understand their business and the impact they have, or can have, on the business. day.

Bibliography

Bannister, Kenneth E., "8 KPI's to Improve Your Lube Program", *Efficient Plant Magazine*, 2019

International Organization for Standardization, ISO 55001: 2014, Switzerland, ISO, 2014
International Council for Machinery Lubrication, ICML 55.1: 2023, USA, Rivers Publishing,

SECTION 7

Lubrication Certification

38

Lubrication Certification

38.1 Lubrication Technician Certification

A major success component of any best practice program is having trained staff who understand the impact of their actions on asset reliability. The past 30 plus years have witnessed a massive shift in the maintenance industry toward a total asset reliability approach. More importantly, this revised approach has embraced and recognized the role and impact Good Lubrication Practices (GLP) now plays in achieving and sustaining asset reliability and lifecycle stewardship. Ironically, lubrication management is a discipline and subject matter known to directly, and indirectly affects up to 70% of all mechanical failures, yet there still is no lubrication specific trade, and likely never will be as vocational schools and apprenticeship programs in their current format continue to shrink and disappear. To meet this challenge many of the world's leading lubrication experts and organizations that include scientists, consultants, practitioners, and suppliers have supported and worked diligently on three fronts to develop industry recognized worldwide lubrication certification programs.

These certification programs have been painstakingly designed to certify qualified individuals to meet and surpass the escalating reliability demands of industry in the proactive field of lubrication application, lubricant selection, and lubricant analysis.

38.1.1 Why certify?

As an individual, lubrication certification is about telling the world that you have worked diligently to study and assimilate lubrication knowledge from an accredited body and domain of knowledge, and have demonstrated that learning by successfully passing a closed book exam that draws questions from that entire knowledge base. In addition, you have also had to meet numerous training and practical requirements to earn the right to take the exam.

ISO 17204, a world standard on the certification process, likens certification to a level of competency that "demonstrates [an individual's] ability to apply knowledge, skills, and attributes" in a chosen field. Individuals embark on the certification process for numerous reasons:

- Personal qualification to set themselves apart from the many others they may have to compete with for an internal or external job posting in the maintenance reliability and lubrication fields. Many potential employees offer substantial pay grade increases for certified personnel. The US lubrication magazine "Lubes & Greases" reported that in a 2018 lubricant sales representative salary survey, certified personnel can earn up to $30,000 more than their non-certified counterparts in the industry!
- Certification may be a current employment requirement. Many lubrication service providers that include lubricant sales, lubrication systems engineering and sales, and oil analysis laboratories now demand their employees be appropriately certified as a condition of sustained employment, or as a basis for hiring and/or promotion.
- Industry maintenance and reliability departments are rapidly recognizing the value of lubrication certified individuals as pivotal members of their reliability team to facilitate and manage proactive approaches and initiatives designed to manage and eliminate preventable mechanical failures.
- Corporations are now turning to certification programs to meet prescribed workforce competence levels while simultaneously raising their level of due diligence.

Whatever the driving factor, the advent of the lubrication professional here, and is reflected in the ever-accelerating number of individuals certifying in a variety of professional lubrication disciplines.

38.2 Certification Choices

Almost every professional group now has a certification process; doctors, lawyers, engineers, and accountants have long had to go through a board certification process to achieve professional status and practice their profession. Similarly, the maintenance industry has in the recent past chosen to recognize its own through a variety of general certification programs. Canada took the lead in the 1990s through the PEMAC (Plant Engineering

Table 38.1 Lubrication certification comparison guide.

Attribute	ICML	STLE
Web Site	www.icmlonline.com	www.stle.org
First Certification	2001	1993
Experience Requisite	1-5 year Discipline dependent	1-5 year Discipline dependent
Formal Training Requirement	0-16 Hours Discipline dependent	0-16 Hours Discipline dependent
Exam Pass Score	70% minimum	70% minimum
Renewal Cycle	3 years	3 Years
Exam Rewrite wait time	1 month	6 month
2023 Exam Cost	\$60 - \$375 USD Discipline and Country dependent	\$450USD member \$675USD non-member
Exam method	Print, On-Line	On-Line
Body of Knowledge	Yes	Yes
Domain of Knowledge	Yes	Yes
Exam Language	English, French, Dutch, German, Spanish, Italian, Portuguese, Japanese, Korean, Mandarin, plus requests	English

Note: Details may change; always refer to websites listed above for current information

and Maintenance Association of Canada) organization with its Certified Maintenance Manager—CMM designation designed to be a national maintenance management certification and delivered through the Canadian provincial technical college system. The Society of Maintenance and Reliability Professionals, a US based international organization, soon followed suit in 2001 with its CMRP (Certified Maintenance Reliability Professional) certification designation. In the specific field of lubrication the US based Society of Tribologists and Lubrication Engineers—STLE, has continued to offer its Certified Lubrication Specialist (CLS) certification—originally designed for lubrication engineers—since 1993.

To meet the practical needs of lubrication practitioners and analysts the International Council of Machinery Lubrication (ICML) was formed in the year 2000 and has aggressively converted lubrication certification into a reliability mindset thanks to a broader range of certification disciplines, and its recognition of "hands-on" lubrication practitioners through its Machinery Lubrication Technician—MLT certifications.

There are now three active certification bodies in the lubrication field that include the previously described STLE, ICML, and more recently, the International Organization for Standardization (ISO), who has worked in close collaboration with the ICML to develop and model their lubrication analyst LCAT certification programs.

STLE (Society of Tribologists and Lubrication Engineers)

The STLE has offered certification since 1993 and currently offers four lubrication certifications, these include:

CLS—Certified Lubrication Specialist

A lubrication specialist is an individual who might be designated as a "Lubrication Engineer" by his/her employer. It is not limited to this designation, as individuals from varied backgrounds, including sales and management can be included in this designation. Typical core responsibilities include:

- Conduct lube surveys
- Evaluate, select, and specify lubricants to use and their purchases
- Troubleshoot and solve lubrication related problems
- Develop in-house quality assurance and lube analysis programs
- Train lubrication staff for any in-house lubrication management program
- Maintain lubrication records, waste and disposal records

OMA I—Oil Monitoring Analyst Level I

The OMA I analyst is a predictive maintenance professional responsible for sampling oil at the equipment level with the ability to review and interpret the oil sample reports and make decisions about the overall care of equipment being sampled.

OMA II—Oil Monitoring Analyst Level II

The OMA II individual can be found in the test laboratory responsible for managing the appropriate tests, interpreting data and supervising the oil analysis program.

CMFS—Certified Metalworking Fluids Specialist

This is a certification for individuals experienced in the specialized field of metalworking and cutting fluids. Similar to the CLS, this individual would be typically responsible for evaluating, specifying, troubleshooting and managing all aspects of metal removal, forming and cutting fluid issues.

All certifications have a published recommended reading list that potential certificate designates are encouraged to read and familiarize themselves with well in advance of examination. A tailored list for each certificate designation is published on the STLE website and can be viewed at www.stle.org/certifications

The CLS program is by far the STLE's most popular certification with over two thirds of certifications listed in this discipline. Because many large North American oil companies and oil analysis laboratories—and their employees—are active members in the STLE, it is understandable that the CLS and OMA designations are popular with this industrial demographic.

The CLS certification requires between 1 to 3 years job experience (discipline dependent) in the field of lubrication; the OMA certification requires one year of laboratory experience backed up with 16 hours of formal oil analysis training, whereas the CFMS certification requires between 3–5 years of experience and 20 hours of training.

STLE certification exams are predominantly performed on-line requiring a 70% or higher mark to be successful. As of 2023, exams are offered in English only. The certification is good for 3 years, and recertification is easily obtained by completing and providing proof of four of twelve options that include professional training program credits, conference attendance, volunteer STLE work, etc. or by writing a recertification exam. No figures are available for the number of current certifications or certified persons.

STLE certification exams are offered at $450USD for members and $625USD for non-members. Retakes are available after 6 months for $225USD for members and $305USD for non-members.

In addition to its certification programs, the STLE offers scholarship and fellowship awards to students interested in pursuing a career in the field of tribology and lubrication engineering.

ICML (International Council of Machinery Lubrication)

In comparison, the ICML—a non-profit organization dedicated to helping lubrication practitioners succeed in their professional careers—has accrued a roster of over 7000 professional certified individuals in 70+ countries worldwide since 2001, in eight occupation-specific disciplines.

In addition, ICML has introduced a number of specialized certification badges for Varnish identification/measurement; Varnish prevention/removal; and a new one for 2023 is the Food Grade Lubrication badge.

Exams are offered in over 120 countries Exam cost at time of publishing are varied depending on the designation and country where the candidate is based. Pricing ranges from $100USD to S275USD for all professional designation courses except the MLE which ranges from $140USD to $375USD. Badge exams range from $60USD to $140USD. Recertifications are required on a 3-year cycle. Pricing varies and exact pricing can be obtained on the ICML website. Note: always refer to the examining body website for up-to-date pricing and conditions.

Machinery Lubrication Technician Level I—MLT I

A typical MLT is a plant technician responsible for daily lubrication tasks that can include oil op up and changes, filter changes, manual greasing, lubricant receiving and general care of lubricating dispensing and transfer equipment

Machinery Lubrication Technician Level II—MLT II

An MLT II designation is designed for the technician or engineer responsible for managing the lubrication program and its staff. The individual would also be responsible for appropriate lubricant selection, lubrication troubleshooting and supporting machine design activities.

Machinery Lubricant Analyst Level I— MLA I

This designation targets in plant technicians associated with performing basic lubrication tasks and machine condition monitoring/analysis. Activities can include oil changes, greasing bearings, lubricant receiving, storage and transfer; basic oil sampling duties and contamination control.

Machinery Lubricant Analyst Level II— MLA II

The MLA II is for the plant technician responsible for the condition monitoring/oil-sampling program that includes such duties as oil sampling, sample management, simple on site testing and managing test results/diagnostics.

Machinery Lubricant Analyst Level III— MLA III

A level III MLA targets technicians and engineers responsible for managing the lubricant analysis function. Responsibilities include team management; test slate selection, setting alarms and limits sampling systems design and advanced diagnostics.

Laboratory Lubricant Analyst Level I— LLA I

The LLA designation targets in house laboratory technicians performing and reporting on basic lubricant sample testing

Laboratory Lubricant Analyst Level II— LLA II

A level II laboratory technician is responsible for daily activities supporting the production of lubricant analysis data for machine condition monitoring performing tasks that include testing, diagnosing lubricant failure mechanisms

Machinery Lubrication Engineer—MLE

The MLE designation is the highest designation offered and specifically targets staff, contract or consulting lubrication engineers responsible for the development and management of all aspects related to the implementation and sustainability of lubrication management programs

The most popular designation with over 55% of certifications is the base MLT certification designation.

The ICML has always prided itself as a lubrication end user/practitioner organization reflecting the needs of the asset reliability industry in management offices, on the shop floor, and in the laboratory. In addition, the ICML now examines and promotes its certifications in ten languages across the world.

ICML certification examinations are proctored exams offered in person and on-line. Similar to the STLE, they require a 70% pass rate to be successful.

Depending on the certification discipline the ICML requires between 1-5 years job experience in the appropriate lubrication field and 16-32 hours of formal lubrication training based on the corresponding bodies of knowledge (depending on the discipline being examined for) as a pre-requisite to take the exam. Initial certification is good for 3 years and recertification is based on a points system accrued based on training, experience and service as with the STLE, or by taking a recertification exam; recertification requirements are an excellent way of ensuring individuals stay current. ICML examination costs are considerably lower at $200 US. Individuals are encouraged, but not required to be an ICML member to certify. See Table 1 for STLE/ICML program offering comparison.

All ICML certifications are in compliance with the ISO 18436-4/5 standard for condition monitoring and diagnostics of machines—requirements for qualifications and assessment of personnel—Part 4—field lubricant analysis and Part 5—laboratory analysis.

ISO (International Organization for Standardization)

The ISO organization offers the following three levels of international standard certification:

Lubrication Analyst ISO Category I— LCAT I

Lubrication Analyst ISO Category II— LCAT II\\

ISO 18436-4 Lubrication Analyst ISO Category III—LCAT III

ISO collaborated closely with ICML to develop their certification designation requirements and Bodies of Knowledge resulting in the current ICML MLA level I and II being adopted by the ISO as their LCAT level II and III certification designations, with the LCAT Level I being derived from the original ICML MLT I curriculum in which the qualification ensures personnel are trained and certified to perform simple tasks related to the proper lubrication of machines according to established and recognized procedures.

Participants who have attended the requisite preparatory formal training for MLT I, MLA I and II are also eligible to write the corresponding ISO LCAT exams with payment of the appropriate examination fee.

Regardless of which certification route you choose, any choice to certify is an excellent choice demanding a lot of preparatory hard work and assimilation of lubrication knowledge prior to writing the examination(s). All certification designations discussed above are industry recognized, and potential designate candidates can be assured their efforts will be rewarded both intrinsically and extrinsically for the rest of their working life.

38.3 Certification Body of Knowledge

Each certification body publishes both a domain of knowledge and a body of knowledge (Learning requirements for the benefit of potential designees and its certified designates. The domain of knowledge (DOK) is recognized as the subject matter knowledge base represented in a library of reference books, texts and papers written by lubrication subject matter experts. This domain of knowledge is then used to develop the body of knowledge (BOK), or areas of knowledge expertise, in each certification level that designees are expected to study and understand for certification purposes. Exam questions for each of the body of knowledge requirements are based on the domain of knowledge reference material.

For the most up-to-date information on the DOK and BOK requirements visit the STLE website at www.stle.org and view the "Career Development" tab. For ICML information, visit www.icmlonline.com and view the "Certifications and Exams" tab.

Appendix

Approximate Color Scale Equivalents

DIN 51-578* ASTM D 1500	Union TAG- (N.P.A.) Robinson Colorimeter Colorimeter (ASTM D 155)	N.P.A. Common Terms
0	22% -	—
0.5	$21^{1}/2$ 1	Lily White
1.0	17% 1	Cream White
1.5	$13^{1}/2$ 1%	—
2.0	11% 2	Extra Pale
2.5	$10^{1}/22^{1}/2$	Extra Lemon Pale
3.0	9% 3	Lemon Pale
3.5	$9^{1}/23^{1}/2$	Extra Orange Pale
4.0	9 3%	Orange Pale
4.5	7 4^{1}A	—
5.0	$5^{1}/24^{1}/2$	Pale
5.5	$4^{1}/24$%3V$45^{1}/2$	Light Red
6.0		
6.5	$2^{1}/_{2}6$	Dark Red
7.0	2 7	Claret Red
7.5	$1^{1}/27^{1}/2$	—
8.0	18	—'

*Color scales for ASTM D 1500 and the German standard DIN 51-578 are the same.

Representative Masses of Petroleum Products

	kg/m^3	m^3/Mg	lb/U.S. gal	bbl/tonne
LPG	542	1.84	4.53	11.60
Aviation Gasoline	707	1.42	5.90	8.90
Motor Gasoline	740	1.35	6.18	8.50
Kerosine	812	1.23	6.77	7.75
Distilate Fuel Oil	843	1.19	7.04	7.46
Lubricating Oil	899	1.11	7.50	7.00
Residual Fuel Oil	944	1.06	7.88	6.66
Paraffin Wax	799	1.25	6.67	7.87
Grease	998	1.00	8.33	6.30
Asphalt	1038	0.96	8.66	6.06

Abridged Gravity, Volume, and Mass Conversion Table

Approximate Conversions at 15°C Except Where Noted

Density 15°C kg/m^3	Specific Gravity 60/60°F	API Gravity 60°F	dm^3/Mg	U.S. gal/Mg	Barrels/Mg	Imp. gal/Mg
1075	1.0757	0.04	930.2	245.77	5.852	204.63
1050	1.0507	3.18	952.4	251.62	5.991	209.50
1025	1.0256	6.46	975.6	257.76	6.137	214.61
1000	1.0006	9.92	1000.0	264.20	6.291	219.97
990	0.9906	11.34	1010.1	266.87	6.354	222.20
980	0.9806	12.80	1020.4	269.59	6.419	224.46
970	0.9706	14.29	1030.9	272.37	6.485	226.78
960	0.9606	15.81	1041.7	275.21	6.553	229.14
950	0.9505	17.36	1052.6	278.11	6.622	231.55
940	0.9405	18.95	1063.8	281.06	6.692	234.01
930	0.9305	20.56	1075.3	284.09	6.764	236.53
920	0.9205	22.22	1087.0	287.17	6.838	239.10
910	0.9105	23.91	1098.9	290.33	6.913	241.73
900	0.9005	25.64	1111.1	293.56	6.989	244.42
890	0.8905	27.40	1123.6	296.85	7.068	247.16
880	0.8805	29.21	1136.4	300.23	7.148	249.97
870	0.8705	31.06	1149.4	303.68	7.231	252.84
860	0.8605	32.95	1162.8	307.21	7.315	255.78
850	0.8504	34.89	1176.5	310.82	7.401	258.79
840	0.8404	36.87	1190.5	314.52	7.489	261.87
830	0.8304	38.90	1204.8	318.31	7.579	265.03
820	0.8204	40.98	1219.5	322.20	7.671	268.26
810	0.8104	43.11	1234.6	326.17	7.766	271.57
800	0.8004	45.29	1250.0	330.25	7.863	274.97
790	0.7903	47.53	1265.8	334.43	7.963	278.45
780	0.7803	49.83	1282.1	338.72	8.065	282.02
770	0.7703	52.19	1298.7	343.12	8.170	285.68
760	0.7603	54.61	1315.8	347.63	8.277	289.44
750	0.7503	57.10	1333.3	352.27	8.387	293.30
725	0.7252	63.61	1379.3	364.41	8.677	303.41
700	0.7002	70.59	1428.6	377.43	8.986	314.25
675	0.6751	78.08	1481.5	391.41	9.319	325.89
650	0.6501	86.16	1538.5	406.46	9.678	338.42
625	0.6251	94.88	1600.0	422.72	10.065	351.96
600	0.6000	—	1666,7	440.33	10.484	366.62

Mass Conversions

	mg	g	lb	kg	short ton	Mg (tonne)
mg	1	0.001	0.000 002	0.000 001	–	
g	1 000.0	1	0.002 205	0.001	–	–
lb	453 592.4	453.592 4	1	0.453 592	0.000 5	0.000 454
kg	1 000 000	1 000.0	2.204 623	1	0.001 102	0.001
short ton	–	–	2 000.0	907.184 7	1	0.907 185
Mg (tonne)	–	–	2 204.623	1 000.0	1.102 311	1

Volume Conversions

	U.S. qt	dm^3 (litre)	U. S. gal	Imp. gal	Barrel	m^3 (kL)
U.S. qt	1	0.946 353	0.25	0.208 168	0.005 952	0.000 946
dm^3 (litre)	1.056 688	1	0.264 172	0,219 969	0.006 290	0.001
U.S. gal	4.0	3.785 412	1	0.832 674	0.023 810	0.003 785
Imp. gal	4.803 802	4.546 092	1.200 950	1	0.028 594	0.004 546
Barrel	168.0	158.987 3	42.0	34.972 30	1	0.158 987
m^3 (kL)	1 056.688	1 000.0	264.172 0	219.969 2	6.289 811	1

Pressure Conversions

	mm Hg	in. $H_2$0	kPa	psi	kg/cm^2	atm.
mm Hg	1	0.535 776	0.133 322	0.019 337	0.001 360	0.001 316
in. $H_2$0	1.866 453	1	0.248 84	0.036 091	0.002 537	0.002 456
kPa	7.500 615	4.018 647	1	0.145 038	0.010 197	0.009 869
psi	51.714 92	27.707 59	6.894 757	1	0.070 307	0.068 046
kg/cm^2	735.559 1	394.094 6	98.066 50	14.223 34	1	0.967 841
atm.	760.0	407.189 4	101.325 0	14.695 95	1.033 227	1

Power Conversions

	W	hp '	kW	Btu/s	kcal/s	MW
W	1	0.001 341	0.001	0.000 948	0.000 239	0.000 001
hp	745.699 9	1	0.745 700	0.706 787	0.178 227	0.000 746
kW	1 000.0	1.341 022	1	0.947 817	0.239 006	0.001
Btu/s	1 055.056	1.414 853	1.055 056	1	0.252 164	0.001 055
kcal/s	4 184.0	5.610 836	4.184	3.965 666	1	0.004 184
MW	1 000 000	1 341.022	1 000.0	947.817 0	239.005 7	1

Energy, Work Conversions

	kJ	Btu©	kcal©	MJ	hp-h	kWh
kJ	1	0.947 817	0.239 006	0.001	0.000 373	0.000 278
Btu	1.055 056	1	0.252 164	0.001 055	0.000 393	0.000 293
kcal	4.184 0_x	3.965 666	1	0.004 184	0.001 559	0.001 162
MJ	1 000.0	947.817 0	239.005 7	1	0.372 506	0.277 778
hp-h	2 684.520	2 544.434	641.615 7	2.684 520	1	0.745 700
kWh	3 600.0	3 412.141	860.420 7	3.60	1.341 022	1

① International Table

② Thermochemical

Length Conversions

	mm	cm	in.	ft	m	km	mile
mm	1	0.1	0.039 370	0.003 281	0.001	0.000 001	–
cm	10.0	1	0.393 701	0.032 808	0.01	0.000 01	–
in.	25.4	2.54	1	0.083 333	0.0254	0.000 025	–
ft	304.8	30.48	12.0	1	0.304 8	0.000 305	0.000 189
m	1 000.0	100.0	39.370 08	3.280 840	1	0.001	0.000 621
km	1 000 000	100 000.0	39 370.08	3 280.840	1 000.0	1	0.621 371
mile	–	–	63 360.0	5 280.0	1 609.344	1.609 344	1

Area Conversions

	mm^2	cm^2	$in.^2$	ft^2	m^2	. acre	ha
mm^2	1	0.01	0.001 550	0.000 011	0.000 001	–	–
cm^2	100.0	1	0.155 000	0.001 076	0.000 1	–	–
$in.^2$	645.16	6.451 6	1	0.006 944	0.000 645	–	–
ft^2	92 903.04	929.030 4	144.0	1	0.092 903	0.000 023	0.000 009
m^2	1 000 000	10 000.0	1 550.003	10.763 91	1	0.000 247	0.000 1
acre	–	–	–	43 560.0	4 046.856	1	0.404 686
ha	–	–	–	107 639.1	10 000.0	2.471 054	1

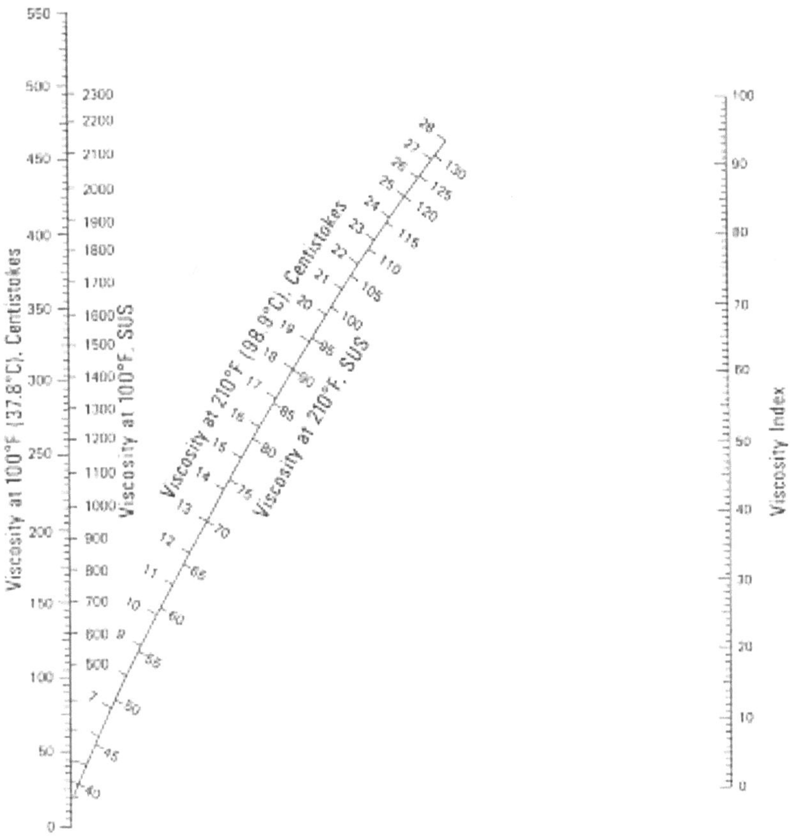

Viscosity Index, 0-100

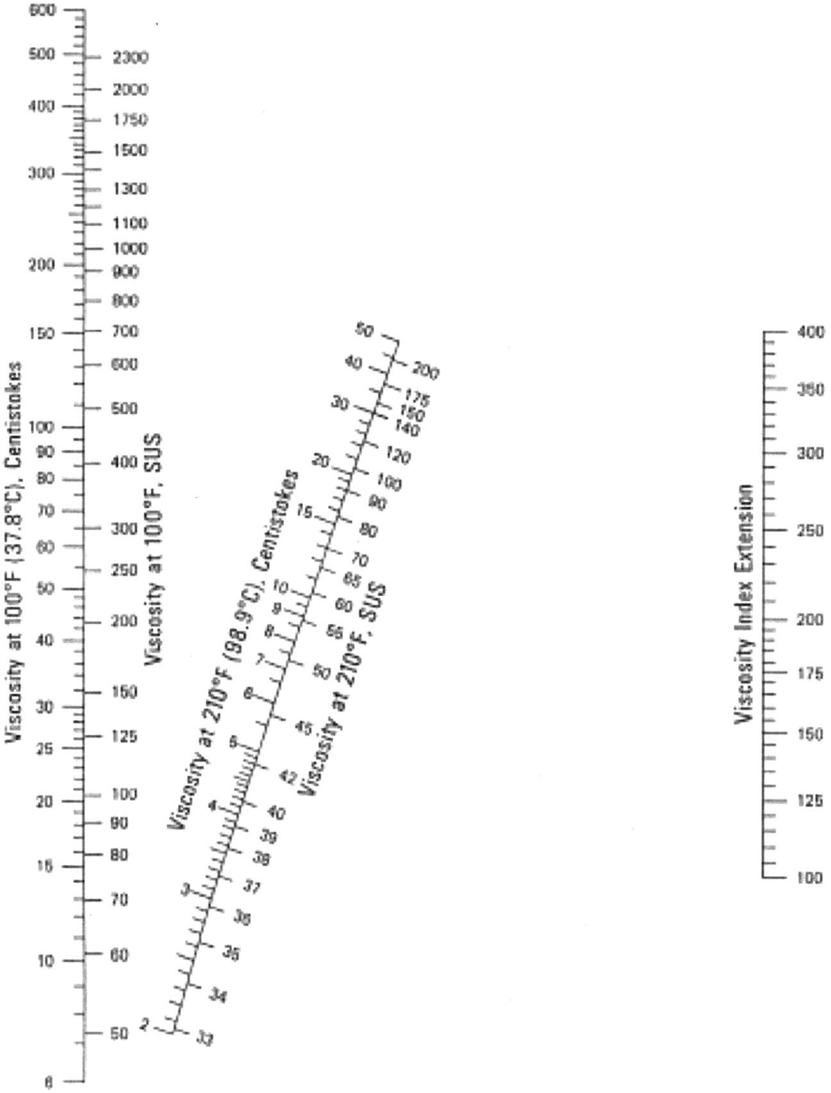

Viscosity Index, 100-400 (Based on ASTM D 2270)

Glossary of Terms

—A—

AAR—Abbreviation for "American Association of Railroads"

Absolute viscosity—A term used interchangeably with viscosity to distinguish it from kinematic viscosity or commercial viscosity. It is occasionally referred to as dynamic viscosity. Absolute viscosity and kinematic viscosity are expressed in fundamental units. Commercial viscosity such as Saybolt viscosity is expressed in arbitrary units.

Acidity—In lubricants, acidity denotes the presence of acid-type constituents whose concentration is usually defined in terms of acid number. The constituents vary in nature and may or may not markedly influence the behavior of the lubricant. (See also Total Acid Number.)

Acid Number—See Strong Acid Number and Total Acid

Additive—A chemical compound or compounds added or to a lubricant for the purpose of imparting new properties or to improve those properties that the lubricant already has.

AGMA—Abbreviation for "American Gear Manufacturers Association," an organization serving the gear industry.

Aluminum Soap Base Greases—Greases containing aluminum soap and mineral oils. They are mainly used in gearboxes for gear lubrication.

Aniline Point—The Aniline Point of a petroleum product is the lowest temperature at which it is completely miscible with an equal volume of freshly distilled aniline.

Anti Friction Bearing—A rolling contact type bearing in which the rotating or moving member is supported or guided by means of ball or roller elements. Does not mean without friction.

Antioxidant—A substance which retards the action of oxidation.

Anti-Stick-Slip Additives—They prevent stick-slip operation, e.g. carriage tracks and guideways in machine tools.

Antiwear Additives—Additives to reduce wear in the mixed friction range:

- Mild additives, e.g. fatty acids, fatty oils
- EP additives, e.g. lead, sulphur, chlorine and phosphorus compounds
- Dry lubricants, e.g. graphite and molybdenum disulphide.

API— American Petroleum Institute

API Gravity—A gravity scale established by the API and in general use in the petroleum industry, the unity being called the "API degree." This unit is defined in terms of specific gravity as follows:

$$\text{sp.gr.}(@60°\text{F}) = \frac{141.5}{(\text{Degrees API} + 131.5)}$$

Degrees API = [141.5/sp.gr.(@60—°F)]—131.5

Apparent viscosity—A measure of the resistance to flow of a grease whose viscosity varies with shear rate. It is defined as the ratio of the shear stress to the shear rate calculated from Poiseulle's equation at a given rate of shear and is expressed in poises.

Aromatics—Unsaturate hydrocarbons with a molecular ring structure (benzene, toluol, naphthalene). Aromatics have poor viscosity temperature properties and affect the oxidation stability of lubricants.

Ash Content—refers to the incombustible residues of a lubricant. The ash can be of different origin: it can stem from additives dissolved in the oil; graphite and molybdenum disulphide, soaps and other grease thickeners are ash producers. Fresh, straight refined mineral oils must be ash free. Used oils also contain insoluble metal soaps produced during operation, incombustible residues from contaminants, e.g. wear particles from bearing components and seals. Sometimes, incipient bearing damage can be diagnosed from the ash content.

Asphaltic—Essentially composed of or similar to asphalt. Frequently applied to naphthenic base lubricating oils derived from crudes that contain asphalt.

ASTM—Abbreviation for "American Society for Testing Materials," a society for developing standards for materials and test methods.

ATF—Abbreviation for Automatic Transmission Fluid. Special lubricants adapted to the requirements in automatic transmissions.

Axial Load Bearing—A bearing in which the load acts in the direction of the axis of rotation.

—B—

Babbitt—A soft, white, non-ferrous alloy bearing material composed principally of copper, antimony, tin and lead.

Ball Bearing—An antifriction bearing comprising rolling elements in the form of balls.

Barium Complex Soap Base Greases—Greases consisting of barium complex soaps and mineral oils or synthetic oils. They are water repellent, retain their consistency, and form a lubricating film with a high load carrying capacity.

Base Oil—is the oil contained in a grease. The amount of oil varies with the type of thickener and the grease application. The penetration number and the frictional behavior of the grease very with the amount of base oil and its viscosity.

Base Stock—A fully refined lube oil, which is a component of lubricant formulations.

Bearing—A support or guide by means of which a moving part such as a shaft or axle is positioned with respect to the other parts of a mechanism.

Bentonites—Minerals (e.g., aluminum silicates) which are used for the production of thermally stable greases with good low-temperature properties.

Bleeding—The oil contained in the grease separates from the soap. This can be caused e.g. by low resistance to working and/or a low thermal stability of the greases.

Block Grease—Generally, a grease of high soap content, which, under normal temperatures is firm to the touch and can be handled in block or stick form.

Bloom—A sheen or fluorescence evident in some petroleum oils when viewed by reflected light.

BMEP—Brake mean effective pressure (in gas engine power cylilnders).

Boundary Lubrication—A condition of lubrication in which the bulk viscosity characteristics of the lubricant do not apply or in which partial contact takes place between the mating surfaces.

Also refers to a thin film, imperfect, or non-viscous lubrication.

Bright Stock—A term referring to high viscosity lubricating oils which have been refined to make them clear products of good color.

By-Pass Filtration—A system of filtration in which only a portion of the total flow of a circulating fluid system passes through a filter or in which a filter, having its own circulating pump, operates in parallel with the main flow.

—C—

Calcium Soap Base Greases—Calcium soap base greases are water repellent and are therefore excellent sealants against the ingress of water. Since their corrosion protection is limited, they are usually treated with corrosion inhibitors. Additive-treated calcium soap base greases are appropriate for rolling mill bearings, even in chocks which are exposed to roll cooling water. Calcium soap base greases are normally suitable for temperatures from—20 to +50°C.

Carbon Residues—The residue remaining after the evaporation of a sample of mineral oil under specified conditions, i.e., Ramsbottom and Conradson.

Centipoise (cP)—A unit of absolute viscosity. 1 centipoise = .01 poise. Former unit for the dynamic viscosity.
1 cP= 1 mPa s

Centistoks (cSt)—A standard unit of kinematic viscosity = 0.10 stoke. Former unit for the kinematic viscosity.
1 cSt= 1 mm2/s

Cetane Number—A number that expresses the ignition quality of diesel fuel and equal to the percentage by volume of cetane (C16H34) in a blend with methyl naphthalene, which blend has the same ignition performance as the test fuel.

Channeling—The tendency of grease to form an unobstructed path or channel following the movement of the rolling elements in a bearing.

Channel Point—Lowest safe temperature at which a gear lubricant can be used.

Characteristics—The following are the most important characteristics of lubricating oils: flash point, density, viscosity at 40°C, setting point, and additive data. Greases are defined by: saponification basis, drop point, worked penetration and, where present, additives.

Circulating Effect—If grease is carried along by rotating parts the rotation causes lumps of grease to be pulled between rolling elements and raceways with a corresponding increase in friction due to grease working. High-speed applications therefore require greases which are not likely to be carried along. The circulating effect depends on the type of thickener, penetration, temperature and the bearing type. Sodium soap base greases tend to participate in the circulating movement.

Circulating Lubrication—A system of lubrication in which the lubricant, after having passed through a bearing or group of bearings, is re-circulated by means of a pump.

Cleveland Open Cup—See Flash Point, Fire Point.

Coefficient of Friction—The ratio of the friction force between two bodies to the normal, or perpendicular, force between them.

Color of Oils—Spent oils are usually judged by their color. However, caution should be exercised in using this criterion because even fresh oil can be more or less dark. Whether the discoloration is due to oxidation can only be confirmed by comparing it with a fresh sample of the same oil type. Contamination by dust and soot may also be the cause of discoloration.

Complex Greases—Besides metal soaps of high-molecular fatty acids, complex soap base greases contain metal salts of low-molecular organic acids. These salts and the soap form a complex compound which outperforms conventional greases as far as thermal stability, water resistance, anti-corrosive action and load carrying capacity are concerned.

Compounded Oil—A petroleum oil to which have other chemical substances added .

Consistency—A term used synonymously with the term Penetration Number of a grease.

Copper Corrosion Test—Method for determining active sulphur in mineral oils (DIN 51759, ASTM D 130-75) and greases (DIN 51 811, ASTM D 130-68, IP154/69).

Corrosion—The attrition or wearing away of a substance by acid or electrochemical action.

Corrosion Inhibiting Greases, Corrosion Inhibiting Oils—They protect corrodible metal surfaces against moisture and atmospheric oxygen.

Cup Grease—An early term for a calcium or lime base grease, practically obsolete now but meant originally to designate a degree of quality suitable for grease cup application, etc.

Cutting Fluid or Oil—Any fluid applied to a cutting tool to assist in the cutting operation by cooling, lubricating or other means.

CVD—Chemical Vapor Deposition—A method of thin coating (3-5 microns) metal parts with metallic alloys through a gaseous medium. The coating adds to the hardness while reducing wear and increasing lubricity of base metal.

—D—

Demulsibility—The ability of a non-water-miscible fluid to separate from water with which it may be mixed. The higher the demulsibility rating, the more rapidly the fluid separates from water. Demulsibility is sometimes expressed as the rate, in cubic centimeters per hour, or settling out of a fluid from an emulsion under specified conditions See Steam Emulsion Number.

Density—The mass of a unit volume of a substance Its numerical value varies with the units used.

Detergent—In lubrication, either an additive or a compounded lubricant having the property of keeping insoluble matter in suspension, thus preventing its deposition where it would be harmful. A detergent may also re-disperse deposits already formed.

Deterioration—is the chemico-physical alteration of lubricants under the effect of atmospheric oxygen, heat, pressure, humidity, metallic debris etc. Deterioration of mineral oils is indicated by a change in color and viscosity and by sludge formation. Grease deterioration: change in color, consistency and structure. Oxidation life test ASTM D-942.

Dewaxing—Process which removes wax from a lube distillate by solvent means (physical separation) or catalytic means (conversion).

Dielectric Strength—A measure of the ability of an insulating fluid to withstand electric stress (voltage) without failure. Fluids with high dielectric strength (usually expressed in volts or kilovolts) are good electrical insulators.

Dispersing—In lubrication, usually used interchangeably with detergent. An additive which keeps fine particles of insoluble materials in a homogeneous solution. Hence, particles are not permitted to settle out and accumulate.

Dispersion lubrication—Here the grease is dispersed in a suitable solvent, e.g., toluol; it is contained in the liquid in a finely dispersed undissolved state. The cleaned dry bearing is dipped into the compound and dried in a dust free environment, a thin grease film remaining on the bearing. These bearings excel by an extremely low lubricant shearing friction.

Distillate—A term applied to a liquid collected when condensing distilled vapors such as naphtha, kerosene, fuel oil and light lubricating oils.

Dopes—see additives.

Doped Lubricants— see Additive-treated lubricants.

Drop Feed Lubrication—A system of lubrication in which the lubricant is applied to the bearing surfaces in the form of drops at regular intervals.

Drop Point—Temperature at which a grease sample, when heated under standard test conditions, passes into a liquid state, flows through the opening of a grease cup and drops to the bottom of the test tube. Grease: DIN51 801T1, ASTMD-566

Dry Lubricants—Substances, such as graphite, molybdenum disulphide, or PTFE suspended in oils and greases, or applied directly.

Dynamic Viscosity—See Absolute Viscosity.

—E—

Elastic Behavior of Greases—The elastic properties of lubricating greases indicate the suitability of a grease for centralized lubrication systems (DIN 51 816T2).

Emcor Method—Testing of corrosion preventing properties of rolling bearing greases according to DIN 51 802.

Emulsibility—Tendency of an oil to emulsify with water.

Emulsifiers—Additives which help to form an emulsion.

Emulsion—Mixture of insoluble substances, usually mineral oils with water, which is activated by emulsifiers.

EP Lubricants—Lubricants that have been fortified with additives that appreciably increase the load carrying properties of the base lubricant, thus reducing excessive wear.

Esters (Synthetic Lubricating Oils)—Compounds of acids and alcohols with water eliminated. Esters of higher alcohols with divalent fatty acids form the diester oils (synthetic lubricating oils). Esters of polyhydric alcohols and different organic acids are particularly heat stable.

Evaporation Loss—Lubricating oil losses occurring at higher temperatures due to evaporation. It can lead to an increase in oil consumption and also to an alteration of the oil properties.

Extreme-Pressure Lubricants—see EP lubricants.

—F—

Fatty Acid—An organic acid of aliphatic structure originally derived from fats and fatty oils.

Fiber Grease—Grease having a distinctly fibrous structure which is noticeable when a sample of the grease is pulled apart. Greases having this fibrous structure tend to resist being thrown off gears and out of bearings.

Filler—Any solid substance such as talc, mica, or various powders, etc., which is added to a grease to increase its weight or consistency.

Filter—Any device or porous substance used as a strainer for cleaning fluids by removing suspended matter.

Fire Point (Cleveland Open Cup)—The flash point of an oil is the temperature to which it must be heated to give off sufficient vapor to form a sustained flammable mixture with air when a small flame is applied under specified conditions.

Flash Point—Flash point is that temperature to which an oil must be heated for sufficient vapor to be given off to form, briefly, a flammable mixture with air. The flash point is one of the characteristics of oils; it is not a criterion for their quality.

Flinger Disk—Disk dipping in lubricant sump.

Flow Pressure—Pressure required to press grease in a continuous stream from a nozzle. It is a measure of the consistency and fluidity of a grease. It is determined according to DIN 51 805.

Foam—A froth produced by whipping air into a lubricant. Foaming in mineral oils should be avoided. Foaming promotes deterioration of the oil. Excessive foaming can lead to an overflow and, consequently loss of oil.

Force Feed Lubrication—A system of lubrication in which the lubricant is supplied to the bearing surface under pressure.

Form Oil—A compound or an oil used to lubricate wooden or metal concrete forms in order to keep cement from sticking to them.

Four Ball Test Rig—Machine for lubricant testing (DIN 51 350, ASTM D 2266-67, ASTM D 25 9669, ASTM D 2783-71, IP 239/73). Four balls are arranged in a pyramid shape, with the upper ball rotating. The load applied can be increased until welding occurs between the balls (welding load). The load, expressed in N, is the four ball welding load. The diameter of the weld scar on the stationary balls measured after one hour of testing is the four ball wear value which is used for wear evaluation.

Fretting Corrosion—A process of mechanical attrition combined with chemical reaction taking place at the common boundary of loaded contact surfaces having small oscillatory relative motion.

Friction—The resisting force encountered at the common boundary between two bodies when, under the action of an external force, one body moves or tends to move over the surface of the other.

Full Flow Filtration—A system of filtration in which the total flow of a circulating fluid system passes through a filter.

—G—

Gear Greases—Gear greases are usually sodium soap based, stringy, soft to semifluid greases (NLGI 0 and 00) for gears and gear motors. Some greases are treated with EP additives.

Gel Greases—They contain an inorganic-organic thickener made up of finely dispersed solid particles; the porous surface of these particles tends to absorb oil. Gel greases are suitable for a wide temperature range and are water resistant. Caution is recommended at high speeds and loads.

Graphite—A crystalline form of carbon either natural or synthetic in origin, which is used as a lubricant.

Gravity—see Specific Gravity, API Gravity

Grease—A lubricant composed of an oil or oils thickened with a soap, or other thickener to a solid or semi-solid consistency.

Grease Service Life—Life of a grease charge determined in laboratory and field tests. The individual life values scatter by 1: 10 even under comparable test and operating conditions.

Gum—A rubber-like, sticky deposit black or dark brown in color, which results from the oxidation of lubricating oils in service.

—H—

HD Oils—Heavy-duty oils are additive-treated engine oils particularly adapted to the rugged conditions in internal combustion engines.

High-Temperature Greases—Lithium greases can be used at steady-state temperatures up to 130°C and bentonite greases up to 140°C. Special MoS_2, silicone, and synthetic greases can be used up to 260°C.

Homogenizing—Final step in grease production. In order to obtain a uniform structure and fine dispersion of the thickener, the grease is thoroughly worked in a special machine.

Hydraulic Fluids—Fire-resistant pressure fluids for hydraulic load transmission and control.

Hydraulic Oil—An oil specially suited for use as a power transmission medium in hydraulically operated equipment. Non-aging, thin-bodied, non-foaming, highly refined hydraulic fluids produced from mineral oil, with a low setting point, for use in hydraulic systems.

Hydrodynamic Lubrication—A system of lubrication in which the shape and relative motion of the sliding surfaces causes the formation of a fluid film having sufficient pressure to separate the surfaces.

Hydrotreating—A process which converts and removes undesirable components with the use of a catalyst.

Hypoid Gear Lubricant—A gear lubricant having extreme pressure characteristics for use with a hypoid type of gear as in the differential of an automobile.

HVI—High Viscosity Index, typically from 80 to 110 VI units.

—I—

ICML—International Council for Machinery Lubrication

Inhibitor—Any substance which slows or prevents chemical reaction or corrosion.

Interfacial Tension (I F T)—The energy per unit area present at the boundary of two immiscible fluids. It is commonly measured as the force per unit length necessary to draw a thin wire or ring through the interface.

Intermediate Base Crude—See Mixed Base Crude.

ISO—International Standards Organization, sets viscosity reference scales.

ISO-equivalent—Consistency of greases at 25°C measured by the penetration depth of a standard cone, after treatment of the grease sample in a grease worker (DIN 51 804).

—J—

Journal Bearing—A sliding type of bearing in conjunction with which a journal operates. In a full or sleeve type journal bearing, the bearing surface is 380. in extent. In a partial bearing, the bearing surface is less than 360° in extent.

—K—

Kinematic Viscosity—The absolute viscosity of a fluid divided by its density. In a c.g.s. system, the standard unit of kinematic viscosity is the stoke and is expressed in sq. cm. per. sec. In the English system, the standard unit of kinematic viscosity is the newt and is expressed in sq. in. per sec.

—L—

Lacquer—A deposit resulting from the oxidation and polymerization of fuels and lubricants when exposed to high temperatures. Similar to but harder than varnish.

Lard Oil—An animal oil prepared from chilled lard or from the fat of swine.

Lime Base Grease—A grease prepared from a lubricating oil and a calcium soap.

Lithium Soap Base Greases—have definite performance merits in terms of water resistance and temperature range. Frequently, they incorporate oxidation inhibitors, corrosion inhibitors and EP additives. Lithium soap base greases are widely used as rolling bearing greases. To a limited extent, lithium greases emulsify with water. They can, to a certain degree, tolerate moisture, but severe moisture or ingress of water should be prevented, because this would cause the grease to become extremely soft and escape from the bearing. Standard lithium soap base greases can be used for temperatures ranging from –35 to +130°C.

Low-Temperature Properties—see Setting Point.

Lubricant—Any substance interposed between two surfaces in relative motion for the purpose of reducing the friction between them. Less exactly, any substance interposed between two surfaces in relative motion to facilitate their action.

Lubricant Analysis Data—The analyzed data are density, flash point, viscosity, setting point, drop point, penetration, neutralization number, saponification number. These physical and chemical properties of lubricants indicate the fields of application of the lubricants. See also Specifications.

Lubricating Greases—Greases are mixtures of thickeners and oils:

- Metal soap base greases consisting of metal soaps as thickeners and oils
- Non-soap greases comprising inorganic gelling agents or organic thickeners and oils
- Synthetic greases consisting of organic or inorganic thickeners and ynthetic oils

LVI—Low Viscosity Index, typically below 40 VI units.

—M—

Mineral Oil—Oils derived from a mineral source, such as petroleum, as opposed to oils derived from plants and animals.

MIL Specifications—Specifications of the US Armed Forces indicating the minimum mandatory requirements for the materials to be supplied. Some engine and machine builders apply the same minimum mandatory requirements to the lubricants. The MIL minimum mandatory requirements are taken as a quality standard.

Mineral Oils—Crude oils and/or liquid oil products.

Miscibility of Oils—Oils of different grades or from different manufacturers should not be mixed. The only exception are HD engine oils which can generally be mixed. If fresh oils are mixed with used oils, sludge can deposit.

Whenever there is the risk of sludge formation, samples should be mixed in a beaker.

Multigrade Oils—Engine and gear oils with improved viscosity-temperature behavior.

—N—

Naphthenic Base Oils—A characterization of certain petroleum product prepared from naphthenic type crudes (crudes containing a high percentage of ring type hydrocarbon molecules).

Neatsfoot Oil—A pale yellow animal oil made from the feet and shinbones of cattle.

Needle Bearing—A bearing comprising rolling elements in the form of rollers that are relatively long compared to their diameter.

Neutralization Number—A term still used in the petroleum industry, but rapidly becoming obsolete in the lubrication field. See Strong Acid, Strong Base, Total Acid, and Total Base Numbers.

Nitration—chemical attack on the lube oil by nitration oxides that are formed in the process of combustion. The nitrogen-bearing products that are formed degrade the lube oil, hasten additive depletion, and contribute to deposit formation. Nitration results from operating at airfuel mixtures that give 1-5% excess oxygen.

NLGI—An abbreviation for "National Lubricating Grease Institute," a technical organization serving the grease industry.

NPRA—National Petrochemical & Refiners Association

—O—

Oil—A viscous, unctuous liquid of vegetable, animal, mineral or synthetic origin.

Oil Ring—A loose ring, the inner surface of which rides a shaft or journal and dips into a reservoir of lubricant from which it carries the lubricant to the top of a bearing by its rotation with the shaft.

Oil Separation—Oil can separate from the greases if they are stored for a longer time or temperatures are high. Oil separation is determined according to DIN 51 817, ASTM D 1742, IP 121/63. For-life lubrication requires a small oil separation rate which must, however, be large enough to lubricate all contact areas.

Operating Viscosity—Kinematic viscosity of an oil at operating temperature. The operating viscosity is termed v. It can be determined by means of the viscosity-temperature diagram if the viscosity values at two temperatures are known. For determining the operating viscosity of oils with average viscosity-temperature behavior, diagrams can be used.

Oxidation—see Deterioration

Oxidation Inhibitors—see Anti-oxidants

Oxidation Stability—Ability of a lubricant to resist natural degradation upon contact with oxygen.

—P—

Pad Lubrication—A system of lubrication in which the lubricant is delivered to a bearing surface by a pad of felt or similar material.

Paraffin Base Oil—A characterization of certain petroleum products prepared from paraffinic type crudes (crudes containing a high percentage of straight chain aliphatic hydrocarbon molecules). Lubricating oils made from these crudes are normally distinguished from similar oils from other crudes (both oils equally well refined) by higher API gravity and higher viscosity index.

Penetration or Penetration Number—The depth, in tenths of a millimeter that a standard cone penetrates a solid or semisolid sample under specified conditions. This test is used for comparative evaluation of grease and grease-like materials. (See Worked Penetration.)

Petrolatum—A jelly-like product obtained from petroleum and having a microcrystalline structure. Often used in rust preventives.

Plain Bearing—Any simple sliding type bearing as distinguished from tapered land, tilting pad, or antifriction bearings, etc.

Poise—The standard unit of absolute viscosity in the c.g.s. system expressed in dyne sec. per sq. cm.

Polyalkylene Glycol—A base fluid prepared by polymerizing one or more alkylene oxides, most usually ethlene oxide and/or propylene oxide.

Polyalphaolefin—A base fluid prepared by polymerizing alpha olefinic hydrocarbons and hydrogenating the polymer.

Polyurea Base Grease—A grease prepared from a lubricating oil and a polyurea thickener.

Pour Point—The pour point of a lubricant is the lowest temperature at which the lubricant will pour or flow when it is chilled without disturbance under specified conditions.

Power Factor—A measure of the dielectric loss, or ability to perform as an electrical insulating oil.

Pour Point Depressant—An additive that retards wax crystallization, and lowers the pour point.

Process Oils—A lube base stock that receives additional processing to impart a very specific hydrocarbon composition in addition to viscometrics. Process oils are not used as lubricants; they are used as chemical components in the manufacture of rubber, plastics, and other polymeric materials.

PVD—Physical Vapor Deposition—A thin metal-plasma coating (2-5 microns) that is applied in a low heat temperature environment (350°F to 600°F) which can be applied to standard metal surfaces to help resist wear while increasing lubricity and hardness.

—R—

Rated Viscosity—This is the kinematic viscosity attributed to a defined lubrication condition. It is a function of speed and can be determined with diagrams by means of the mean bearing diameter and the rotational speed. By comparing the rated viscosity v1 with the operating viscosity v the lubrication condition can be assessed.

RCFA—Root Cause Failure Analysis

Refined Oils—A positive resistance to aging of lubricating oils is obtained by refining the distillates. Unstable compounds which can incorporate sulphur, nitrogen, oxygen and metallic salts are removed. Several refining processes are used, the most important being the treatment with sulphuric acid and the extraction of oil-insoluble unstable compounds with solvents.

Relubrication Interval—Period after which lubricant is replenished. The relubrication interval should be shorter than the lubricant renewal interval.

Ring Lubrication—A system of lubrication in which the lubricant is supplied to the bearing surfaces by an oil ring.

R&O—An additive inhibitor package which contains Rust and Oxidation Inhibitors.

Roller Bearing—An antifriction bearing comprising rolling elements in the form of rollers.

Rust Prevention Test (Turbine Oils)—A test for determining the ability of an oil to aid in preventing the rusting of ferrous parts in the presence of water.

—S—

SAE—An abbreviation for "Society of Automotive Engineers," an organization serving the automotive industry.

SAE Numbers—Numbers applied to motor, transmission and rear axle lubricants to indicate their viscosity range. Conversion of the SAE values for engine oils are indicated in DIN 51511, and for automotive gear oils in DIN 51512.

Saponification Number—The state of straight oil deterioration can be assessed by means of the saponification number. It is expressed in milligrams of potassium hydroxide which are required to neutralize the free and bonded acids contained in one gram of oil.

Saybolt Furol Viscosity—The time in seconds required for 60 cubic centimeters of a fluid to flow through the orifice of a Saybolt Furol Viscometer at a given temperature under specified conditions The orifice of the furol viscometer is larger than that of the universal viscometer, the former instrument being used for more viscous fluids.

Saybolt Universal Viscosity—The time in seconds required for 60 cubic centimeters of a fluid to flow through the orifice of the Standard Saybolt Universal. Viscometer at a given temperature under specified conditions. (ASTM Designation D 88-56.)

Seals, Seal Compatibility—The reaction of sealing materials with mineral oils and grease differs widely. They can swell, shrink, embrittle or even dissolve, operating temperatures and lubricant composition playing a major role. Seal and lubricant manufacturers should be consulted for seal compatibility.

Sediments—Sediments are usually formed by soot and dirt particles. They are caused by oil deterioration, mechanical wear, excessive heating, too long oil renewal intervals. They settle in the oil sump, in the bearings, in filters, and in lubricant feed lines. Sediments are hazardous to the operational reliability.

Semi-fluid Greases—These are lubricating greases of semi-fluid to pasty consistency, e.g. aluminum, calcium and sodium soap base greases with a mineral base oil of a viscosity > 70 mm2/s at 40°C. To improve their load carrying capacity, semi-fluid greases which are generally used for gear lubrication, can be doped with EP additives or solid lubricants.

More generally, any substance in which the force required to produce a deformation depends both on the magnitude and on the rate of deformation.

Setting Point—The setting point of a lubricating oil is the temperature at which the oil ceases to flow if cooled under specific conditions. The low-temperature behavior of the oil slightly above the setting point may be unsatisfactory and must therefore be determined by measuring the viscosity.

Shear Stress—The force per unit area acting tangent to the surface of an element of a fluid or a solid.

Silicone Oils—Synthetic oils which are used for special operating conditions. They have better physical data than mineral oils, but have poor lubricating properties and a low load carrying capacity

Sleeve Bearing—A journal bearing, usually a full journal bearing. (See journal bearing.)

Sludge—Insoluble material formed as a result either of deterioration reactions in an oil or by contamination of an oil, or both.

Air and water can effect the formation of oxidation material and polymerizates in mineral oil products. They settle as sludge.

Slushing Oil—An oil or grease-like material used on metals to from a temporary protective coating against rust, corrosion, etc.

Sodium Soap Base Greases—These greases adhere well to the bearing surfaces and form a uniform and smooth lubricating film on the functional surfaces. They are more prone to emulsifying with water than lithium soap base greases, i.e. they are not water resistant.

The grease is able to absorb minor quantities of water; larger amounts of water would liquefy the grease and make it run out of the bearing. Sodium soap base greases have poor low-temperature properties. They can be used for temperatures ranging from –30 to +120°C.

Solid Foreign Particles—All foreign contaminants insoluble in naphtha and benzene. Solid foreign particles in oils are evaluated according to DIN 51 592, in greases to DIN 51 813.

Solid Lubricants—see Dry Lubricants

"Soluble" Cutting Oil—A mineral oil containing an emulsifier which makes it capable of mixing easily with water to form a cutting fluid.

Solvates—Mineral oils refined with solvents.

Solvency—Ability of a fluid to dissolve organic materials and polymers, which is a function of aromaticity.

Specifications—Military and industrial standards for lubricants which stipulate physical and chemical properties as well as test methods.

Specific Gravity—The ratio of the weight in air of a given volume of a material to the weight in air of an equal volume of water at a stated temperature.

Sperm Oil—A fixed nondrying pale yellow oil obtained from the head cavities and blubber of the sperm whale. Formerly used as an oil additive but now prohibited from use by law in the United States.

Spindle Oil—A light-bodied oil used principally for lubricating textile spindles and for light, high speed machinery.

Splash Lubrication—A system of lubrication in which parts of a mechanism dip into and splash lubricant onto themselves and/or other parts of the mechanism.

SSU—An abbreviation for Saybolt Seconds Universal used to indicate viscosity, e.g., SSU @ 100°F. Also SUS

STLE—Society of Tribologists and Lubrication Engineers

Stability—Ability of a lubricant to resist natural degradation reactions upon exposure to UV radiation, heat, or oxygen.

Static Friction—The friction between two surfaces not in relative motion but tending to slide over one another. The value of the static friction at the instant relative motion begins is termed break-away friction.

Strong Acid Number (S A N)—The quantity of base, expressed in milligrams of potassium hydroxide, required to titrate the strong acid constituents present in 1 gram of sample.

Strong Base Number (S B N)—The quantity of acid, expressed in terms of equivalent number of milligrams of potassium hydroxide, required to titrate the strong base constituents present in 1 gram of sample.

Sulfurized Oil—Oil to which sulfur or sulfur compounds have been added.

Surface Tension—The tension exhibited at the free surface of liquids, measured in force per unit length.

Suspension—Colloidal suspension of solid particles dispersed in liquids, e.g. the oil-insoluble additives in lubricants.

Swelling Properties—The swelling properties of natural rubber and elastomers under the effect of lubricants are tested according to DIN 53 521.

Synthetic Ester—Oil Molecule prepared by reacting an organic acid with an organic alcohol and possessing some lubricant properties

Synthetic Hydrocarbon—Oil Molecule prepared by reacting paraffinic materials.

Synthetic Lubricant—A lubricant produced from materials not naturally occurring in crude oil by either chemical synthesis or refining processes. Lubricants produced by chemical synthesis; their properties can be adapted to meet special requirements: very low setting point, good V-T behavior, small evaporation loss, long life, high oxidation stability.

—T—

Tacky—A descriptive term applied to greases which are particularly sticky or cohesive.

Tallow—Animal fat prepared from beef and mutton.

Thermal Conductivity—Measure of the ability of a solid or liquid to transfer heat.

Thickener—Thickener and base oil are the constituents of lubricating greases. The percentage of the thickener and the base oil viscosity determine the consistency of the grease.

Thixotropy—The property of a grease to become softer when mechanically stressed and to return to its original consistency when left to rest. Preserving oils with special additives are also thixotropic.

TOST—Turbine Oil Oxidation Stability Test, ASTM D-943

Total Acid Number (TAN)—The quantity of base, expressed in milligrams of potassium hydroxide, that is required to titrate all acidic constituents present in 1 gram of sample.

Total Base Number (TBN)—The quantity of acid that is required to titrate all basic constituents present in 1 gram of sample.

Turbine Quality—Lube base stocks suitable for turbine applications, finished with severe hydrotreating. TQ base stocks exhibit improved oxidation stability over other base stocks.

—U—

Unworked Penetration—The penetration at 77°F of a sample of grease that has received only the minimum handling in transfer from a sample can to the test apparatus and which has not been subjected to the action of a grease worker.

—V—

Varnish—When applied to lubrication, a deposit resulting from the oxidation and polymerization of fuels and lubricants. Similar to but softer than lacquer.

Viscometer—Viscosimeter—An apparatus for determining the viscosity of a fluid.

Viscosity—That property of a fluid or semi-solid substance characterized by resistance to flow and defined as the ratio of the shear stress to the rate of shear of a fluid element. The standard unit of viscosity in the c.g.s. system is the poise and is expressed in dyne sec. per square centimeter. The standard unit of viscosity in the English system is the reyn and is expressed in lb. sec. per square in.

Viscosity Classification—The standards ISO 3448 and DIN 51519 specify 18 viscosity classes ranging from 2 to 1,500 mm2/s at 40°C for industrial liquid lubricants

Viscosity Grade—Any of a number of systems that characterize lubricants according to viscosity for particular applications, such as industrial oils, gear oils, automotive engine oils, automotive gear oils, and aircraft piston engine oils.

Viscosity Index (VI)—A measure of a fluid's change of viscosity with temperature. The higher the viscosity index the smaller the change in viscosity with temperature.

Viscosity Index Improver—Additive that increases lubricant viscosity index, necessary for formulation of multi-grade engine oils.

Viscosity-Pressure Behavior—Viscosity of a lubricating oil as a function of pressure. With a rise in pressure the viscosity increases.

V-T Behavior—The viscosity-temperature behavior refers to the viscosity variations with temperatures. The V-T behavior is good if the viscosity varies little with changing temperatures.

—W—

Water Content—If an oil contains water, the water droplets disrupt the lubricating film and reduce lubricity. Water in oil accelerates deterioration and leads to corrosion. The water content can be determined by distillation or by settling in a test tube; due to its higher specific gravity the water settles at the bottom. Samples of emulsifying oil must be heated. A small amount of water (0.1% or less) is identified by a crackling noise which is produced when the oil is heated in a test tube. A higher water content will cause the oil to boil over.

Water Resistance—The water resistance of greases is tested according to DIN 51807 (static test); it is not indicative of the water resistance of the grease when used in the field. The test merely shows the effect which static, distilled water has on an unworked grease at different temperatures.

Water Separation Ability—Ability of an oil to separate water. The test is carried out according to DIN 51589.

"Wetting Bearings"—The pre-lubrication of bearing surfaces prior to starting a machine that has been idle for an elongated time period. Prevention of possible Brinell damage to bearing components upon sudden dry start of a machine.

Wet Fuel—Gas containing heavier products, such as ethane or propane, and having a heat content greater than normal (above 1000 Btu/cubic foot).

White Oils Light—Colored and usually highly-refined mineral oils, usually employed in medicinal and pharmaceutical preparations and as a base for creams, salves, and ointments, but also used as lubricants.

Worked Penetration—The penetration of a sample of lubricating grease immediately after it has been brought to 77°F ±1°F and then subjected to 60 strokes in the ASTM standard grease worker.

Index

Symbols
4-Ball EP test 119

A
abrasive wear 13–14, 485, 486
absolute viscometer 633, 634
absorption 33, 96, 98, 130, 132, 153
accelerated aging 431
accelerated leakage test 156
AC circuit 136
acidity 31, 122, 124, 129, 131, 163, 656
acid test results 124
active chemicals 52
AC transmission 136
additives 12, 14–15, 19–21, 25, 27, 32–34, 44, 50–53, 59, 68, 71, 75–76, 79–82, 88, 89, 94–95, 98–99, 112–113, 118, 121, 129, 134, 142–143, 153, 159, 163, 172, 174–175, 177, 180, 188, 190–191, 194–195, 199–200, 210–213, 223–224, 227, 232–234, 262, 267–268, 272, 340, 360, 427–430, 437–438, 440, 480–482, 487, 491, 500, 517, 545, 567, 583, 590, 606–607, 610, 616–617, 637–638, 643, 656, 671
  depletion 31–32, 34, 89, 124, 163, 396, 490, 534, 628, 634–638, 657
  for heavier loads 50
  formulation 52, 72–73, 80, 89, 160, 175, 183, 188, 190, 201, 430, 632
  loss 3, 7, 13, 20, 31, 34, 38, 47, 51, 57, 92, 94, 110, 127, 136, 141, 144, 167, 177, 198, 211–212, 280, 336, 341–342, 350, 364, 382–383, 385, 390, 395–396, 401, 403, 419–420, 445–446, 482, 487, 501, 517, 527, 533, 549, 590, 606, 610, 680, 697, 715, 718–719
  monitoring 80, 142, 314, 333, 356, 368, 371, 454, 557, 569, 603, 615–618, 623, 627, 631–634, 637–639, 648, 680, 689, 693–697, 704, 717, 728, 730–731
  package 22, 29, 32, 72–73, 79–83, 139, 219, 236, 243, 578, 590, 606, 619, 704–706
  stability 23, 68, 73, 75, 81, 101, 105, 108, 116, 128–130, 132, 135, 138–139, 157, 160, 168, 171, 174–175, 183, 190, 195, 217, 222–223, 226–227, 230, 232–233, 237–238, 242, 247, 253, 257, 266, 453, 457, 460, 466, 468, 490–491, 494, 523, 540, 631–632, 634, 636–637, 643, 674
adhesive wear 14–15, 267, 485

adiabatic compression 31
advantages of a lubricating oil 75
after-drip 360
agents 12, 20, 30, 50–53, 70, 81, 98, 101, 138, 163, 233–234, 269–70, 540, 548–549
hydraulic oils 51, 77, 135, 174–176, 225–226
properties 12, 22–25, 30, 50, 55, 69–73, 75, 79–80, 85, 87, 89, 99, 101–103, 109–110, 112, 118–122, 124, 139, 142, 146, 153–154, 164, 167–168, 175, 177, 183–186, 207, 211, 223–224, 226–228, 230, 232, 239, 242–243, 249, 261, 267–268, 367, 388, 425–428, 430, 437, 453–454, 460, 472, 475, 481, 484, 486, 492, 517, 583, 590, 598, 604, 619, 629, 631– 632, 637, 646, 648, 674
protection 13, 27, 34–35, 52, 68, 70, 75–77, 107, 146, 160, 163, 165, 168, 171, 174–175, 177, 183–184, 190–192, 194, 200–201, 210, 216–217, 225, 232, 236, 246, 252–253, 258, 268, 270, 275, 359–360, 392, 394–396, 403–404, 417, 429, 456, 460, 462, 466, 468–469, 471, 481, 493, 534, 539, 551, 560, 565, 584, 589, 606, 664, 667, 669, 672–673, 709
turbine oils 87, 101, 124, 128, 138–140, 224, 634
aggregate adsorption 33
AGMA lubricants 480
air bearings 65
air entrainment 87–88, 112–113, 160, 175, 236
air-oil lubrication 336–341, 438, 441
systems 336
air-oil method 436
air-oil valve assembly 338, 340
air-operated injector pump 361
air stripping 644
units 644
alkalinity 122, 124, 234
alkalinity agents 234
alloys 55, 98–99, 132, 169, 175, 269, 454, 467
aluminum-complex thickener 192, 193
American Gear Manufacturers Association (AGMA) 167
analytical ferrography 648–650
aniline point 88, 114
animal/vegetable base oil 23
annular plate springs 469
annular springs 470
anti-oxidants 195
anti-wear 15, 51, 81, 119–120, 163, 165, 174–177, 183, 190, 200, 213, 217, 233, 249, 258, 487, 493, 500
API-610 513
API gravity 112–113
apparent viscosity (mPa·s) 27, 604
application limits for greases 509
application methods for dry lubricants 276
application of lube oil analysis 616
applications for PAOs and diesters 247
applicator tubing 436
aquatic toxicity 92–93
aromatic compounds 97
aromatics 72–73, 88, 91, 97, 138, 142–43

ash content 89
ASIM D 1218 137
asperities 7, 11, 12, 20, 35, 53, 507
ASTM colors 137
ASTM cone penetrometer 114
ASTM copper strip classification 100
ASTM D 86-67 106
ASTM D 88 145
ASTM D 92 109
ASTM D 96, D 95, and D 473 154
ASTM D 97 134
ASTM D 130 98, 491, 492
ASTM D 216-54 106
ASTM D 217 and D 1403 114
ASTM D 286 90
ASTM D 287 113
ASTM D 322 104
ASTM D 323 143
ASTM D 566 and ASTM D 2265 108
ASTM D 567 and D 2270 147
ASTM D 611 and ASTM D 1012 88
ASTM D 664 and D 974 121
ASTM D 665 139–140, 163, 492
ASTM D 721 test 127
ASTM D 850-70 106
ASTM D 877 and D 1816 102
ASTM D 892 112
ASTM D 924 136
ASTM D 942 • 1P142, D 1402, and D 1261 130
ASTM D 943 128, 138–139
ASTM D 971 116
ASTM D 1078-70 106
ASTM D 1263 156
ASTM D 1401 and ASTM D 2711 100
ASTM D 1742 127
ASTM D 2155 90
ASTM D 2272 138–139
ASTM D 2500 94, 135
ASTM D 2699 and D 2700 124
ASTM engine 125
ASTM method D 1298 102
A Strategy for Tribology in Canada 597
atmospheric contamination 623, 625
atmospheric/vacuum distillation process 72
atomization 170
auto-ignition temperature 90–91, 109–110
automated centralized lubrication 327
automated lubrication 312, 317, 330, 332, 347, 375, 690
automated single point lubrication (SPL) devices and systems 317
automatic lubrication systems 329–330, 333
automatic particle counters 640
automotive gasoline 106, 107, 200
automotive hypoid rear axles 99
auxiliary pump 380
aviation gasoline 107, 126

B

babbitt 12, 38, 448, 453
babbitted liner 457
Baldor Electric 510–511
barium 82, 88
based on synthetic oils 428
base oils 25, 27, 69, 71, 73, 79, 82, 159–160, 173, 195, 221–224, 430, 517, 583, 606
base oil viscosity 27, 427, 500, 578
basic gear designs 478

Bearing 3, 4, 8–9, 12–13, 15, 17, 20–22, 25, 29–32, 35, 39–47, 53–55, 57, 59–60, 65–67, 70, 80–82, 85, 99, 124, 129–133, 145, 153, 156, 159–170, 172, 175, 179, 189, 197–198, 211, 223–224, 234–236, 244, 246, 249, 252–253, 255, 258, 261–262, 279–290, 292–296, 301–306, 308, 311–313, 317–321, 323, 326–327, 330, 332–333, 336–337, 339–341, 344, 346, 350, 352–359, 361, 363–364, 372–376, 380–381, 384, 386–388, 390, 392, 394, 396, 401, 403, 411, 425–446, 448–460, 474, 482, 497–510, 512,–517, 521–528, 531, 539, 551, 559, 569–570, 599–600, 604–606, 620, 631, 651–652, 662, 680, 688–694, 697–698, 707, 717, 719
bearing protector 525
Belleville washers 469
benefits, of synthetic lubricants 217, 232, 238, 241, 244, 339, 363, 401, 493, 517, 553, 617–618, 627, 652, 657–659, 686–688
Beta ratio 409–411
bevel gears 168, 212, 487
bi-directional running 448
bill of materials 556, 577, 703
binder system 275
biodegradability 77, 80, 91–93, 195, 224–225
biodegradable 92, 187, 189, 192, 194, 541
biodegradation 91–94, 108, 194
  test 91–111, 113–116, 118–133, 136–146, 150, 153–157, 163, 179, 183–184, 194
biological activity 606
biological growth 598, 606
black oil 527
blends of the synthetic lubricants 230
block penetration 115
blowers 330, 336, 363, 494–496
body of knowledge 732
bore options 458
bore profile 457
  failures 31, 171, 234–236, 249, 279–280, 330, 361, 364, 499, 502, 605, 627, 629, 679–680, 691–693, 713, 717, 719–720, 725–726
  housing pressure 445
  housing seals 319–320
  inches 286, 296–297, 368, 398, 479, 513–514, 524–525, 541, 564
  isolators 392
  longevity 693
  lubrication 39
  metals 12–13, 21, 33, 52, 82, 89, 99, 110, 129, 133, 139, 160, 169, 172, 188, 190, 194, 219, 223, 236, 267–269, 405, 463, 466, 468, 470, 539, 615, 647, 651
  operating temperatures 28, 48–49, 110, 163, 224, 248, 431, 435, 437, 440, 504, 637
  protector seal 525
  reconditioning service 507
  replacement costs 236, 549
  surface temperatures 448, 454
  temperature limits 70, 275
boron compounds 210
boundary lubrication 32, 35, 43–44, 50–53, 71, 120, 483–484, 644
bowl mill wear 237

Boyle's law 31
breakdown voltage 103
break-in 11–12, 628
breathers 21, 180, 182, 319, 321, 402, 416–418, 421, 585, 599, 662, 689–690, 706
bung location 566
burner fuel 107

C
cage designs 515
can seamer oil 192
carbon deposits 31, 252–253
carbon-forming tendencies 249
carbonization 637
carbon particles 103
carcinogens 142–143, 188
catalytic metals 129, 160, 169, 190
CEC L-33-A-94 primary biodegradation test 92
center-pivoted pads 448
central generator, for oil mist systems 368–369
centrifugal compressor 257
centrifugal forces 33, 215, 493
  on grease 21, 153, 295
centrifugal pump 164, 501, 528
centrifugation 33, 274, 590
centrifuges 215, 493, 607, 609–610
centrifuging 584, 606
cetane number 107
chain 82, 138, 193, 227, 233, 242, 393, 537–542, 668, 697, 707
  components 21, 65, 66, 75, 95–98, 110, 134–135, 144, 167–168, 174, 178–180, 213, 215, 236, 242, 258, 269, 275–276, 321, 335, 359, 363–364, 366, 373, 379, 415, 435, 443, 452, 461, 463, 465–466, 469, 471–472, 474, 485, 488, 493, 515, 538, 567, 583, 605, 610, 615, 619, 621, 643, 653, 674, 719
  test rig 253, 295
changing oil properties, monitoring of 631
channeling characteristics 509
chattering action 53
check valves 337, 434
chemical activated SPLs 322–323
chemical separation 610
  principle 607–610, 627, 641
chlorine 52, 224
choice of lubricant 41, 384, 534
chromatography 96–97, 637
Cincinnati Milacron test (ASTM D 2070-91) 637
circulating lubrication systems 135, 333
circulating oil lubrication 435, 437, 440
circulating oil system 20, 57, 161, 215, 234, 380–381, 493
circulation lubrication center 333
classification equivalents 149
classifications 24, 147, 187, 262, 718
classification systems 148
clay/silica gel analysis 97, 138
clean oil sampling 625
clearance 39, 40–41, 47, 65, 70, 105, 140, 317, 343, 389–390, 453, 492, 503, 506, 692
closed bottle test 92
closed cup 111
closed gears 207, 475
closed-loop oil mist system 370
close-fitting seals 504–505
cloud point 94–95, 134–135
CLS—certified lubrication specialist 728

CMFS—certified metalworking fluids specialist 728
coalescers 608–610
coalescing 606
filter 21, 33, 140, 173, 180–181, 199, 260, 280, 321, 374, 376, 380, 38–384, 402, 403–415, 417–418, 548, 562–563, 567, 572–574, 576, 582, 585, 590, 608–610, 615, 617, 620, 639, 641, 647, 651, 687, 691, 694, 697–698, 707, 715, 718, 730
coating costs 274
coaxial distribution delivery tubes 338
coefficient of friction 4–7, 44–47, 55, 223, 238
cohesive lubricants 66, 70–71, 267
coil springs 469
coke deposits 250, 253
cold startup 244, 254
cold-weather performance 249
collection container 366
color 95, 123, 143, 174, 192, 264, 267, 295, 299, 313, 379, 397, 399–400, 418, 420, 562, 568, 571–572, 574–578, 584, 588, 604, 632, 635, 637, 646, 648, 663
color-bodies, in oxidized oils 635
color code 577
colorimetric method 123
color scale comparison 95
color tests 95
combination friction 9, 10
combination thrust/radial tilt pad bearing 455
commercial oils 100
compatibility 196, 211, 214, 224, 225, 226, 230, 239, 389, 453, 456, 466, 468, 481, 488, 534, 549
complex greases 500
composition analysis 95
composition of bearing metals 13
compounded oils 211, 481, 565
compressors 161, 163–164, 224, 227, 231, 249–250, 257–260, 273, 359, 494, 549, 567, 598, 636
lubricants 4, 10, 19–24, 27, 29, 32–33, 50, 52, 55, 59, 60, 65–67, 69–71, 73, 79, 81–82, 85, 99, 102, 119–121, 138, 159, 160, 162–164, 167–168, 172, 175, 181–183, 186–188, 190, 192–200, 207, 210,–214, 219, 221–227, 230–232, 237–239, 241–243, 246–247, 249, 252–255, 257–258, 267–269, 272–276, 279, 313, 336, 368, 373, 379, 387–389, 391, 414–415, 425, 441, 458–459, 461–463, 466–467, 469, 471, 473–474, 480–482, 486–488, 490–491, 509, 516–517, 534–535, 539, 545–557, 559–569, 571–572, 575–579, 581, 583–586, 588, 597–598, 602, 608, 621, 629, 631–633, 638–639, 643, 652–653, 657, 661–663, 665, 667, 669–670, 674, 679, 686–691, 707, 717, 719, 728
operation 14, 22, 38, 46–47, 49, 54, 87, 89, 94, 104–105, 131, 134–135, 139, 150–151, 153, 154, 163, 168–170, 173, 177, 194, 199, 211, 213–214, 234–235, 237, 239, 250, 260, 262, 294, 332, 347–348, 360, 369, 389, 392–394, 396–397, 415–416, 425, 428, 436, 443, 452, 456, 460–461, 463, 471–472, 486–491, 493, 497,

501, 547, 557, 559, 564, 574–575, 584, 610, 627, 637–638, 651, 653, 665, 679, 685, 688
concentration in oil at different temperatures 644
condemnation limits 606
condensation settling 33
condition-based oil change 635
conductive pastes 267
congealing point 155, 156
of petroleum wax (ASTM D 938) 127, 155
connecting ducts 436
consistency 57, 98, 108, 114–115, 124, 128, 133, 261–262, 266, 427–428, 430, 431, 619, 626, 656, 694
classes 93, 107, 204, 379, 427, 627, 658
of greases 115, 127, 130, 132, 216, 294, 312, 427, 429
constant level oiler 320, 322
contacts 54, 442, 453, 466–468, 514, 638, 651
containers with open spouts 570
contaminant severity factor (CSF) 640–641
contamination 14–15, 17, 20–21, 31, 33, 57, 60, 71, 82, 89, 93, 95, 101, 103–104, 110, 113, 133, 136–137, 140, 142, 154–155, 178–181, 187, 198, 272, 275, 280, 283, 294, 300–301, 304–305, 332, 364, 370, 373–375, 380, 383, 396, 402, 408, 411–412, 414–418, 425, 440, 532–533, 545, 553, 559–560, 565–568, 570–572, 574–575, 577–578, 581, 584–585, 593, 597–601, 603, 605, 608, 610, 615, 617–618, 622–623, 625, 628–629, 631–632, 638–639, 641, 643–646, 657–658, 686–688, 693, 717, 730
effects 3, 8–10, 19, 21, 31, 52, 94, 103, 112, 147, 153, 192, 255, 266, 268, 301, 516–517, 569, 603, 606–607, 643–644, 648, 672
factor 104, 118, 129, 136–137, 144, 150, 156, 207, 250, 261, 373, 385, 391, 427, 443, 452–453, 476, 509, 554, 588, 627, 640–641, 657, 726
monitoring 80, 142, 314, 333, 356, 368, 371, 454, 557, 569, 603, 615–618, 623, 627, 631–634, 637–639, 648, 680, 689, 693–697, 704, 717, 728, 730–731
of oil 28, 38, 40, 41, 47, 49, 51, 54, 58, 80, 87–88, 100–101, 104, 108, 112, 117, 124, 127, 133–134, 140–141, 145, 167, 170, 210–212, 216, 261, 284, 317, 345, 360–361, 363–365, 369, 374–375, 387, 402, 411–412, 414, 435–436, 439–441, 452, 457, 480, 482–483, 492–493, 513, 515, 522, 527, 565–566, 571, 575, 588, 591, 600, 610, 615–616, 619–620, 623, 626–627, 629, 631, 645–646, 652, 658, 701
sources 6, 21, 23, 137, 159, 179, 180, 194, 265, 492, 566, 599, 644, 647
continuous lubrication 343, 432
controllers 330, 337, 392–393

conventional, "one-way" oil mist
system 22, 72, 103, 109, 124, 170–171, 177, 216, 221, 223, 235, 239, 241–250, 253–254, 361, 364, 366, 445, 448, 493, 509, 515
converging geometry 452
conveyor ropes 534
coolers 235, 492, 651
cooling of the oil 436
copper strip corrosion 98–99, 214, 491
test 91–111, 113–116, 118–133, 136–146, 150, 153–157, 163, 179, 183–184, 194
Corona 102–103
corrosion, of bronze/of copper 3, 16–17, 21, 60, 75–76, 82, 85, 98–99, 101, 119, 124, 132, 163–164, 172, 180, 183–184, 214, 217, 226, 233, 267–268, 270, 272, 275, 425–426, 429, 453, 456, 460, 462, 466, 468–469, 472–473, 491–493, 500, 533–534, 539–540, 583, 598, 604–606
corrosion products 605
corrosion test 99, 491
corrosive acids 21, 30, 82, 180
corrosive wear 15, 17
cost justification 610
approach 6, 60, 93, 179, 312, 330, 336, 375, 411, 457, 499, 502, 546, 564, 586, 592, 635, 640, 651–652, 657, 681–683, 686, 700, 702–703, 705, 709, 725
for synthetic lubricants 243
of oil mist 361, 364–365, 441, 513
cost savings through power loss reduction 517
coupling grease 215, 216, 493
couplings requiring grease lubrication 215, 494
crackle test 645
for water contamination 645
cradle-to-cradle lubricant management program 545
crankcase dilution 104–106, 133
critical loading conditions 440
cross-flow 503–504, 508
grease lubrication 59, 215, 261–262, 282, 285, 306, 312, 315, 329, 333, 336, 361, 364, 426, 442, 460, 494, 504, 524, 528
cross lubricant contamination 374
crude scale waxes 127
crystalline structure 134, 135
cylinder lubrication 259

D

DCF return 365
Dean & Stark apparatus 645
decomposition 32, 141, 635, 637–638
dedicated filter cart pumping system 572
deep groove ball bearings 440, 514
deformation of the contact zone 484
degradation 3, 13–14, 20, 33, 92, 137, 164, 168, 227, 253, 321, 583, 585, 618, 632–633, 643
degradation of lubricating oil 632
degrees of rusting 139
demisting system for the textile industry 363

demulsibility 100–101, 102, 160, 163, 167–168, 175, 214, 490, 492–493
characteristics 493
test 493
density 28, 102, 164, 216, 363, 367, 493, 619, 620–621, 631, 637, 647–649, 656
depletion of oxidation 634–635
depth fluid filter—oil 406
designing and preparing a lubricant storage area 559
detergency 89
detergent 21, 33, 82, 88–89, 133, 199, 233, 567
engine oils 133
dibasic acid esters 227
dielectric loss 136
dielectric strength 102–104, 118
diesel fuel 107, 199
diester base stock 252
diester-base synthetics 254, 257
diester blend 223
diester lubricants 249, 254
diesters 209–210, 223–224, 246–247, 257, 480–481
differential pressure 406
regulating valve 346
DIN 51 381 TUV impinger test 87
dip-feed lubrication 214, 490
directed lubrication 445–449
direct reading ferrograph 648
dirt and contamination 639
discounted rate of return 365
dispensing equipment 115, 556–557, 566–569
dispersants 82, 199, 233
dispersion of air bubbles 87
dissolved water 604, 609–610
distillate fuels 94, 135
distillation 72, 104–108, 144, 146, 154, 180, 611–612
temperature 146
distilled vapors 610
distressed oils 617
distribution header system design 368
distribution of hydrodynamic pressure 40
DN value 336, 443–444, 505
domain of knowledge 725, 732
dosing modules 329, 332
double-acting compressor cylinders 256
double-helical, low-speed gear 477
double-shielded bearings 502–503, 506
double thrust bearing 445–446, 448
double V-rings 516
drain leg 363, 369
drain lines 366, 620
drain plug 405, 498, 503, 505, 507–508
drive chains 537
drop melting point of petroleum wax (ASTM D 127) 155
dropping point 85, 108, 265, 428
drop-tube vacuum samplers 622
drum covers 565
drums, for oil storage 330, 379, 545, 560, 562, 565, 578, 584, 588–589, 662
dry friction 8, 19–20
drying oils 566
dry lubricants 71, 273–276, 458–459, 461
dry plant or instrument air 512
dry-running bearings 458–459
dry sliding bearings 456
dry sump 364, 512

oil mist 361–364, 512
dual-header lubrication system 332
dual-line systems 332
dual supply line parallel system 349
DVGW approval 464
dynamic compressors 166
dynamic 93, 128, 130, 132, 139, 166–167, 319, 392, 456, 531, 533, 541

E

ecotoxicity 91–92, 108, 194
effective area 43
effect of friction 7
effect of load on fluid friction 47
effect of speed 43
effect of viscosity 38
efficiency factors 47
efficiency of mineral hydrocarbon oil compared to 487
elastic deformation 15, 54
elastohydrodynamic (EHD) lubrication 25, 53, 77, 484
elastomer compatibility 230
electrical contacts 466
electrical switches 466–467
electric motor rolling element bearings 513
electric motors 169, 235, 249, 330, 361, 434, 499, 501, 504–505, 512–513, 515–517
electro-chemical activated SPLs 323
electrode filter spectroscopy 647
electro-mechanical activated SPLs 325
electrostatic coating 275
elemental analysis 646, 649
elemental spectroscopy 635, 638, 646–648
elevated temperatures 227, 427, 491, 604
embeddability 453
emulsibility 52, 75, 100–101
emulsification 100–101
emulsified water 399–400, 411, 604–605, 607, 610, 643, 645
emulsion 82, 100–101, 183, 219, 567, 604
end points 123
energy consumption 22, 193, 223, 232, 280, 336, 339
energy savings 60, 246, 254–255, 257
potential 257
engineered distribution devices 312
entrained air 31, 87, 180, 234, 636
entrained gases 607
environmental claims 194–195, 197
environmentally friendly 182, 192, 194, 197–198, 366, 662
lubricants 192, 194, 197–198
environmental toxicity 92–94
EP additives 33, 51–52, 99, 191, 210, 428–430, 438, 440, 491
EP additives in oils 438
EPA shake flask test 92
EP gear oil 214, 243–246, 490
equilibrium temperature 433
equivalent viscosities 210, 481
evaporation 33, 90, 105, 107–108, 110, 137, 144, 146, 257, 637
loss 110, 144
excessive inherent viscosity 48
excess of lubricant 426
exit hole for the grease 433
extreme-pressure additive 192
extreme-pressure conditions 51
extreme-pressure protection 191, 246
extreme-temperature additive 52

F
face seals 516
failure to trip 605
fatigue wear 15
fatty oils 66–67
fault detection 615, 658
FDA 142–143, 175, 187–190
FDA Method 142
feed rate 235
feedstock 72, 73
ferritic 405
ferromagnetic debris 649
ferrous density analysis 631, 647–649
ferrous particle concentration 647
ferrous particle counter 631, 648
FIFO 181, 564, 582, 584–585, 589
fill and drain plugs 505
filled coupling 215, 493
film strength 59, 81, 146, 180–181, 223–224, 227, 252, 258
  additives 50
film thickness 14, 20, 35, 43, 167, 212, 245, 255, 274, 479, 480, 483, 598
filterability 163
filter/dryers 610
filtergram 649
filters 33, 95, 169, 181–182, 190, 333, 334, 402, 406–408, 411, 415, 417–418, 574, 611, 620, 629, 638, 641, 644, 656, 662, 689–690, 706, 715
filtration 33, 82, 133, 154, 254, 406, 407, 410, 411, 413–414, 517, 559, 563, 590, 618, 639, 641, 658, 687, 697
fire point 109–110, 168
flash and fire points, open cup 109–111, 168
flash point 28, 90–91, 107, 109, 111–112, 169, 183, 226, 232, 248, 578, 629
  closed cup 111
flash temperature theory 485
flinger discs 524–526
flinger ring 435
flooded lubrication 445–447
fluid contamination 178–179, 618, 629, 717
  analysis 629
fluid environment severity 627
fluid film 6, 9, 19, 36, 38, 43, 46–47, 50, 53–54, 65
fluid friction 5, 13, 20, 31, 41–43, 46–47, 70, 145, 280, 282, 336, 456, 498
fluid properties analysis 629
fluid wear debris 629, 631
fluid wear debris analysis 631
fluid wedge 40
fluorescent indicator analysis 97
flush rule 624
flush-through arrangements 502
foaming 30, 87–88, 112–113, 257, 400, 567
  characteristics 112
foam resistance 112, 183
foam suppressants 234
food additives 142–143, 360
Food and Drug Administration (FDA) 142, 187
food grade lubricants 187
forced feed lubrication 484
four-ball EP test 120–121
four-ball wear test (ASTM D 2266) 5' 85, 113, 120
frame 93, 332, 386, 497, 680, 701, 705
  gas turbine 448

free water 104, 603, 604–611, 644
in lubricating oil 644
frequency 66, 93, 98, 108, 285, 302, 311, 314, 359, 406, 440, 453, 454, 500, 509, 510, 574, 626, 627, 635, 694
frequency of re-lubrication 108
fretting corrosion 267, 473
friction 3,–10, 12–15, 19–20, 22, 31, 35, 38–39, 41–47, 51–55, 65, 70, 75–77, 82, 99, 119, 131, 145, 173, 194, 199, 204, 212, 219, 223, 225, 234, 238, 254–255, 257, 267–268, 271–272, 275, 280, 282, 284, 298–300, 302, 311, 329, 332, 336, 364, 389, 393, 443, 456–458, 460–461, 466, 469, 472, 474, 486–487, 490, 497–498, 516–517, 532–534, 538, 598, 606, 693–694
frictional drag 45, 53
frictional heat 7–8, 11, 29, 32, 47, 178, 389, 505, 605
frictional power losses of industrial equipment 517
friction-locking conditions 474
full-fluid film 36, 38, 43, 47, 50, 53, 54
furnace air preheaters 234
FZG Test 119, 210, 481

G

galvanized containers 567
gas 5, 19, 22, 65–66, 73, 96–97, 163, 224, 226, 231, 256, 258, 322–324, 389, 448, 464, 598, 705
chromatography 97
gaseous lubricants 65
gases, compatibility with synthetics 65, 143, 199, 227, 231, 389, 416, 463, 550, 598, 607, 610, 667, 705
gas liquid chromatography 97
gas turbine 448
gearboxes 217, 626, 689
gear coupling lubrication 214–215, 493
geared blowers 494
gear lubricants 207, 212–213, 487
gear lubrication 167, 224, 475, 484
gear motor, double-stage spur gear 489
gear oils 101, 129, 147, 214, 224–225, 237, 243–244, 247–248, 250, 490–492
gears 10, 21, 41, 50, 53, 75, 119, 167–168, 191, 207–208, 211–214, 225, 236, 246, 276, 284, 329, 435, 457, 475–476, 478, 483, 486–488, 490, 492, 495, 521, 619, 651, 656, 680
gear unit 208, 212, 214, 238, 457, 476–477, 483
gearwheel 486
general purpose R&O oils 159
Girth gear drive 486
G level 216, 493
gold-plated plug contact 469
good lubrication practices (GLP) 49, 725
grades 24, 27, 79, 115, 149, 159, 161, 164, 165, 167, 168, 172, 175, 200, 204, 225, 226, 243, 253, 255, 632
graphite 55, 71, 267, 275
gravity 28, 31, 102, 112–114, 134–135, 138, 145, 265, 296–297, 317–318, 321, 325, 380, 403–405, 521, 541,

603–604, 606–607, 609, 629, 633, 656
feed oil 135
settling 603
grease lubrication 59, 215, 261–262, 282, 285, 306, 312, 315, 329, 333, 336, 361, 364, 426, 442, 460, 494, 504, 524, 528
greases 70, 77–78, 82, 115–116, 127–128, 130, 132, 141, 153–154, 156, 190, 214–217, 261–263, 266, 275, 294, 296, 298, 312–313, 425, 427–431, 434, 438, 458, 490, 493, 499–500, 505, 509, 548, 569, 575, 588, 683, 726
based on synthetic oils 428
for rolling bearings 427
separation 332
service life 431
Group I Base Oils 221
Group II Base Oil 73
Group III Base Oil 73
Group IV Base Oils 73, 222
Group V Base Oils 73, 222

H

handling 103, 119, 155, 181, 546, 553, 557, 559–562, 564, 566, 577–578, 623, 653, 661–664, 673, 679, 686, 688
handling lubricants 661
hard particles 641
health and safety 336, 339, 553, 569, 572, 706
heat loss 211, 482
heat transfer fluids 167–169
heat-transfer oils 110, 129
heavily loaded bearings 429
heptane 125, 126
hermetic sealing 320, 321, 569
Hertzian contact area 25
high flank load 486
high shock loads 483, 538
high speed coupling grease 216
high-speed operation 436
high-temperature screw compounds 268
high viscosity index 110, 152, 163, 194, 222, 244–245, 439, 492–493, 540
high-viscosity oil 47
hydraulic applications 165, 175
hydraulic fluids 87, 93, 99, 101, 129, 173–174, 178–180, 182–183, 195–196, 521, 572, 616, 634, 637–639, 653
hydraulic jacking provisions 449
hydraulic oils 51, 77, 135, 174–176, 225–226
hydraulic power 165, 180
hydraulic pumps 177
hydraulic systems 76, 87, 135, 161, 165, 174–175, 177–179, 182, 190, 231, 249, 376, 396, 626, 658, 686, 698
hydraulic thrust metering systems 448
hydrochloric acid (HCl) 123, 141
hydrodynamic bearing clearances 459
hydrodynamic film 167, 445, 452
hydrodynamic lubrication 15, 35–37, 40–41, 49–50, 172, 484
hydrodynamic oil film bearing 452
hydrodynamic sliding bearings 455–457
hydrometer 102, 114
hydrophilic materials 117–118
hydrostatic sliding bearings 455
hydroxyl ions 122

I
impurities 23, 154, 590, 641, 671
indoor storage 566, 584
industrial pumps 164
industrial revolution 23, 358
infant oils 628
infrared spectroscopy 634, 645
inherent biodegradability 92
initial lubrication 170, 505
injector pump systems 360
inner reservoir space 506
insoluble resins 133, 134
instability 453–454
insufficient reservoir fluid level 87
insulating oils 99, 101, 103, 124, 136
integral sight glasses 320
interfacial tension 104, 116
internal combustion engine 88, 104
internal reservoir design 368
interpreting test results 650
ionic species 606
ionization 102, 122
ISO 18436-4 Lubrication Analyst ISO Category III—LCAT III 732
ISO 55001—Asset Management Standard 679
ISO cleanliness level 413, 579
ISO contaminant code (ISO 4406) 640
iso-octane 125, 126
isoparaffins 91

J
jet of oil 436
Jost Report 3, 597
journal bearing design requirements 41
journal bearings 39, 452–453, 497
justification for use of oil mist 365

K
Karl Fischer titration 645
kinematic viscometer 633
kinematic viscosity 26, 152, 439–440, 632–633
Kluberplast sheet
knock ratings 126

L
L10 grease 431
labyrinth seal 21, 389–390, 503
Lambda 35–36
laminar flow 41–42
large motors 507
lead-based additives 429
lead-based compounds 429
leading edge bearings 450
leakage characteristics 157
leakage tendencies 156
lemon bore 457
leveling links 445
level switch 369, 395
liability for the lubricant 425
life-time lubricated, “sealed” bearings 504
lifetime lubrication 70, 213, 272, 458, 488
lifting chains 537
line losses 136
lip seal 389–391
liquid lubricants 66, 70, 441
liquid vapor pressure 143
lithium base greases 430–431
live zone 620
load-carrying ability 50, 53, 119, 214, 490
load cells 448–449
load wear index 121
lobes 457
long-chain paraffin molecules 242

low consistency 431
lower pour points 28, 254
lower temperature limit 428
low pour points 160, 175
  advantage 254
lubed-for-life bearings 505
lube oil contamination 598
lube oil system 212, 483, 605–606
lubricant certificate of analysis 579
lubricant consolidation 60, 239, 280, 312, 412, 499, 548, 550–551, 557, 576, 653, 661, 686–688, 690
  program 312
lubricant contamination 374, 570
lubricant disposal costs 652
lubricant film regimes 35
lubricant of the polar type 50
lubricant purchase policy/program 546
lubricant selection 57, 164, 207, 209, 475–476, 479, 499, 597, 725, 730
  for closed gears 475
lubricants in gear 211, 482
lubricant supply 43
lubricant transfer policy 571
lubricant types, for gears 70, 75, 80, 209, 223, 480, 548, 557, 710
lubricated-for-life anti-friction bearings 131
lubricated-for-life bearings 169
lubricating and assembly paste 267
lubricating grease, advantage 85, 108, 114–115, 127, 131, 192, 262, 427, 458
lubricating grease, disadvantage 85, 108, 114–115, 127, 131, 192, 262, 427, 458
lubricating greases 70, 77–78, 141, 153, 214, 261, 427, 458, 490
lubricating oils 25, 30, 52, 68–69, 71, 75–76, 78, 109–110, 112, 128, 134–135, 140, 145, 147, 150, 199, 275, 548, 604, 639, 643
lubricating pastes 70, 267, 270, 461
lubricating waxes 70, 78, 268–269
lubrication certification programs 725
lubrication delivery system 59, 282, 284, 371, 379, 698, 719
lubrication of chains 537
lubrication of large open gears 486
lubrication of screws 461
lubrication of small geared blowers 494
lubrication of small gears 488
lubrication of worm gears 487
lubrication program 60, 198, 280, 312, 411, 533, 534, 548, 574, 679, 682–683, 685, 693, 698, 717, 719–720, 730
lubrication system 41, 60, 160, 163–164, 199, 235, 259, 262, 280, 289–290, 315, 322, 325, 327, 329, 332, 334–335, 337–341, 343–344, 347, 358, 363–364, 366, 371, 373, 375–376, 383–384, 386–387, 392, 393, 401–405, 417–418, 458, 485, 516, 567, 597, 600, 656, 665, 680, 690–693, 697, 703
Lubrication Systems Company 367, 369–370, 495

M

machine elements 11, 269, 388, 455, 472, 475

Machinery Lubrication Technician Level II—MLT II 730
Machinery Lubrication Technician Level I—MLT 730
Macpherson contamination curve 600–601
maintenance costs 330, 516
manifold 318, 338, 370–371
manifold block 338, 370
manometric respirometry test 92
manual drill, double-stage gear 213
marine environment 493
mass spectrometry 96, 97
mass "spectroscope" 97
maximum knock 125
McCoy's patented lubricator 317
mean Hertz load 121
mean sliding speeds 446, 448
measuring filter efficiency 409
mechanically stable greases 427
mechanical powered pump units 371
melting point (plateau) of petroleum wax (ASTM D 8) 12, 108, 155–156, 268, 500
mercury 22, 116, 244
metal-containing pastes 267
metallic derivatives 88
metallic soap thickened greases 427
metal-to-metal contact 10, 20, 39–40, 46, 49, 53, 119, 150, 163, 252, 425, 453, 483–484, 493
metering plate facing grease reservoir 503
methods employed to remove water 607
methods of supplying lubricant 484
methyl orange indicator 123
microcrystalline wax 127
micro-polish 223
mild EP additives 51
mineral oils, for gears 66–67, 69–70, 82, 98, 112, 141–142, 152, 173, 189, 195, 209–211, 222–223, 232, 241–242, 252–255, 437, 480–481, 517, 635
mineral powders 71
miniaturized, closed-loop lubricator 363
minimum kinematic viscosity 440
minimum load 119, 443, 515
miscibility 269, 430, 434
of greases 430
mist generator 366–369
mist manifolds 362, 366, 370
mixed film (mf) 13, 35, 43–44, 46, 81
mixed friction 268, 456–457, 486
regime 457
mixers 188
mixtures of compatible greases 430
modified silicone system ("Radix") 515
modified Sturm test 92
modified Uniontown procedure 126
moisture contamination 15, 17, 585, 643
molybdenum disulfide 264
monitoring oil contamination 603, 638
MoS2 71, 267, 275, 481
motor octane number 124–126
motor oils 49, 101, 104–105, 110, 200
movable contacts 466
MSDS 653, 670–671
multi-channel controllers 330
multi-grade viscosity oils 26, 200
multiple boundary lubrication 52

multiple row bearing or paired
bearing arrangements 514
multiple spline shaft 473
multipoint systems 358

N
naphthalenes 91, 138
naphthenic oils 28, 128, 153
National Lubricating Grease
Institute 115, 262, 427
scale 427
natural base oils 195
naval gearbox 446
neutralization number 104, 118,
121–122, 129
nipple 287, 290, 291, 292, 294,
295, 299, 300, 302, 304, 305,
306, 317, 326, 433, 577
NLGI grades 115
NLGI grease grading system 115
non-magnetic debris 649
non-tacky wax film 268
non-toxic 192, 194–195
NUTO®FG 175
NUTO® H 175–176

O
octane number 124–126
offset pivots 448, 450
oil analysis 31, 80, 179, 259,
311, 399–400, 413, 555, 603,
615–619, 623–624, 626–627,
629–633, 637, 639–640,
645–647, 651–654, 657–658,
686, 689–690, 726, 728–729
data interpretation 654
software 631
tests, selection of 629
oil bath lubrication 435, 440, 541
oil bath/oil spray 436
oil carryover 258, 454
oil change frequency 406
oil change interval 631
oil change procedures 258
oil consumption 105, 193, 200, 222,
364, 631
oil cooler 20
oil cup 287–288
oil degradation 632–633
oil drain intervals 257
oil drums 588
oil feed rate 235
oil flow meter groups 335
oil gallery 484
oiliness 50–53, 81, 234
oiliness additives 51, 234
oiliness agents 50–53
oil jet lubrication 441
oil jet method 436
oil level 105, 339, 416, 435–436,
523–525, 527
oil lubrication 58–59, 287,
335–341, 402, 426–427, 435,
437–438, 440–441, 522, 690
oil mist 235–236, 250, 259,
361–370, 441–442, 512–517,
524, 526–527, 541, 667
distribution piping 366
for electric motors 512
generator 366
lubricants 526
lubricated bearings 364
manifold 362
system 235, 365
oil odor 637
oil oxidation 117, 138, 140, 190,
629, 634–635
oil purification 610
oil purifier 610
oil reservoirs 32, 384, 396

oil ring 436, 524–525
Oil-Rite Corporation 320–322, 361–362
oil sampling frequency 627
oil sampling intervals 628
oils containing additives 437
oil separation in grease storage 127
oil separation test 216
oil supply tank 366
oil velocity 446
oil versus grease 57
oil wiper 484
OMA II—Oil Monitoring Analyst Level II 728
OMA I—Oil Monitoring Analyst Level I 728
one-time lubrication 214, 490
on-site moisture analyzers 631
on-site oil analysis 631, 633, 639
on-stream purification 612
open bearings 503, 507
open, centralized oil lubrication systems 109–110, 168–169, 180, 213, 244, 298–299, 304–306, 308, 321, 329–330, 332, 335–336, 346, 363–364, 380, 387, 396, 398, 415–417, 460, 462, 486, 488, 497–498, 502–507, 509, 532–533, 535, 541, 555, 562, 567, 569–570, 586, 592–593, 661
open fill-ports 570
operating environments 196, 330
operating temperature ranges for greases 429
operating temperatures 28, 48–49, 110, 163, 224, 248, 431, 435, 437, 440, 504, 637
operating viscosity 47–48, 178
organic esters 227
origin of synthetic lubes 224
O-ring 276
outdoor storage 563–565, 592
overall bearing friction 45
overflow space 505
over-greasing 330
over lubricated 281, 498, 691
oxidation 30–33, 73, 75, 81, 101, 103–105, 108, 117–119, 124, 128–135, 137–141, 147, 159–160, 163–164, 168, 175, 190, 209, 212, 214, 217, 222–223, 225, 232–233, 235, 242–243, 248, 249, 253–254, 257–258, 332, 387, 428, 460, 468, 473, 480, 483, 490–492, 494, 500, 503, 523, 567, 583, 628–629, 631, 633–635, 637, 643
oxidative degradation 164
oxygen absorption rate 132

P

pad materials 448
PAO-based synthetic EP oils 244
PAO-based synthetic lubes 517
PAO/diester blends 223
paper chromatography (blotter spot test) 637
paper machine drying sections 333
paper machines 332
paper mill 244, 334
Paraffinic base stocks 134
paraffinic oils 28, 128, 152, 153
partial film lubrication 43
partial lubrication 12–13, 43, 47, 70–71, 78
particle contamination 625, 638–639, 643–644
particle count 629, 640–641, 643, 648

particle counter 631, 639, 642, 648–649
particle counting 629, 631, 639, 648
particle counts 625, 631, 640
particle count trend graphs 643
particle count trends 641
particle quantifier 648
particle scrubbing 33
particle-size error 647
Parvan 271
PAS 55 (publically available specification #55) 679
PDI system 345–346, 348
penetration number 114
penetrometer 114
Pensky-Martens 111
Pensky-Martens Closed Tester 111
pentane insolubles 133
performance characteristics 75, 100, 107, 119, 137, 190
performance numbers 126
peroxides 138
petroleum-based EP gear oil 243–244
petroleum base oil 23
petroleum oils 24, 98, 99, 122, 136, 195, 224, 226, 249, 254, 261, 492
petroleum solvent 111
petroleum wax 127, 155
phosphate esters 227, 635
phosphorus 52, 210, 224, 480, 490
pH reading 123–124
pH scale 122
pH value 122–123
pinion 208, 475–476, 479–480, 484, 486
pitch line speed 212, 476, 483
pitch line velocity 208, 477, 484
pitting 15–17, 51, 191, 483, 485
pivot position 448
plain bearings 9, 273, 284, 455, 501
plug contact 469
p-naphtholbenzene 123
pneumatic components 471
pneumatic cylinder 471
pneumatic powered pumps 373
pneumatic valves 471–472
polar characteristics 51–52
polar compounds 72–73, 97–98, 643
polar contaminants 638
polar molecule 51
polar rust inhibitor 140
polyalphaolefin 223, 226, 242–243, 250
  diester blend 223
polyglycols 225, 227
polymeric materials 448
polymerization of the base oil 635
polynuclear aromatics content 142
polyurea 428, 500, 505
  grease 428
pore blockage-type particle counter 642
porosity of sintered metal sliding bearings 460
portable mist density monitor 363
positive displacement blowers 494–496
positive displacement injector 337, 344–345, 347, 385
potentiometric method 122
pour depressant 135
pour point 23, 28, 69, 72, 82, 110, 134–135, 221, 224–226, 243–245, 254
  advantage 254
  depressants 243
powder metallurgy (P/M) bearings 169–171

power factor 104, 118, 136, 137
power loss 212, 445–446, 487, 517
power transmissions 475
practical dispensing equipment 569
precipitation analysis 96, 98
precipitation number 491
premium EP gear oil 243
premium-quality, multi-purpose greases 175
pressure balanced 320
  constant level oiler 319
  device 320
  lubricator 320
pressure build-up in the oil film 452
pressure differential 407
pressure sensing switch 329
pressure switch 329–330, 345, 348, 394, 396
primary biodegradability 91
prime operating influences 509
proactive maintenance 615, 617–619, 627, 629, 639, 644, 646, 658–659
process-dependent dilution 598–599
product identification labeling 554
progressive divider style distribution block 312, 322
prolonged worked penetration 115
protective equipment 663, 672–673
protective film 40, 233, 268
protector seal 525
proximity probes 448
PTFE 55, 71, 267, 272, 275, 340, 391
PTFE based tribo-system lubricants 272
pulverizing mills 236
pumpability 192, 265, 434, 534
pumping action 81, 391, 393, 426, 442, 505
pumping stations 329, 338
pumps 31, 58, 161, 164, 169, 175, 177, 178, 199, 224, 231, 235–236, 249, 258–260, 273, 318, 330, 333–334, 343, 364, 371, 373, 380–381, 393, 403, 513, 521–522, 528, 548, 577, 585, 598, 605, 625, 656
pumps—mechanical and pneumatic activated types 371
pungent odors 635
purchase specifications 549
pure mist 364
pure oil mist 512, 516
purge mist 364

R

Radicon Worm Gear Test 492
RBOT test 139
RCFA (root cause failure analysis) 549
reciprocating air compressors 250
reciprocating compressors 250, 257, 259, 636
reciprocating pumps 164
re-classifiers 235, 236, 366
reconditioning service 507
reconditioning service, for bearings 507
recycled oil 549
reduced maintenance 252, 255
reduced valve overhauls 250
reduction in energy cost 238
reduction of operating temperature 631
reductions in bearing life 430
Redwood and Engler methods 145
reference properties 632

refractive index 97, 137–138
re-greasing electric motor
 bearings 505
regulating valve 346
Reid vapor pressure 144
related damage 605
relative motion 65, 233, 452
release agents 269, 270
re-lubrication 108, 425, 509, 535
 frequency 509
 intervals 431–433
 procedures 433
remote-controlled shut-off
 valve 329, 335
replenishment grease 433
research octane number 124–126
reserve EP capability test 214, 490
resistance to oxidation 129–130,
 164, 168
retentive properties of grease 261
road octane number 124–126
rolling bearings 75, 275, 425–428,
 431
 contact 53–54
rolling friction 9–10, 15, 538
rotary bomb oxidation
 test(RBOT) 138–139
rotary compressors 258, 260
rotary lobe blowers 363
rotary lobe machines 495
rotating bomb oxidation test 629
rotating speed 514
rotation 38, 170, 181, 427, 452,
 503, 526–527, 582, 583–586, 589
rough surface finish 485
RPVOT (rotary pressure vessel
 oxidation test) 31
rubbing contact 33
rust inhibiting properties 427–428
 properties of a grease 428
rust inhibitors 192, 233, 428
rust particles 140
rust preventatives 113
rust-preventive characteristics 139
rust protection 160, 175, 190–191
rust test, ASTM D 665 140, 163

S

SAE (Society of Automotive
 Engineers) 27, 147, 161, 200
Safematic 329, 334–336
sample data density 619
sample data disturbance 619, 623
sample trap 620
sampling oil, different options 728
sampling point 619, 621, 656
sampling valves 622
sap number 140–141
saponification 140, 261
 number 140
saturation point 180, 607, 643
scoring 51, 99, 119, 121, 140, 485,
 491, 640
screw compound 268
screw compressors 259
screw connection 461
screw lubricants 462
screw pastes 268
scuffing (adhesive wear) 14–15,
 191, 484–485
sealed bearings 501, 505
sealing and drainage issues 515
seals 21, 32, 59, 179, 182, 208,
 211–212, 217, 273, 276, 280,
 294–295, 319–320, 332–333,
 388–391, 416, 425, 442, 471,
 478, 481, 483, 497, 504–505,
 516, 527, 565, 581–582, 598,
 651, 719
sediment 154–155, 600

selecting oil analysis tests 630
selection criteria 269–270, 273, 338, 541
selection of an oil 47, 438
self-lubricating bearings 169, 171
sensory indications 651
sensory inspection information 651
separation 29, 33, 35, 52, 85, 97, 114, 127–128, 160, 215–217, 224, 332, 493, 607–611, 693
series progressive divider systems 350
service level agreement (SLA) 550, 581, 714–715
service temperature range 275
severe ambient environment 247
severe EP application 99
Sharples high speed centrifuge 216
shear stability 217, 266, 494
shear stress 42
sheet bearing 459
shielded, grease-lubricated ball bearing 153, 501–503, 505–507
shield facing the grease cavity 502
shipping container 560, 562
shut-off valve 320, 329, 335
signal devices 392, 394
silica gel analysis 97, 138
silicones 230
silver strip corrosion test 99
single line resistance (SLR) systems 328, 337, 341–342, 344–345, 373, 385, 396
single-shield bearings 502
single weight oil 27
sintered bearings 456, 460
sintered metal 170, 460
SKF 428, 431–432, 644
sliding friction 6–7, 9–10, 14–15, 46, 275
sliding load 8, 10, 37, 215, 493
slow-speed bearings 13
sludge 30–31, 81, 104, 128, 138, 160, 163, 168, 171, 190, 199, 222, 225, 227, 398, 491, 598, 605, 623, 634–635, 641, 656
slumpability 265
small gear lubricants 213
small gears 213–214, 488, 490
snap acting valve for draining 370
soft sludge deposits 491
solenoid dump valve 605
solid friction 5–7, 9, 41, 43, 46, 53
solid kinetic friction 53
solid lubricants 70–71, 269, 274–275
solids contamination 179, 380, 600–601, 629
solid static friction 53
solubility limit 604
solvent-diluted oil 649
solvent extraction 69, 72, 154
solvents 88, 99, 106, 107, 111–112, 133, 137–138, 144, 593, 610, 611, 666–667
SOPs (standard operating procedures) 679
sources of metals in oil 647
Spartan Synthetic EP 244, 250
specific gravity 28, 31, 112–114, 138, 296–297, 629, 633
speed factor (n-dm) 427
speed limits for conventional greases 509
spherical roller 426, 434, 436, 440–443, 509
  thrust bearings 426, 440–443
spill control 562, 564
splash lubrication 484
splash system 484

split inner ring bearings 442
spread in volatility characteristics 106
springs 38, 276, 443, 468–470
sputter test 645
stainless steel tubing 366, 370
standardized reference temperature 439
static sampling 622, 624
static sumps 523
steam cylinder oils 101
steam turbine oils 87, 101, 128, 140
steam turbines 163, 224, 259, 389
steel industry rolling mills 333
steel mills 110, 333
steel plants 184, 333
stick-slip lubrication 52
STLE Society of Tribologists and Lubrication Engineers 727, 728
stock rotation procedure 582
stop-and-start operation 49
storage temperatures 566
straight mineral oils 98, 112, 152, 210, 437, 481
stray mist 516
strength-enhancing additives 272
stripping units 644
sulfated ash 89
sulfur 20, 23, 52, 98–99, 221, 224, 491
surface fatigue 485
surface tension 81, 82, 116, 234
surfactants 234
SUS (Saybolt Universal Seconds) 25–26, 445, 633
switches 324, 337, 349, 367, 393, 395–396, 466, 467
synergistic additive systems 223
Synerlec 223
Synesstic 252–253, 255–256
Synesstic synthetic lubricants 255
synthesized hydrocarbons 209–211, 246–247, 480–481
synthetic additive technology 223
synthetic base oil 24, 184, 221–223, 260, 490
synthetic hydrocarbon fluids 226
synthetic lubricant 22–23, 73, 221–223, 230, 232, 236, 238–239, 249, 251, 487
synthetic lubricants 24, 70, 73, 193–194, 210, 212–213, 221–224, 230–231, 238–239, 241–243, 252, 254–255, 257–258, 481, 487, 516
synthetic oils 66, 69, 101, 212, 219, 260, 427–428, 437, 483, 487–488, 517
systems leaks 87

T

tackiness agents 234
tag closed tester 111
target cleanliness 625, 639–640
task 213, 279, 295, 374, 460, 472, 488, 510, 547, 551, 571, 679–680, 688, 710
TEFC motors 516
TEFC vs. WPII construction 515
temperature guidelines for premium grade R&O oils 170
temperature measurement 454
temperature peaking 433
temperature range 27–28, 49, 80, 82, 108, 211, 223, 227, 265–266, 275, 391, 425, 427–429, 482
  for greases 429
temperature rise 211, 339, 482, 497, 500, 506
temperature sensors 420, 448

Teresstic 162
test for crankcase dilution 105
tetraethyl lead 125–126
the contamination effect 597
thermal failure 30–31, 636–637
thermal stability 160, 174, 222, 226, 253, 466, 468, 636–637
thermocouples 454
thickener 77, 108, 192, 193, 216, 217, 261, 266, 427, 428, 430, 493, 500
thickness of the film 40, 452
thin-film lubrication 174, 177, 190
through-flow oil mist 515
throw-off system 484
thrust bearings 37–39, 426, 432–433, 440–443, 445, 448–449
  selection 450
thrust measurement 454
tilting pad radial bearings 452
tilting pad thrust bearings 445, 449
Timken machine 119
Timken OK load 85, 119, 210, 243, 246, 481
tin base babbitt 453
titration 122–123, 645
  end points 123
toluene insolubles 133
toothed gear systems 486
tooth flanks 214, 486, 490
tooth-type coupling 216, 495
total acid number 123, 629, 635
toxicity 73, 92–94, 133, 164, 195, 227, 581, 661, 671
training and education 658
transfer/filtration equipment 563
transformer oil 135, 591, 604
trending, depletion of oxidation inhibitors 615, 618, 635, 652, 656–657, 694
tribo-corrosion 468, 472–473
tribology 3, 4, 10, 79, 597, 729
tribo-system coatings 71, 273–274, 457, 461, 469
tribo-system materials 70–71, 271–273, 457–458
tribotechnical data 78
tricresyl phosphate 51
trip cylinder 605
trip piston 605–606
tumble processing 274
turbine oils 87, 101, 124, 128, 138, 139, 140, 224, 634
turbines 13, 60, 163, 224, 226–227, 259, 336, 389
TUV impinger test 87
twin filters 333
types of lube oil contamination 598
typical inspections 85, 162, 175–176, 192, 244–245, 251, 269, 271

U

ultimate biodegradability 91–92
ultraviolet (UV) absorption Analysis 96–98, 142–143
under lubricated 282, 691
undissolved suspended water 643
undisturbed penetration 115–116
uniform thermal failure 637
uninhibited oils 101
Uniontown Procedure 126
United States Department of Agriculture (USDA) 187
unworked penetration 115
upper temperature limit 428

USDA 175, 187, 190, 192, 194
USDA H-1 175
USP/NF TESTS 141
USS 491
U-tube kinematic 633
UV absorbance 142–143

V

vacuum applications 272, 275
vacuum dehydrators 644
vacuum distillation tower 72
vacuum oil purification 610
vacuum oil purifier 610
vacuum pump 610, 621, 625
vacuum sampler 622–624
valve components 463, 465
valves and fittings 462
vane pump 174, 177, 382
vapor lock 107, 144, 360
vapor pressure 106, 143–145, 168, 227
vapor-to-air ratio 90
varnish tendency 636–637
vertical shafts 428, 431, 442
vibration amplitude 457
virgin oil sample 31, 619
viscometer 145, 146, 633–634
viscosity 13–14, 22, 23, 25–28, 31, 35, 37–38, 41–51, 53–54, 59–60, 65, 69, 71–72, 75, 79–82, 85, 94, 105–106, 110–111, 128, 134–135, 138, 145–152, 155–156, 160–161, 163–164, 167, 169, 172–173, 175, 177–179, 181, 183, 192, 194, 198–200, 204, 207–208, 211–212, 216, 218, 221–222, 226–227, 230, 232, 234, 236–238, 241–245, 248, 250, 252–253, 255, 257, 261–262, 326, 335, 368, 385, 407, 414, 427–428, 438–440, 443–445, 450, 454, 456–457, 476, 478–480, 482–485, 491–493, 500, 517, 525, 540, 548–549, 575, 578–579, 581, 583–584, 588, 597–598, 604, 621, 629, 632–635, 637, 639, 656, 666, 693
  changes 633
  classification equivalents 149
  classifications 147
  classification systems 148
  grades 200, 204
  improvers in gear drives 211
  ratio 440
  requirement 439
  selection chart 443
  stability 23, 68, 73, 75, 81, 101, 105, 108, 116, 128–130, 132, 135, 138–139, 157, 160, 168, 171, 174–175, 183, 190, 195, 217, 222–223, 226–227, 230, 232–233, 237–238, 242, 247, 253, 257, 266, 453, 457, 460, 466, 468, 490–491, 494, 523, 540, 631–632, 634, 636–637, 643, 674
  temperature relationship 234
viscous fluid 452
visual inspection 132
volatility 28, 73, 102, 104–107, 109–110, 143–144, 146, 164, 168, 222, 230, 232, 235, 242–244, 248–249, 492
voltametry 635
V-rings 516

W

warm, high precipitation climate 26, 107, 151, 253, 507, 554
waste oil management 590, 674
water 21, 25–26, 28, 30, 32–33, 52, 81–82, 85, 91–94, 100–101, 103–105, 112–114, 116–117, 122, 128–129, 137–140, 142–143, 153–155, 160, 163–165, 173, 179–181, 183–184, 187, 191–192, 197, 210, 219, 224–227, 233, 261, 266–269, 273, 283, 296, 321, 363, 366, 368, 380, 388, 399–401, 411, 426, 429, 431, 464, 466, 470, 473, 490, 492–493, 507, 540, 545, 560, 565, 571, 582–585, 589, 590, 597–598, 603–611, 627, 638, 641, 643–645, 656, 670, 691
  absorbent filters 644
  accumulation 369, 516, 643, 656
  and sediment 154, 155
  concentration in oil at different temperatures 644
  condemnation limits 606
  contamination 104, 140, 179–180, 545, 560, 584–585, 603, 605, 608, 645
  content 645
  ingression 604, 606, 607, 644
  removal 493, 607, 609
  resistance 153, 192, 266
  soluble 224–225, 227
  washing 33
  washout 85, 153
wax crystals 72, 82, 94, 134, 242, 256
wax emulsions 268–269, 458
wax film 268–269
wax melting point 155
wax-plugging 250
wear 3–4, 7, 10, 12–15, 17, 19–21, 26, 35, 41, 44, 51, 53, 59, 75–77, 81, 85, 87, 89, 105, 113, 119–121, 129, 133, 140, 150, 163, 165, 167, 173–177, 179–181, 183–184, 190, 194, 200, 212–213, 215–217, 224, 225, 232–233, 235–238, 244, 249, 252, 254, 257–258, 267–268, 272, 274–275, 280, 283, 303, 336, 373, 389, 391, 404–405, 409, 425, 460, 463, 466, 468–469, 473, 475, 483–488, 493, 500, 516, 532–534, 538–540, 598–599, 615–617, 619–620, 622, 629, 631, 646–649, 652–653, 656–658, 672–673, 689, 694, 717
wear curve of mineral hydrocarbon oil compared to synthetic oils 488
wear debris analysis 616, 631
wear particle analyzer 648
wear particle detection 646
wear particles 13, 233, 404, 473, 646–648, 656
wear prevention 233
wear protection 75, 165, 174, 177, 184, 194, 216–217, 225, 232, 236, 252, 258, 268, 460, 493
wear rates 121
weather-protected (WPII) motors 516
weep hole 515
weld point 121
wet sump 336, 364
wheel bearing grease leakage 156

wheel gears 492
white metal 8, 448, 454
white mineral oils 141–142, 189
worked penetration 85, 115–116
work flow 697
workhorse lubricants 159
worm gear lubricants 212
worm gears 10, 212–213, 225, 487–488
WRC (Water Research Council) 466

X

XP-motor drains 515

Z

ZDDP 33, 51, 81, 634–635, 638
Zerk style grease fitting 304, 306
zinc additives 567
zip-lock sandwich bags 625
ZN/P Curve 45–47

About the Author

Kenneth E. Bannister is a UK technical apprenticed and accredited mechanical design engineer, credited on several engineering patents, two of which involved tribology aspects in their design. Ken is also a CMRP - Certified Maintenance Reliability Professional and since 1988 has consulted worldwide helping clients implement practical and meaningful asset management, reliability, and lubrication management programs. Ken is one of a handful asset management consultants holding expertise and accreditation in the field of tribology, lubrication failure management and industrial lubrication application as a designated professional MLE - Machinery Lubrication Engineer.

Ken was the first consultant to assist a company through the ISO 55001 asset management certification process in North America and was the key architect in the development of the ICML 55® world lubrication standard. More recently Ken was a contributing author and senior editor for the compilation of the ICML 55.0, ICML 55.1, and ICML 55.2 standards documents.

Practical Lubrication for Industrial Facilities – Fourth Edition, is Ken's fourth industrial lubrication book. In addition, Ken also provided the lubrication section for the *Machinery's Handbook* and has published other books on energy management and predictive maintenance. Throughout his career Ken has published over 650 articles and white papers for numerous international maintenance magazines, with over half dedicated to the field of practical lubrication.

Ken was a founding board member of the Plant Engineering Maintenance Association of Canada (PEMAC) and currently sits on the board of directors for the International Council for Machinery Lubrication (ICML) responsible for the ICML 55® world lubrication standard.